A Systems Approach to Managing the Complexities of Process Industries

A Systems Approach to Managing the Complexities of Process Industries

Fabienne Salimi
Process Safety Expert, ADEPP Academy

Frederic Salimi
Process Safety Expert, ADEPP Academy

ELSEVIER

Elsevier
Radarweg 29, PO Box 211, 1000 AE Amsterdam, Netherlands
The Boulevard, Langford Lane, Kidlington, Oxford OX5 1GB, United Kingdom
50 Hampshire Street, 5th Floor, Cambridge, MA 02139, United States

British Library Cataloguing-in-Publication Data
A catalogue record for this book is available from the British Library

Library of Congress Cataloging-in-Publication Data
A catalog record for this book is available from the Library of Congress

ISBN: 978-0-12-804213-7

Publisher: Joe Hayton
Acquisition Editor: Kostas Marinakis
Editorial Project Manager: Jennifer Pierce
Production Project Manager: Sruthi Satheesh
Designer: Mark Rogers

Typeset by MPS Limited, Chennai, India

Contents

Acknowledgments

I was always amazed about the wonders that man-kind can do when they join their forces to turn a dream to a reality. When I was a child, I was thinking "What if" the brains of all people on the world are interconnected to make a huge brain? My friends and mentors smiled and did not take this question seriously.

Today, technology gives us the required tools to think more seriously about this question. The IIoT and augmented reality are already there and stimulate imagination of many of us.

As the health, safety, and environmental practitioners, we understand that the complex solutions engineered to meet fundamental human needs. They should be safe and not harm people and environment. But what does complexity mean in process industry and how can we manage it?

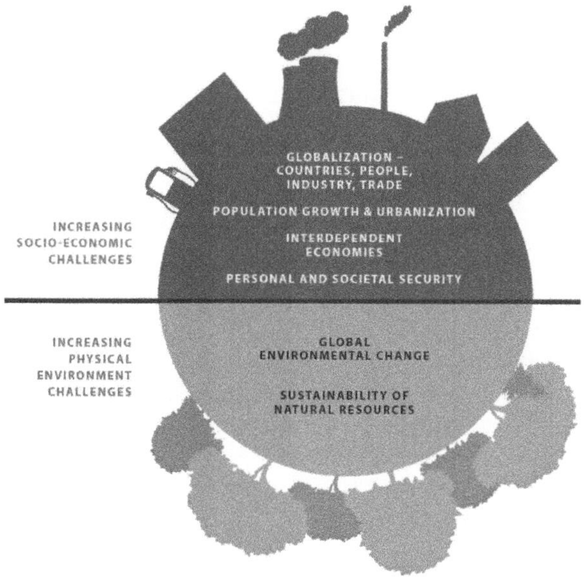

INCOSE Vision 2025 - Page 7.

Since 1994, the authors of this book have reflected on this question and developed a tool called ADEPP (Analysis & Dynamic Evaluation of Project Processes). We acknowledge the importance of technical and financial supports of the French innovation organizations in particular "Institute Français du Pétrole (IFP) in development of version 1 of ADEPP."

In 2014 the authors created a partnership with the following companies to develop the prototype of version 2 of ADEPP. We acknowledge their trust and importance in crystallization of the ideas that we express in this book:

- Hydraulic Analysis Ltd + Software Simulation Ltd (United Kingdom)
- VMG (Europe Branch in Spain)
- Antea (Italy)
- Human Focus (United Kingdom)
- EON Reality (France)

We also acknowledge the positive reception and constructive discussions with the following system engineering and complexity experts who inspired and encouraged us:

- Alan Hayman (System and Complexity Thinking Consultant)
- Frank Verschueren (Engineer Inspector for FOD WASO/Belgian fed government COMAH regulations)
- Matthew E. Weilert (Systems Thinking & Engineering Consultant)
- Gene Bellinger (Systems Thinking & Engineering visionary)
- Joss Colchester (director and developer of complexity academy, i.e.—an excellent source of high-quality system and complexity courses for the beginners)
- Mark Simpson (Product Manager at Siemens)

We acknowledge the usage of the valuable books and documents published by the following organizations for promotion of system engineering, system thinking, complexity thinking, quality and risk thinking, and IIoT:

- INCOSE (International Council on Systems Engineering)
- SEBoK (Systems Engineering Body of Knowledge)
- LNS Research
- Skybrary—Aviation Safety
- NASA (National Aeronautics and Space Administration)
- MIMSOA (Maintenance Information Management Open System Alliance)
- Health & Safety Executive (United Kingdom)
- Energy Institute
- IOGP (International Oil & Gas Producers)

Finally, we express our gratitude towards all those individuals and companies who directly or indirectly contributed in our education and professional understanding. May this book pay a humble tribute to their efforts.

Fabienne-Fariba Salimi & Frederic Salimi

CHAPTER 1

Perspective

1.1 UNDERSTANDING A QUESTION IS HALF AN ANSWER!

"Management System" is a structured and documented set of interdependent practices, process, and procedures used by the managers and the workforce at every level in a company to plan, direct, and execute activities as shown in Fig. 1.1.

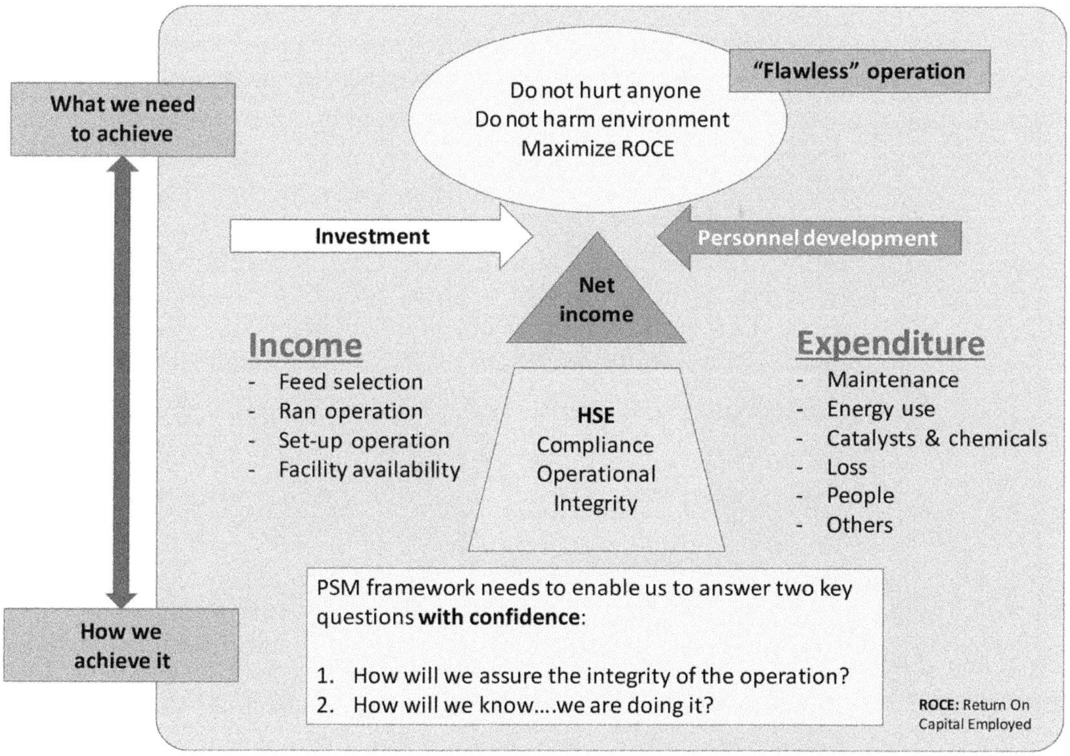

FIGURE 1.1
Process safety management in context. *Figure reproduced from webpage https://www.energyinst.org/technical/PSM published by the Energy Institute.*

A Systems Approach to Managing the Complexities of Process Industries.
DOI: http://dx.doi.org/10.1016/B978-0-12-804213-7.00001-3

Since, first days of quality management system in the late 60s, we have come a long way in improving quality, occupational health & safety, technical safety, and environmental management systems. For decades, these regulatory management systems were taken as the constraints to production and profitability of the businesses. This perception creates the conflicts and unsatisfactory results. In this regard the Energy Institute states:

> ..., most well-run organisations can tell you how many incidents they had yesterday; however, our real challenge is to be able to answer the question "How likely am I to have an incident-free day tomorrow"?

Energy Institute continues,

> We have all seen the typical banner statements 'zero harm', 'flawless operation', 'target zero', 'incident free', 'nobody gets hurt'; but the two key questions for executives and managers at all levels are:
>
> 1. How will we assure the integrity of the operation?
> 2. How will we know we are doing it?
>
> All too often the first two words used to answer these questions are *"I think...."*; in reality, this means *"I dont' know"*! Recent events have shown that such answers are no longer acceptable and that, from top to bottom, organisations need to be able to answer these two key questions with absolute *confidence*.

Not only, the quality and health, safety, environment (HSE) practitioners but also the operational teams feel that the "management systems" do not work as they are advertised. But why is it so? And what can we do about it? The other side of the coin is that we often talk and analyses the failures, but we do not look at success very frequently. Why despite the flaws in the management systems, are the operations performed safely and reliably?

Socrates said that *"Understanding a question is half an answer"*!

To answer these questions, we need to understand the word "System" in "process safety management system" with the mind of a system engineer.

This book aims to raise the awareness of the HSEQ practitioners, managers and operational personnel in process manufacturing to the required system engineering skills. We will clarify how the relatively new ISO/IEC 15228 and ISA-95 (IEC/ISO 62264) standards are embedded in the operational excellence guidances and how they can smooth the journey of the process facilities toward the digital transformation. Then the most practical methods will be introduced to assess and manage the complexities of their day-to-day tasks, configuration management, and the strategic decision makings.

1.2 PROCESS SAFETY MANAGEMENT IN CONTEXT OF THE OPERATIONAL EXCELLENCE

Today the management systems philosophies are refined and go beyond inspection, focusing on the strategies that incorporate processes and people to the physical assets management to achieve the operational excellence.

Operational management system (OMS) is the consolidation of the company's knowledge and requirements into a single framework to manage assets and activities safely and responsibly. It includes the company's policies, standards, practices, procedures, and processes. This "corporate memory" is organized within the System's Elements and Expectations, which are designed to ensure the control measures are complete and robust.

The OMS framework applies to the all the management systems including:

- Production Operations Management,
- Reliability and Asset Integrity Management,
- Quality Operations Management,
- Inventory Operations Management, and
- Regulatory Compliance Management

Management system is a structured and documented set of interdependent practices, process, and procedures used by the managers and the workforce at every level in a company to plan, direct, and execute activities.

Operating covers, the design, implementation, and control of activities that convert resources into products and services to fulfill a company's business strategy. The word "operating" refers to the entire lifecycle of a company's activities and products. In this context, "operating" applies to every upstream or downstream company activity, from engineering to decommissioning, throughout the entire value chain and lifecycle of the business and its products.

In 2011, International Oil & Gas Producers issued the IOGP 510 which is a new Operating Management System Framework to help companies define and achieve performance goals and stakeholder benefits while managing the broad and significant range of risks inherent in the oil and gas industry. This guideline and its supplement IOGP 511 can be applied to the other process industry sectors such as hydrocarbon processing, chemical, pharmaceutical industries too. Fig. 1.2 illustrates the four fundamentals and ten elements of the OMS framework.

FIGURE 1.2

The OMS framework—four fundamentals underpin 10 elements. *From IOGP 510.*

IOGP 510 suggests a generic framework which offers an integrated approach and the flexibility to address some or all the wide range of risks, impacts or threats related to occupational health and safety; environmental and social responsibility; process safety, quality, and security. The degree of integration and the scope of an OMS will be determined by individual companies and will differ depending on their activities, organizational structure and management system maturity as shown in Fig. 1.3.

At the facility level the office should provide information about new customer orders, raw materials that have been ordered, specific customer demands for products, and so on. The shop floor will also have to send information to the office. For example, information about the status of orders, about the exact amounts of raw materials that were used in the production process and so on. Although they speak different languages, both levels should communicate with each other as shown in Fig. 1.4.

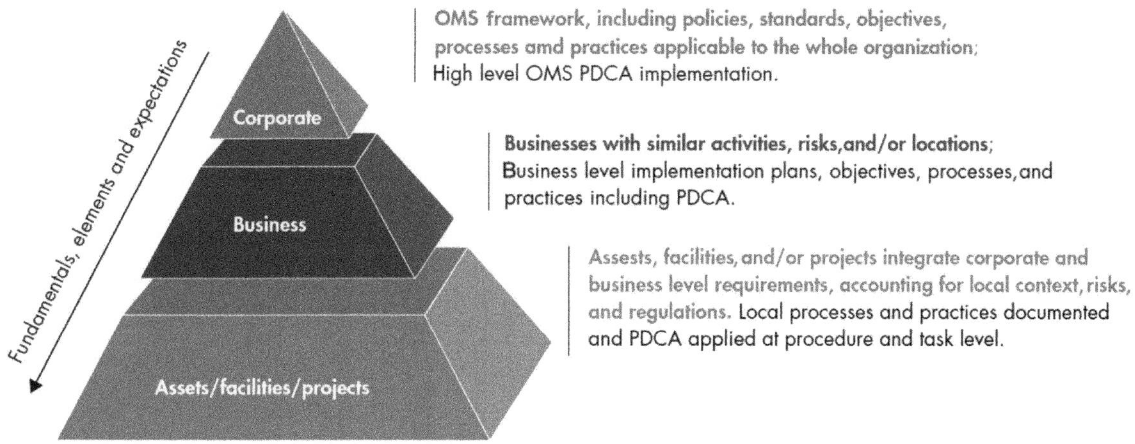

FIGURE 1.3
Hierarchy of the OMS implementation. *PDCA*, Plan, Do, Check, Act. *From IOGP 510.*

With the appearance of new technologies, it is getting easier to automate the exchange of information between the office and the shop floor. An automated interface between enterprise and control systems can lead to a lot of advantages. Relevant information becomes accessible at the right time and the right place to the right person. The company has access to the real-time information such as information about raw materials and end products, which enables optimum usage of storage capacity.

ISA-95 (IEC/ISO 62264) is an international standard which has been developed to address the problems encountered during the development of automated interfaces between enterprise and control systems. This standard applies to all industries, and in all sorts of processes, such as batch, continuous, repetitive, or discrete processes.

The Part 1 of the ISA-95 standard defines a functional hierarchy model. Each level provides specialized functions and has characteristic response times, as shown in Fig. 1.5.

Level 0 defines the actual physical processes.

Level 1 defines the activities involved in sensing and manipulating the physical processes. Level 1 typically operates on time frames of seconds and faster.

Level 2 defines the activities of monitoring and controlling the physical processes. Level 2 operates on time frames of hours, minutes, seconds, and subseconds.

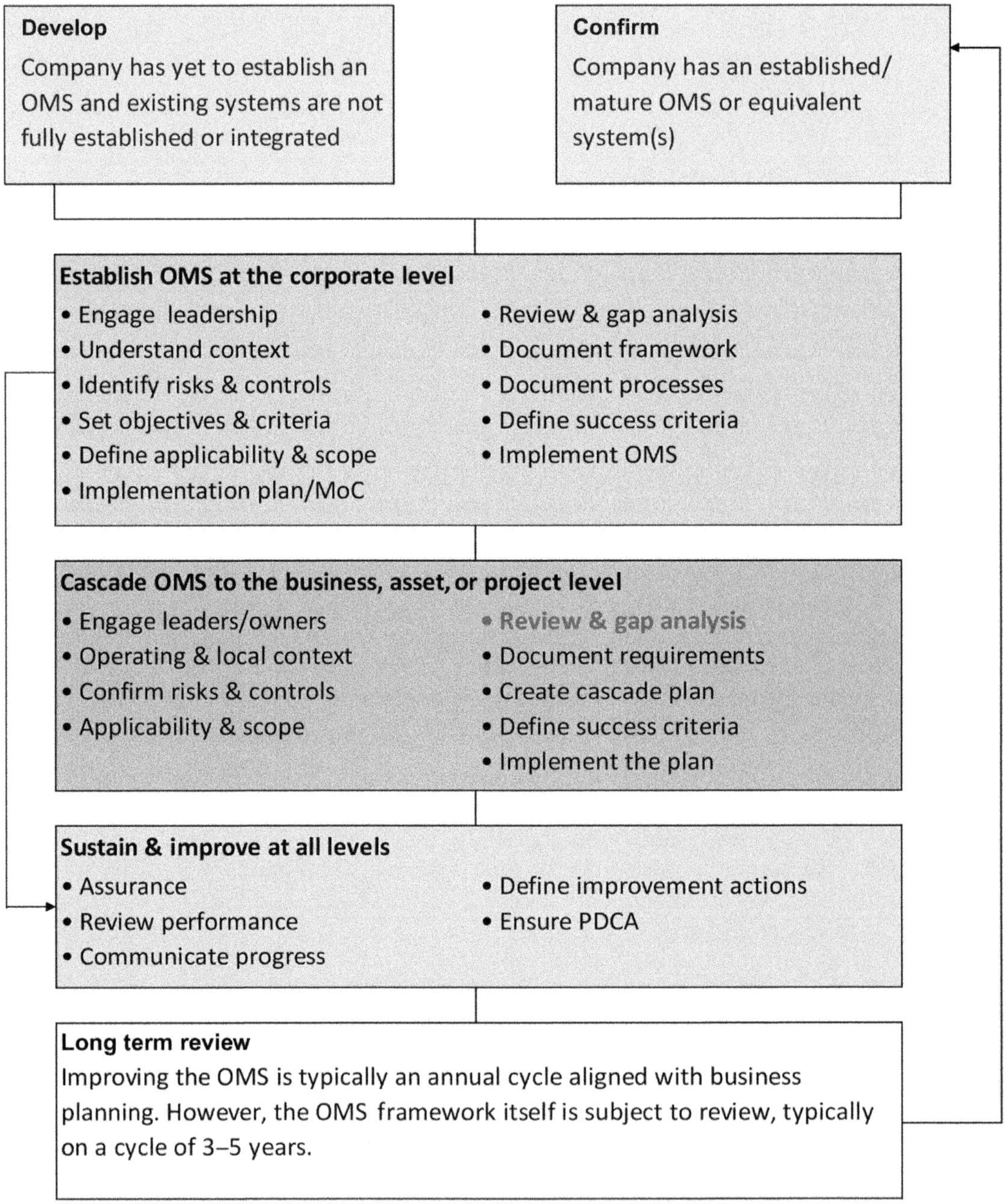

FIGURE 1.4

Establishing and sustaining an OMS flow chart. *From IOGP 510.*

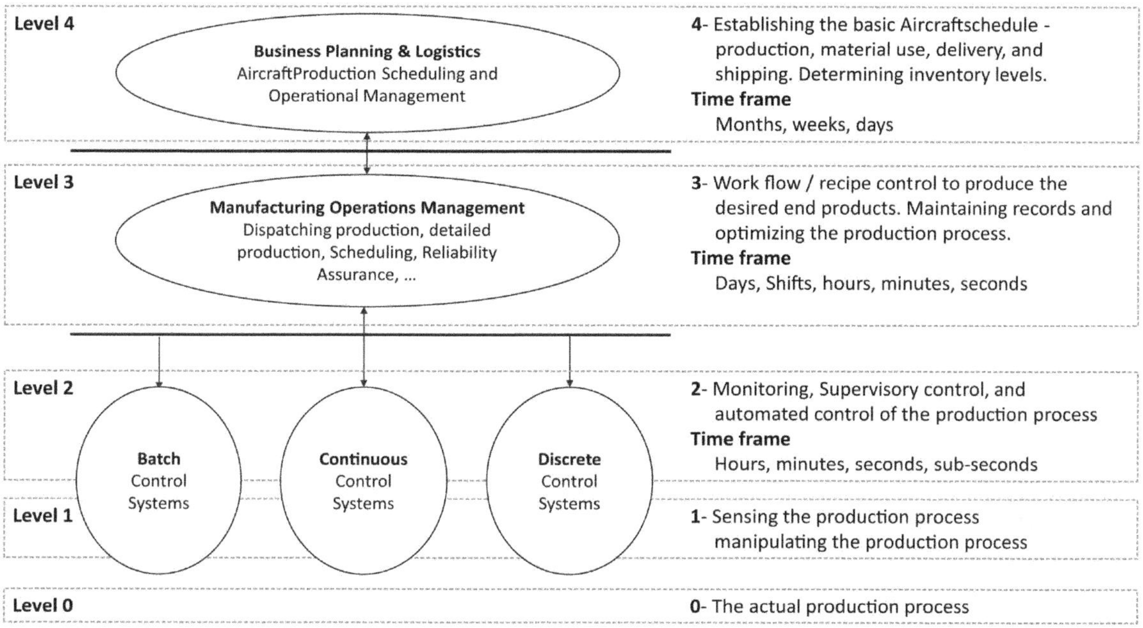

FIGURE 1.5
ISA-95 multilevel functional hierarchy of activities.

Level 3 defines the activities of workflow to produce the desired end products. It includes the activities of maintaining records and coordinating the processes. Level 3 typically operates on time frames of days, shifts, hours, minutes, and seconds.

Level 4 defines the business-related activities needed to manage a manufacturing organization. Manufacturing-related activities include establishing the basic plant schedule (such as material use, delivery, and shipping), determining inventory levels, and making sure that materials are delivered on time to the right place for production. Level 3 information is critical to Level 4 activities. Level 4 typically operates on time frames of months, weeks, and days.

The Level 5 can be added to capture Quality Governance and Planning and then added the value chain as quality management occurs across the lifecycle. Level 5 determines the strategy for Operational Excellence, Knowledge Retention, and Quality and Risk Management. This Level 0-5 framework is applicable to entire value chain as shown in Fig 1.6.

This framework is valuable because it provides a temporal perspective which includes both enterprise quality and functional quality. In a single

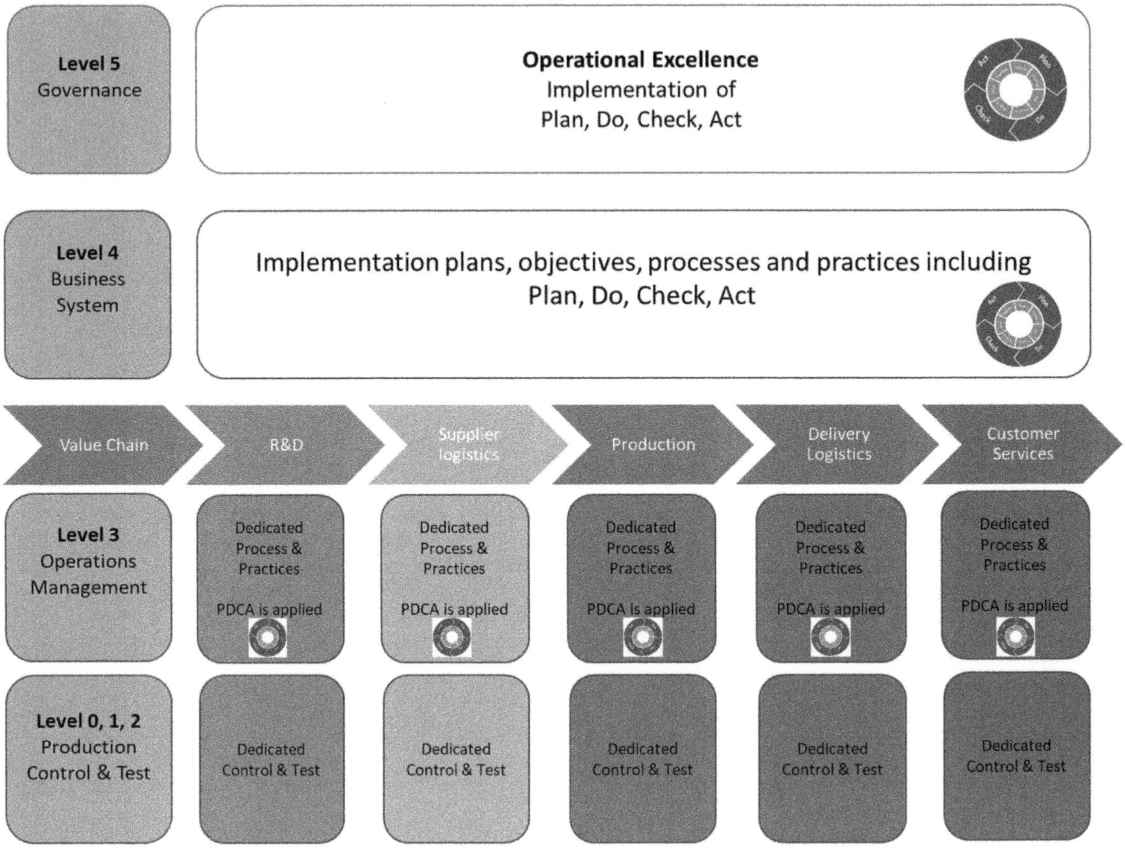

FIGURE 1.6
Adaptation of the ISA-95 framework to understand the total quality management system.

framework, it represents strategy and management down to operations and real-time asset performance. The connected devices and analytics capture the connection to Industrial Internet of Things (IIoT).

System engineering is the foundation of the operational excellence standards and guidelines. In the following sections, we highlight how the system engineering is applied in operating management systems of the process facilities.

1.3 REGULATORY COMPLIANCE MANAGEMENT SYSTEM

The broad footprint of management of regulatory compliance means that many areas of the enterprise can be significantly affected. Failures in

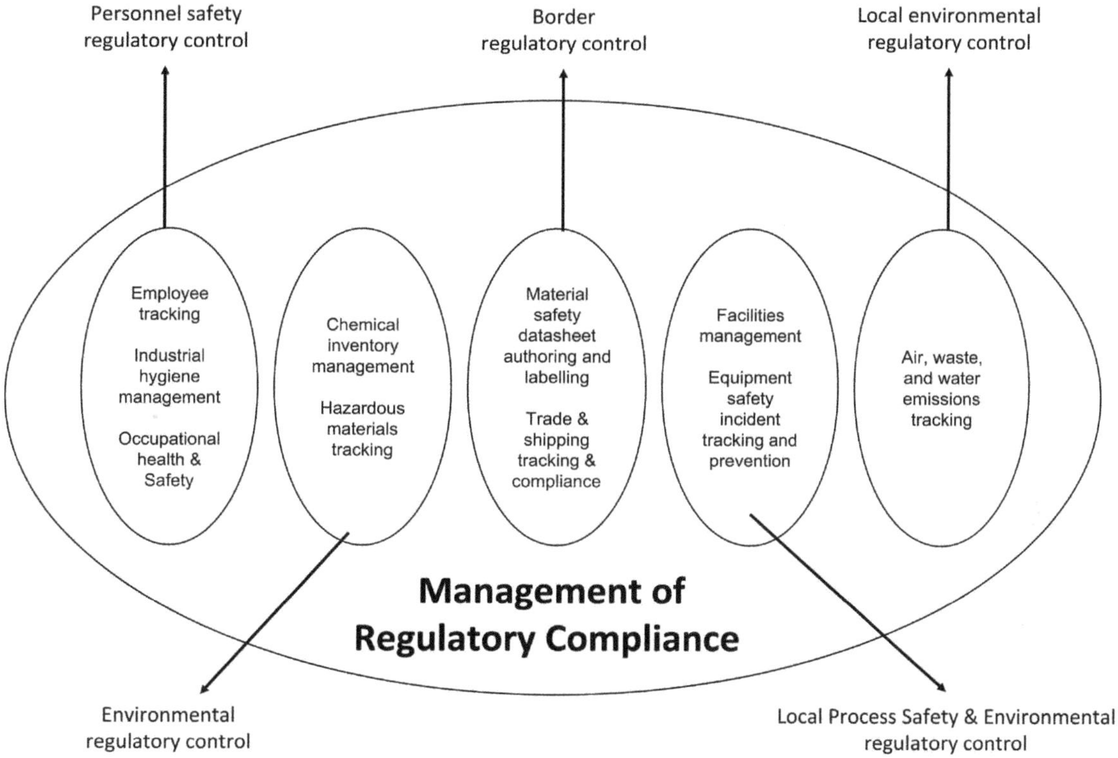

FIGURE 1.7
Functions in management of regulatory compliance.

regulatory compliance can stop production, force product recalls, and potentially cause safety problems. Where management of regulatory compliance activities involves the quality and safety of production, then the activities are in the scope of manufacturing operations. Fig. 1.7 breakdowns the most important regulatory compliances and the general activities associated with them.

Fig. 1.8 highlights the requirements of SEVESO III for process safety management (PSM) systems and Fig. 1.9 compares the structure of the quality management, environmental and occupational health & safety management systems.

The local or activity specific regulatory compliances should be considered case by case.

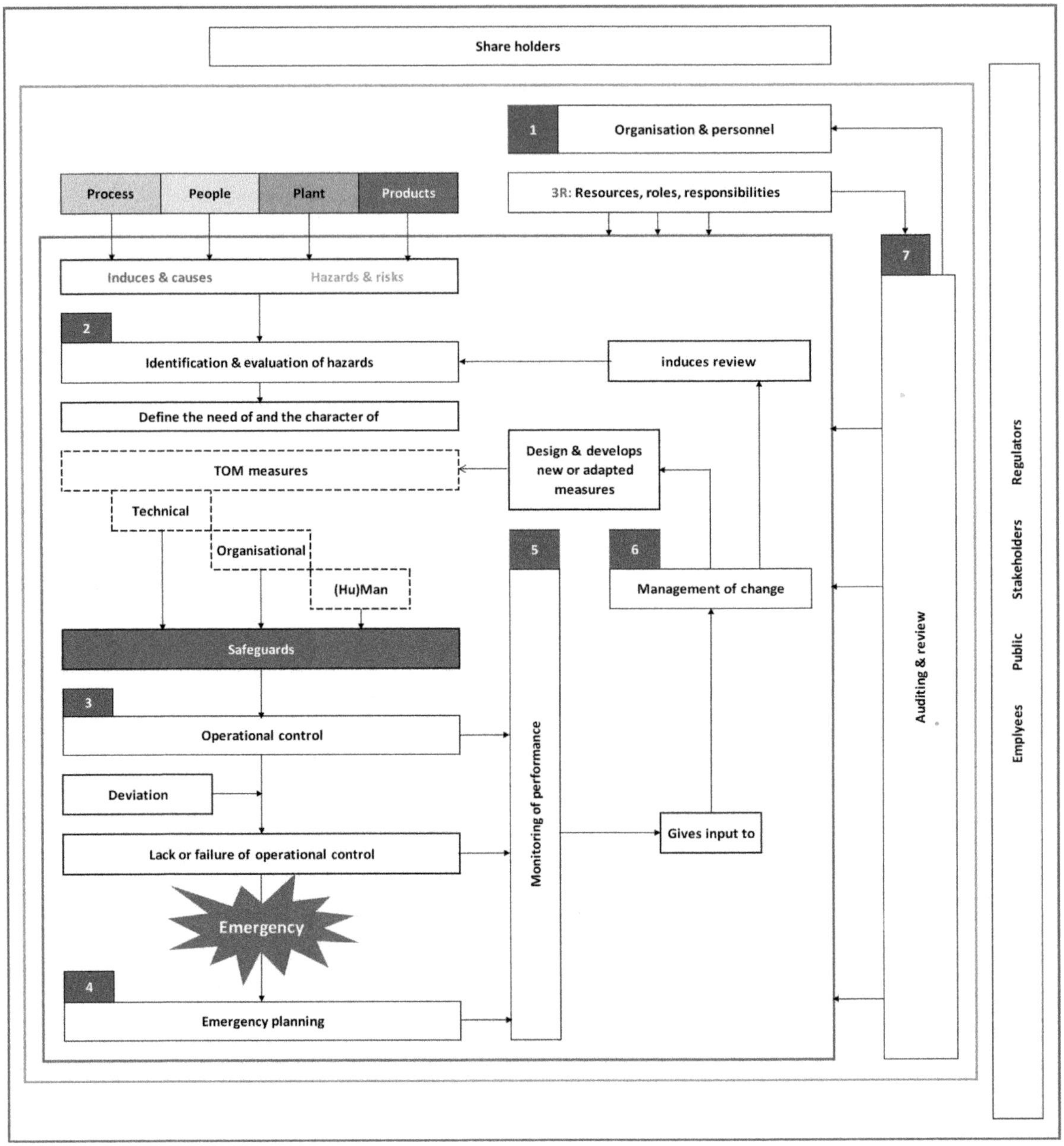

FIGURE 1.8
SEVESO III regulatory compliance framework.

ISO 9001		ISO 14001		OHSAS 18001	
0.1	Introduction		Introduction		Introduction
0.2	General				
0.3	Process approach				
0.4	Relationship with ISO 9004				
	Compatibility with other management systems				
1	Scope	1	Scope	1	Scope
1.1	General				
1.2	Application				
2	Normative reference	2	Normative references	2	Normative references
3	terms and definitions	3	Definitions	3	Definitions
4	Quality management system	4	Environmental management system requirements	4	OH&S management system requirements
4.1	General requirements	4.1	General requirements	4.1	General requirements
4.2	Documentation requirements (title only)				
4.2.1	General	4.4.4	Documentation	4.4.4	Documentation
4.2.2	Quality manual				
4.2.3	Control of documents	4.4.5	Control of documents	4.4.5	Control of documents
4.2.4	Control of records	4.5.4	Control of records	4.5.4	Control of records
5	Management responsibility (title only)	4.4.1	Structure and responsibility	4.4.1	Structure and responsibility
5.1	Management commitment	4.4.1	Structure and responsibility	4.4.1	Structure and responsibility
5.2	Customer focus	4.3.1	Environmental aspects	4.3.1	Hazard Identification, Risk Assessment & Determining controls.
		4.3.2	Legal and requirements	4.3.2	Legal and requirements
		4.6	Management Review	4.6	Management Review
5.3	Quality policy	4.2	Environmental policy	4.2	OH&S policy
5.4	Planning	4.3	Planning	4.3	Planning
5.4.1	Quality objectives	4.3.3	Objectives, targets and programme(s)	4.3.3	Objectives, targets and programme(s)
5.4.2	Quality management system planning	4.3.3	Objectives, targets and programme(s)	4.3.3	Objectives, targets and programme(s)
5.5	Responsibility authority and communication (title only)	-	-	-	-
5.5.1	Responsibility and authority	4.4.1	Resources, roles, responsibility and authority	4.4.1	Resources, roles, responsibility and authority
5.5.2	management representative	4.4.1	Resources, roles, responsibility and authority	4.4.1	Resources, roles, responsibility and authority
5.5.3	Internal communication	4.4.3	Communication	4.4.3	Communication, Participation and Consultation
5.6	Management review	4.6	Management review	4.6	Management review
5.6.1	General				
5.6.2	Review input				
5.6.3	Review output				

FIGURE 1.9

Comparison of the regulatory management systems.

When policies and procedures for management of regulatory compliance do not exist on a company-wide basis, then compliance control can be regarded as a manufacturing operations activity, for manufacturing compliance.

Management of incidents, deviations, corrective actions, and preventative actions is often associated with maintenance of regulatory compliance or with continuous improvement processes. These activities are also often performed in conjunction with other Manufacturing Operations Management (MOM) activities.

Incidents are the unexpected events related to maintaining plant operations, safety, regulatory compliance, or security. Incident management involves investigation to determine the root cause of the incident and may lead to preventive actions to prevent future incidents.

Incidents and response to them should be recorded as part of incident management system.

EXAMPLE 1: An unexpected release of a chemical into the environment may generate an incident, and the incident report may have to be sent to the appropriate regulatory agency.

EXAMPLE 2: An unexpected pump failure from a newly installed pump may generate an incident, and the incident response may be to investigate and potentially change the supplier.

Deviations are the measured differences between an observed value and an expected or normal value, or an anomaly from a documented standard or process. Deviation management involves the determination of the root cause of the deviation and may lead to corrective actions to remove the source of the deviation.

Deviations and response to them should be recorded.

Maintaining plant operations often requires that corrective actions, in response to an incident, deviation, or failure. Clear, appropriate, and implementable corrective actions should be identified at the conclusion of any investigation. Tracking and follow-up should be managed to ensure that the corrective actions are implemented and verified.

The root cause of the incident and the corrective actions should be recorded.

EXAMPLE 1: Corrective actions may include improving procedures, adding maintenance procedures for equipment, or implementing retest or revalidation procedures.

Preventative actions are managed in a similar fashion, to prevent possible future incidents or deviations.

EXAMPLE 2: Batch cycle times on a process cell may not meet the rated value, and this is identified as a deviation; then, a preventive action is created to reduce the batch cycle time.

Recommended actions are managed in a similar function. Recommended actions are predefined sets of actions to occur in the event of an incident or deviation.

1.4 COST OF NONCOMPLIANCE

Industrial facilities are created to satisfy human needs. Today, working in a safe workplace is a fundamental human right, and any business activities must be embedded in the current social, physical, cultural, and economic environment.

Management of the social responsibilities and liabilities post a major accident can be very complicated. The major disaster of Erika oil tanker in 1999 and BP Horizon in 2009 are the examples of these complex situations.

In many cases the complacency or haste of decision makers is the leading cause of the major accidents. Very often the cost of eliminating the technical causes is much less than the financial cost of the accident consequences. Table 1.1 summarizes the cost of nonquality of a few major accidents.

Table 1.1 Cost of Accident Versus Cost of Eliminating the Cause of Accident

Major Accident	Cause	Consequence
1986: Space Shuttle Challenger	Break down of an O-ring Cost of redesigning O-rings: a few hundred thousand dollars	Human: 7 fatalities Financial: 1 billion USD Environmental: Minor Reputation: Inestimable
2009: Toyota	Problematic floor mat and defective breaks leading to unintended acceleration	Human: 52 dead and 38 injured Financial: 5.5 billion USD Environmental: Minor Reputation: Inestimable
2010: BP Horizon	Weak Cement around well The cost of checking the cement: 188,00 USD Time: 10 hours	Human: 11 fatalities Financial: 10 billion USD Environmental: Inestimable Reputation: Inestimable—losing 1/3 of BP value in stock market forcing BP to sell some of its assets worldwide

The managers carry the responsibility for ensuring that the equipment is competitively priced and that its safety integrity is adequate in operation. They should apply a systematic approach to ensure that optimum solutions are implemented to consider the complexity of the system and balance the equilibrium between the cost and safety.

A Study on 319 major industrial accidents which were recorded per the UNEP-specified criteria concludes:

- Although the number of major industrial accidents is higher in *developed countries* than in *developing ones,* the number of deaths and injuries is considerably less. Very probably, this fact is the result of better enforcement of *safety regulatory legislation* in developed countries.
- Another effect of better enforcement of safety regulatory legislation is the fact that it seems that during the *last two decades,* the number of major industrial accidents *is decreasing in general.*

Fig. 1.10 demonstrates two other important facts:

1. Comparison between BP Horizon and Piper Alfa shows that:
   a. *Asset loss*: 100% damage of both BP Horizon and Piper Alfa offshore platforms

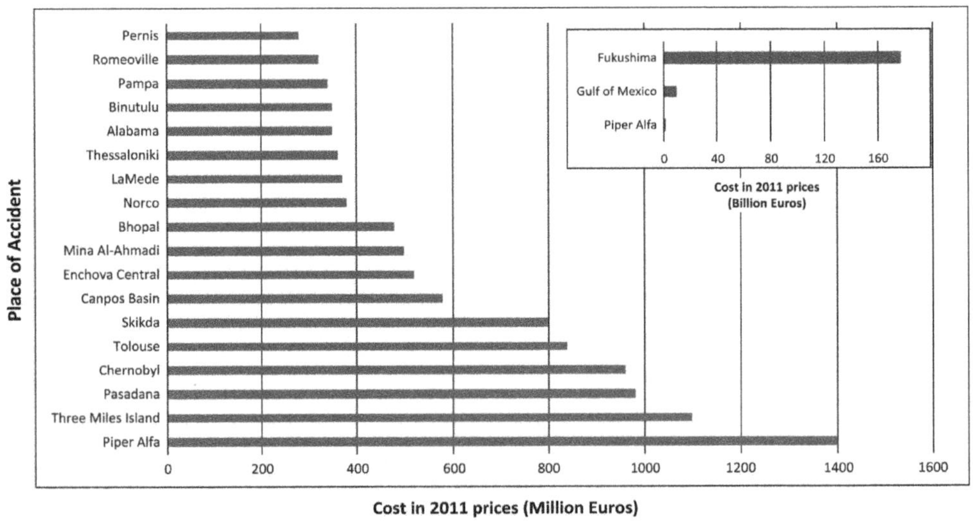

FIGURE 1.10
Cost of some the major accidents.

b. *Human loss*: BP Horizon (11 fatalities) versus Piper Alfa (167 fatalities)
c. *Environmental damage*: BP Horizon (inestimable) versus Pipe Alfa (relatively limited)

Safety case regulations and risk-based approach came into force after Pipe Alfa disaster. This comparison shows that the safety regulations have been effective in protecting lives of BP Horizon personnel. On the other hands, due to the application of the novel technology on much more challenging environmental conditions, the environmental damages of BP Horizon have been much more sever and affected the areas beyond the USA boarders up to the African coasts.

2. The Fukushima major accident demonstrates that when natural events combine with industrial accidents, the losses can be much more devastating. Climate changes cause the more sever natural extrems. Decision makers should bear in mind that the original design basis and safety factors of the ageing plants may not be sufficient to cope with the actual environmental conditions. The cost of the Fukushima accident is about 160 times more than Chernobyl.

Safety and quality are the two faces of a coin. If a defect or noncompliance in the process production leads to toxic and/or flammable materials, then a major accident occurs. Fig. 1.11 demonstrates how quality assurance cycle is related to the risk-based PSM. Without a robust quality management system an effective safety management system cannot be implemented.

Cost effectiveness is the outcome of a realistic and wise balance between opposite spending: the cost of the good quality (or the cost of conformance-immediately and exactly measured) and the cost of poor quality (or the cost of nonconformance-latent and unpredictable extent). As Fig. 1.12 shows the cost of good quality affects:

- Costs for investing in the prevention of nonconformance to requirements.
- Costs for appraising a product or service for conformance to requirements.

The cost of poor quality affects the internal and external costs resulting from failing to meet requirements.

Internal failure costs are costs that are caused by-products or services not conforming to requirements or customer/user needs and are found before delivery of products and services to external customers. They would have

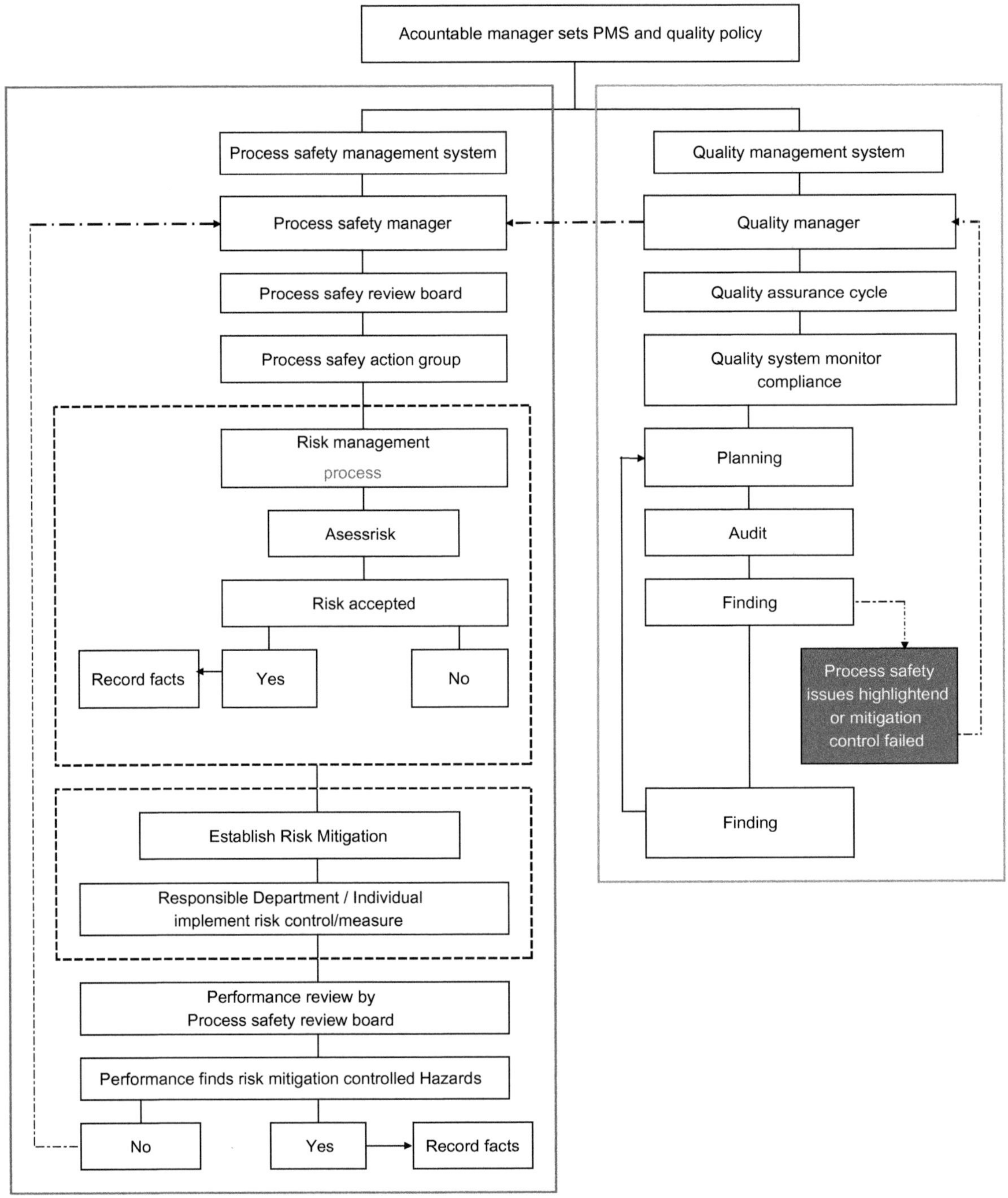

FIGURE 1.11
Integrated quality and process safety management systems. Inspired by: Safety management systems—guidance to organizations.

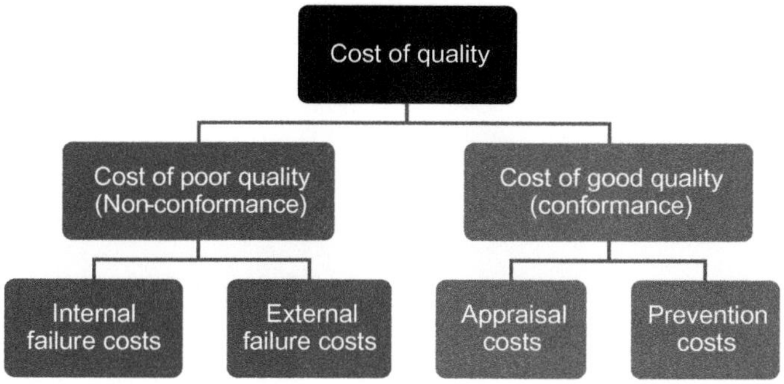

FIGURE 1.12
Cost of quality.

otherwise led to the customer not being satisfied. Deficiencies are caused both by errors in products and inefficiencies in processes.

External failure costs are costs that are caused by deficiencies found after delivery of products and services to external customers, which lead to customer dissatisfaction.

Prevention costs are costs of all activities that are designed to prevent poor quality from arising in products or services.

Appraisal costs are costs that occur because of the need to control products and services to ensure a high-quality level in all stages, conformance to quality standards and performance requirements.

The total quality costs are then the sum of these costs. They represent the difference between the actual cost of a product or service and the potential (reduced) cost given no substandard service or no defective products.

Many of the costs of quality are hidden and difficult to identify by formal measurement systems. The iceberg model is very often used to illustrate this matter: Only a minority of the costs of poor and good quality are obvious—appear above the surface of the water. But there is a huge potential for reducing costs under the water. Identifying and improving these costs will significantly reduce the costs of doing business.

A general study made by UK Health & Safety Executive into the cost of accidents showed that the costs of error rectification far exceeded those that would have been incurred if a systematic approach had been employed from the outset. Fig. 1.13 summarizes the typical insured and uninsured cost associated to an accident.

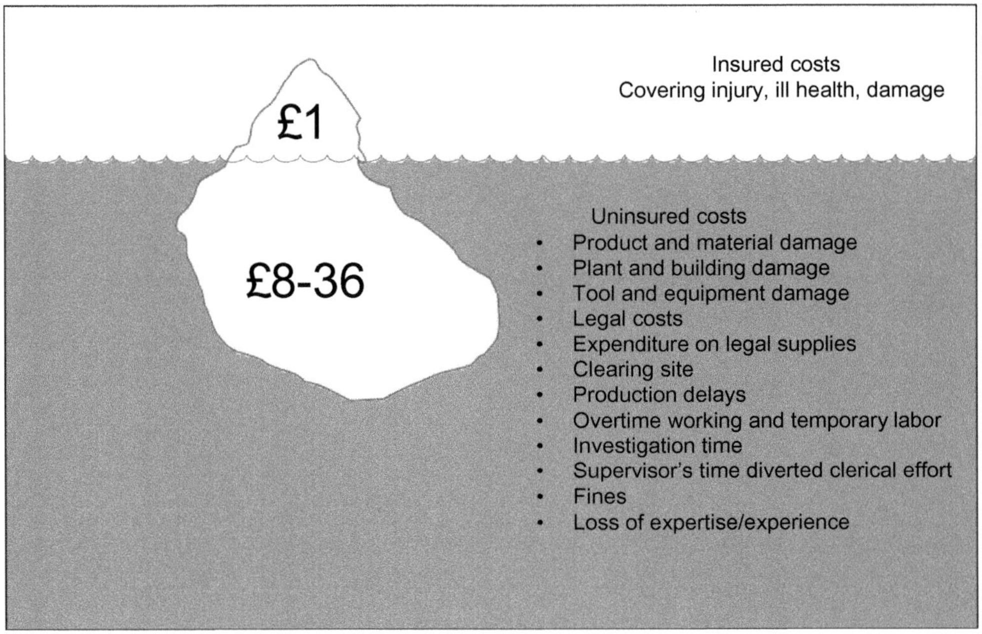

FIGURE 1.13
Iceberg model for the major accident costs. *From Out of control—Why control systems go wrong and how to prevent failure.*

1.5 PROCESS SAFETY VERSUS OCCUPATIONAL SAFETY

In Part 3-Section 10.6 of ISA-95 the typical health and safety activities listed as follows:

1. Handling, classification, packaging, and labeling of hazardous substances including safety data sheets.
2. *Disaster planning including emergency planning and response, and fire safety.*
3. Hazard communication in the form of warning signs, training, and advice.
4. Occupational health surveillance in the form of occupational exposure controls (including chemical, physical, biological agents, and noise).
5. Medical surveillance of personnel.
6. *Process safety in the form of machinery safety, lifting equipment, pressure systems, confined space entry/work permits/access control.*
7. *Management of functional safety.*

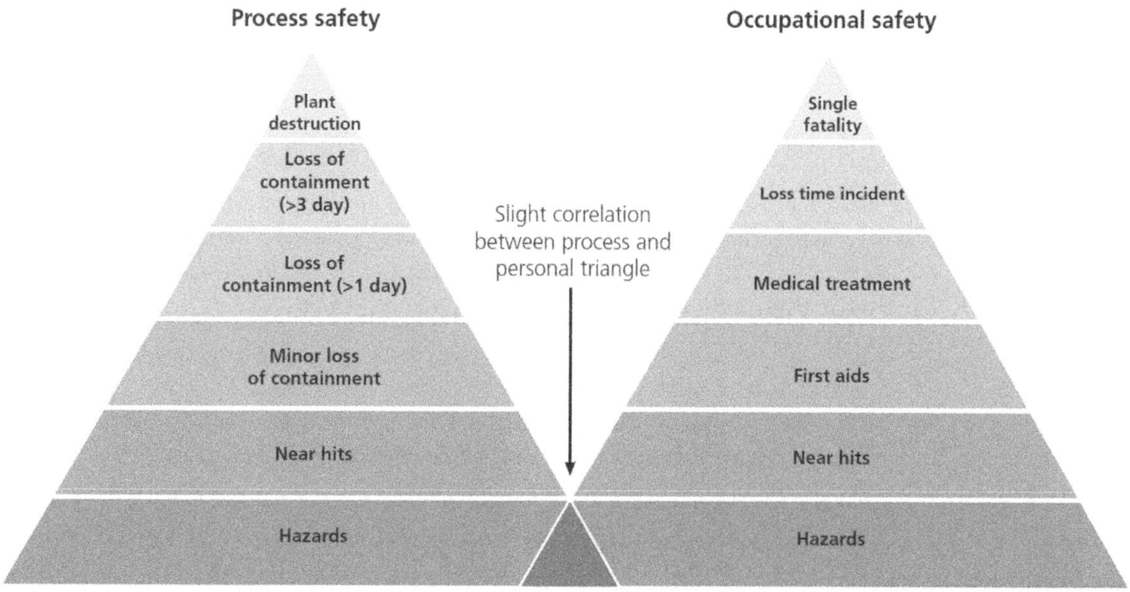

FIGURE 1.14
Process safety and occupational safety. *From Energy Institute-HUMAN FACTORS BRIEFING NOTE No. 20.*

8. Electrical safety.
9. Ergonomics including office work, manual handling of loads, and the like.
10. First aid.

This list mixes the material, occupational health & safety and process safety together. Many people are confused in the same way and ask the HSE practitioners "what is the need for PSM when our HSEMS is already in place?"

The likelihood and the extent of consequences of the occupational safety hazards differ significantly from the process safety hazards. In other words:

- *Occupational safety*—focuses on protecting the safety, health and welfare of people at work (sometimes is called "Personal safety").
- *Process safety*—focuses on the major accident hazards associated with releases of energy, chemicals, and other hazardous substances.

Process safety is a blend of engineering and management skills focused on preventing catastrophic accidents and near hits, particularly, structural collapse, explosions, fires, and damaging releases associated with a loss of containment of energy or dangerous substances such as chemicals and petroleum products. These engineering and management skills exceed those required for managing workplace safety as it impacts people, property and the environment. Fig. 1.14 compares the process safety and the occupational safety indicators.

1.6 PROCESS SAFETY INDICATORS

API 754 introduced a four-tier model for implementation of process safety key performance indicators (KPIs) in the process industry. The model is illustrated by the pyramid diagram in Fig. 1.15 that also shows the need for higher numbers of KPIs at the more leading levels.

The four tiers expressed as a triangle to emphasize that statistically larger data sets are available from the KPIs at the lower tiers. This approach mirrors the now-familiar personal accident triangle shown in Fig. 1.16 based on insurance claim work in 1931 by W. Heinrich and refined in 1969 for safety by Bird & Germain.

Tier 1 and Tier 2 (T1 and T2) are well-defined KPIs based on the recording of process safety events (PSEs) that involve loss of process containment (LOPC) that either exceed gas or liquid release thresholds or result in serious consequences such as injury or fire.

In contrast, Tiers 3 and 4 (T3 and T4) provide an intentionally broader concept, with the aim of encouraging companies to introduce a range of more leading KPIs that are typically defined locally at the facility or asset level, or in some instances across a business or company, to monitor the effectiveness of barriers that are specifically designed as risk controls at the operating level.

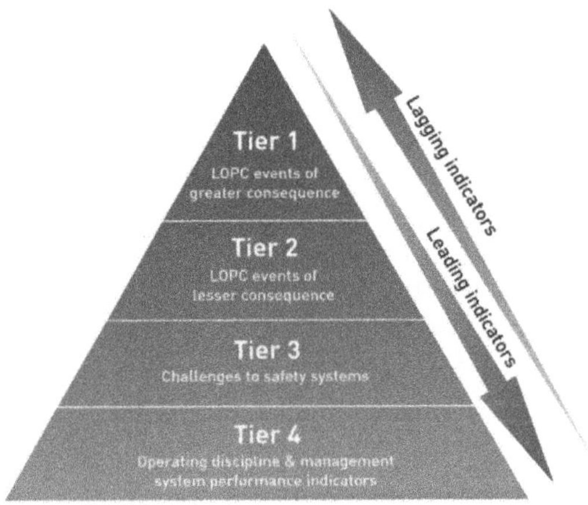

From ANSI/API RP 754 [4]

FIGURE 1.15
Process safety indicator pyramid per API 754.

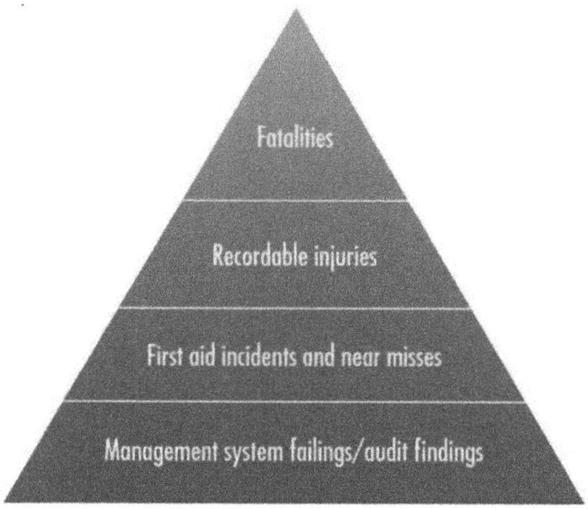

FIGURE 1.16
Occupational safety indicator pyramid.

1.7 WHAT DO WE MANAGE, "SAFETY PROCESSES" OR "PROCESS SAFETY"?

"Process Safety" in "Process Safety Management" is another confusing term. "Process Safety" and "Safety Processes" cover very different scopes. The same confusion can occur in using the following terms:

- *Process manufacturing* is the branch of manufacturing that is associated with formulas and manufacturing recipes. It can be contrasted with discrete manufacturing, which is concerned with discrete units, bills of materials, and the assembly of components.
- *Manufacturing processes* are the steps through which raw materials are transformed into a final product. The manufacturing process begins with the creation of the materials from which the design is made. These materials are then modified through manufacturing processes to become the required part. Manufacturing processes can include treating (such as heat treating or coating), machining, or reshaping the material. The manufacturing process also includes tests and checks for quality assurance during or after the manufacturing and planning the production process before manufacturing.

The "*Process Safety* Management" as we know is, in fact, the "*Safety Processes* Management." The safety processes may or may not be relevant to the chemical engineering and unit operation processes. A process engineer may have no expertise in the activities such as the "permit to work (PTW)" or management of the subcontractors.

1.7.1 Process Safety Engineering

"Process Safety" engineering aims to reduce the risk of an undesirable process events such as the overpressure, overtemperature, overflow, vacuum, under-temperature, low level to as low as reasonably practicable. The safety measures beginning by inherently safer design to emergency response systems are in place to achieve this goal.

HAZard & OPeratability (HAZOP) studies identify the credible undesirable events.

Then, process safety engineers implement the required protection layers using the layer of protection analysis (LOPA) as follows:

1. Inherently safer design
2. Basic Process Control Systems
3. Critical Alarms in compliance with EEMMU 191 and ISA-84.0 guidelines
4. Safety Instrumented Systems (SIS)

   API 14C (ISO 10418) provides the prescriptive recommendation for primary and secondary protection of the conventional oil & gas equipment. API 14C has been developed for offshore facilities, but today it is applied for both onshore installation use this guideline too. We believe that with some customization for the reactors or specific equipment, the approach of API 14C is very useful for evaluation of the process safeguarding requirements of petrochemical and refineries processes too.

   None of the safety barriers is 100% effective. The required safety integrity level of the instrumented-based safety functions is determined and assured by application of the international standards of IEC 61508 and IEC61511.
5. Secondary process safeguards such as relief valves or dikes around the storage tanks are in place to minimize the risk when the primary instrumented-based process safeguards failed to protect the process against an undesirable process event. Fig. 1.17 illustrates how the protection layers reduce the initial risk of the tolerable risk.

Traditionally the requirements of the nonprocess emergency response systems including F&G detection, ESD, active and passive fire protection and EER systems were determined by the loss prevention engineers. The external specialist consultants perform the risk-based studies.

Today, process safety engineering is considered as a new discipline. The "process safety engineer" should cover both process and nonprocess hazard identification, risk assessment, and safety barrier management.

Bow-tie or Swiss cheese method is another common technique that the process safety engineers use to determine the safety critical elements (SCEs).

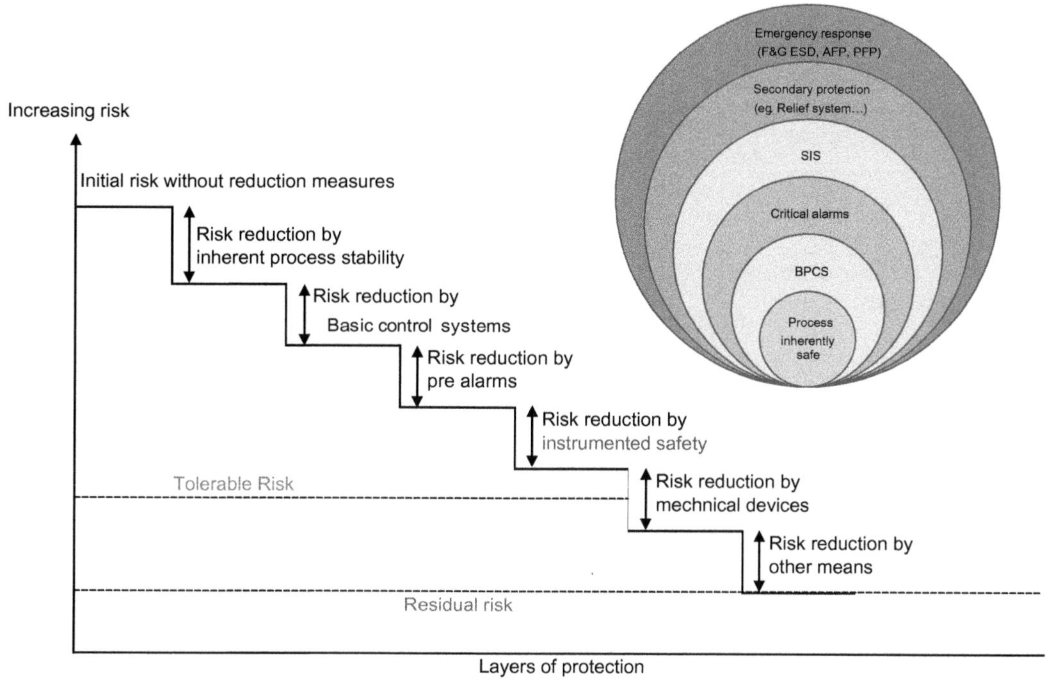

FIGURE 1.17
Layers of protection analysis (LOPA).

Barrier thinking is a useful concept to manage the safety barriers which could be the hardware or human intervention.

A barrier is defined as a functional grouping of safeguards, such as primary containment, process equipment, engineered systems, operational procedures, management system elements, or worker capabilities designed to prevent loss of process containment (LOPC) and other types of asset integrity or PSEs, and mitigate any potential consequences of such events. A set of barriers is also often referred to as a risk control system.

As illustrated in Fig. 1.18, barriers are put in place to manage the risk of a hazard being released resulting in an unintended event, such as LOPC, which could cause harmful consequences. A properly functioning barrier will either stop the event from happening (a prevention barrier) or reduce its consequences (a mitigation barrier).

Hardware barriers include the activities necessary to assure that they continue to meet the performance standards set at the design stage, while the asset's workforce provides human barriers that respond and act to manage the potential cause or threat of an event. The elements of the asset's management system then provide the necessary support processes to ensure the barriers are effective throughout the asset's life (Fig. 1.19).

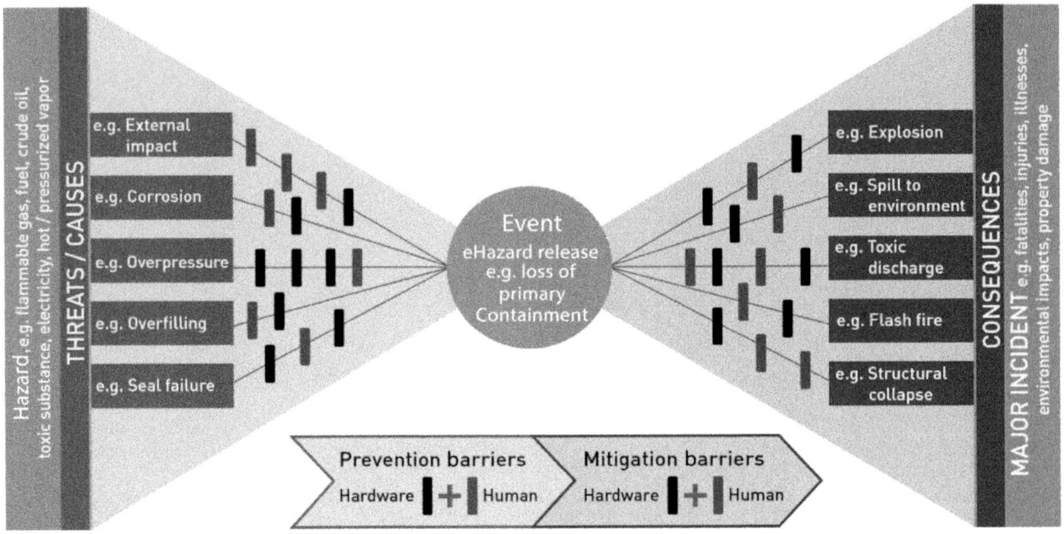

Hardware barriers	Human barriers
Category 1: Structural integrity Category 2: Process containment Category 3: Ignition control Category 4: Detection systems Category 5: Protection systems–including deluge and firewater systems Category 6: Shutdown systems–including operational well isolation and drilling well control equipment Category 7: Emergency response Category 8: Life- saving equipment– including evacuation systems.	Category 1: Operating in accordance with procedures including: • Permit to work • Isolation of equipment • Overrides and inhibits of safety systems • Shift handover, etc. Category 2: Surveillance, operator rounds, and routine inspection Category 3: Authorization of temporary and mobile equipment Category 4: Acceptance of handover or restart of facilities or equipment Category 5: Response to process alarm and upset conditions (e.g. outside safe envelope) Category 6: Response to emergencies.

FIGURE 1.18
Example of bow-tie model. *(From: IOGP-456).*

For example:

- in a pressure vessel engineered with sufficient integrity and controls to prevent release of hydrocarbons, the definition of a hardware barrier includes the management of that barrier or the tasks and activities necessary to assure that it continues to meet the requirements of the performance standard. The management system includes a description of the maintenance and inspection process designed to support this.

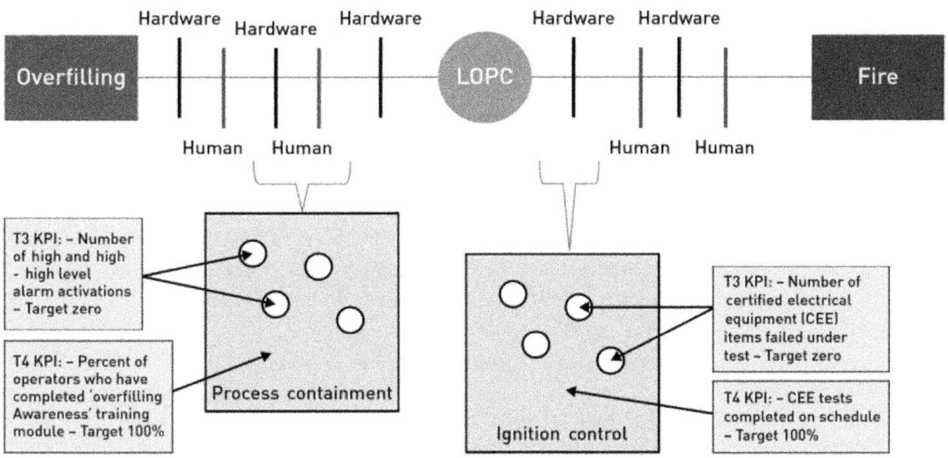

Scenario: An asset has experienced several minor LOPC events caused by overfills of condensate within, or close to, classified hazardous areas where electrical equipment is routinely used. The hazard is flammable liquid hydrocarbon and the threat to the barriers is overfilling, with the potential consequence of a fire leading to a major incident. The asset identified two key barriers which were a concern, process containment,and ignition control, and decided to monitor and assess these barriers using two pairs of T3 and T4 KPIs to provide dual assurance.

FIGURE 1.19
Selection of key barrier for application of dual assurance. *(From: IOGP-456).*

- an operator monitoring the filling of a tank can respond to an alarm by implementing a procedure to prevent an overfill. The effectiveness of this human barrier relies on the discipline and knowledge to operate the plant in accordance with the procedures. Failure to do so may be attributable to an underlying management system failure such as competency management, unclear procedures or under-resourcing.

The KPIs are in place to proactively track and identify the flaws in the safety barrier which are analogue to the "holes in the cheese" and then based on KPI reports the required actions should be taken to eliminate or minimize these defects. Two categories of KPIs can be distinguished. API 754, categorizes the process event KPIs in four categories. The first three categories measure outcomes (i.e., unintended events or effects) and the fourth category of KPIs measure inputs that sustain barriers. IOGP 556 calls this approach as "Dual Assurance."

1.7.2 Management of the Safety Processes

The ethical, legal, and financial imperatives motivate the organization to adopt a more holistic and systematic approach to assuring the integrity of their operations. Process Safety Management (PSM) is a risk-based framework which defines the key Safety Processes to be managed by organizations to assure the integrity of their operations.

Technical, maintenance, operational, human, and organizational factors are incorporated in the PSM framework.

PSM can be defined as

> *...a businesslike approach to safety. It is a systematic, explicit and comprehensive* process *for managing safety risks. As with all management systems, a* safety management system *provides for* goal setting, *planning, and measuring performance. A safety management system is woven into the fabric of an organisation. It becomes part of the culture, the way people do their jobs.*

For the purposes of defining PSM, safety can be defined as

> *... the reduction of risk to a level that is as low as is reasonably practicable (ALARP).*

The guideline of Center for Chemical Process Safety (CCPS) *for risk-based process safety (RBPS) provides a high level framework.* Fig. 1.20 illustrates the Energy Institute version of the CCPS PSM framework.

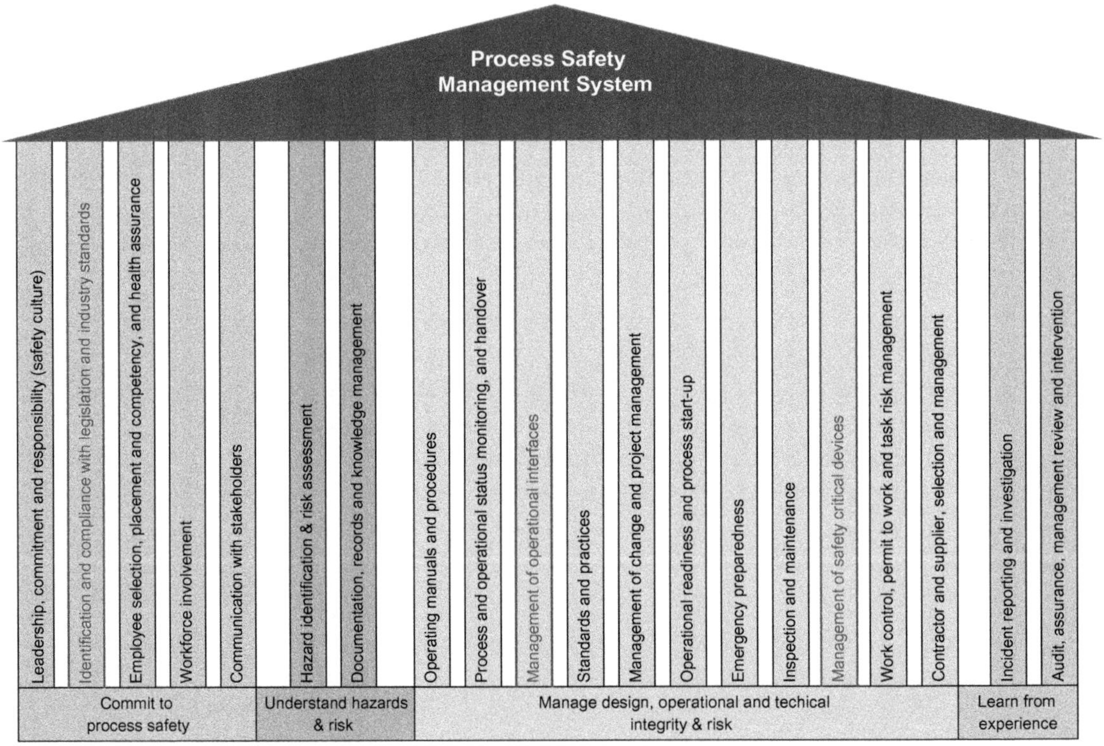

FIGURE 1.20

Energy institute process safety management elements.

Implementation of an effective PSM cannot be achieved if the impact of the other management systems is neglected. Fig. 1.21 illustrates the relationship between the PSM and the other management systems.

The word "process" in PSM does not refer to the chemical engineering processes or process safety engineering that we explained in the previous section. The focus of the PSM is on the "processes" as defined in the *Systems Engineering* standard entitled "ISO/IEC 15288—Guidance for System Lifecycle Processes." Therefore without having a good understanding of system engineering implementation of an effective PSM cannot be possible.

The processes can be used by organizations (for example functional organizations and projects) that play the role of acquirer, supplier (for example main contractor, subcontractor, service provider) or management to fulfill responsibilities about the system-of-interest.

A process is an integrated set of activities that transform inputs (for example a set of data such as requirements) into desired outputs (for example a set of data describing the desired solution).

Fig. 1.22 illustrates example inputs and outputs of a process for engineering a system. The inputs can be either converted to desired outputs or can enable

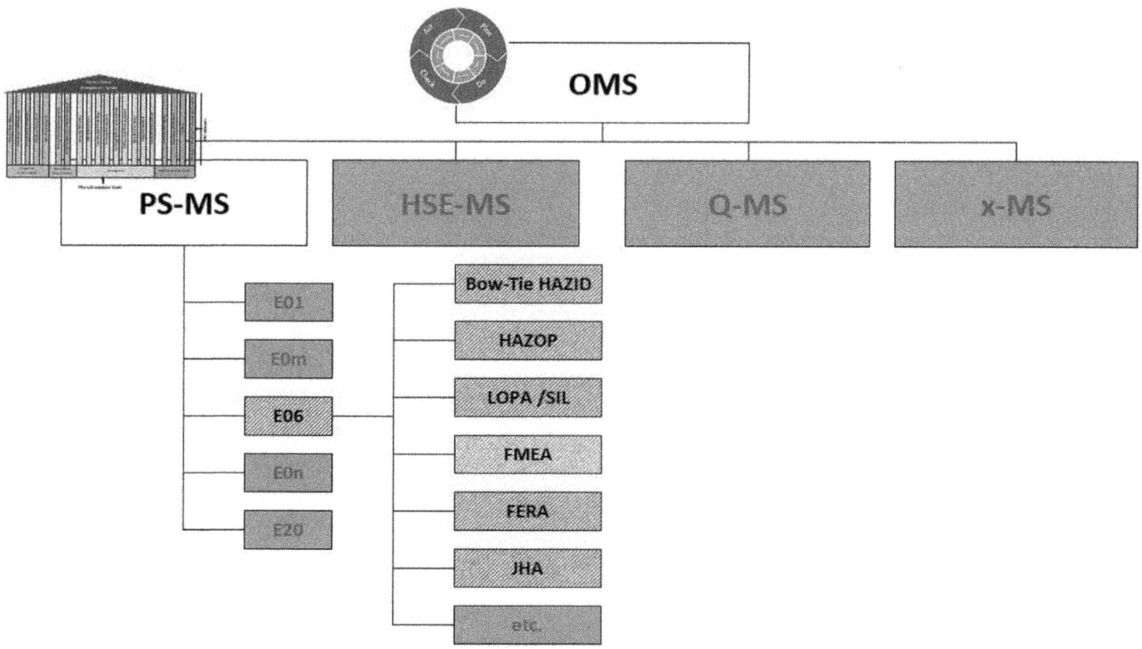

FIGURE 1.21
Operational Excellence Roadmap.

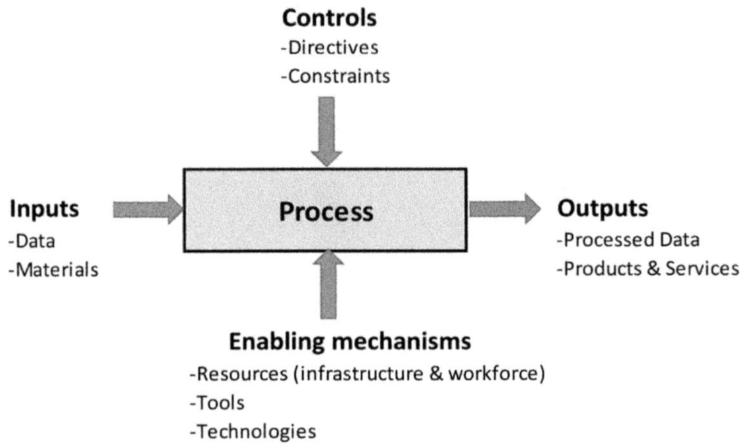

FIGURE 1.22
Example process inputs and outputs according to ISO/IEC 15288.

or control the conversion. Each set of these process inputs and outputs needs to be defined and managed.

Processes can be controlled by organizational or enterprise management directives and constraints and by governmental regulations and laws. Examples of such controls on a process include:

1. The project agreement.
2. The interfaces with other systems for which the project is responsible (see Fig. 1.23).
3. The applicable system lifecycle stage or stages.
4. The organization or enterprise that has project responsibility.

Each process can have a set of process enabling mechanisms as follows:

1. The workforce that performs the tasks related to the process.
2. Other resources required by the process such as facilities, equipment, and funds.
3. Tools (for example software and hardware, automated, manual) required for performing the process activities.
4. Technologies needed by persons performing the activities including methods, procedures, and techniques.

1.7.2.1 Lifecycle Processes

ISO/IEC 15288 describes four groups of system lifecycle processes—agreement, enterprise, project, and technical. Each process has a specific purpose, a set of expected outcomes and a set of activities.

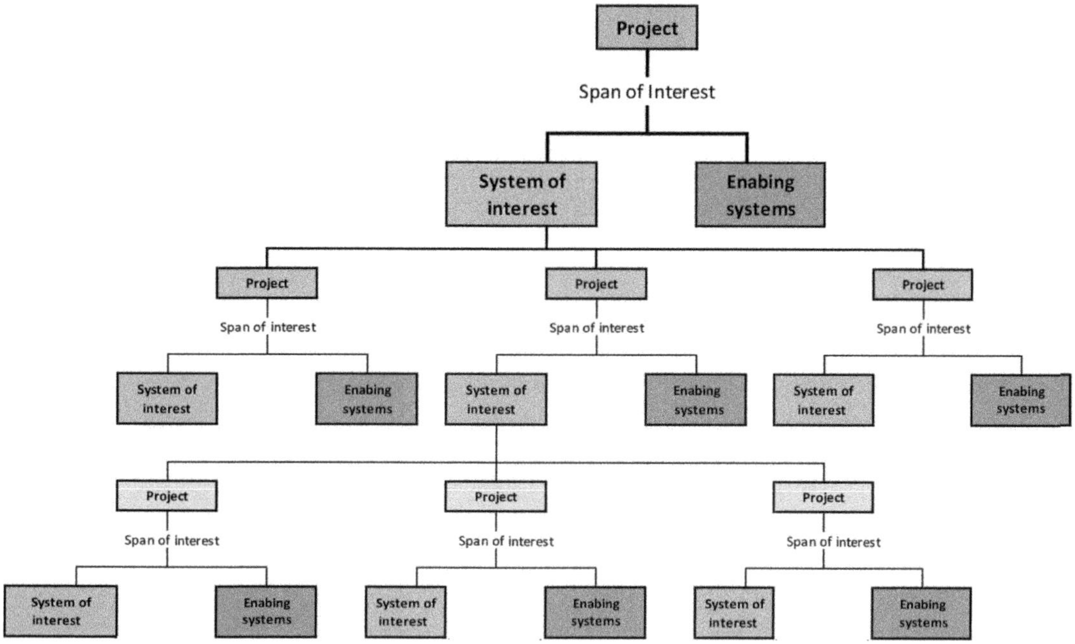

FIGURE 1.23
Hierarchy of projects.

1.7.2.2 Agreement Processes

The agreement processes are applicable for establishing the relationship and requirements between an acquirer and supplier. The agreement processes provide the basis for initiation of other project processes to enable arriving at an agreement to engineer, utilize, support, or retire a system and to acquire or supply related services.

The agreement processes can be used for several purposes such as listed below.

1. To form and ensure completion of an agreement between an acquirer and a supplier for work on a system at any level of the system structure.
2. To establish and carry out agreements to acquire a system or related enabling system services.
3. To obtain work efforts by consultants, subcontractors, functional organizations, projects, or individuals or teams within a project.
4. To provide the basis for closing an agreement after the system has been delivered or work has completed and payment made.

1.7.2.3 Enterprise Processes

Enterprise processes are for that part of the general management that is responsible for establishing and implementing projects related to the products and services of an organization. Thus the enterprise through these distinct processes provides the services that both constrain and enable the projects, directly or indirectly, to meet their requirements.

The enterprise processes are not necessarily the only processes used by an enterprise for governance of its business. For example enterprises also have processes for managing accounts receivable, accounts payable, payroll processing, and marketing. These processes are not within the scope of the International Standard.

For multiple projects involved in or interfacing with an enterprise, or for a teaming arrangement among external organizations, other enterprise processes can be appropriately established or the processes can be appropriately tailored.

To perform these processes, it is not intended that a new organizational unit or discipline within an enterprise be created. Identified and defined roles, responsibilities, and authorities may be assigned to individuals or existing committees or established organizational units. When necessary, however, a new enterprise unit can be formed.

The enterprise processes have specific objectives to fulfill such as listed below.

1. Provide the proper environment so that projects within the organization can accomplish their purpose and objectives.
2. Ensure that there is an orderly approach to starting, stopping, and redirecting projects.
3. Ensure that organizational policies and procedures are defined that set forth the processes of the International Standard and that are applicable to projects within the enterprise.
4. Ensure that appropriate methods and tools are selected and provided to projects so that they can complete process activities efficiently and effectively.
5. Ensure that projects have adequate resources for the project to meet cost, schedule, and performance requirements within acceptable risks and that human resources are appropriately trained for completing their responsibilities.
6. Ensure that project work products for delivery to customers are of a suitable quality.

1.7.2.4 Project Processes

The project processes should be used to manage technical process activities and to assure satisfaction of an agreement. Project processes are performed to establish and update plans, to assess progress against plans and system requirements, to control work efforts, to make required decisions, to manage risks and configurations and to capture, store, and disseminate information.

Outcomes from performing the project processes help in the accomplishment of the technical processes.

The project processes apply to engineering projects that are most often part of larger projects. When that is the case, the appropriate project processes are performed at each level of the system structure. These processes also apply when performing enterprise processes or carrying out the activities related to a lifecycle stage, including utilization, support, and retirement.

When several projects coexist within one enterprise, project processes should be defined to allow for the management of the resources and performance of the multiple projects.

1.7.2.5 Technical Processes

The technical processes are applicable across all lifecycle stages. The following technical processes should be performed to engineer a system.

1. Stakeholder requirements definition process.
2. Requirements analysis process.
3. Architectural design process.
4. Implementation process.
5. Integration process.
6. Verification process.
7. Transition process.
8. Validation process.

These processes should be performed to satisfy the entry or exit criteria of a system lifecycle stage or set of stages. For example, they may be used during early system lifecycle stages to create a feasible system concept, determine technology needs and establish future developmental costs, schedules, and risks. During mid-system lifecycle stages the technical processes may be used to define and realize a new system. During later system lifecycle stages they may be used on legacy systems to make technology refreshments or technology insertions, as well as to correct variations from expected performance during production, utilization, support, or retirement.

The other three technical processes (operations process, maintenance process, and disposal process) can be used during any system lifecycle to accomplish the objectives of a lifecycle stage and support the technical processes used for engineering a system. The operations process and the maintenance process can be performed, as applicable, to support a particular version of a system. The disposal process can be performed to deactivate legacy systems, to dispose of legacy systems and to safely dispose of unwanted by-products from system use.

1.7.3 Technical Process Model

Fig. 1.24 provides a model for the application of the technical processes. This model includes only the technical processes that are primarily used for engineering a *system-of-interest*. The operation process, support process, and disposal process are not shown in Fig. 1.24. These three processes should be used as appropriate to provide inputs to the stakeholder requirements definition process. The requirements could be in the form of acquirer requirements such as

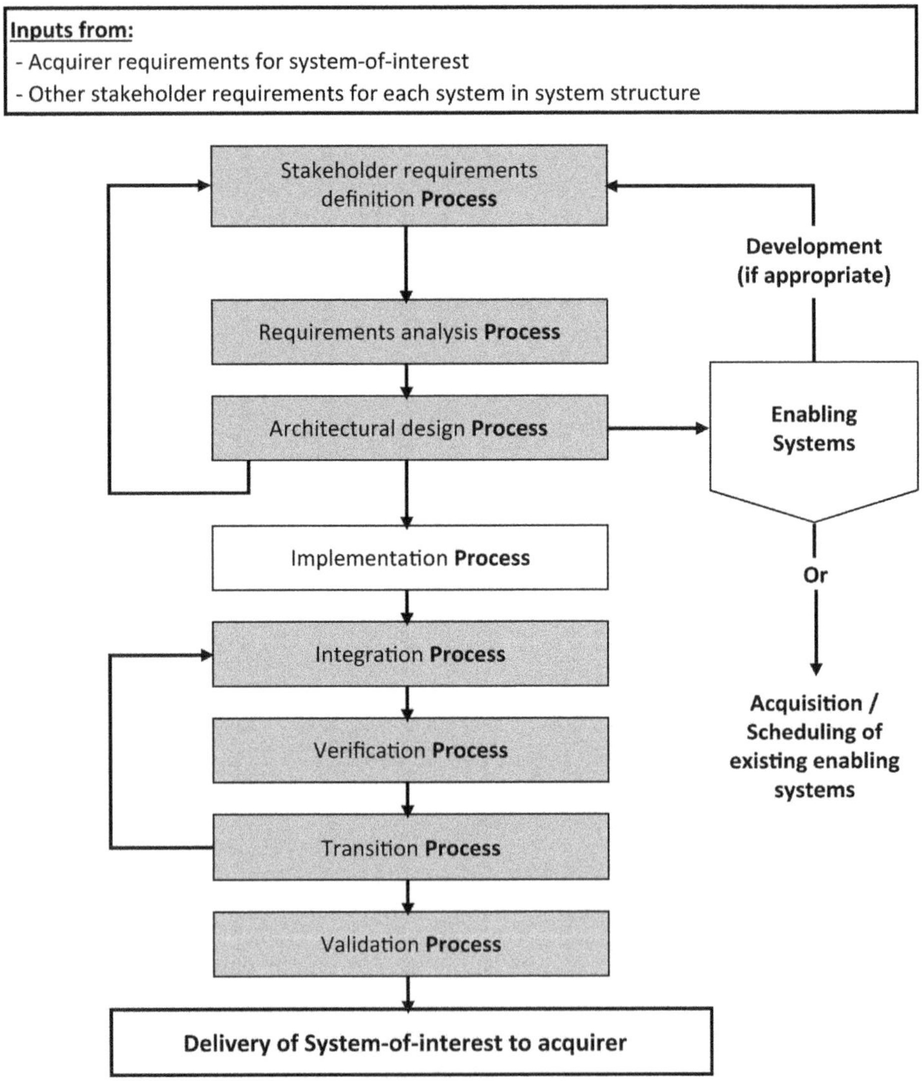

FIGURE 1.24
Application of technical processes to engineer a system-of-interest with the enabling systems.

operability, supportability, and disposability or in the form of other interested party requirements such as for enabling systems to provide related services.

The stakeholder requirements definition process, requirements analysis process, and architectural design process are used to design the solution for each system in the system structure. Application of these processes can be highly iterative to arrive at the desired design solution.

The implementation process, integration process, transition process, and validation process are used to realize the architectural design solution for each system in the system structure. These processes can be highly iterative too.

For each architectural design solution in the system structure the enabling system requirements related to the system should be identified. The enabling system requirements should be satisfied either by engineering the enabling systems that need to be developed or by acquisition or scheduling the existing and available enabling systems.

1.7.4 System Lifecycle Model

Within a lifecycle stage, processes are performed as required to achieve stated objectives. The progression of a system through its life is the result of actions managed and performed by people in one or more enterprises using the processes selected for a lifecycle stage.

ISO/IEC 15288 considers a six stages lifecycle model with the "enterprise view" and the "engineering view" as shown in Fig. 1.25.

The order of use of the lifecycle processes is influenced by multiple factors such as social responsibilities, world trade laws, organizational cultures,

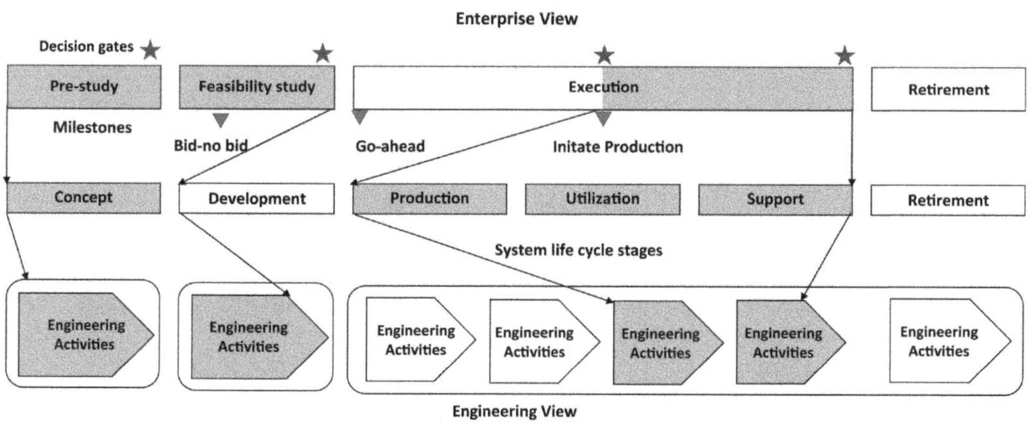

FIGURE 1.25
Enterprise and engineering views related to the representative system lifecycle model.

and technical considerations. Each of these factors can vary during the life of a system. A manager of a system lifecycle stage typically selects the appropriate set of lifecycle processes to meet the exit criteria and other stage objectives. For example, during any of the later lifecycle stages a manager can use the operation process, maintenance process, and disposal process to manage the system while it performs its required functions or is serviced to meet system requirements. During earlier lifecycle stages the same processes can be used to help managing the development of the system as well as affect the disposal of waste products or work products that are no longer needed.

To determine which processes to select and apply during a system lifecycle stage a manager is guided by the purpose and outcomes for each of the stages. The selection of the appropriate processes enables the system's progression through its lifecycle to be managed. The system lifecycle model of Fig. 1.25 can be considered as an illustration of an orderly passage associated with a system going from one stage of life to another. Both the enterprise and engineering views can be helpful in enabling this passage.

An enterprise (for example an automobile company or medical equipment supplier) or a domain group of an organization (for example a government defense agency or industry group) often has a unique view the system lifecycle to control the passage from one system lifecycle stage to the next. The enterprise view illustrated includes management-focused stages that are used to form both milestones and decision gates. The enterprise uses these milestones and gates as decision points where investment decisions can be made as to whether a system should be continued to the next stage or be modified, be canceled or retired or have the plans for the next stage revised before approval. These milestones and decision gates can be used by enterprises to contain the inherent uncertainties and risks associated with costs, schedule, and functionality when a system is created or utilized.

In order to meet the exit criteria of a decision gate a system has to be appropriately engineered and the appropriate work products need to be produced to provide decision-making information and required deliverables. Thus planned engineering activities need to take place during each system lifecycle stage to obtain the outcomes and meet the purpose of the stage or a set of stages.

The engineering view of Fig. 1.25 provides an example framework of engineering activities required to meet the criteria of management decision gates and related system lifecycle model milestones.

Engineering is involved with a system in the early lifecycle stages (concept and development) when it is being studied, defined, and created. Reengineering is involved in later stages (production, utilization, support, retirement) when unwanted and unexpected variations come about due to design errors or failures or new requirements are provided because of technology, competition, or threat system changes.

To engineer a feasible system solution during the concept stage a system structure needs to be sufficiently defined and evaluated. This should be done to assure that system requirements are met and that the costs and risks are understood for the feasible system concept selected. When a parts list is an exit criterion (for example required as part of a proposal or to prepare a creditable cost proposal), sufficiently detailed engineering should be done to ensure that the parts list is complete and that the costs and risks are understood.

To engineer a system solution during the development stage a system needs to be designed with appropriate detail from the system-of-interest level down through successive system levels until a system element can be made, bought, reused or implemented by software. Each system should be verified that it meets its specification requirements included in configuration descriptions from architectural design, and validated that it meets the acquirer and other interested party requirements. Each system elements need to be transitioned to the acquirer where can be assembled and integrated into a higher-level system that is verified, transitioned, and validated. This action continues through successive levels upward to realize the desired system-of-interest.

This approach whether applied to the concept stage or the development stage is typically called top down and bottom up engineering and describes one block of the engineering activities. The top down, bottom up approach is illustrated in Fig. 1.26 and is called the "Vee" diagram or model. This

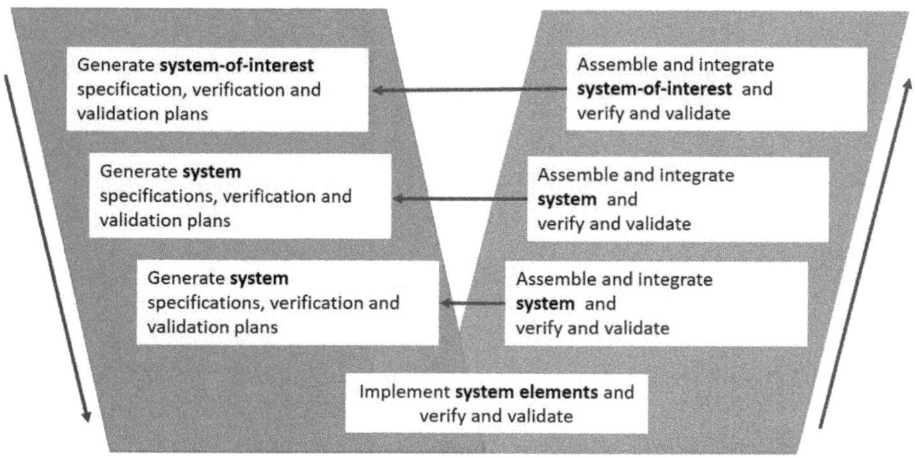

FIGURE 1.26
The engineering "Vee" model.

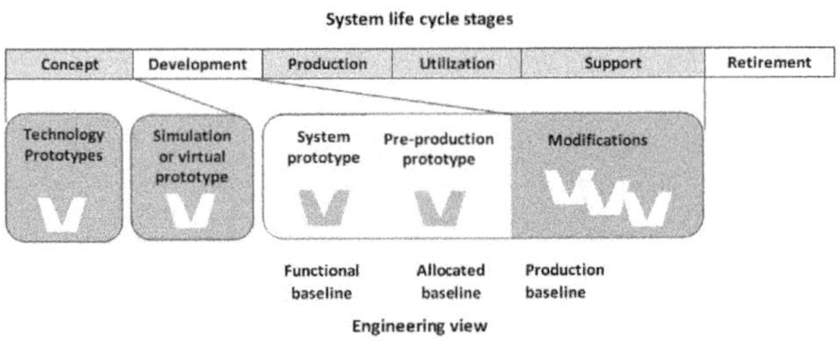

FIGURE 1.27
Engineering view with engineering "Vee" models.

figure reflects the work products and actions expected from the recursive application of the processes in Fig. 1.24 to define and realize the system structure.

Reengineering efforts to correct variations or failures and to meet changed requirements are typically initiated at a system level within the system structure and below the level of the system-of-interest. The same general engineering approach using the "Vee" model is appropriate. In this case, however, the system affected is the place in the system structure where the reengineering effort begins. The requirements for the change are analyzed as to how they could have an impact on interfacing and interacting systems and the performance of the system-of-interest. Then the stakeholder requirements definition process, requirements analysis process, and architectural design process are used downward through successive levels of system structure to define architectural solutions. After the system elements are implemented using the implementation process, the integration process, verification process, transition process, and validation process can be used upward through successive levels to the system-of-interest. This approach is often called middle-out engineering.

The engineering "Vee" model is used in each system lifecycle stages as appropriate to meet stage entry or exit criteria or to meet the enterprise view milestone or decision gate requirements as shown in Fig. 1.27.

1.7.5 Process Versus Procedure

The language of the 2000 revision to the ISO 9000 series significantly moves away from procedure to process and the new concept that the results an organization achieves are the product of the interaction between its processes and not its procedures.

Identifying and managing critical business processes is a vital factor in the effective management of successful organizations. This appears to be a fairly

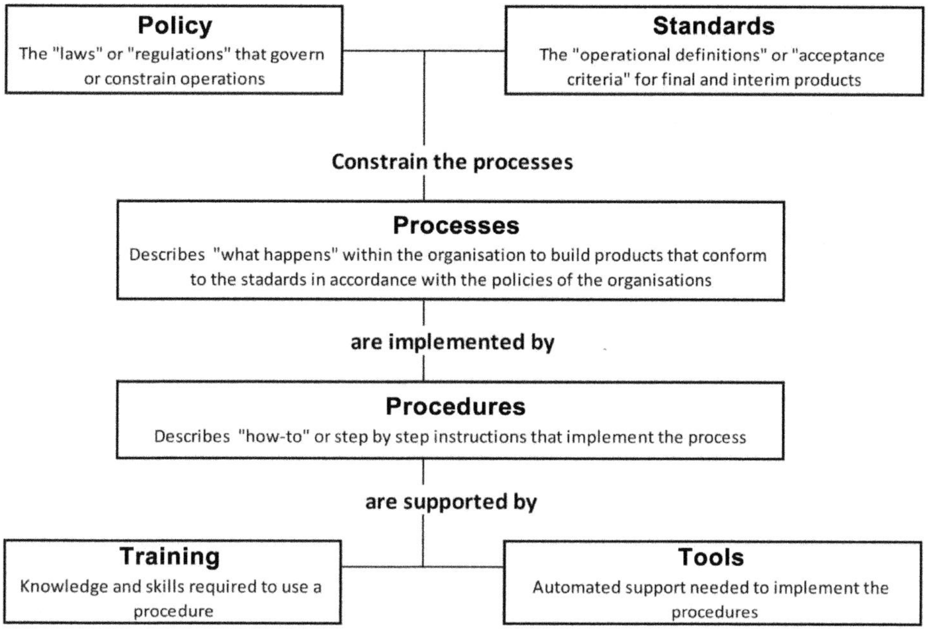

FIGURE 1.28
Process and procedure roles.

obvious statement. At the heart of the business excellence model there is a strong beat generated by the emphasis on process management. Within the context of quality management standards, and more specifically ISO 9000, "procedure" is a key word which has acquired a particular meaning over the years (Fig. 1.28).

1.7.5.1 Procedures

In its simplest form a procedure is a way in which one works to accomplish a task. It can therefore be a sequence of steps that include preparation, conduct, and completion of a task. Each step can be a sequence of activities and each activity can be a sequence of actions. The sequence of steps is critical to whether a statement or document is a procedure or something else. Specifications, contracts, and records are not procedures as they do not tell us how to do anything. These describe the outputs resulting from carrying out procedures or tasks, leaving us to decide any further actions necessary to use these outputs. The output will more than likely be used as inputs to other procedures.

We need procedures when the task we have to perform is complicated or when the task is routine and we want it to be performed consistently. Hence procedures are intended to make something happen in a certain way. If we are not concerned about how something is done and are interested only in the result we do not produce procedures but issue instructions such as "post the letter," "repair the spin drier" or "recruit another person." These are the work instructions as they intend us to do "quantitative" work without telling how to do it or the "qualitative" standard to which the work should be carried out. Instructions are not procedures unless they follow in a sequence and enable us to perform a task.

A set of self-assembly instructions is a procedure as it tells how to proceed to assemble the product. But the wording on the label telling us not to put hot objects on the surface is an instruction or a warning (a special type of instruction).

As procedures are normally used by people they are designed with a user in mind. The user is usually an individual or a group of individuals, although procedures can cover a sequence of steps each of which is performed by different individuals or groups. The perceptions of procedures vary considerably depending on the context in which they are created and used. Any sequence of steps, no matter how simple or complicated, can be expressed as a procedure that is intended to cause someone to act in a certain way to accomplish a task. The key is that the steps follow a sequence. A random collection of statements is not a procedure unless we rearrange these in a sequence that enables someone to proceed.

1.7.5.2 Processes

Processes produce results by converting, transforming, or simply using inputs to create outputs. An input could be material, information, people or a set of conditions and these are passed through a sequence of stages during which they are either used, transformed, or their status changed to emerge as an output with different characteristics. Hence, processes act upon inputs and are dormant until the input is received.

At each stage the transformation tasks may be procedural, but may also be mechanical, chemical, etc. Inherently processes do not normally recognize departmental or functional boundaries (but are often hindered by them) nor the boundaries between customers and suppliers.

Each process has an objective with both quantitative and qualitative measures of its outputs directly related to its objectives. The transformation or process stages are designed to ensure the combination of resources achieves the objectives—the desired outputs. Of course, this means that the process has to receive the right inputs to deliver the desired outputs. Also the correct resources should be applied at the right stages, in the correct quantities and in the right manner.

Table 1.2 Procedures Versus Processes

Procedures	Processes
Procedures are driven by completion of the task	Processes are driven by achievement of the desired outcome
Procedures are implemented	Processes are operated
Procedures steps are completed by different people in different departments with different objectives	Process stages are completed by different people with the same objectives—departments do not matter
Procedures are discontinuous	Processes flow to conclusion
Procedures focus on satisfying the rules	Processes focus on satisfying the user
Procedures define the sequence of steps to execute a task	Processes transform inputs into outputs through use of resources
Procedures are driven by humans	Processes are driven by physical forces some of which may be activated by humans
Procedures may be used to process information	Information is processed by use of a procedure
Procedures exist they are static	Processes behave they are dynamic
Procedures cause people to take actions and decisions	Processes cause things to happen

It is true that a process can be illustrated as a sequence of steps just as a procedure is illustrated, but the similarity ends there. Table 1.2 compares the key features of procedures and processes.

To make a transition away from managing procedures towards process management an organization must answer whether it has:

- clearly defined what its objectives are and how it will measure and review the success of achieving those objectives
- evaluated the impact of those objectives on the interested parties, the stakeholders
- designed the critical, end-to-end processes necessary to deliver the objectives
- assessed and provided the resources, skills, and competence to make the processes work

The change in language from procedure to process is not about perception or semantics. To remain competitive the processes should be designed to add value consistently.

1.7.6 Efficiency Versus Effectiveness

Productivity is determined by looking at the production obtained (effectiveness) versus the invested effort in order to achieve the result (efficiency); in other words, if we can achieve more with less effort, productivity increases (Fig. 1.29).

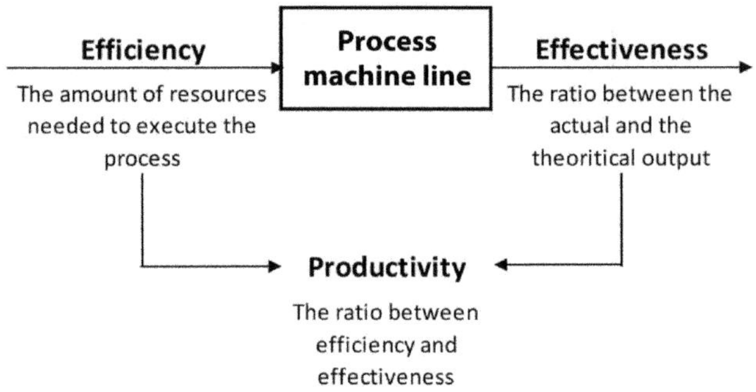

FIGURE 1.29
Productivity is the ratio between efficiency and effectiveness.

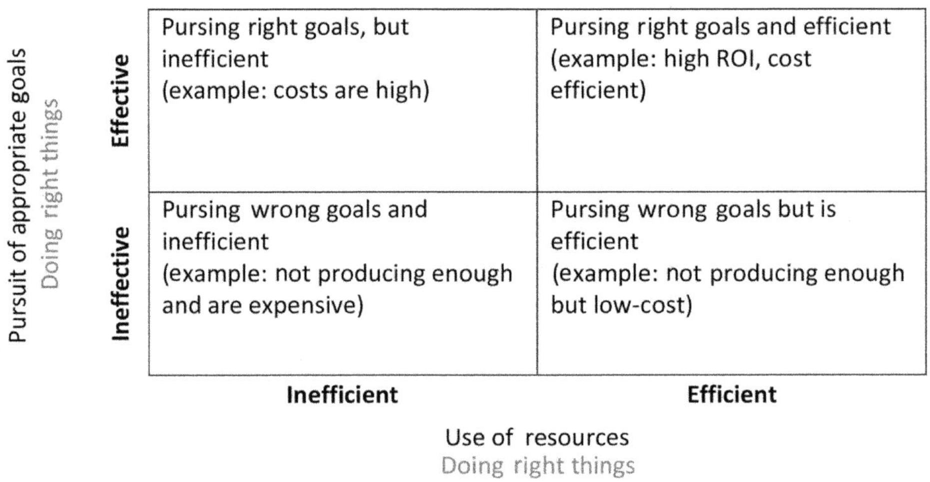

FIGURE 1.30
Effectiveness and efficiency matrix.

For example Amec Foster Wheeler reports that following implementation of their lean "More4Less" program, they achieved 30%–60% cost saving for modifications with accelerated delivery times of more than 50%.

Effectiveness and efficiency are two fundamental goals; all businesses over the world are pursuing, however, more often, there are lack of clarity upon their true means and how to achieve them accordingly (Fig. 1.30).

Business must ensure both efficiency and effectiveness, and with agility if they are to be successful. Efficiency is *doing things right* with minimum inputs and resources (do it right the first time) and effectiveness is *doing the right thing* by following the principles and leading in the right directions. Leaders focus on effectiveness, to ensure business having the vision and well-defined goals to reach it; while managers focus on efficiency, "efficiency" is the relationship between how much time (or labor) you expected or planned to expend, versus what the actual was. If you expend less time or labor than expected, you were efficient.

Companies often talk about employee effectiveness and efficiency when brainstorming ways to improve business. While they sound similar, effectiveness means something entirely different than efficiency. An effective employee produces at a high level, while an efficient employee produces quickly and intelligently. By combining effectiveness and efficiency a company produces better products faster and with fewer resources.

To improve effectiveness, companies must take the initiative to provide thorough performance reviews, detailing an employee's weakness through constructive criticism. Managers must make it a point to address effectiveness and explain how an employee's performance affects the company as a whole. To avoid a workplace full of ineffective employees, companies must hire high-performing employees by weeding out candidates at the recruiting level. Employees are often ineffective because they do not care about their work or because they do not possess the skills to contribute. By interviewing candidates, calling references and conducting tests, companies can bring on employees with skills better suited for performing at a high level.

Employees and managers are often inefficient because they either do not know how to be efficient or do not have the necessary tools to perform tasks efficiently. Ways to improve efficiency include meeting with managers and employees to outline ways to implement efficiency in the workplace and asking for opinions on what the workplace is missing. For example, a small business that lacks an employee email system prevents managers from communicating with employees efficiently.

Quality is doing the right thing right, the first time which means operation with zero backlog and wastage and highest customer satisfaction rate. Generally, you have to assure the effectiveness first, and then make the effect more efficient. Efficiency means a way to measure how good you are in what you are doing in terms of available resources.

Most of the organizations think of efficiency as the most beneficial (profitable) means of doing business. Distributing workloads and delegating is often necessary, but many business practices add complexity that inevitably leads to losses, through additional expenses, waiting, bureaucracy, etc. Often such unnecessary complexity becomes obstacle for businesses to achieve *agility to response* to the changes, or lack of flexibility to make alternative options to do the work. At higher mature level, agility is the ultimate goal for business to response to change with speed, and out-beat competitors with capacity.

Effectiveness is a product of *wisdom* which enlarges both the range of consequences considered in making a decision and the length of time over which the decision is believed to have possible consequences. By taking long- as well as shortrun consequences into account, wisdom prevents sacrificing the future for the present. For example, our technology enables us to keep terminally ill people alive at great cost. But is this the right thing to do in the long run? Is it wise? Might the same resources be better used elsewhere? Wisdom is required for the effective pursuit of ideals, and therefore is required of leadership. Leaders must also have a creative and recreative role in the pursuit of ideals, and these are esthetic functions.

1.8 PROCESS INDUSTRY VERSUS DISCRETE MANUFACTURING

All manufacturing processes can be broadly categorized into two groups: discrete parts assembly manufacturing and process industry manufacturing. Assembly manufacturing generally consists of the manufacture of individual parts and components and then welding, bolting, or otherwise fastening them together into a finished product. Examples include automobiles, aircraft, motorcycles, cell phones, computers, power tools, television sets, and hair dryers.

Process industries are characterized by processes including chemical reactions, mixing, blending, extrusion, sheet forming, slitting, baking, and annealing. Finished products can be in solid form packaged as rolls, spools, sheets, or tubes; or they can be in powder, pellet, or liquid form in containers ranging from bottles and buckets to tankcars and railcars. Examples include automotive and house paints, processed foods and beverages, paper goods, plastic packaging films, fibers, carpets, glass, and ceramics. The outputs may be sold as consumer products (e.g., food and beverages, cosmetics) but more often become ingredients or components for other manufacturing processes (Fig. 1.31).

Process industry and discrete parts assembly manufacturing operations are different and have different challenges. The differences are profound enough that the application of system engineering and other industrial engineering tools must be approached quite differently (Fig. 1.32).

The difference between assembly and process industries has often been characterized as discrete versus continuous processing, but that is a profound over simplification. While many of these processes are continuous (e.g., oil refining, manufacture of bulk chemicals), many are batch chemical (house paints, industrial lubricants) or what could be considered mechanical batching (e.g., rolls of paper, tubs of fiber) and become discrete later in the

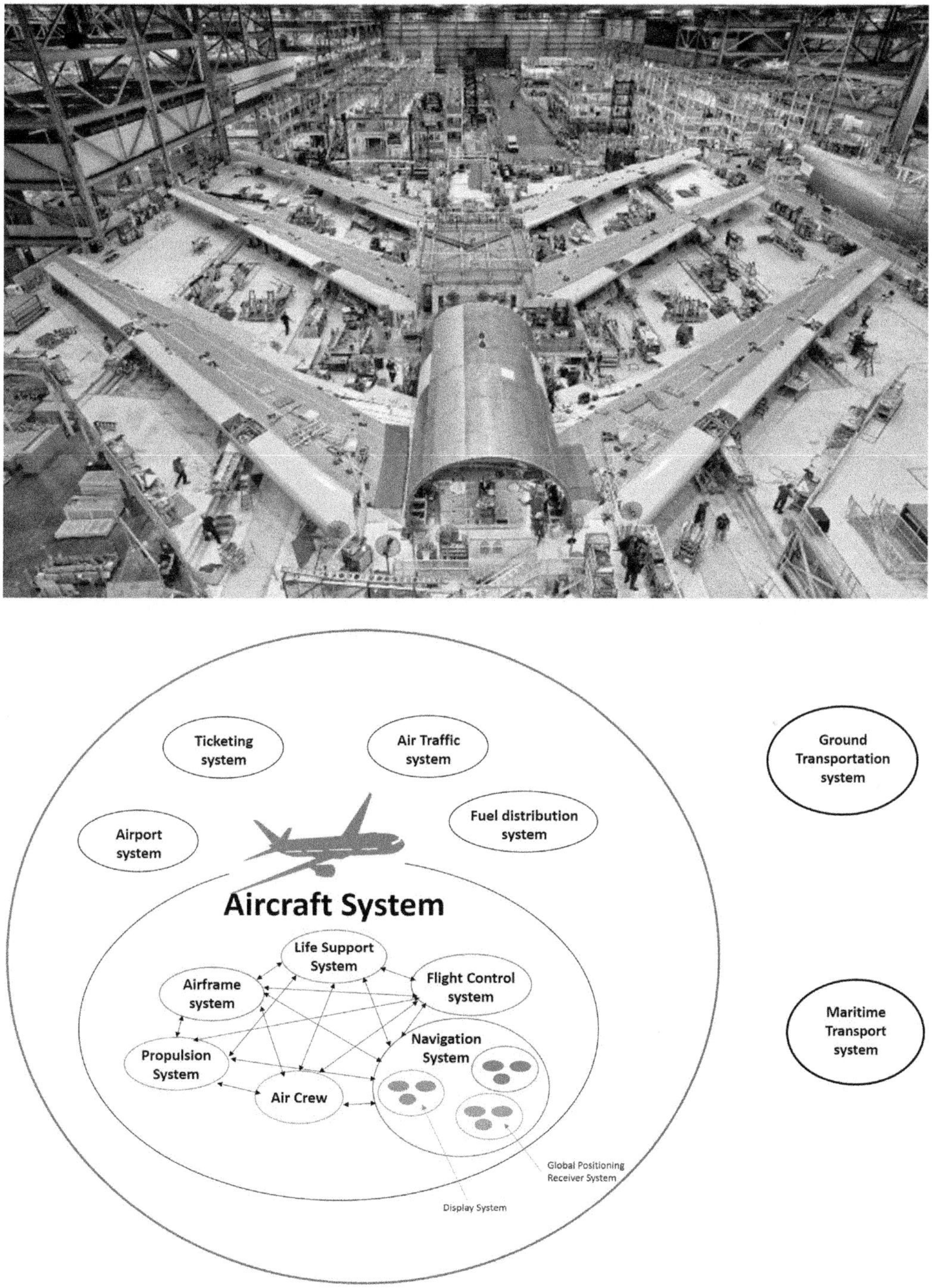

FIGURE 1.31

Air transport system. Example of "System of Systems" in Manufacturing industry.

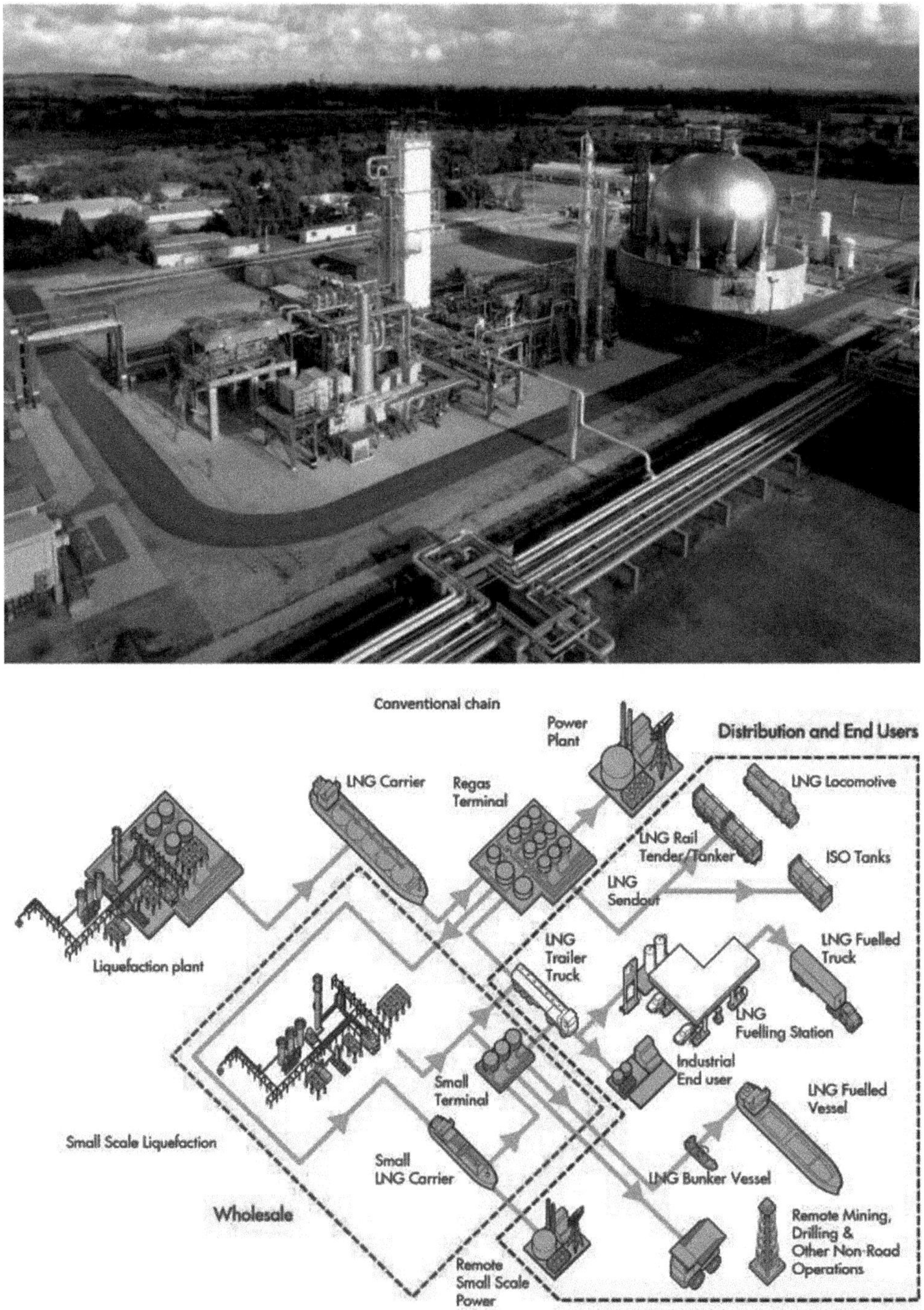

FIGURE 1.32
LNG production & distribution system. Example of the "System of Systems" in Process industry.

process (e.g., tubes of toothpaste, rolls of carpet, buckets of paint, jars of mayonnaise, boxes of cereal).

A better characterization of the difference would be that the number of different part types converges as material flows through an assembly operation, while the product variety increases as material flows through a process operation. Assembly manufacturing starts with a very large number of raw materials and ends with a small number of finished product stock keeping units (SKU), while process operations are the opposite; very few raw materials become highly differentiated as material flows through the process, ending with a large number of finished SKUs. Table 1.3 summarizes the contrast between discrete part assembly manufacturing and process industry.

1.9 APPLICATION OF SYSTEM ENGINEERING IN PROCESS INDUSTRY

Traditional systems engineering was seen as a branch of engineering in the classical sense, that is, as applied only to the *physical systems,* such as spacecraft and aircraft. More recently, systems engineering had evolved to take on a broader meaning especially when *humans* are seen as an essential component of a system. Fig. 1.33 illustrates how the human organizations are embedded in the aircraft and its enabling systems. The other systems are the "enabling systems".

With the introduction of the international standard ISO/IEC 15288 in 2002 the systems engineering discipline was formally recognized as a preferred mechanism to establish an agreement for the creation of products and services to be traded between two enterprises—the acquirer and supplier.

ISO/IEC 15288 considers two specific kinds of systems: *systems-of-interest* and *enabling systems.* There is a relationship between these two kinds of systems. Each system-of-interest has its associated set of enabling systems needed for the system-of-interest to be created, utilized and retired from use during its lifecycle.

Consistent with the broader scope of systems engineering the Systems Engineering Body of Knowledge (SEBoK) has defined three types of systems engineering:

1. *Product systems engineering* is the traditional systems engineering focused on the design of physical systems consisting of hardware and software.
2. *Enterprise systems engineering* pertains to the view of enterprises, that is, organizations or combinations of organizations, as systems.

Table 1.3 Comparison Between Assembly Manufacturing and Process Industry Business Characteristic

Business Characteristic	Assembly Manufacturing	Process Industries
Examples	Automobiles, aircraft, cell phones, computers, power tools, industrial equipment, home appliances	Oil & gas, petrochemical, refineries, chemical, paints, paper & plastic sheet goods, food & beverages, fibers, yarns, carpets, metals & ceramics
Production	Parts or assemblies Bill of material (BOM)	Ingredients, formula, recipe (process bill of materials)
Repetitiveness	Unit	Bulk—ability to adjust batch sizes depending upon available inventory of ingredients/other materials
Parts	The assembly products can be broken down to the parts and reassemble using the repaired parts or new spare parts	In process manufacturing, after a product is produced it cannot then be broken down into its component parts
	An assembly line is a manufacturing process in which interchangeable parts are added to a product in a sequential manner to create an end product	An off-spec product may or may not be processed again
Process flow model	"A" processes Part variety convergence Many raw materials Little final differentiation 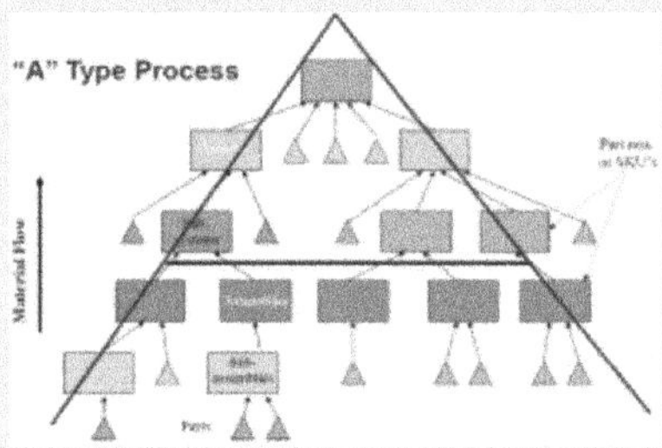	"V" processes Material variety divergence Few raw materials High final differentiation 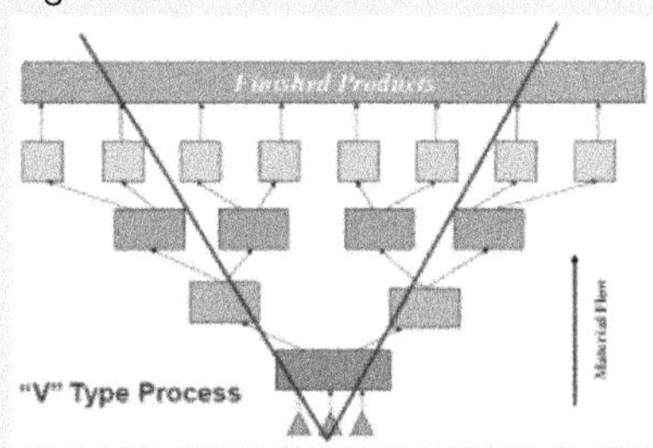
Primary economic drivers	Labor productivity Inventory reduction	Asset productivity Inventory reduction Increased throughput Reduced yield losses
Primary rate limiting factor	Labor	Equipment
Tools/ techniques	Value stream mapping 5S Standard work Poka-yoke SMED One-piece flow Cellular manufacturing Production leveling Mixed model production Autonomation Synchronize flow to TAKT Pull systems	Supply chain mapping Value stream mapping 5S Standard Work Poka-yoke SMED Flow determined by equipment size Cellular manufacturing Product wheels Autonomation Synchronize flow to TAKT Pull systems

Continued

Table 1.3 Comparison Between Assembly Manufacturing and Process Industry Business Characteristic *Continued*

Business Characteristic	Assembly Manufacturing	Process Industries
Batch logic influenced by	Machine set-up time Transportation lot size	Batch size by: Vessel size Roll length and width Bale size campaign size by: Changeover time EOQ
Set-up issues	Time to replace, reset tooling	Time to clean out process vessels Time to reset, stabilize temperatures Time for pressures to equillibrate Time to get properties on aim after
Cellular manufacturing implementation	Group technology physical work cells	Group technology virtual cells
Production leveling techniques	Control market demand Mixed model production Heijunkaa	Product wheels: Batch sequence optimization Batch length optimization Heijunkaa

[a]*Heijunka (English: Production smoothing or leveling): Heijunka (hi-JUNE-kuh) is a Japanese word for leveling. It is part of the lean methodology of process improvement that helps organizations match unpredictable customer demand patterns and eliminate manufacturing waste by leveling the type and quantity of production output over a fixed period of time.*

3. *Service systems engineering* is about the engineering of service systems. A service system is conceived as serving another system. Most civil infrastructure systems are service systems.

In the process industry the notion of the "product systems engineering" can be ambiguous. The process plant is the final product in the project phase. The EPC contractors shall integrate the physical subsystems and assure the integrity of the overall plant system, as shown in Fig. 1.34.

An empty process plant like an aircraft is made of the dissociable parts and can be considered as a discrete assembly product.

Modularization of the production units reduces the cost of the EPC projects but adds the interfaces which increase the complexity of manufacturing and assembly of the process units.

The advances in technology enable us to realize the very high capacity plants in the tough environmental conditions. These projects are called *megaprojects* and need a significant infrascrture and supply chain configurations. They

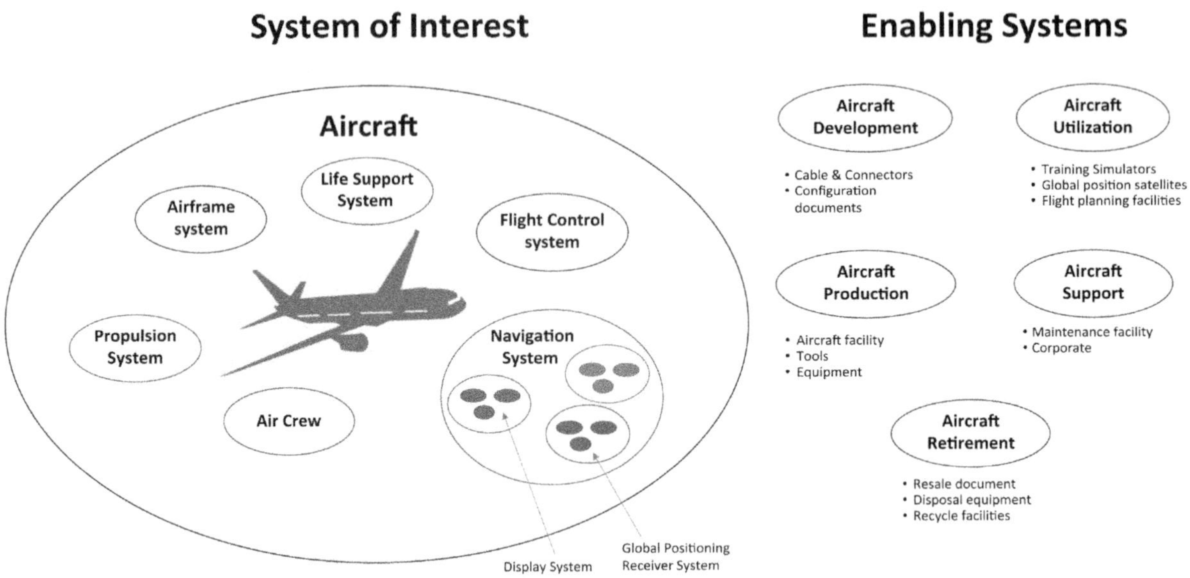

FIGURE 1.33
Example aircraft system-of-interest and its enabling systems. *From: ISO/IEC 15288 in 2002.*

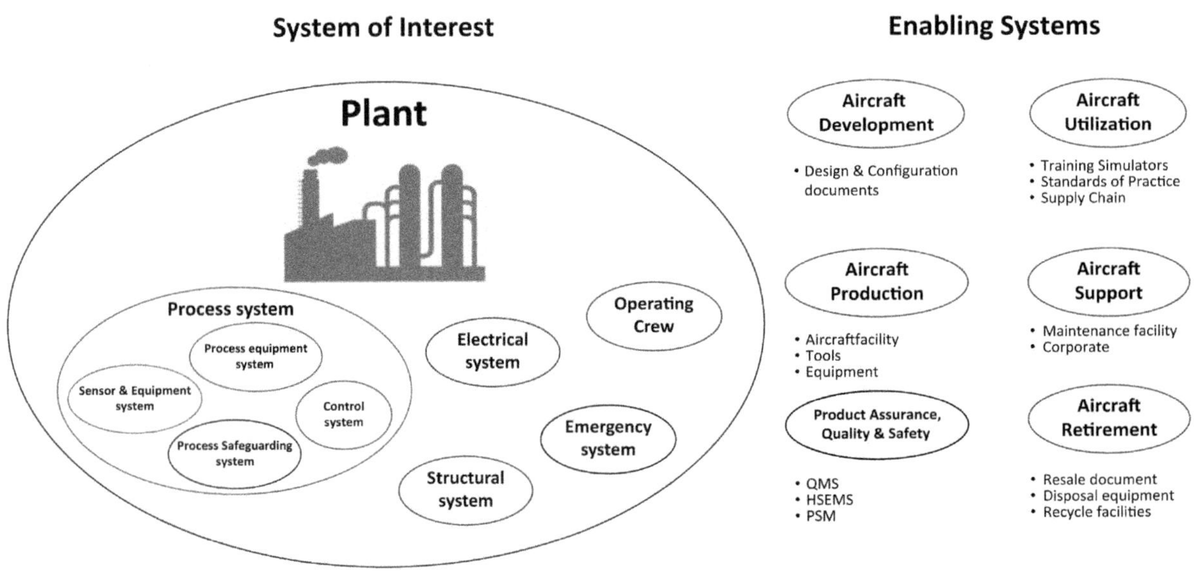

FIGURE 1.34
System-of-interest and its enabling systems for a process plant according to ISO/IEC 15288.

attract a lot of public attention because of substantial impacts on communities, environment, and budgets, and many of them costs more than US$1 billion. In theory the higher capacities mean the lower production costs, but it also means coping with much higher complexity during design and realization of the project.

Ernst & Young published a report in 2014, revealing that almost *two-thirds* (64%) of multibillion-dollar, technically and operationally demanding megaprojects continue to exceed budgets, with three-quarters *(73%)* missing project schedule deadlines. The report "Spotlight on megaprojects" examines the performance of 365 megaprojects and the impact on the oil and gas industry of these overruns (Fig. 1.35).

Modularization and megaprojects increase the need for understanding the complexity management techniques in the system engineering of the process manufacturing plants.

In conventional system engineering, after testing and iteration of the prototype, the new product is industrialized and produced in series (Fig. 1.36).

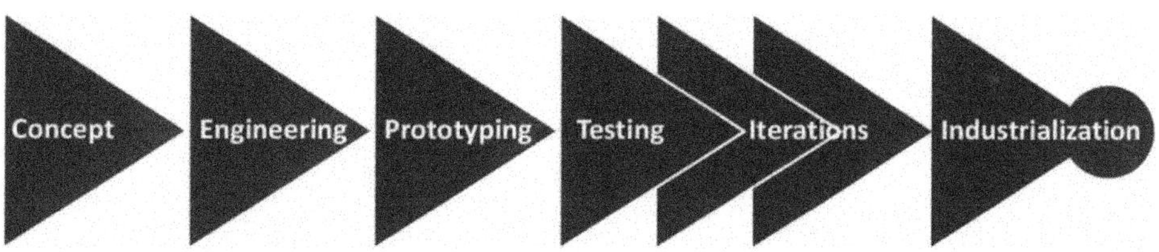

FIGURE 1.35
Lifecycle of a new discrete assembly product.

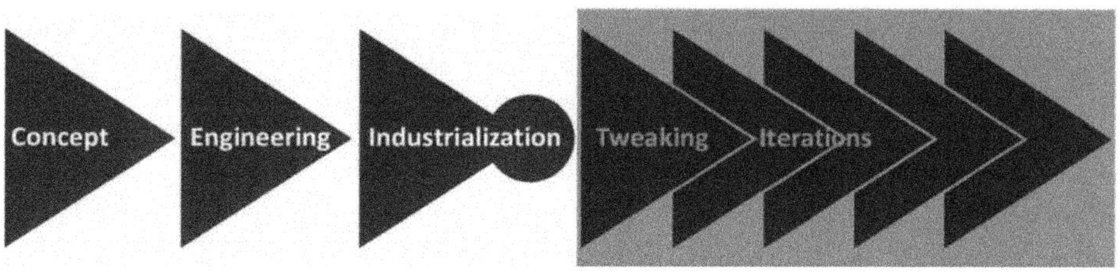

FIGURE 1.36
Lifecycle of a new process plant.

The process plants are the *unique products* because even for similar process licenses when the location, layout, and capacity change, the entire configuration of the facility changes. Therefore industrialization starts soon after the detailed design. Optimization of the plant is performed after industrialization by the "Tweaking and Iterations" of the industrialized process plant (Fig. 1.37).

The consequences of this approach are as follows:

- The costs are unpredictable:
  - Magnitude of deficiencies
  - Number of iterations
  - Standstill time (while upgrading)
- The final results will always be a compromise:
  - Final performance will depend on the initial philosophy of the concept/prototype
  - Depending on the additional costs or investments one is willing to make, the resulting performance will be "as good as it gets."
- The industrialized solution will most probably be conservative, with little innovations.

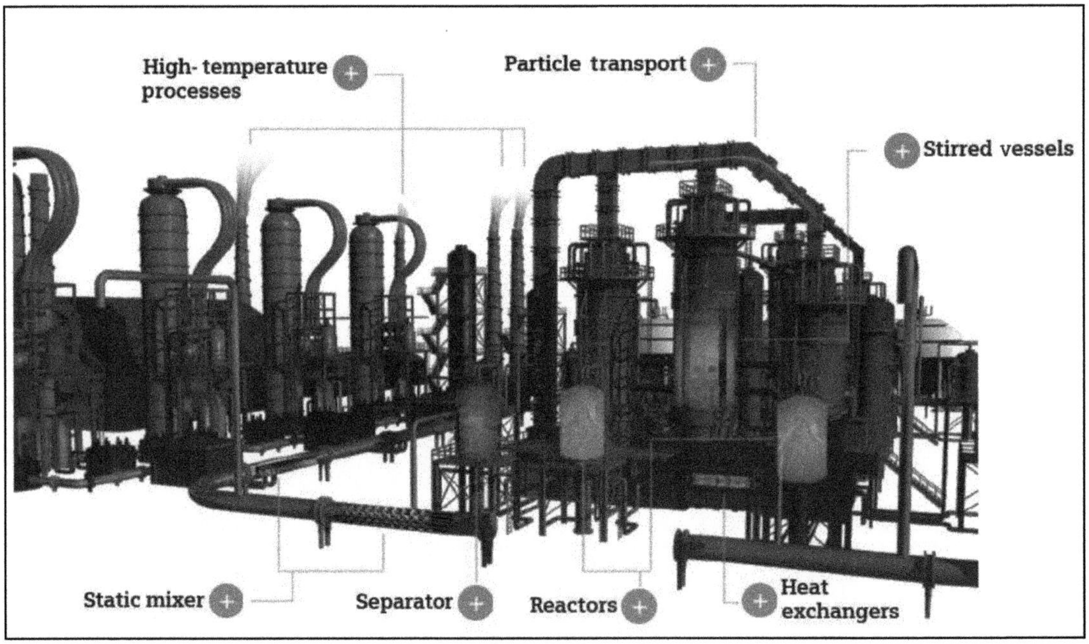

FIGURE 1.37
Virtual prototyping and CFD modeling enables the cost effective "Tweaking & Iteration."

Virtual prototyping and simulation is an effective solution to fulfill the requirements of the process plants for "Tweaking and Iterations" phases. With this approach:

- Concepts and solutions can be new, unknown, risky, and innovative:
  - Big failures are allowed in virtual prototypes
  - Designers can develop up to the limits, and learn from it
  - Crazy ideas are allowed (and might just work)
- The final design will be optimized:
  - Virtual iterations are fast and inexpensive
  - Visualizations (surface plots, flow trajectories, iso-surfaces) give more insight into a design then measurement ever will
  - Use of parameter optimization

Fig. 1.38 compares the virtual prototyping and simulation phases of the process plants with the physical prototyping of a new discrete assembly product. These features for the system engineering of the process manufacturing plants are not covered by ISO/IEC 15288.

"Structural," "electrical," and "emergency response" systems are essential to the integrity and safe functions of the process system. They can be considered as the independent systems, but their failure will lead to the safe or unsafe failures of the "process" functions. For example:

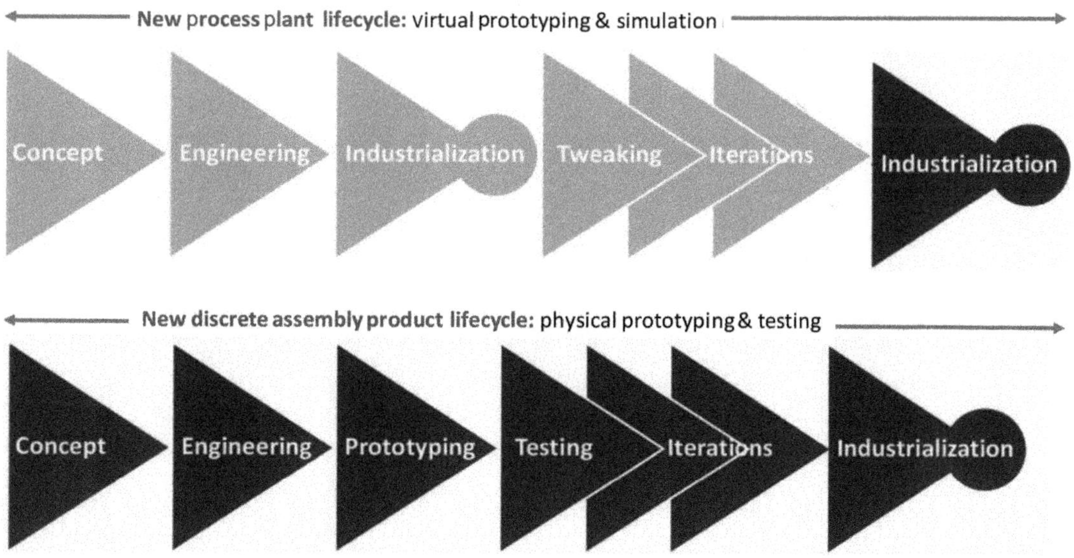

FIGURE 1.38
Comparison of the prototyping solutions of the process and discrete manufacturing plants.

- If the structure of the unit is buckling the pipework system will be under stress and hazardous materials may leak from the pipe flaws or loosen flanges.
- If the electrical system fails all the rotatory equipment will stop, and safe shutdown of the plant will be initiated.
- If there is a fire and emergency firefighting system fails to operate then the fire will escalate, and process equipment may totally be damaged.

The structural and process equipment systems interact continuously. For example, when a relief valve is opened, vessel structures are subjected to vibration, an emergency blowdown cause the cold stress in the equipment and associated pipework, or the land movements may cause the buckling and stress to the process equipment affecting the performance and safety of the unit operation.

Fig. 1.39 illustrates the effects of the internal and external environments on a process column and Fig. 1.40 illustrates the mechanical stress distribution on

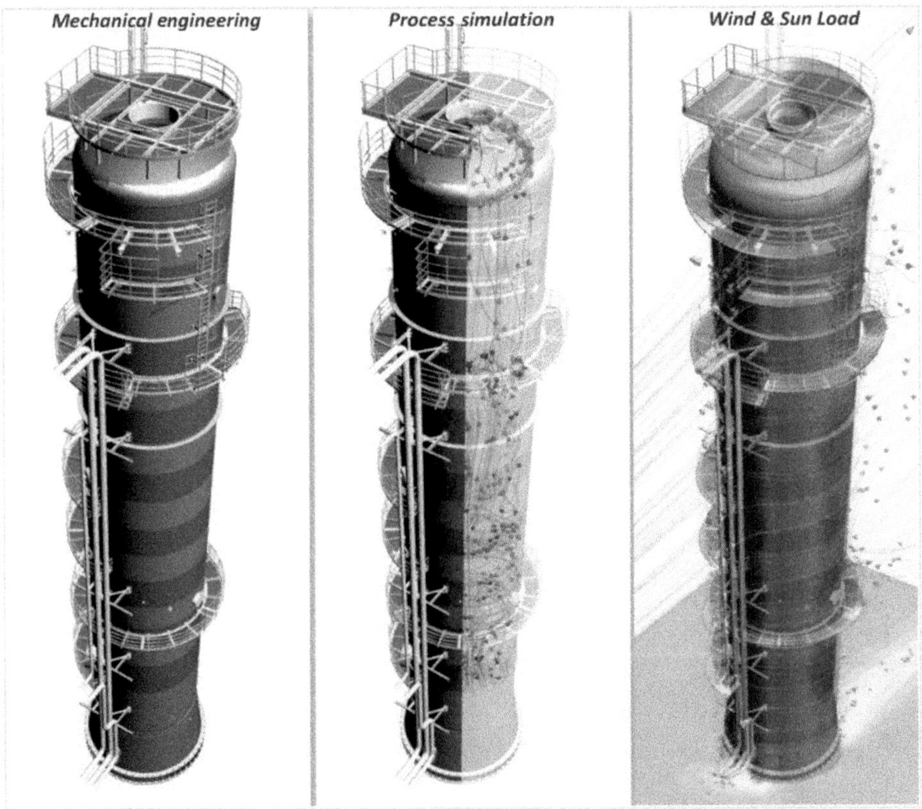

FIGURE 1.39

Internal process materials and natural environment interacts continuously with the mechanical and structural parts of process units.

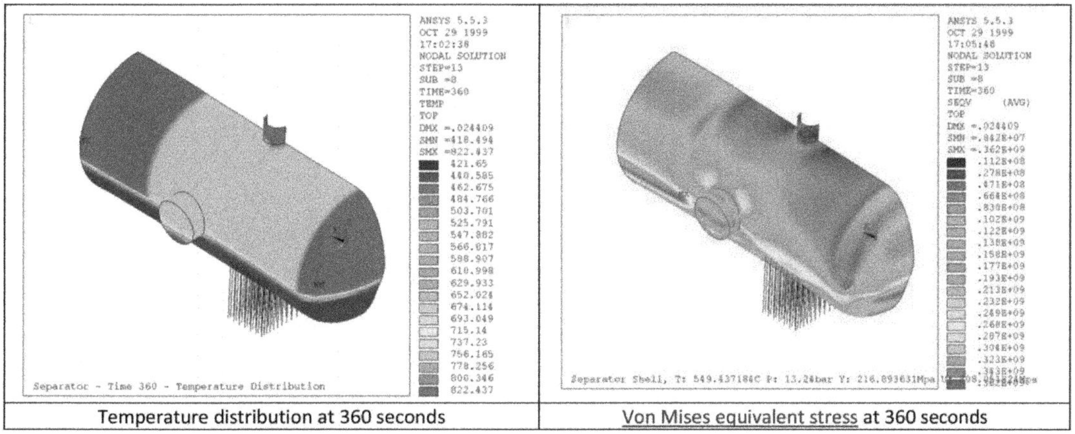

FIGURE 1.40

Response of pressurized process vessels and equipment to fire attack.

the surface of a pressure vessel which is attacked by a fire. API 579 provides the guidance for the fitness for service study of these cases.

Fig. 1.41 illustrates the interaction of the process plant systems. Each of these systems has their own lifecycle processes. If there is any changes in the configuration of one of these systems, the effects of this change on the configuration and perforamce of the other systems should be verified to assure the integrity of the whole system.

A perfect plant without the materials in the process equipment serves to nothing. During operation the "process materials streams system" is our primary "system of interest." The other systems are the "enabling systems". The other systems such as process equipment and control systems are the enabling systems. In other words the process flow diagrams such as Fig. 1.42 illustrate how the process related system of interest and its enabling systems are configured.

"Materials in movement" in the process equipment systems are homogeneous and cannot be dissociated into the parts. Their state and conditions are the functions of the upstream and downstream process conditions, i.-e., temperature, pressure, etc. Process conditions are controlled by the combination of the human intervention and control systems. If the process conditions exceed the design limits product can be irreversibly damaged.

Sometimes the material components or state phases are in the equilibrium, and the materials system shows the self-regulation characterstics against the process conditions changes. For example, to control a distillation column at least five control loops should be provided. If these control loops are optimized individually,

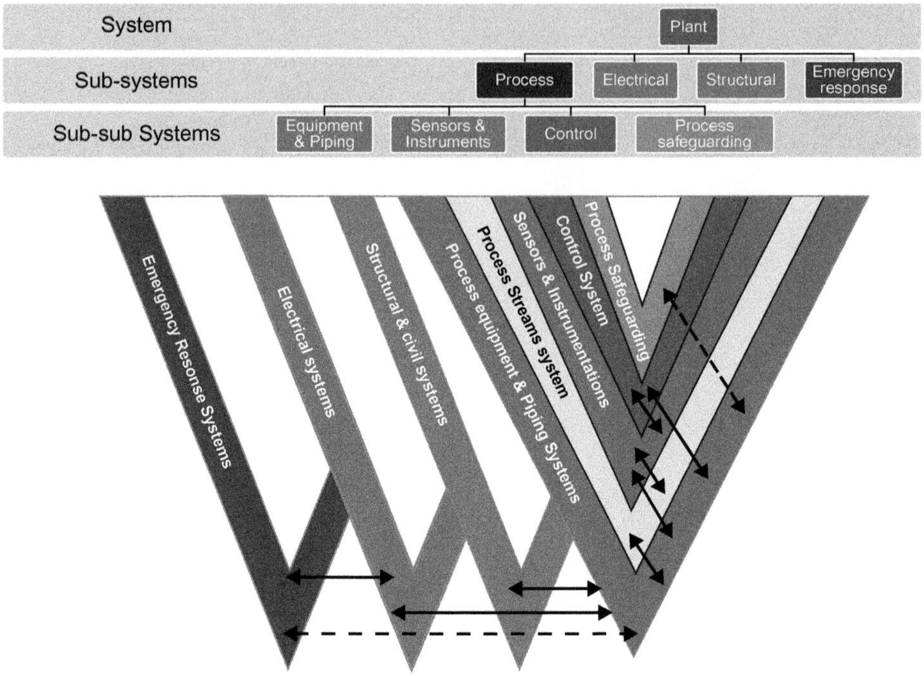

FIGURE 1.41
Process plant subsystems and "vee" diagrams.

it does not mean that the whole distillation column is optimized. Thermodynamic equilibrium creates the dependency between liquid and vapor phases. When one phase is subjected to change, the other phase changes too. This can have the positive or negative effect on the performance of the control system.

Process equipment system contains and provides the necessary conditions for the transformation of the raw materials into the final products. Sensors and control systems interact *continuously* with the process equipment and material systems. "Process safeguarding system" not only protects the "process equipment system" against the undesirable process events but also protects people and the environment against the major accident outcomes. Process safeguarding system like emergency response system is initiated *on demand*.

ISA-95 (IEC/ISO 62264) is an another international standard which has been developed about a decade ago to address the problems encountered during the development of automated interfaces between enterprise and control systems. This standard applies to all industries, and in all sorts of processes, such as batch, continuous, and repetitive or discrete processes.

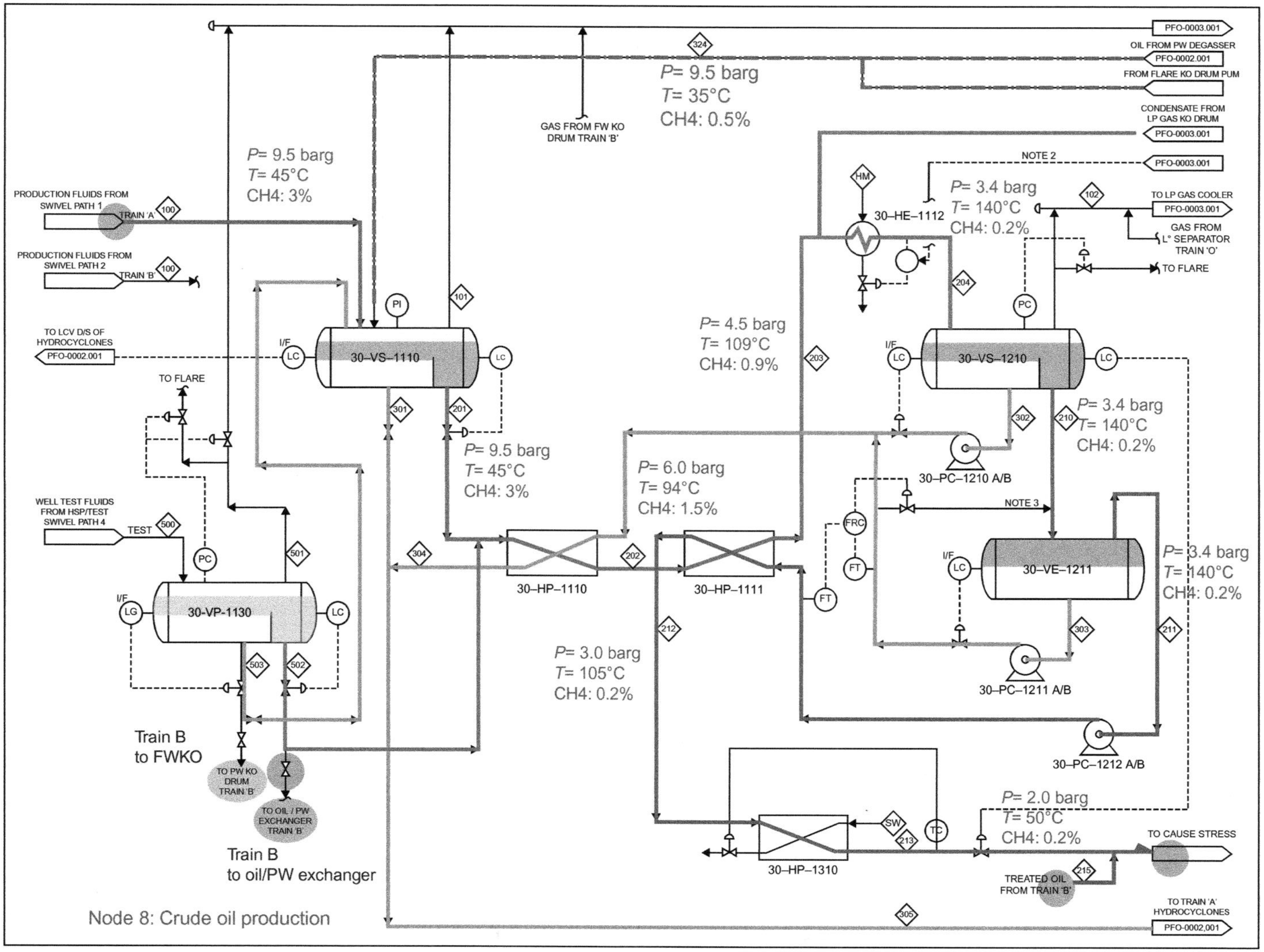

FIGURE 1.42

Transformation of the raw material to products is simulated with process simulators.

ISA-95 distinguishes between the "process equipment" and other "physical asset" systems. This concept has been illustrated in Figs. 1.43 and 1.44.

Physical asset is defined as a tangible, man-made object that has a specific function, normally within a broader system. ISO 55000, focus on the physical asset integrity management system. The management of physical assets and asset systems is inextricably linked to the other categories of assets. The critical interdependencies are illustrated in Fig. 1.45. Although human factors such as leadership, motivation, and culture are not directly addressed within the scope of ISO 55000, they are critical to the successful achievement of optimized and sustainable asset management and require due consideration. This is applicable to the organization's owners, managers, employees, contractors, and suppliers and is considered as the elements of PSM systems (see Fig. 1.20).

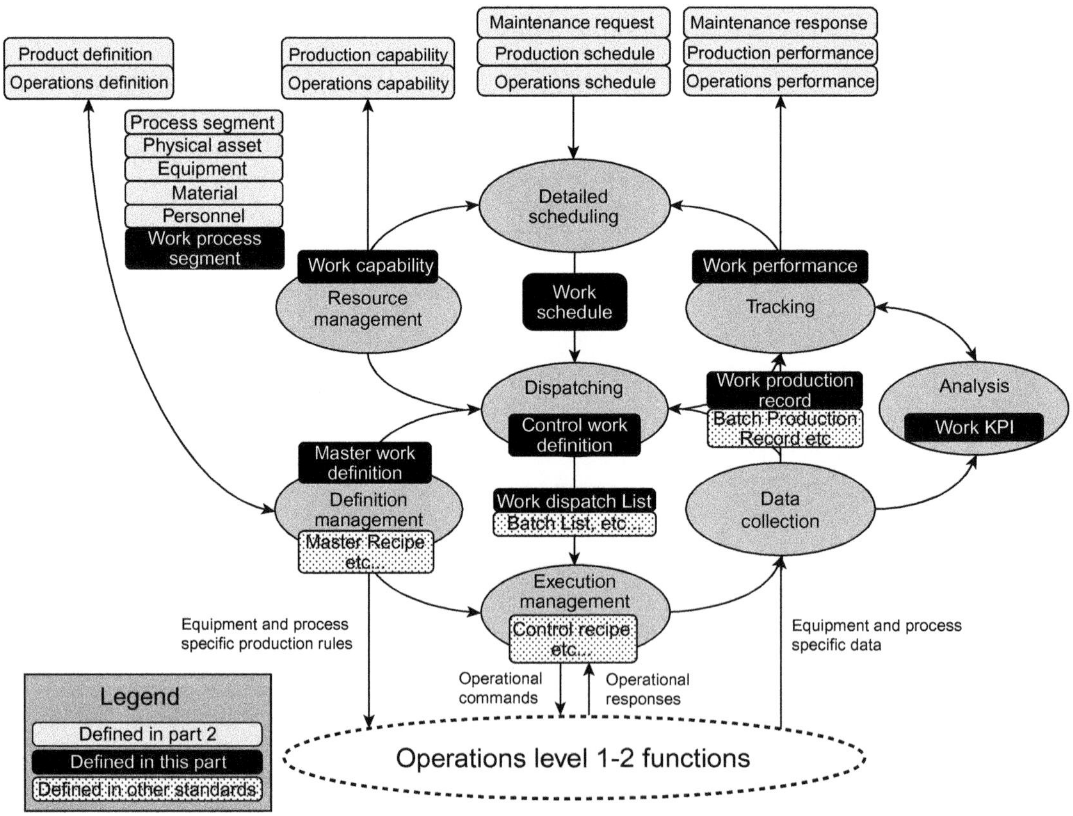

FIGURE 1.43

ISA-95 work information models for operations management.

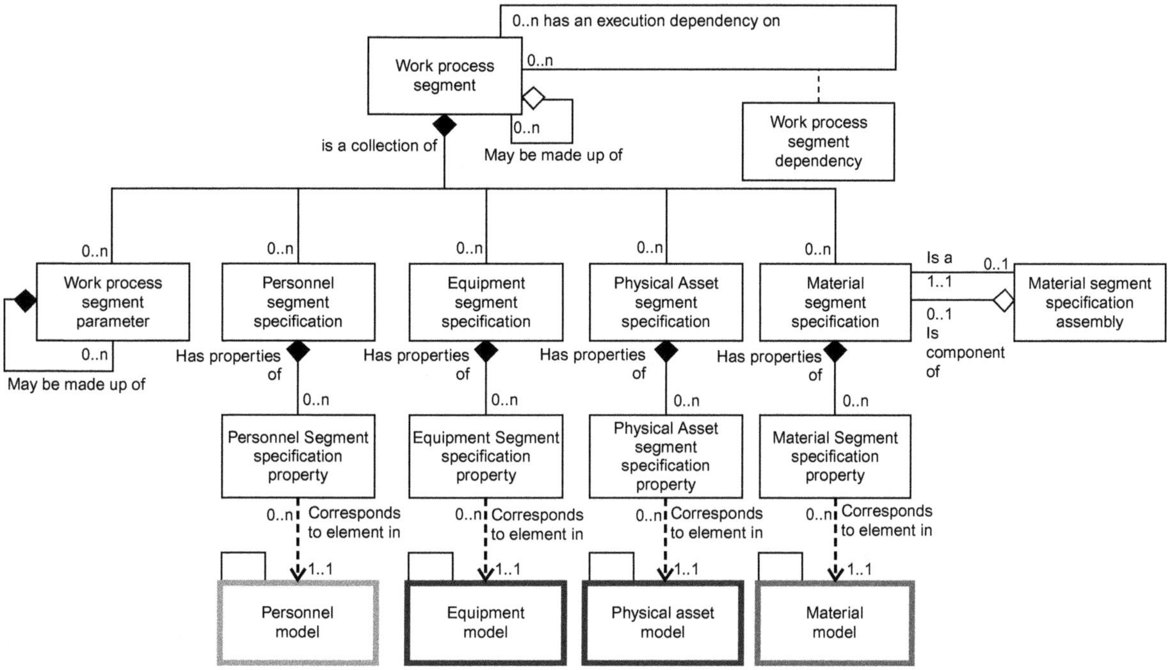

FIGURE 1.44
ISA-95 Work process segment model.

Some of the physical assets are SCEs. These elements shall comply with the regulatory requirements. Fig. 1.46 summarizes how the SCEs are identified and Fig. 1.47 categorizes the SCEs for the oil & gas facilities.

In the process industry the project managers play the "product system engineer" role during EPC phase.

During operation, process engineers play the role of "product system engineer." When the raw materials or capacity change, the process engineers should optimize process system for the new conditions. Process engineer should coordinate with the instrumentation and control engineers to assure the **performance** of the sensors and control enabling systems on one hand and coordinate with Maintenance engineers assure the **integrity** of the physical assets on the other hand.

Plant managers have a few or no degree of freedom to change the plant system. They have double role of "Enterprise System Engineering" at the plant level and "Service System Engineering" in supply chain context.

Quality and HSE assurance systems are among the enabling systems too. They deal with the regulatory requirements and obligations. Fig. 1.48 illustrates the types of the "Systems in Operational Environment."

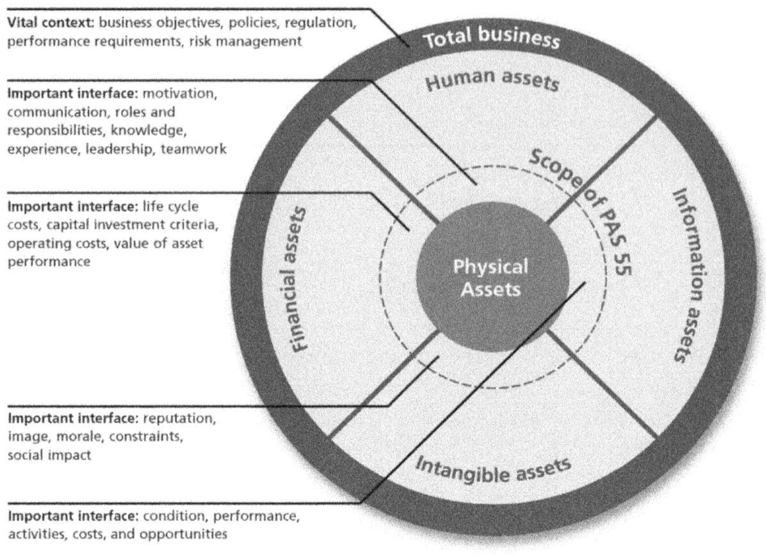

FIGURE 1.45

ISO 55000 Physical assets system in relation to the other categories of assets. *Reproduced with permission from BSI Standards limited (BSI).*

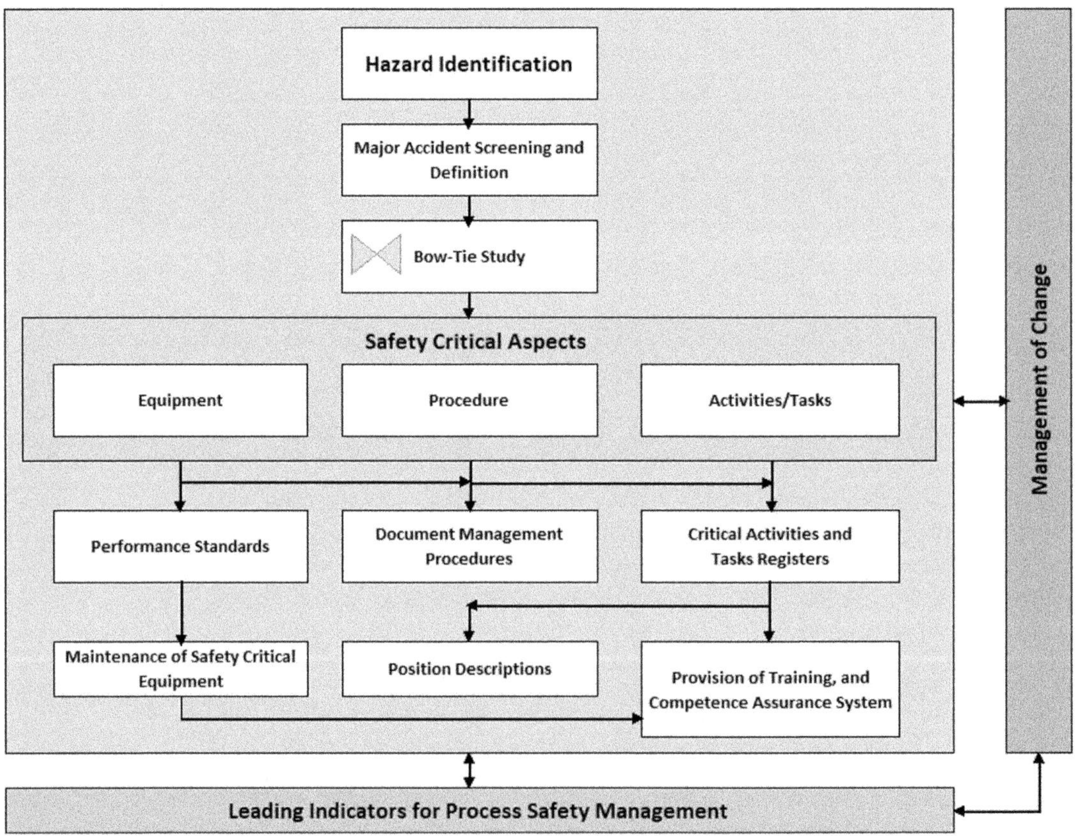

FIGURE 1.46

Safety critical system management. *From: Bow-Tie diagrams in downstream hazard identification and risk assessment.*

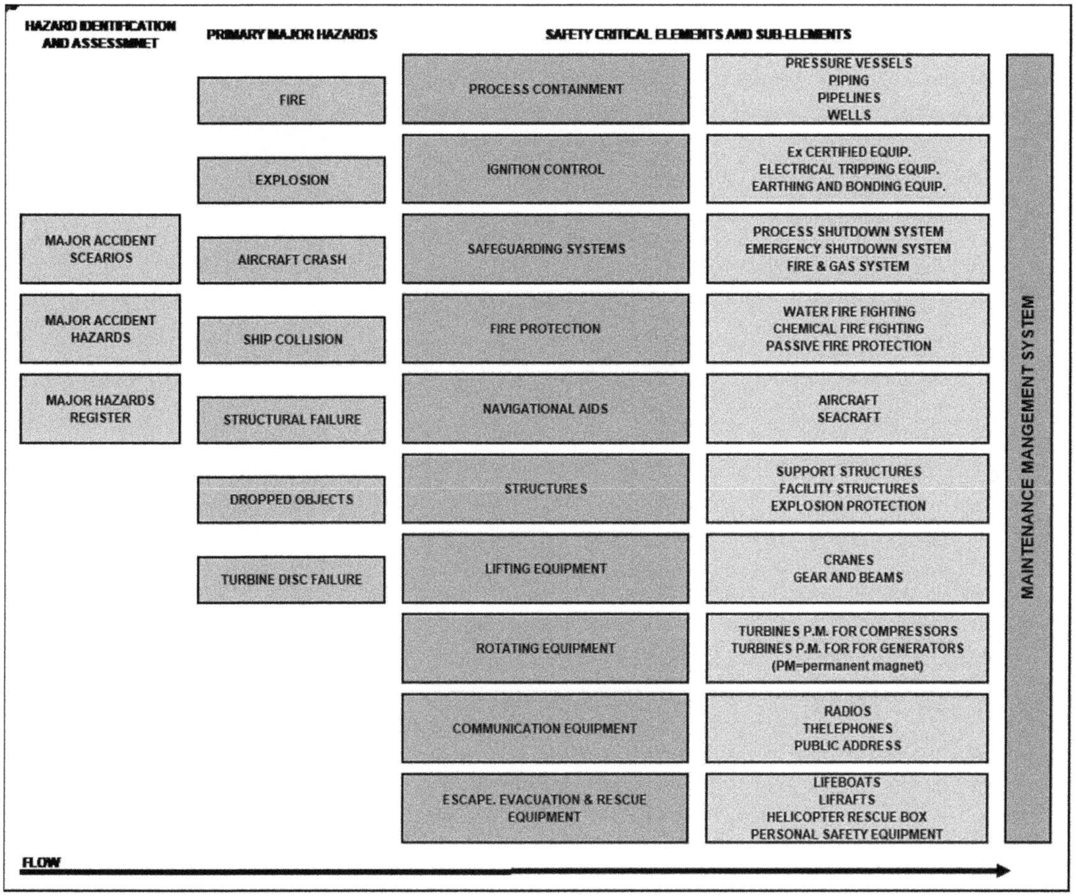

FIGURE 1.47

Safety critical elements for the offshore oil & gas facilities. *From: Energy Institute Guideline (ISBN 978 0 85293 462 3).*

Conclusion: To implement an effective "Process Safety Management System," we should understand that it is not a "Standalone Closed System" but one of the "Systems in Operational Environment" of the production plant which is our main "System of Interest."

1.10 ESSENTIAL SKILLS TO COPE WITH THE CYBER-PHYSICAL SYSTEMS

The 21st century is moving at a tremendous pace with technology taking us places, both personally and professionally, that most of us would not have envisioned 10 years ago. From a manufacturing perspective, these advances

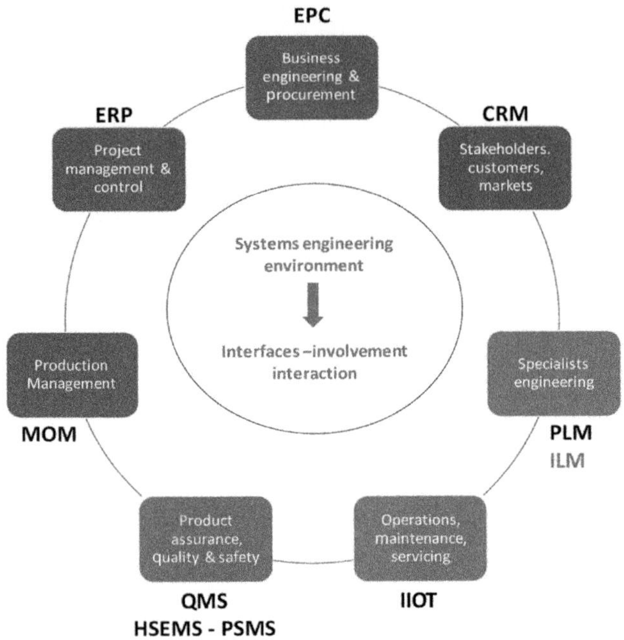

FIGURE 1.48

System engineering environment (Ref. Prof. H. Stoewer). *Inspired from: INCOSE Handbook-2006, Figure 4-1: context of systems engineering technical processes.*

are heralding another industrial revolution which has been named Industry 4.0. Fig. 1.49 summarizes the industrial revolution timeline.

Today, control systems can autonomously operate manufacturing equipment within clearly defined parameters. The transition to the cyber-physical system (CPS) is fluid because the lines between the virtual, digital, and real worlds are disappearing before our very eyes. A CPS responds to changes; it can digitally process tasks in a split second, and it can convert the inputs into commands.

The traditional manufacturing world is converging with the digital manufacturing world to enable organizations to digitally plan and project the entire lifecycle of products and production facilities. This approach offers tremendous benefits to the manufacturing processes, with significant cost benefits and robust growth.

The IIoT is everywhere now and connect millions of devices, machines, sensors, and systems throughout the world. The collaborative interfaces and successive break-up of the traditional automation pyramid and substitution with networked, decentralized, or partially self-organizing services are the great opportunities for improving the quality and safety management system frameworks. Fig. 1.50 lustrates the evolution of the traditional automation pyramid to the decentralized or partially self-organizing services.

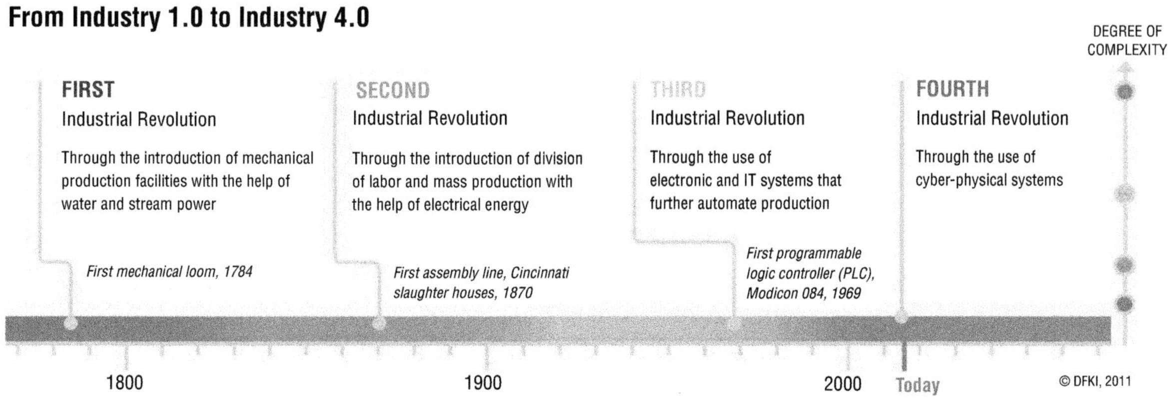

FIGURE 1.49
Industrial revolution timeline.

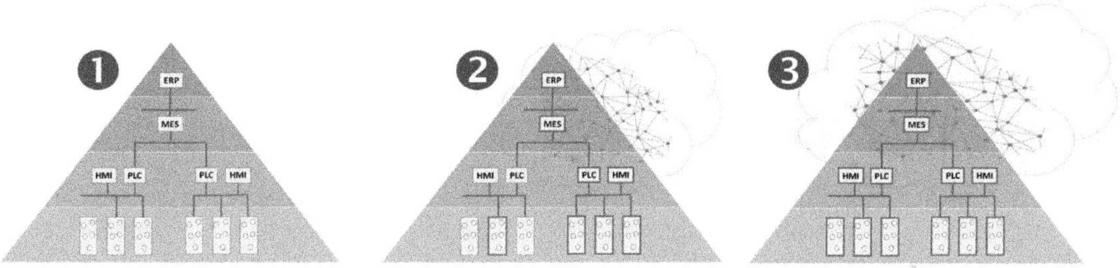

FIGURE 1.50
Evolution of the traditional automation pyramid to the decentralized or partially self-organizing services. *From: http://www.dlg.org/fileadmin/downloads/food/Expertenwissen/e_2015_5_Expertenwissen.pdf.*

A decentralized, self-organizing system can respond independently and adeptly to unexpected events. The result is the resilient factory—one that is error-tolerant, agile, and able to learn.

With these available advanced technologies, dream of the integrators for implementation of an effective and easy to use integrated management systems can become a reality. Integration of the management systems which can be achieved only if the following notions are fully understood and democratized throughout the organizations:

- *System engineering* is an *interdisciplinary* field of engineering that focuses on how to design and manage complex engineering systems over their *lifecycles*. Issues such as requirements engineering, reliability, logistics, coordination of different teams, testing and evaluation, maintainability, and many other disciplines necessary for successful system development, design, implementation, and ultimate decommission become more difficult when dealing with large or complex projects.

- *System thinking* is a way of thinking about, and a language for describing and understanding, the forces and interrelationships that shape the behavior of systems. This discipline helps us to see how to change systems more effectively and to act more in tune with the natural processes of the natural and economic world.
- *Complexity thinking*: Russell L. Ackoff quoted: "Problems that arise in organizations are almost always the product of interactions of parts, never the action of a single part. Complex problems do not have simple solutions. The search for simple –if not simpleminded– solutions to complex problems is a consequence of the inability to deal effectively with complexity."
- *Risk thinking*: Risk awareness is growing among quality and HSE managers. They recognize that risk is not limited to negative possibilities. Companies can also use risk-based thinking to pinpoint opportunities, which represent the positive side of risk. The "Plan-Do-Check-Act (PDCA) cycle" is the link between the operational excellence and risk management and can assure the effective process improvement.
- In the context of ISO 9001:2015, *risk-based thinking* replaces what was called the preventive action in the previous standard version. ISO's risk-based thinking requirements center on incorporating risk into decision-making, without exactly formalizing how to do it. Areas, where risk appears in the new standard requirements, include organizational context, leadership, planning, operation, performance evaluation, improvement.
- *Risk management* is the identification, assessment, and prioritization of risks (defined in ISO 31000 as the effect of uncertainty on objectives) followed by coordinated and economical application of resources to minimize, monitor, control the probability, and impact of unfortunate events or to maximize the realization of opportunities. Risk management's objective is to assure uncertainty does not deflect the endeavor from the business goals.
- *Resilience engineering* is a new way of thinking about safety.

  The traditional view of safety (known as "Safety I") is to prevent things from going wrong. However, the new view ("Safety II") is that safety is the ability to succeed under varying conditions. Resilience engineering aims to increase the number of things that go right, rather than to reduce the number of things that go wrong.
- It is concerned with building systems that are resilient to change. By analyzing what goes right, resilience engineering attempts to understand normal performance, so that work can be made better and safer.

- Success has been ascribed to the ability of groups, individuals, and organizations to anticipate the changing shape of risk before damage occurs; failure is simply the temporary or permanent absence of that.
- IT, OT, ET integration: Technology now makes the integration of "Engineering," "Operational," and "Information" technologies possible. Successful integration enables the companies to break down barriers between horizontal and vertical lifecycle departments, phases of business, and chain values.
- *Big data management*: Collecting more and more data without processing them to knowledge present no value. Today's OMS captures more data than ever, allowing companies to leverage sophisticated reporting and business intelligence.

With these skills and using the right tools the user-friendly and easy to use workflows and applications can be designed. Then, all workforce of business can be involved and contribute in the design and application of the operational excellence.

The new versions of ISO 9001, ISO 14001, and ISO 45001 and ISA-95 provide the guidance to achieve this goal.

1.11 WHY DOES COMPLEXITY MATTER?

Enterprise systems are inherently complex, often involving many business processes, people, and organizations across a company. Given this built-in complexity, it is no surprise that failures abound; it is amazing these systems function at all. Fig. 1.51 illustrates the complexity of a production plant which is a "system of the systems" operated by the "team of the teams" with the "organisation of the organisations."

Systems engineering deals with work-processes, optimization methods, and risk management tools in such projects. It ensures that all possible aspects of a project or system are considered, and integrated into a whole.

Fig. 1.52 illustrates the system fundamentals and elements of the engineered systems. A process plant consists of many closed and open systems which interact with other systems in a shared environment affected by the behaviors, properties, and functions characterized by *emergence* and *complexity*.

An open system is defined by the interactions between system elements within a system boundary and by the interaction between system elements and other systems within an environment. Closed systems have no interactions with their surrounding environment.

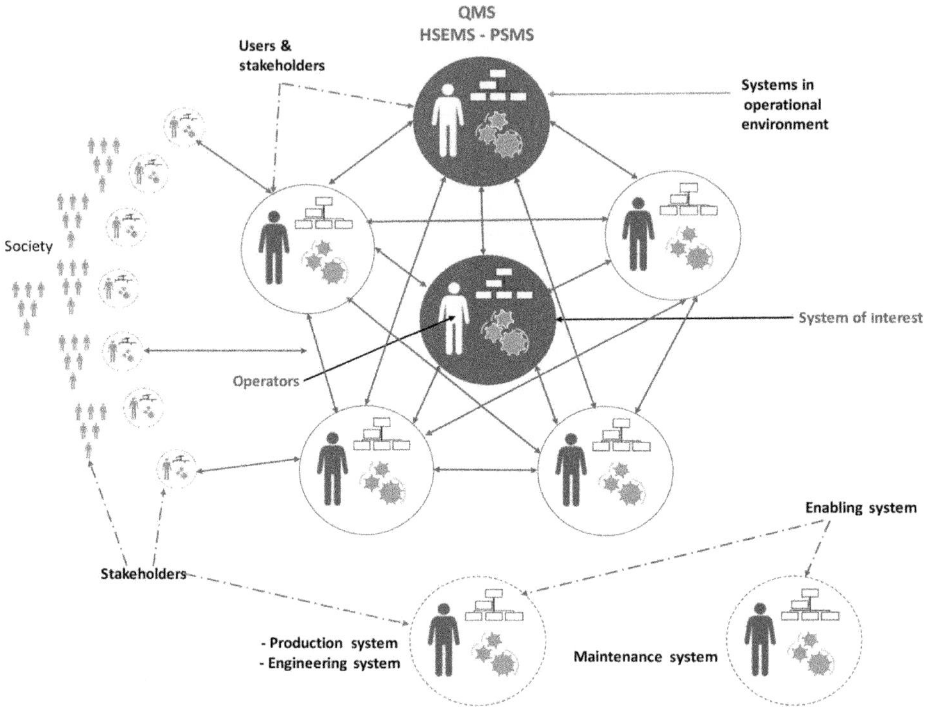

FIGURE 1.51
Requirements elicitation captures the needs of stakeholders, operators, and users across systems boundaries (INCOSE 2002).

Emergence is a consequence of the fundamental system concepts of *holism* and *interaction*. System wholes have behaviors and properties arising from the organization of their elements and their relationships, which only become apparent when the system is placed in different *environments*.

The stem of the word *complexity*, i.e., *Complex* is composed of the Latin words *com* (meaning: "together") and *plex* (meaning: woven). This concept is different from *Complicated* where *plic* (meaning: folded) refers to many layers. A *complex* system is thereby characterized by its interdependencies, where as a *complicated* system is characterized by its layers.

Neil Johnson states that "even among scientists, there is no unique definition of complexity — and the scientific notion has traditionally been conveyed using particular examples..." Ultimately he adopts the definition of "complexity science" as *"the study of the phenomena which emerge from a collection of interacting objects."*

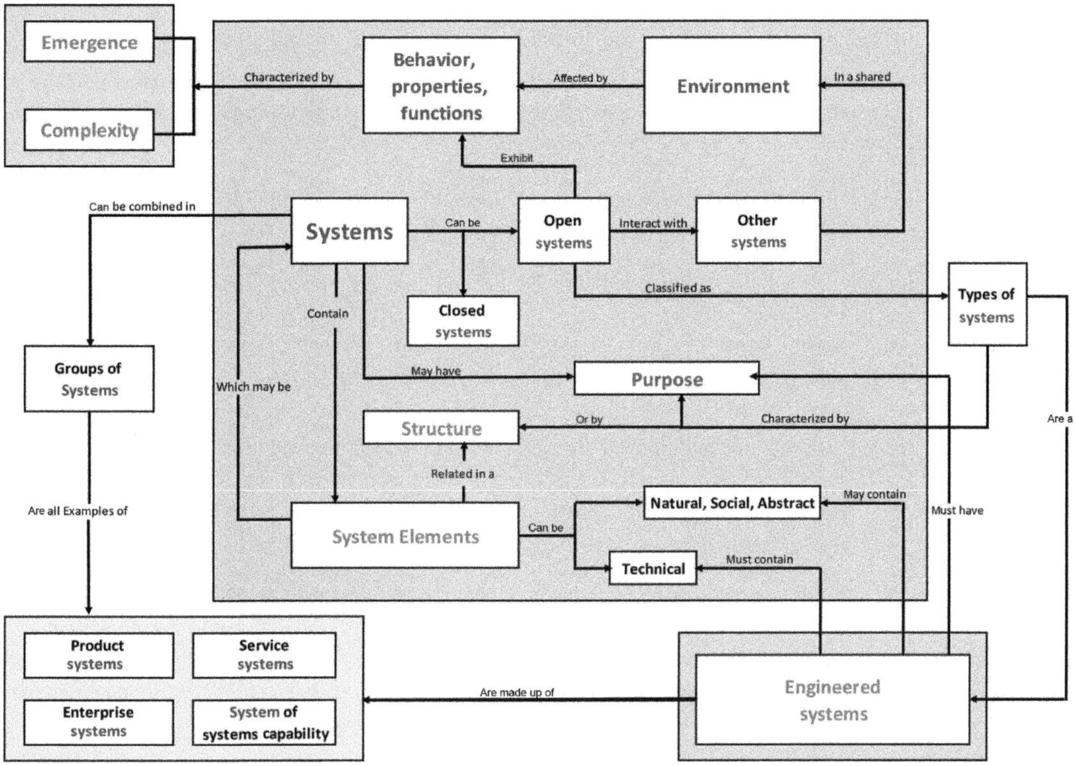

FIGURE 1.52
System fundamentals and engineered systems (SEBoK v.01).

Emergence and complexity concepts represent many of the challenges which drive the need for *systems thinking* and an appreciation of systems science in system engineering.

Russell L. Ackoff quoted: "Problems that arise in organisations are almost always the product of interactions of parts, never the action of a single part. Complex problems do not have simple solutions. The search for simple –if not simpleminded– solutions to complex problems is a consequence of the inability to deal effectively with complexity."

Complexity is an attribute of the *technical system* being developed but also of the problem space (including *people* and *organizations*), and the *environment*. Complexity is associated with the *number of parts, diversity, dynamism* and with *emergence*. It is a challenge to systems engineers not to *over-simplify* in pursuit of representations and capabilities that can be understood and controlled; the right level of complexity is essential.

Although the meaning of complexity varies from "confusion" to "measurable characteristics" of technical systems, it is most useful to systems engineers to identify characteristics that can be resolved and whose resolution will improve the development and operation of modern systems. Complex systems engineering requires both a shift in thinking and an expanded set of tools and techniques.

Discrete high-tech and hazardous manufacturing such as space, aeronautics, railway, and automotive projects define clearly their complexity assessment and management tools and techniques in the deliverables of their projects and operation.

Application of the complexity assessment and management are not explicit and straight forward in the process manufacturing such as oil & gas, petrochemical, and chemical industries. The nuclear industry is an exception and very often inspired the other process industry sectors. Also the manufacturers of the specialized packages such as compressors, gas turbines, and subsea equipment may apply the complexity science in design and operation of their products. However, these closed systems are only a few parts of the entire plant open system.

System engineering, system thinking, and complexity thinking skills are essential to design and operate an effective PSM system. To guarantee success, not only the HSE practitioners but also everybody in the managerial and operational teams should be trained for these skills and apply them in their day to days tasks and decision makings.

The Cynefine framework is a quick and easy technique for complexity assessment and management of a given context. It can be combined with any qualitative decision making or risk-based activity such as HAZID, HAZOP, JHA, PTW, MOC, etc.

Cynefin framework is divided into five domains: obvious (*simple*), complicated, complex, chaotic, and disorder. This framework strives to make sense of the prevailing environment. It distinguishes the decision-making models in two groups: "categorization models" and "sense-making models". In a categorization model the framework precedes the data. In a sense-making model the data precedes the framework.

Fig. 1.53 summarizes how Cynefin framework is applied for the complexity assessment of the self-determined learning processes. From this assessment, we understand that procedural "learn then work" is not effective if we intend to promote the creativity and innovation in a particular activity. On the other hand, giving the possibility of "working to learn" will increase the risk of failures and business interruptions. The company should make sure that everything is in place to be "safe to fail."

This book aims to raise the awareness process manufacturing personnel and consultants to the importance of the system engineering, system thinking,

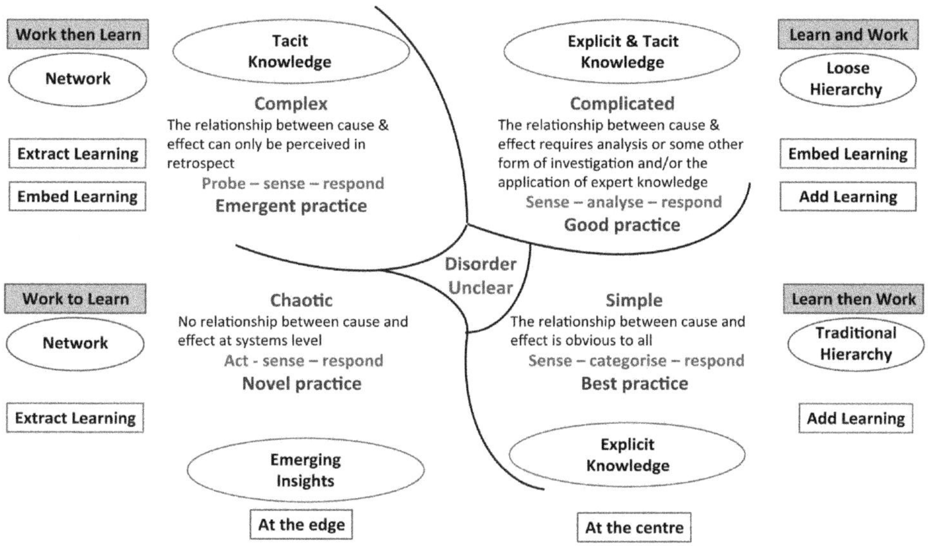

FIGURE 1.53

Cynefin framework for self-determined learning complexity level assessment. *From: http://colabria.com/heutagogy/.*

and complexity thinking in the implementation of their OMS. Without these skills a thorough risk assessment is not achievable. We will scrutinize the complexity issues associated with the PSM to provide the insights to the HSE practitioners to choose the optimal approach for implementation of PSM in their organization.

1.12 BARRIER THINKING & COMPLEXITY

As an example, let us assess the causes the relief valve (PRV) failure and the impact of this failure on a gas release incident (Fig. 1.7). In a thorough bow-tie, all the technical, organizational and human causes and safety barriers during the lifecycle of PRV should be considered.

None of the safety barriers is 100% effective. The reliability of a safety barrier can be increased by different techniques such as intrinsically safer systems, redundancy, shorter time of the test, etc. can be used to increase the integrity level of the individual safety barriers. The bow-ties of Fig. 1.54 illustrates the relationship of the preventive and mitigation measure for the individual top events such as "Gas Release," "Overpressure," "PRV failure," "Design Failure," etc.

In Fig. 1.55 we have considered only one of the safety barriers of in each bow-tie of Fig. 1.54 and related them to each other. We assumed that the barriers are 100% independent and neglected the common cause failures.

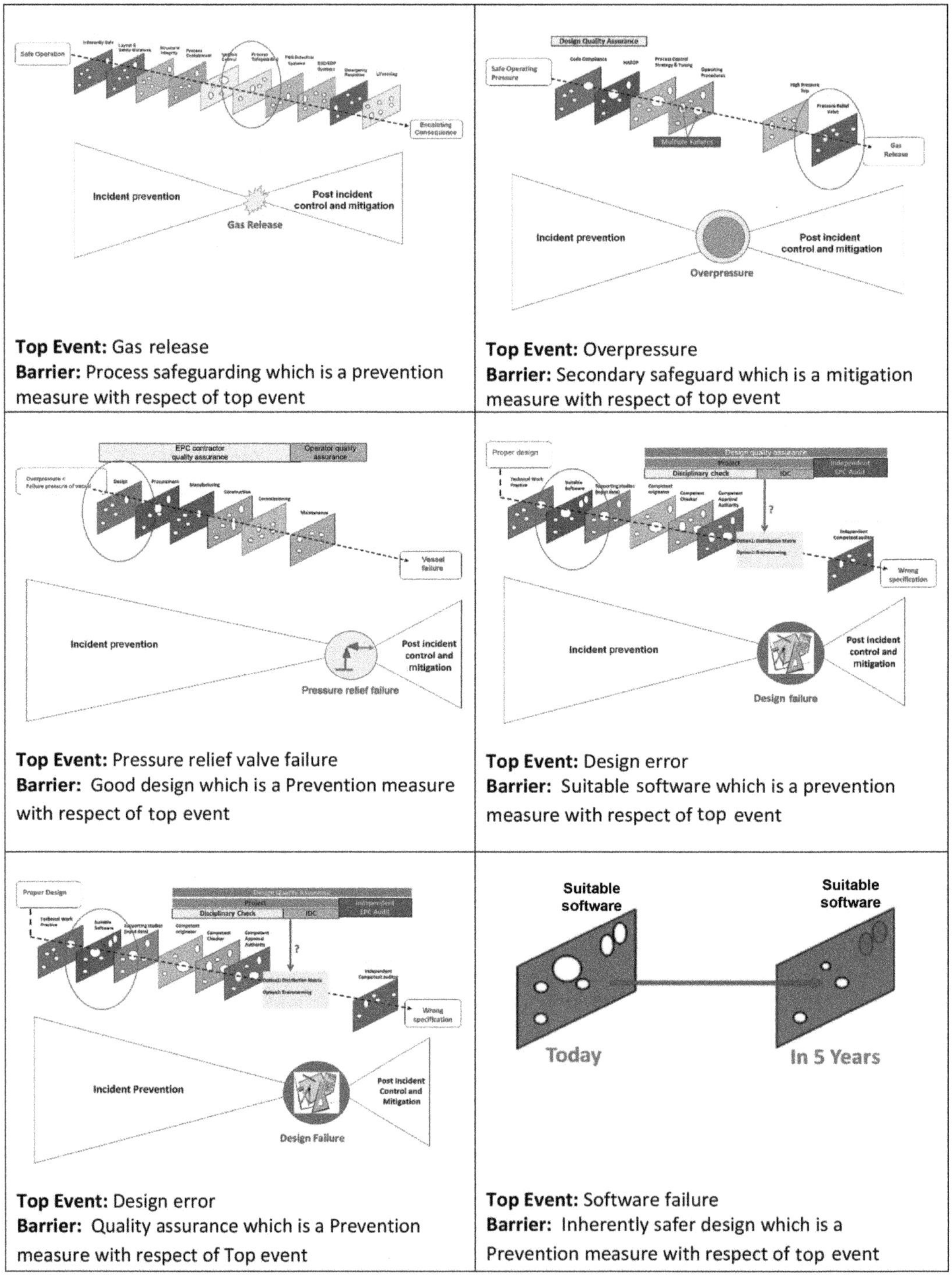

FIGURE 1.54
PRV bow-tie assessment.

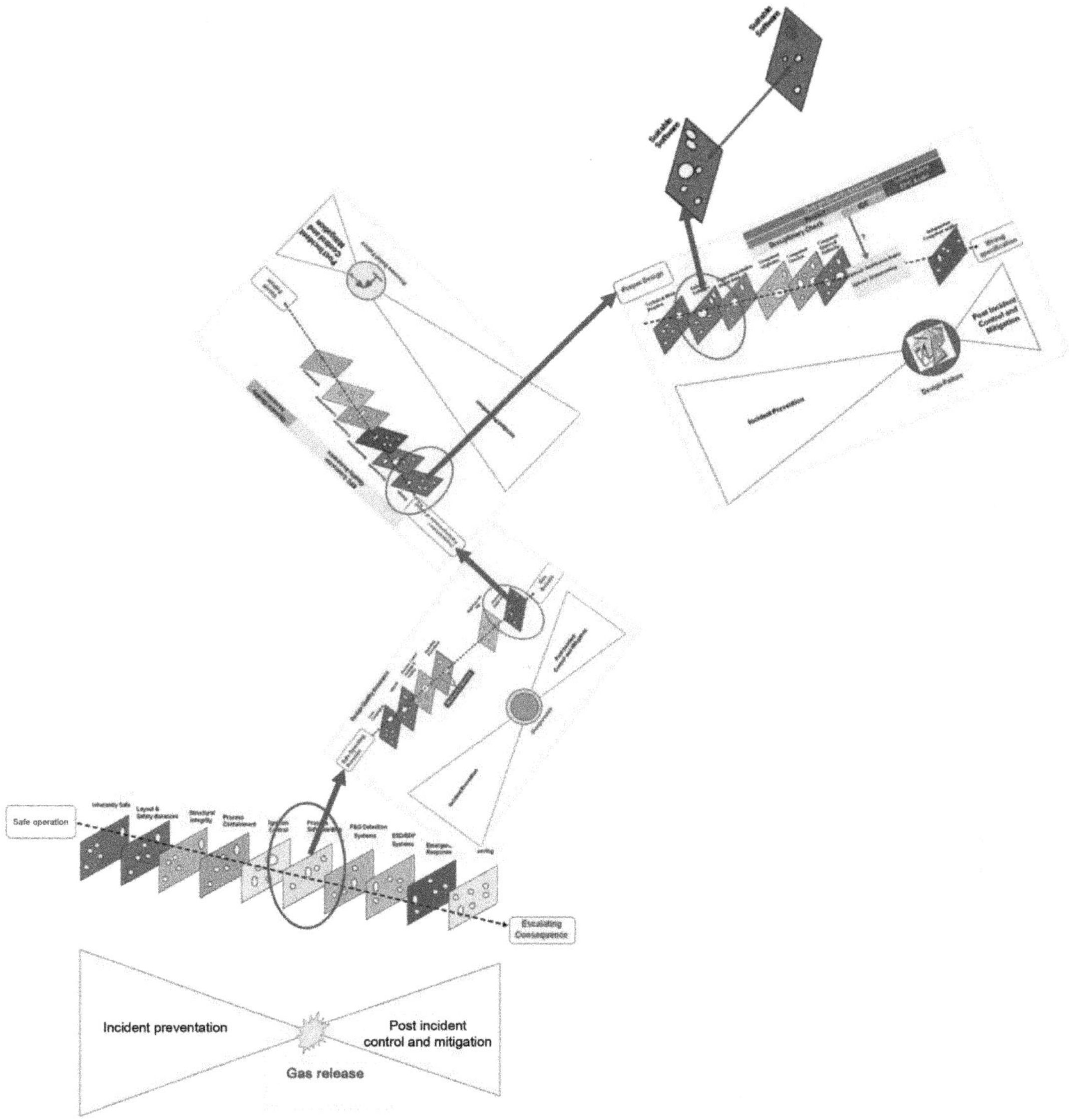

FIGURE 1.55
Combined bow-ties for PRV failure assessment.

For the real systems which all the barriers should be scrutinized to the sub-layers up to individual element, the combined bow-tie and Swiss cheese model can become complex very quickly (Fig. 1.55).

Considering all these facts take us to believe that a realistic bow-tie assessment looks more like a Croissant with interrelated *resilient*, random, and

FIGURE 1.56
The safety critical elements are not always independence. A croissant in its whole is much better representative the system of the safety critical elements in a real plant.

enclosed hollow spaces rather than the rigid sliced Swiss cheese with perfect and independent holes.

No matter how and the croissant is made or subjected to the shocks the croissant will be acceptable as far as the composition and texture of pastry substance, as well as the number and configuration of the holes, remain at defined range (Fig. 1.56).

In a real production facility the combination of technical, organizational and human factors assures the ability to perform in a resilient manner.

1.13 CHANGE MANAGEMENT & COMPLEXITY

The world that enterprise lives in it is anything but static. Technology and people change continuously and cause the constant moving of the things around all the time. Enterprise management system should continually cope with both shallow and deep conflicts between the designed features and the reality. ISO 10007 for configuration management aims to provide the guidance to minimize the risk of the gap between design requirements and the reality of the plant (Fig. 1.57).

The novel conditions are common, and things that we have not seen before do occur relatively frequently. Therefore lots of out of the box adapting and tailoring activities are required to make systems working.

In 2002 the McKinsey Quarterly study of 40 companies found that 58% of change initiatives failed to reach their goals. We need to understand how organizations work before we can effectively change them (Fig. 1.58).

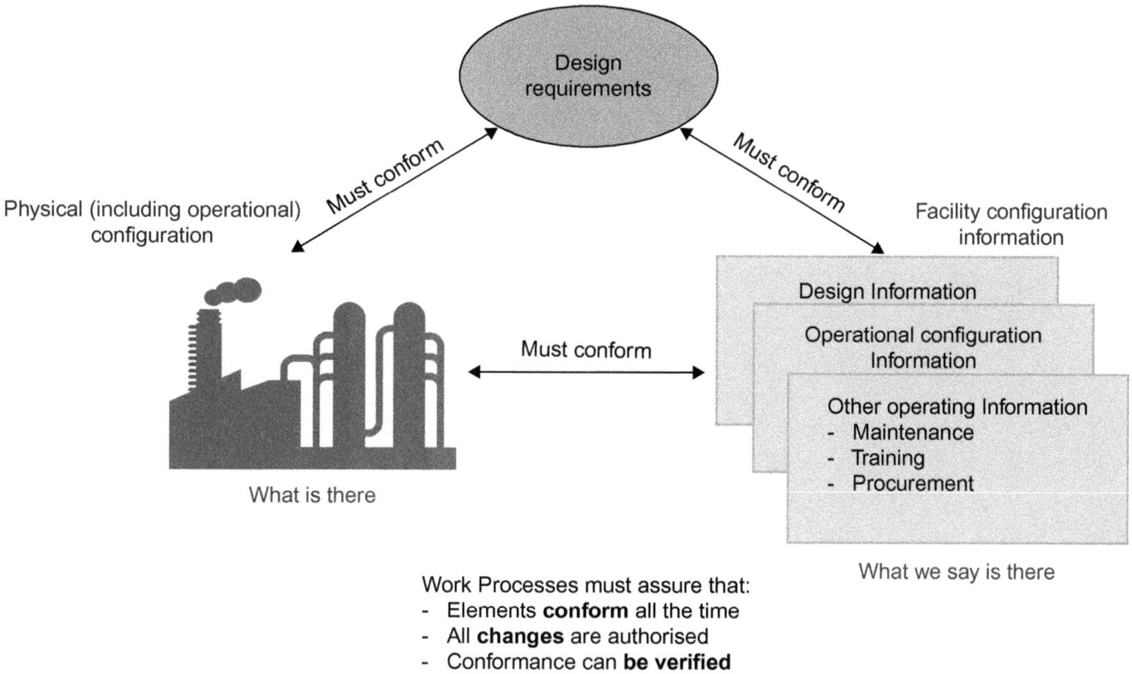

FIGURE 1.57

Relationship among design requirements, documentation, and physical configuration. *Inspired from: Configuration management in nuclear power plants.*

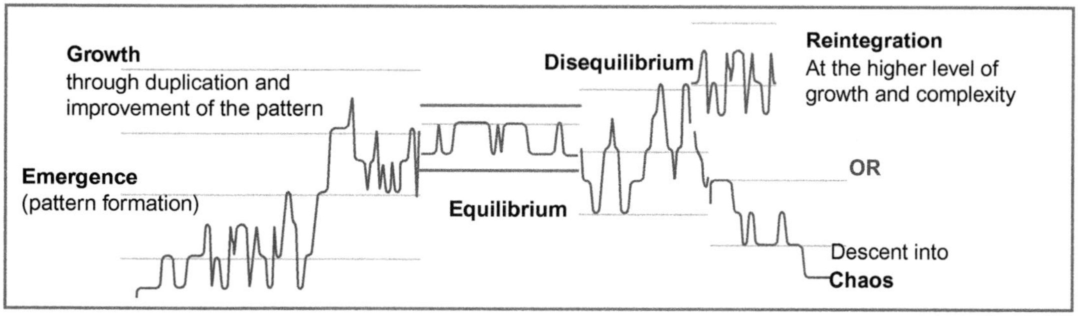

FIGURE 1.58

Change cycle. *Inspired from: complex adaptive systems theory-12613245196525-phpapp02.*

Change is only made possible by people who by their very nature are complex, unpredictable, dynamic, and resistant to engineering. Organizations depend on people. Therefore to improve operational systems we should understand the complexity of people individually and organization environment.

The definition of insanity is *"doing the same thing twice and expecting a different result,"* but this simple wisdom rule does not apply to the complex environment in which, *"doing the same thing twice* will *give a different result."*

Also, we always say that *"You can't fix what you can't measure"* but *"You* can *intervene in a complex environment, even though you* can't *measure it reliably."*

Complexity is an attribute of the *technical system* being developed but also of the problem space (including *people* and *organizations*), and the *environment*. Complexity is associated with *size, diversity, dynamism* and with *emergence*. It is a challenge to systems engineers not to *over-simplify* in pursuit of representations and capabilities that can be understood and controlled; the right level of complexity is essential.

OMS including the PSM are complex by their very nature. They encompass a great number of diverse and dynamic technical, organizational, and people simultaneously. Over simplification of the monitoring and control of them to the check list-based audits and gap analysis of the sample cases will lead unsatisfactory results and inappropriate prioritization and decision making.

1.14 COMPLEXITY AND DECISION MAKING AND COMPLEXITY

In the context of ISO 9001:2015, *risk-based thinking* replaces what was called the preventive action in the previous standard version. ISO's risk-based thinking requirements center on incorporating risk into decision-making, without exactly formalizing how to do it. Areas, where risk appears in the new standard requirements, include organizational context, leadership, planning, operation, performance evaluation, and improvement.

Fig. 1.59 from ISO 31000 illustrates the risk management steps. The trouble with the conventional risk assessment approach is that we do not just "find" causes; we tend to "create" them, and when none can be found, we use the "act of God" opt-out clause. This approach is a social process, which changes over time just as thinking and society change: from the end of the Second World War until the late 1970s, most accidents were perceived as a result of *technical failure*. The Three Mile Island accident (March 1979) saw the emphasis begin to shift from technical to *human failure*. With the challenger disaster in 1986 the cause identified was not solely technical or human but *organizational failure*.

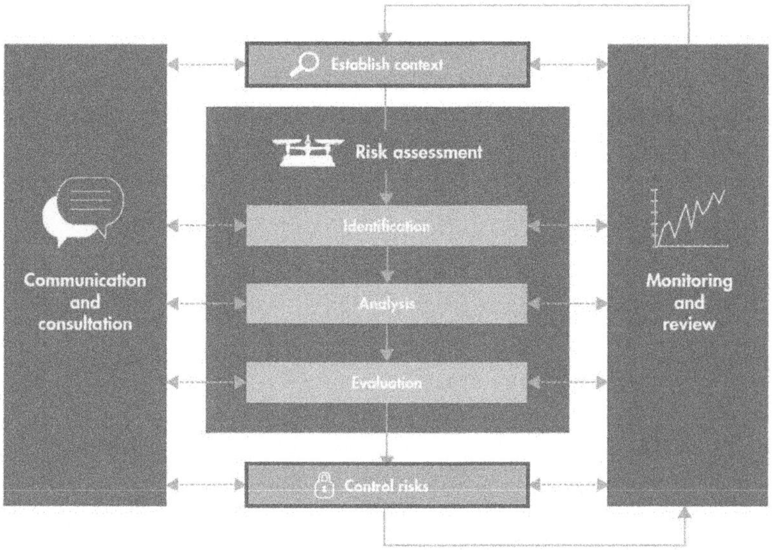

FIGURE 1.59
Risk management according to ISO 31000 (page 8).

In the world of accident investigation, it has been expressed as the "What-You-Look-For-Is-What-You-Find" or WYLFIWYF principle. In other words, our (current) understanding of the world heavily influences what we pay attention to or notice, both the larger "picture" and the nuances. In consequence of that it also excludes from our awareness that which we are not prepared for, that which we do not expect, and that which we are otherwise unwilling to "see."

Understanding the complexity level of the context is the key to an effective risk assessment and risk reduction strategy. Over simplification of the complex contexts to simple leads to wrong conclusions and blindness to new ways of thinking.

The Cynefin Framework developed by David J. Snowden is a useful tool to determine the complexity level of a given context. This framework strives to make sense of the prevailing environment. It distinguishes the decision-making models in two groups: "categorization models" and "sense-making models." In a categorization model the framework precedes the data. In a sense-making model, the data precedes the framework.

Cynefin framework is divided into five contexts domains: obvious (*simple*), complicated, complex, chaotic, and disorder.

Snowden says failure to recognize a situation's context may result in disorder. He refers to disorder as "not knowing which space you are in." He goes on to state that this is the place "where we are most of the time" because people "interpret each situation according to their preference for action." Typically a certain blindness is caused by our preferences and our experiences.

There is a "cliff" between the ordered domains (including obvious and complicated) and chaos. Chaos may result from either deliberate unethical behavior or failure to recognize complicated or complex situations. In the latter cause, complacency may cause one to over-simplify and misinterpret a problem causing an already complicated situation to become chaotic. Once one falls into chaos, it is difficult to recover. Therefore decision maker should manage in the complicated and complex spaces to avoid the cliff (Fig. 1.60).

Complex and chaotic contexts unordered: "there is no immediate apparent relationship between cause and effect." A complex context as a place where "cause and effect are only obvious in hindsight, with unpredictable emergent outcomes."

Making decisions in a complex context calls for leaders to probe, sense, and respond in order to discover an emergent practice. Of these three actions the key to success in a complex context is effective probing. Probing is considered as conducting "safe-to-fail experiments" (not fail-safe experiments). If a solution does not work, leaders should get rid of it. If it succeeds, they should amplify it.

Table 1.4 summarize the Cynefin framework for decision making of the leaders. Consultants should help the leaders to determine the complexity level of their context and then set the effective strategy to achieve the required results.

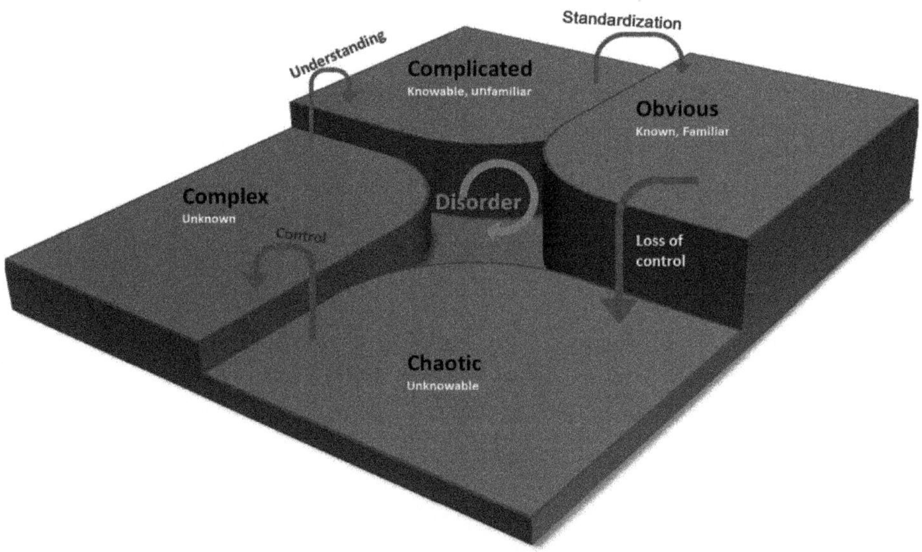

FIGURE 1.60
Cynefin frame work domain.

Table 1.4 The Cynefin Framework for Decision Making

	The Context's Characteristics	The Leader's Job	Danger Signal	Response to Danger Signals
Simple	• Repeating patterns and consistent events • Clear cause and effect relationships evident to everyone; *right answer exists* **Known knowns** Fact-based management	• Sense, categorize, respond • Ensure that proper process are in place • Delegate • Use the best practices • Communicate in clear, direct way • Understand that extensive interactive communication may not be necessary	• Complacency and comfort • Desire to make complex problem simple • Entrained thinking • No challenge of received wisdoms • Overreliance on best practice if context shifts	• Create communication channels to challenge orthodoxy • Stay connected without micromanaging • Do not assume things are simple • Recognize both the value and limitations of best practices
Complicated	• Expert diagnosis required • Clear cause and effect relationships discoverable but not immediately apparent to everyone; *more than one answer is possible* **Known unknowns** Fact-based management	• Sense, analyze, respond • Create panels of experts • Listen to conflicting advice	• Expert overconfident in their own solutions or in the efficiency of past solutions • Analysis paralysis • Expert panels • Viewpoints of no experts excluded	• Encourage external and internal stakeholders to challenge expert opinions to combat entrained thinking • Use experiments and games to force people to think outside the familiar
Complex	• Flux and unpredictability • No right answers; *emergent instructive patterns* • Many competing ideas • A need for creative and innovative approaches **Unknown unknowns** Pattern-based leadership	• Probe, sense, respond • Create environment and experiments that allow patterns to emerge • Increase levels of interaction and communication • Use methods that can help generate ideas: open up discussion (as through large group methods); set barriers; stimulate attractors; encourage dissent and diversity; and manage starting conditions and monitor for emergence	• Temptation to fall back into habitual, command-and-control mode • Temptation to look for facts rather than allowing patterns to emerge • Desire for accelerated resolution of problems or exploitation of opportunities	• Be patient and allow time for reflection • Use approaches that encourage interaction so patterns can emerge
Chaotic	• High turbulence • No clear cause and effect relationship, so *no point in looking for right answers* • Many decisions to make and no time to think • High tension **Unknowables** Pattern-based leadership	• Act, sense, respond • Look for what works instead of seeking right answers • Take immediate action to reestablish order (command and control) • Provide clear, direct communication	• Applying a command-and-control approach longer than needed • "Cult of the leader" • Missed opportunity for innovation • Chaos unabated	• Set-up mechanisms (such as parallel teams) to take advantage of opportunities afforded by a chaotic environment • Encourage advisers to challenge your point of view once the crisis has abated • Work to shift the context from chaotic to complex
Disorder[a]	• You do not know where you are **Confused** Pattern-based leadership	• Gather more info on what you know or identify what you do not know • Get enough info to move to a more defined domain	• Complacency and choosing the easiest or wrong domain	• Choose the right experts • Implement a proper big data management system

[a] *This section has been developed by the author of this book.*

1.15 DIGITAL TRANSFORMATION AND COMPLEXITY

Conway's law suggests that effective systems design emerges from system-oriented organizations which implement the three simple rules:

1. facilitate communications
2. streamline controls, and
3. simplify paperwork.

The way to effective systems engineering management is not "in the direction of formal, formidable, massive documentation. It does, however, reside in the direction of creating a total environment which is conducive to the emergence and effectiveutilisation of creative and inventive talents oriented toward achieving a system approach with a minimum of management encumbrances."

Collecting just more and more data has been proven to be inefficient and improve very little the overall system. The next level of big data queries will examine the associational connections between many different data elements that come from a diversity of sources, and they will use associational practices similar to those used by the human brain.

Today, companies old and new are transforming themselves to extract value from traditional industrial sectors by delivering digital, data-rich services. By taking advantage of the rapid explosion of sensors, ultra-low cost connectivity, and data storage together with powerful analytics (commonly referred to as IIoT) these value-added services can produce business outcomes for customers and produce incremental revenue for the company. Indeed, customers are coming to expect not only best-in-class physical products from manufacturers but also new ways that they can drive increased levels of productivity and business impact (Fig. 1.61).

FIGURE 1.61
Digital twin of a gas turbine with augmented reality visualization.

FIGURE 1.62
Structured System Engineering helps users to navigate at different complexity level of the digital twin.

A key innovation in delivering a radically new approach to industrial systems in a digital world is the Digital Twin. The Digital Twin provides knowledge, insights, and outcomes about its physical twin performance and operation (past and present) and allows for predictions and simulation of future operation (Fig. 1.62).

Digital Twins consists of aggregated data from asset model, asset data, contextual data (such as environmental measurements), and reference data together with intelligence built using a variety of analytics and models. Advanced techniques such as AI or machine learning can also be added to the picture. In a nutshell the Digital Twin is a container for all knowledge and expertise about a specific asset class, designed to deliver insights and outcomes that are essential to building apps about performance, optimization, and business transformation.

High-technology manufacturers that use engineering and manufacturing facilities frequently battle with consistency of execution because processes and approaches to manufacturing vary depending on where you are in the world. A system "learning" mechanism that can ingest and develop best practices from the most highly skilled people across all of these operations and then infuse the intelligence into machine-aided decisioning and automation processes for manufacture can improve their overall production results.

Investment in the IIoT is expected to top $60 trillion during the next 15 years and by 2020, over 50 billion assets will connect to the Internet.

Dive into a fully-immersive exploration of the IIoT in this visit to an undersea oil and gas site. Then lift off in your own helicopter for further examples of the IIoT in action. You will respond to early warning alerts and race to repair a pipeline breach in the freezing deep sea darkness, and you will navigate around weather turbulence to perform an emergency rescue. Thanks to cutting-edge analytic software, you will get real-time insight and guidance to complete these critical tasks.

IIoT lets developers connect machines and devices, manage industrial data, and transform the way that industries work—from oil and gas to healthcare to manufacturing to aviation—and everything in between. The VAR platforms offer the safe and pleasant interactive experiences for the users of the IIoT solutions. VAR is used to stimulate the creativity of the personnel through the collaborative projects. Involvement of the greater number of the personnel in the collaborative activities enhance the adaptive self-organized characteristics of the executive teams which increase the resilience of the activity and the business as a whole.

1.15.1 Solutions for Complexity

Throughout its history, systems engineering has been the primary method for engineering in the face of complexity. As the complexity of systems and their contexts has grown, systems engineering methods and tools have increasingly fallen short of what is needed in the face of this reality.

A common approach has been to seek clever ways to *simplify*, or reduce, the subjective complexity so that the problem and the system are understandable. Scientific advances have, in fact, often come from elegant simplifications that model the important variables or forces that dominate behavior. However, this is not always possible—complexity often cannot be simplified away without *losing the essence* of the problem or possible solutions. Further, this simplification leads to an inability on the part of the solution to be able to engage with the complexity that remains despite our preference to assume it away.

Ross Ashby's law of requisite variety shows that a system controller must have at least as many degrees of control as the degrees of freedom in the environment to be controlled. If a system operates within an environment of human processes, as in today's air traffic or process plant control, then the system solution must have sufficient complexity to do so. In such a system, it is difficult and even dangerous to ignore the complexity.

Therefore systems engineers should acknowledge, understand, and learn to work better with complexity as follows:

- the first step—identify the kinds of complexity in a system and its environment.

- A second step—create appropriate new ways to think about complexity that guides the approaches used.
- A third step—evolve and publicize methods to deal with different types of complexity in various situations.

Working with complexity will never be a trivial task that can be reduced to *following a checklist.*

The complex, adaptive neural network of our human brains will always be needed to supplement documented lessons of any sort.

Decision makers should acknowledge that the world is complex and that our orderly, simple views are inadequate. Their business must grow toward a new appreciation of the implications of complexity—in their systems and ecosystems (both technical and socio-political), and in the interplay between the two—to be successful.

1.15.2 Consultant Role in Performance Improvement of the Complex Systems

Michel Baudin in his blog says:

Most of the people who call themselves consultants are really *contractors*, hired to produce outputs that are occasionally needed, but not often enough to justify hiring employees. The contractors are a temporary extension of the workforce, and their knowledge and skills leave with them at the end of their engagements.

Then, there is a second category of consultants who are really *trainers*, hired to help an organization comply with external mandates, such as regulations from the local regulatory bodies, ISO 9000 certification, or certification as a Lean supplier. These consultants coach clients in the art of passing a particular kind of audit and are needed as a cost of doing business. Their emphasis is not on improving performance but on providing auditors with what they need to see to check-mark the items on their lists. This is the realm of named, formal processes where no deviation from a standard is tolerated. They transfer knowledge, but this knowledge is procedural, and usually has no value beyond ensuring certification.

The last category are consultants who are brought in to help clients *improve performance*. There, what matters is less the ideas they already have than their ability to come up with new ones, as well as ways to implement them that are feasible in the existing *organization*. It is similar to solving a Harvard Business School case, with the differences that it is a real situation and that the issues are *technical* as well as *managerial*.

Top management has identified shortcomings in the company's operations, feels they must be remedied to remain competitive, but has not found the ability to do it with internal resources. The external consultant brings in a fresh pair of *trained* eyes the experience of similar situations, and a kit of tools, either mastered personally or made available through colleagues. The consultant's *generic abilities* must then be melded with the *specific business* and *technical* knowledge available internally to come up with innovative, ad-hoc solutions.

This kind of work requires direct observation of the operations, communication with people at all levels in the client organization, and analysis of the company data. It cannot be done rigidly or dogmatically. The solutions will vary, as will the means of implementing them. It requires us to be open-minded and evaluate ideas on their own merits from wherever they may come.

In the next chapters, we try to give the necessary background to readers to evaluate their capabilities to improve the performance of the management systems and accompany their clients in their journey towards implementation of the effective integrated operating system and digital transformation of the business.

Literature

1. INCOSE Vision 2025
2. CCPS Vision 20/20
3. Out of Control—Why Control Systems Go Wrong and How to Prevent Failure
4. N.G. Leveson, Engineering a Safer World. Systems Thinking Applied to Safety, MIT Press, Engineering Systems Series, 2011. ISBN 978-0-262-01662-9, Jan 2012.
5. A New Accident Model for Engineering Safer Systems.
6. Venkatasubramanian, V. Systemic Failures: Challenges and Opportunities in Risk Management in Complex Systems., doi:10.1002/aic.12495. Published online November 30, 2010 in Wiley Online Library (wileyonlinelibrary.com).
7. Applying System-Theoretic Accident Model and Processes (STAMP) to Hazard Analysis.
8. Applying STAMP in Accident Analysis.
9. A Systems Theoretic Approach to Safety Engineering.
10. King, P. L., Kroeger, D.R., Foster, J.B., Williams, N., Proctor, W. Making Cereal – Not Cars, The Process Industries – Challenges Facing Industrial Engineers. IE PID Article – Final to Publish 9-09, Oct. 9, 09.
11. The Cynefin Framework for Decision Making.

Blog

1. MIT Article Comparing Lean, TQM, Six Sigma, "And Related Enterprise Process Improvement Methods".

Handbook

1. INCOSE Handbook-2002
2. INCOSE Handbook-2006
3. INCOSE Handbook-2012
4. INCOSE Handbook-2015
5. Guide to the Systems Engineering Body of Knowledge (SEBoK)
   - SEBoK v. 1.7
   - SEBoK v. 1.6
   - SEBoK v. 1.5.1
   - SEBoK v. 1.5.
   - SEBoK v. 1.4
   - SEBoK v. 1.3.1
   - SEBoK v. 1.3
   - SEBoK v. 1.2
   - SEBoK v. 1.1
   - SEBoK v. 1.0

Standards

1. ISA-95: Enterprise-Control System Integration
2. ISO/IEC 15228: Systems and Software Engineering – System Life Cycle Processes
3. API 754: Process Safety Performance Indicators for the Refining and Petrochemical Industries
4. ISO 9000: Quality Management Systems
5. ISO 14000: Environmental Management
6. OHSAS 18000 Replaced by ISO 45000
7. ISO 10007: Quality Management Systems – Guidelines for Configuration Management
8. ISO 55000: Asset Management
9. ISO 31000: Risk Management – Principles and Guidelines
10. EEMMU 191: Alarm Systems – A Guide to Design, Management and Procurement
11. ISA 84.0: Guidance on the Evaluation of Fire and Gas System Effectiveness
12. IEC 61508: Electronic Functional Safety
13. IEC 61511: Functional Safety – Safety Instrumented Systems for the Process Industry Sector
14. API 14C (ISO 10418): Petroleum and Natural Gas Industries – Offshore Production Platforms – Analysis, Design, Installation and Testing of Basic Surface Safety Systems

Guidance

1. IOGP 510: Operating Management System Framework for Controlling Risk and Delivering High Performance in the Oil and Gas Industry.
2. IOGP 511: OMS in Practice. A Supplement to Report No. 510, Operating Management System Framework (Jun 2014).

3. IOGP Report 456: In November 2011, Process Safety – Recommended Practice on Key Performance Indicators [1], Which Built on Report 415, Asset Integrity – The Key to Managing Major Incident Risks [1], 2008.
4. IOGP Report 556: Process Safety – Leading Key Performance Indicators. Supplement to IOGP Report 456.
5. Safety Management Systems – Guidance to Organisations.

Regulations

1. SEVESO III: The Seveso Directives are the main EU legislation dealing specifically with the control of on-shore major accident hazards involving dangerous substances.
2. OSHA PSM: American regulations for prevention and control of major accident hazards which intends to provide a generic, non-exhaustive overview of a particular standards-related topic. It does not itself alter or determine compliance responsibilities, which are set forth in OSHA standards themselves and the Occupational Safety and Health Act.
3. PFEER: Prevention Fire & Explosion Emergency Response a UK regulation which is on force since 1995 and call the offshore Oil & Gas facilities to determine and manage the safety critical elements.

CHAPTER 2

Fundamentals of Systemic Approach

2.1 SYSTEMIC VERSUS SYSTEMATIC

It is very easy to distinguish what is systemic and what is systematic. Perhaps the confusion between the two words only arose because of semantics, spelling, and sound. The two words have almost the same letters nevertheless. It is just the addition of the two letters "a" and "t" for systematic, which make it longer. The pronunciation is also the same but there is obviously an additional syllable for the word systematic.

By definition the two adjectives differ. Systematic means something is well organized or arranged according to a *set of plan* and or is *grouped into systems*. Conversely, systemic means something matters to the *entire system*. This describes something that belongs to, works together with, or can affect the entire body or *system as a whole*. The two words further amplify its definition when used to describe other terms across various fields of study. For business, there are systemic and systematic risks and also systemic and systematic circulations in the field of system engineering and science.

2.2 CRITICALITY OF THE SYSTEMIC, SYSTEMATIC CHANGES

A system engineer should continually deal with the fear of

- *systemic uncertainty* in unforeseen events and unintended consequences,
- losing or even reinforcing excessive *systematic control*,
- *change in systems*; an undue ultimate optimism in old or new systems.

To assure the positive systemic, systematic and systems changes, scope of change should be categorized and assessed as shown in Table 2.1.

A Systems Approach to Managing the Complexities of Process Industries.
DOI: http://dx.doi.org/10.1016/B978-0-12-804213-7.00002-5

Table 2.1 Critical Changes Features and Pitfalls

Type of Change	Location of Change	Primary Intent	Risk or Traps	Some Keywords
Systemic	Complex realities or situation	Make simple and manageable the complex web of realities for improving situations	Seeing a mess as simple problem-solving, i.e., reductionist thinking rather than as improvement resolution	Complexity, feedback, emergence, uncertainty, autonomy
Systematic	Stakeholders	Developing mutual understanding and shared practice	Fixing people as objects for purposive endeavors rather than as purposeful subjects	Perspectives, praxis, learning, stakeholding
Systems	Conceptual worlds	Improvement of situations and emancipation through reflective practice	Complacency and obsession with "systems," e.g., as holistic devices, rather than as temporary pragmatic constructs	Judgements, boundaries, reframing, critique

2.3 SYSTEMATIC VERSUS SYSTEMIC FAILURE

"Systematic failure" is occurring in a deterministic way when given conditions are encountered. It should not be confused with systemic failures which occurs "at system level." ISO/TR 12489:2013 defines the systematic and systemic failures as follows.

Systematic failure is a "failure that consistently occurs under particular conditions of handling, storage, use or maintenance."

The cause of a systematic failure is found in the specification, design, manufacture, installation, operation, or maintenance.

Corrective maintenance without modification will usually not eliminate the cause of a systematic failure. A systematic failure can be reproduced by deliberately applying the same conditions, e.g., in verifying the failure cause. Systematic failures are nonrandom failures.

In operation a systematic failure is a manifestation of a systematic fault, i.e., a preexisting state of the system.

The software systematic failures called "bugs," are example of systematic failures: they are due to preexisting bugs (i.e., faults) and they occur when the input data activate them. Fig. 2.1 illustrates a systematic failure.

Systemic or holistic failure is a "failure at system level which cannot be simply described from the individual component failures of the system." In other words, "The whole is more than the sum of its parts."

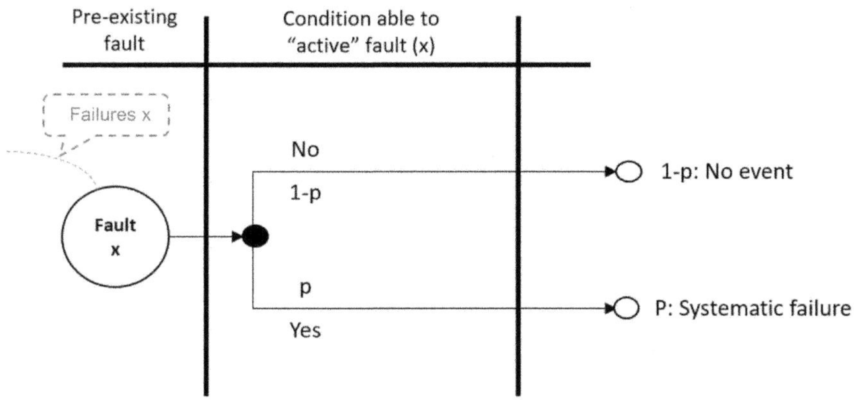

FIGURE 2.1
Illustration of a systematic failure.

Components have only failure modes. Those failure modes become dangerous, safe or spurious only when the components are implemented into a safety "system." Therefore dangerous, safe, or spurious failures are typical systemic failures. For example, the failure "fail to close" of a valve is dangerous only if it belongs to a safety system closing this valve on demand. Otherwise this failure mode does not matter (Fig. 2.2A and B).

The cause and effect of the systemic failures can be assessed by fault tree analysis as tool of the hard system thinking and structural model for the soft system thinking.

Hard system thinking applies to the predicable systems such as a safety system and soft system thinking applies to the complex systems which the behavior of system can be dynamic and unpredictable such as human factors or organizational issues.

2.4 WHAT IS A SYSTEM?

2.4.1 Background

The term "system" comes from the Latin word *systēma*, "whole compounded of several parts or members, system," literary "composition."

The 20th century philosophe *Marshall McLuhan* interpret "System" as *"something to look at."* For example in Fig. 2.3, depending our focus on rescue (left image) or traffic (right image), system of interest and enabling systems for the same event will be different. You must have a very high visual gradient to have systematization. In philosophy, before Descartes, there was no "system." Plato had no "system." Aristotle had no "system."

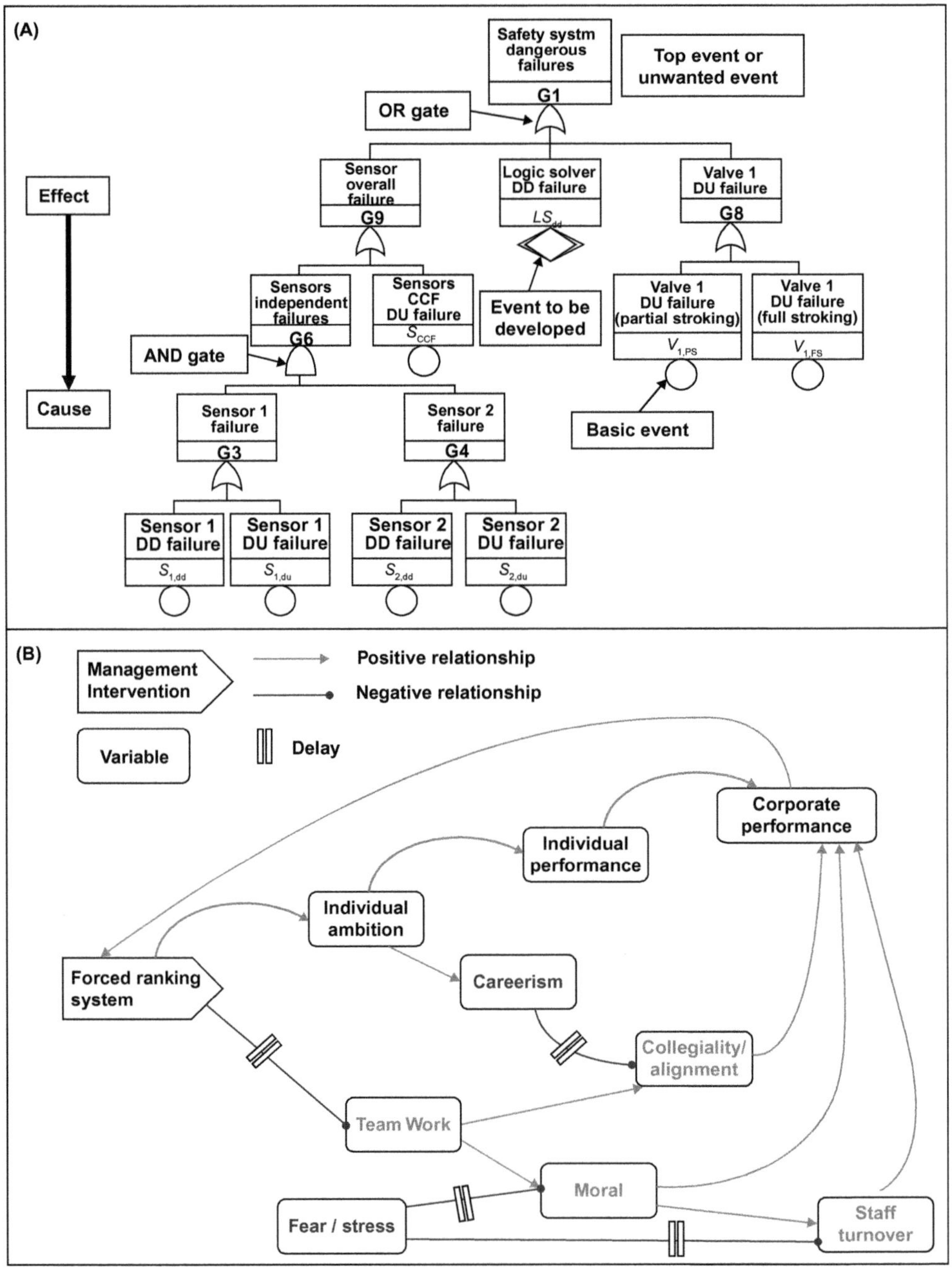

FIGURE 2.2

(A) Fault tree for failure analysis of the hard system thinking ISO/TR 12489:2013. (B) Structural model for the failure analysis of the soft system thinking.

FIGURE 2.3
A helicopter view is necessary to put order in the road transport system after a chaotic car accident.

In the 19th century the French physicist Nicolas Léonard Sadi Carnot, who studied thermodynamics, pioneered the development of the concept of a "system" in the natural sciences. In 1824 he studied the system which he called the *working substance* (typically a body of water vapor) in steam engines, in regards to the system's ability to do work when heat is applied to it. The working substance could be put in contact with either a boiler, a cold reservoir (a stream of cold water), or a piston (to which the working body could do work by pushing on it). In 1850 the German physicist Rudolf Clausius generalized this picture to include the concept of the *surroundings* and began to use the term "working body" when referring to the system.

The biologist Ludwig von Bertalanffy (1901–72) became one of the pioneers of the general systems theory. In 1945 he introduced *models, principles, and laws that apply to generalized systems or their subclasses, irrespective of their particular kind, the nature of their component elements, and the relation or "forces" between them.*

Norbert Wiener (1894–1964) and Ross Ashby (1903–72), who pioneered the use of mathematics to study systems, carried out significant development in the concept of a *system*.

In the 1980s John H. Holland, Murray Gell-Mann and others coined the term "complex adaptive system" at the interdisciplinary Santa Fe Institute.

2.4.2 System Definitions

There are different definitions of a system. A system can be broadly defined as an integrated set of *elements* that accomplish a defined objective. Every system is delineated by its spatial and temporal boundaries, surrounded, and influenced by its *environment*, described by its structure and purpose and expressed in its functioning.

The organized, purposeful structure of a system consists of interrelated and interdependent elements (components, entities, factors, members, parts, etc.). These elements continually influence one another (directly or indirectly) to maintain their activity and the existence of the system, to achieve the goal of the system.

People from different engineering disciplines have different perspectives of what a "system" is. For example, software engineers often refer to an integrated set of computer programs as a "system." Electrical engineers might refer to complex integrated circuits or an integrated set of electrical units as a "system." Quality engineers, define system as a set of detailed methods, procedures and routines created to carry out a specific activity, perform a duty, or solve a problem. As can be seen, "system" depends on one's perspective, and the "integrated set of elements that accomplish a defined objective" is an appropriate definition. All systems have:

1. inputs, outputs, and feedback mechanisms,
2. maintain an internal steady-state (called homeostasis) despite a changing external environment,
3. display properties that are different than the whole (called emergent properties) but are not possessed by any of the individual elements, and
4. boundaries that are usually defined by the system observer.

Fundamental properties of a system described in the "system engineering" include *togetherness, structure, behavior,* and *emergence*. In this context *the essence of a system is "togetherness," the drawing together of various parts and the relationships they form in order to produce a new whole.*

The whole has the properties that none of the parts has. For example:

- a "car" can move you from point A to B but the car motor alone cannot.
- "You" have life, your cut hand cannot write anything.

2.4.3 Different Types of System

The different types of system can be distinguished. The system engineering strategy for each of them can be different. For example, ISO 55000 divides the total business to different assets as shown in Fig. 2.4.

Physical or abstract: Physical system is tangible entities that may be static or dynamic in nature. Abstract system is conceptual or nonphysical. The abstract is conceptualization of physical situations.

Subsystem and super system: Each system is part of a larger system. For example, as shown in Fig. 2.5, the dynamic performance is a subsystem of a car while the

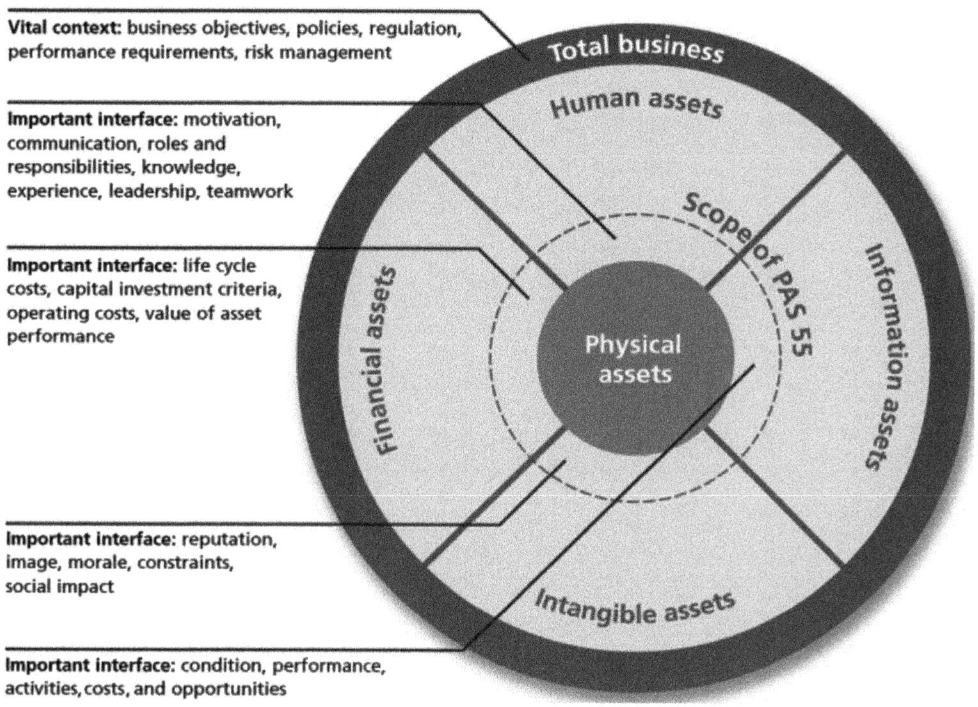

FIGURE 2.4
ISO 55000 Physical assets system in relation to the other categories of assets.

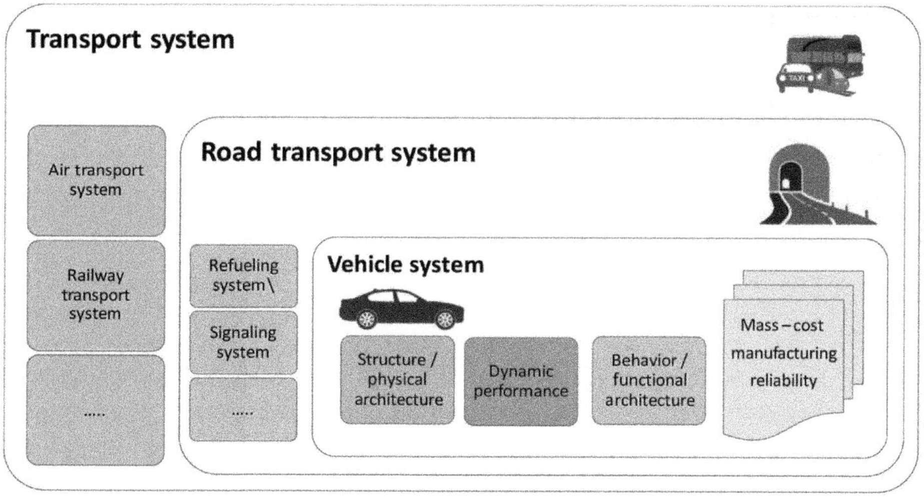

FIGURE 2.5
Subsystems and super systems of a car system.

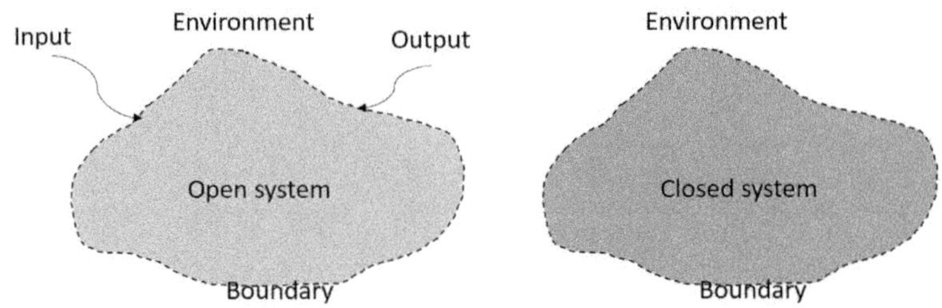

FIGURE 2.6
Open versus closed system.

refuelling system is a part of the road transport system which is the super system with respect to car. The business firm is viewed as the system or total system when focus is on production, distribution of goal, and sources of profit and income.

The total system consists of all the objects, attributes, and relationship necessary to accomplish an objective given a number of constraints. Subsystems are the smaller systems within a system. Super system denotes extremely large and complex system.

Open and closed: An open system continually interacts with its environment. It receives input from the outside and delivers output to outside. A closed system is isolated from environment influences.

For *closed systems* all aspects of the system exist within this boundary. This idea is useful for abstract systems and for some theoretical system descriptions.

The boundary of an *open systems* defines those elements and relationships which can be considered part of the system and those which describe the interactions across the boundary between system elements and elements in the environment. Fig. 2.6 illustrates the open system versus a closed system.

Natural and man-made system: A system which is made by man is called man-made system. Systems which are in the environment made by nature are called natural system. SE Body of Knowledge (SEBoK) defines the three open system domains as follows:

1. *A natural system* is one whose elements are wholly natural.
2. *A social system* includes only humans as elements.
3. *An engineering system* is a man-made aggregation which may contain physical, informational, human, natural, and social elements; it is normally created for the benefit of people.

Fig. 2.7 provides an example of the overlap between these three domains, which raises the needs for the abstract management and assurance systems such as quality and safety management systems.

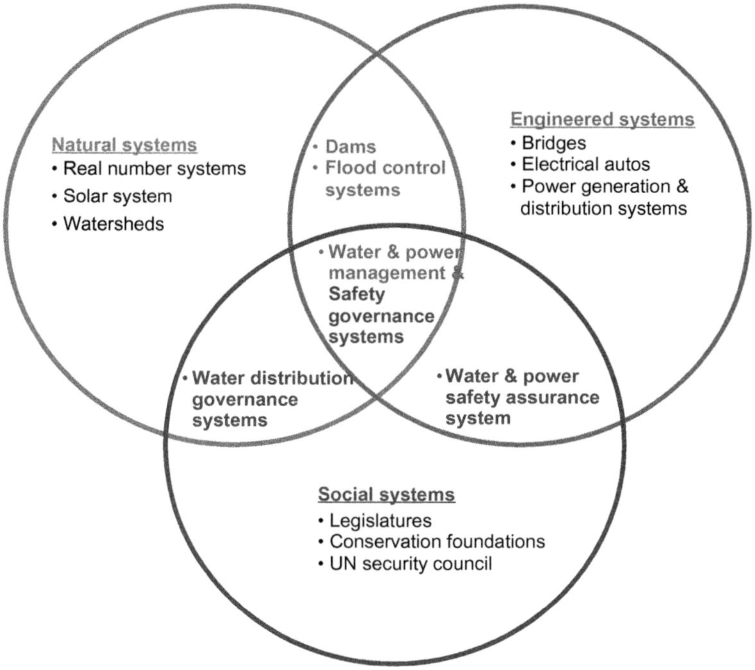

FIGURE 2.7

System boundaries of engineered systems, social systems, and natural systems.

Permanent and temporary system: A permanent system is a system enduring for a time span that is long relative to the operation of human. Temporary system is one having a short-time span.

For example, the permit to work (PTW) management system is a permanent subsystem of process safety management (PSM) system but the PTW of a given maintenance or modification job forms a temporary system and team for the period of job; system will be retired and team will be demobilized when the job is completed.

Deterministic and probabilistic: A deterministic system is one in which the occurrence of all events is known with certainty. If the description of the system state at a particular point of time of its operation is given, the next state can be perfectly predicted. If we get the description of the system state at a particular time, the next state can be easily predicted. A clock, a car, and a compressor are the deterministic systems.

In a probabilistic system, unlike in a deterministic system, what has just occurred is not always an accurate predictor of what will transpire next. The weather, for example, is a probabilistic system, in that future events can only be imperfectly predicted.

The theory of probabilistic systems by extension leads to probabilistic analysis and forecasting. A probabilistic system must be analyzed according to the various possible outcomes and their relative probability of occurrence.

Human judgment under uncertainty has been a major focus of behavioral research in the last quarter of the century. A central motivation for this work has been the desire to compare human judgment against the "gold standard" of the normative probability theory and decision theory in order to identify places where human decision making could be improved.

Numerous studies have shown that people usually rely on *rules of thumb*, or *heuristics*, in their estimations of the likelihood of uncertain events. These heuristics make estimation of probabilities computationally tractable, but often result in systematic biases and inconsistencies. A concise summary of the position that is known today as the "heuristics and biases" school of thought is given by Tversky and Kahneman.

The *subjective assessment* of probability resembles the subjective assessment of physical quantities such as distance or size. These judgments are all based on data of *limited validity*, which are processed according to heuristic rules. For example, the apparent distance of an object is determined in part by its clarity. The more sharply the object is seen, the closer it appears to be. This rule has some validity, because in any given scene the more distant objects are seen less sharply than nearer objects. However, the reliance on this rule leads to *systematic errors* in the estimation of distance. Specifically, distances are often overestimated when visibility is poor because the contours of objects are blurred. On the other hand, distances are often underestimated when visibility is good because the objects are seen sharply. Thus the reliance on clarity as an indication of distance leads to common biases. Such biases are also found in the intuitive judgment of probability.

It is these biases that are believed to provide the most information about cognitive processes used in judgment. Designing experiments that lead to errors in human reasoning and observing the nature of these errors allows to infer the heuristic procedures that most likely led to them. Fig. 2.8 illustrates the causes of the cognitive failures.

Man-made information system: It is generally believed that the information reduces uncertainty about a state or event. An information system is the basis for interaction between the user and the analyst. It determines the nature of relationship among decision makers.

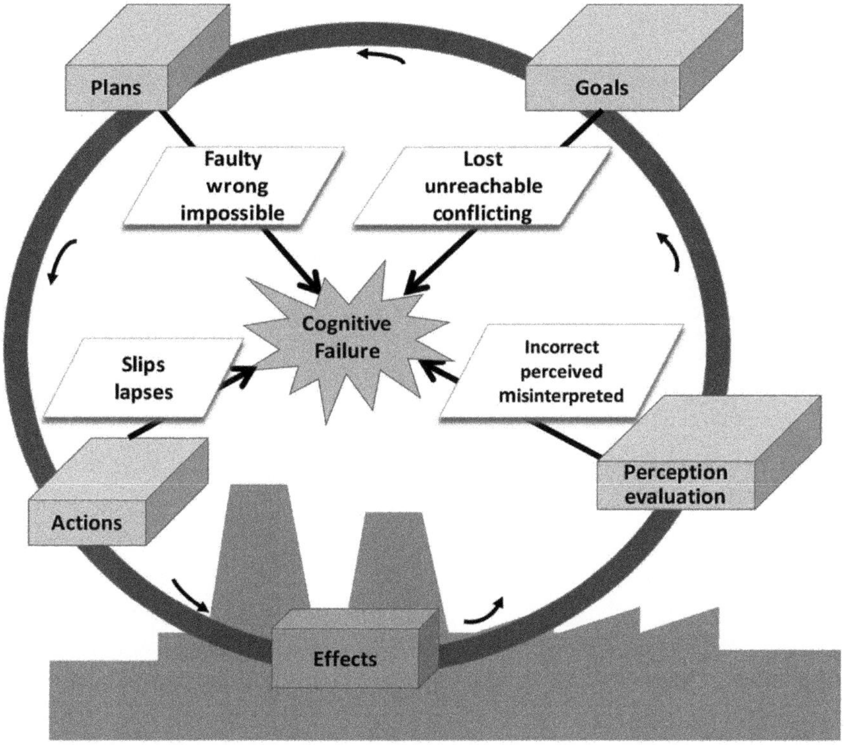

FIGURE 2.8
Causes of the cognitive failure.

An information system may be defined as a set of devices, procedures, and operating system designed around user-base criteria to produce information and communicating it to the user for planning control and performance.

Very often, organizations believe that as much as 50–70% of the data they store has no value and is not required for legal or regulatory purposes. Because data storage is relatively cheap, it is easier to let dead data accumulate than it is to deal with it. But we should not confuse "storage" cost with all-in data compliance expense. Fig. 2.9 shows the required tools for the information management values.

Today, with the advanced web technology and cloud computing, the digital transformation impose itself to all type of businesses. Big data, cyber security, and legacy management are the key issues to be considered in the modern information systems.

The collaborative aspects of web-based applications provide the powerful tools to involve not only all workforce but also the relevant external

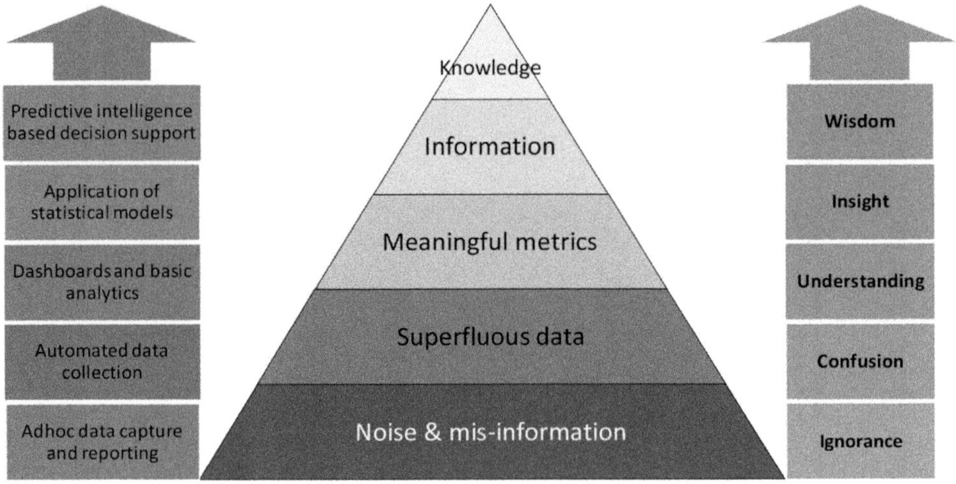

FIGURE 2.9
Information management values and tools.

identities such as subcontractors, regulatory bodies, etc. in the information management system.

Industrial Internet of Things (IIoT) and connectivity of the IT and OT enables us to design the self-regulated and self-organized information and knowledge management systems. Fig. 2.10 provides a high-level view of general enablers (GEs) in the data/context management.

Management system: it is a structured and documented set of interdependent practices, process, and procedures used by the managers and the workforce at every level in a company to plan, direct, and execute activities. The most common management system are listed as follows. Each of these management Systems generates the considerable amount of structure and unstructured data which should be processed to become the meaningful information to make the better decision. Very often the information from one management system interact with the information from/to other management systems.

- Manufacturing operation management system
- Inventory management system
- Product lifecycle management (PLM)
- Supply chain management system
- Operational excellence
- Quality management system
- Configuration management system
- Environmental management system

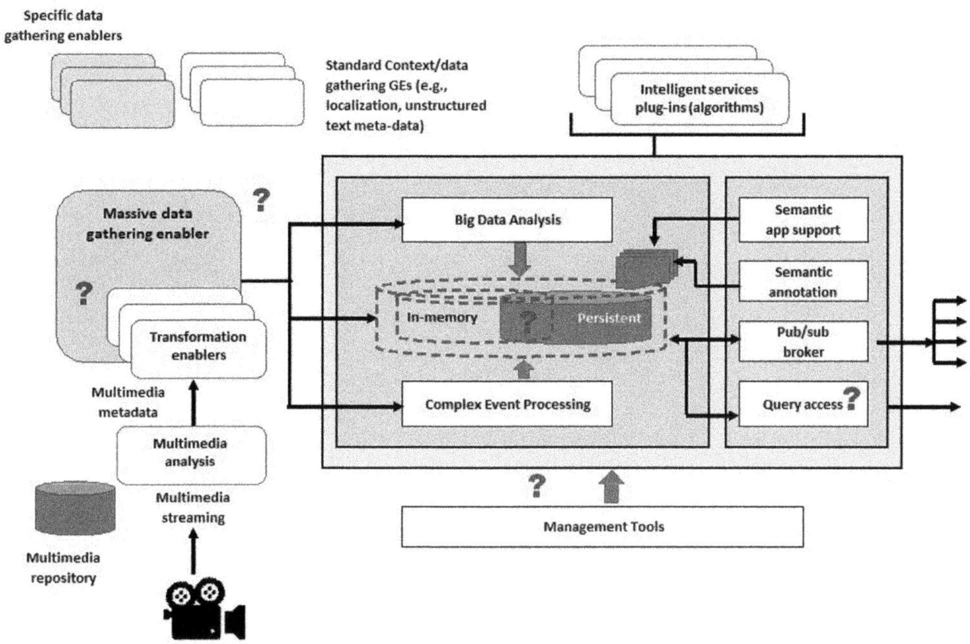

FIGURE 2.10
High-level view of general enablers (GEs) in the data/context management chapter.

- Occupational health & safety management system
- Asset integrity management system
- PSM system
- Security management system
- *Installation Lifecycle Management (ILM)*
- *IIOT*

Enterprises that are critically dependent upon complex physical assets have historically focused integration efforts on two major horizontal layers; real-time control and business information systems. Experts within these two areas seldom work directly with each other and do not focus on integration between the layers, which has resulted in a significant vertical information gap. This gap is compounded when O&M processes, systems and people are not efficiently integrated with each other, resulting in a corresponding horizontal information gap. Together, these gaps create an empty space in the very center of enterprise process and information integration.

Operational inefficiencies coming from the lack of integration have been overlooked or minimized due to a general lack of interdisciplinary understanding. Overall optimization requires proper integration of O&M processes, systems, and people. MIMOSA is working on effective solutions to eliminate these impediments to efficiency.

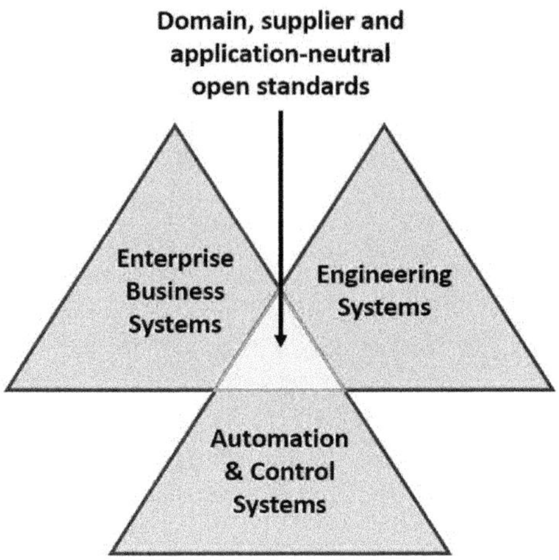

FIGURE 2.11
MIMOSA system integration model.

MIMOSA is a not-for-profit trade association dedicated to developing and encouraging the adoption of open information standards for operations and maintenance in manufacturing, fleet, and facility environments. MIMOSA's open standards enable collaborative asset lifecycle management.

Historically, the O&M community has also lacked tight alignment with the life-cycle engineering community. This has led to a series of poorly connected activities with highly suboptimal results including data quality problems and the loss of configuration control for complex physical assets including plants, platforms, and facilities. The effect has been a loss of integrity management for the digital asset which makes integrity management for the physical asset much more difficult.

Working in close cooperation with groups such as POSC Caesar Association and Fiatech, MIMOSA is helping to establish the basis for a more integrated approach to Critical Infrastructure Management, holistically combining full life cycle engineering with O&M activities. This supports more sustainable approaches to both integrity management and risk management. Fig. 2.11 illustrates the MIMOSA system integration model.

Organizations such as Standards Leadership Council (SCL) goes further in uniting the leaders of the upstream oil and gas industry standards organizations in a collaborative environment promoting the adoption of open and freely-available standards. Fig. 2.12 summarizes the SCL members.

The mission of the SCL is to speak with one voice to endorse and promote the adoption of standards and to focus on the semantics and workflow mapping to harmonize the standards landscape for the benefit of the industry.

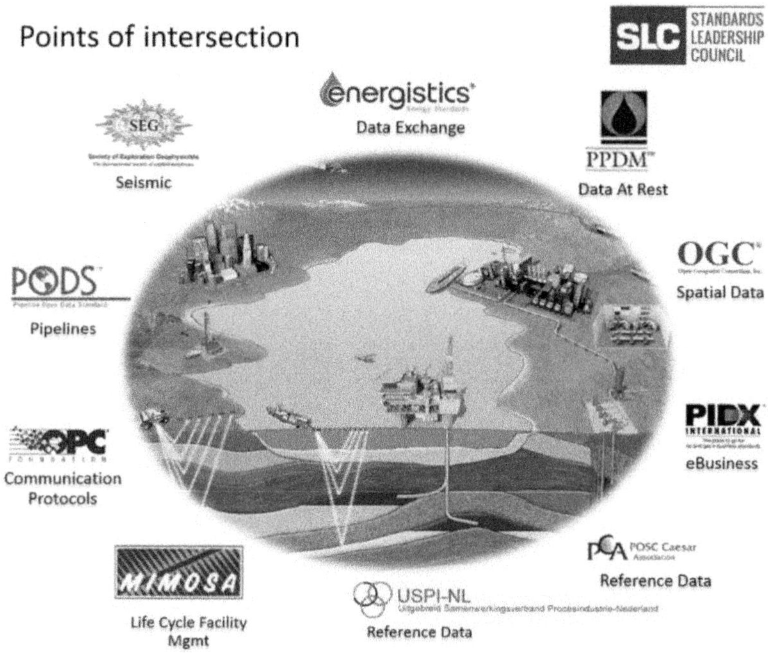

FIGURE 2.12
Standards Leadership Council—Member Organizations.

IIoT provides the required tools for the effective integration of "Operational Technology (OT)," "Information Technology (IT)" and "Engineering Technology (ET)."

Product lifecycle management is used by discrete manufacturing since 1985. The authors of this book believe that with the combination of PLM and IIoT the new concept of "Installation Lifecycle Management (ILM)" has already been found. Fig. 2.13 illustrates the possible configuration of an ILM. However, we have a long way to democratize this concept and implement it in the process manufacturing plants.

2.4.4 Environment Definition

Environment is defined as the surroundings (natural or man-made) in which the system-of-interest is utilized and supported; or in which the system is being developed, produced, or retired.

Technical processes enable systems engineers to coordinate the interactions between engineering specialists, systems stakeholders and operators, and manufacturing. They also address conformance with the expectations and

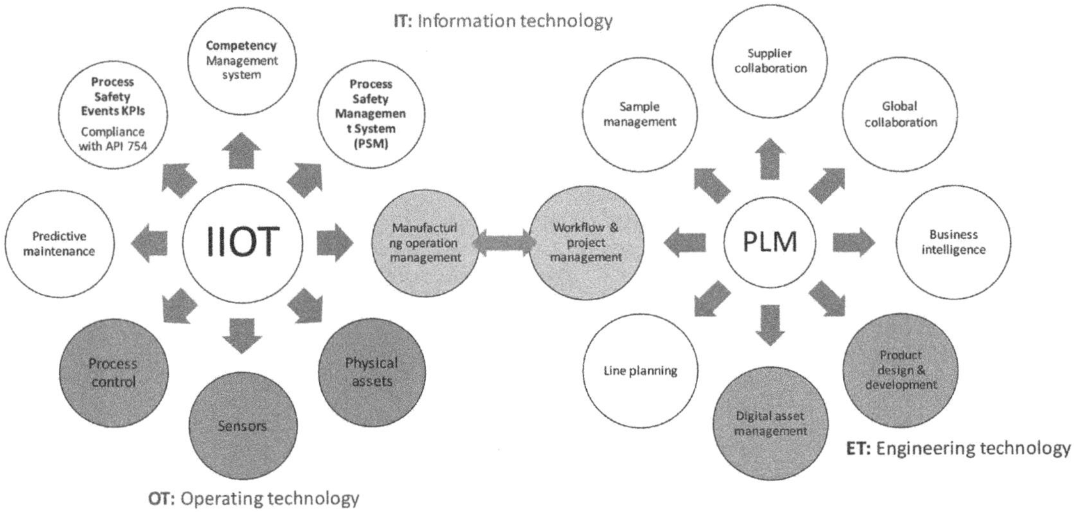

FIGURE 2.13
Possible configuration of installation lifecycle management (ILM).

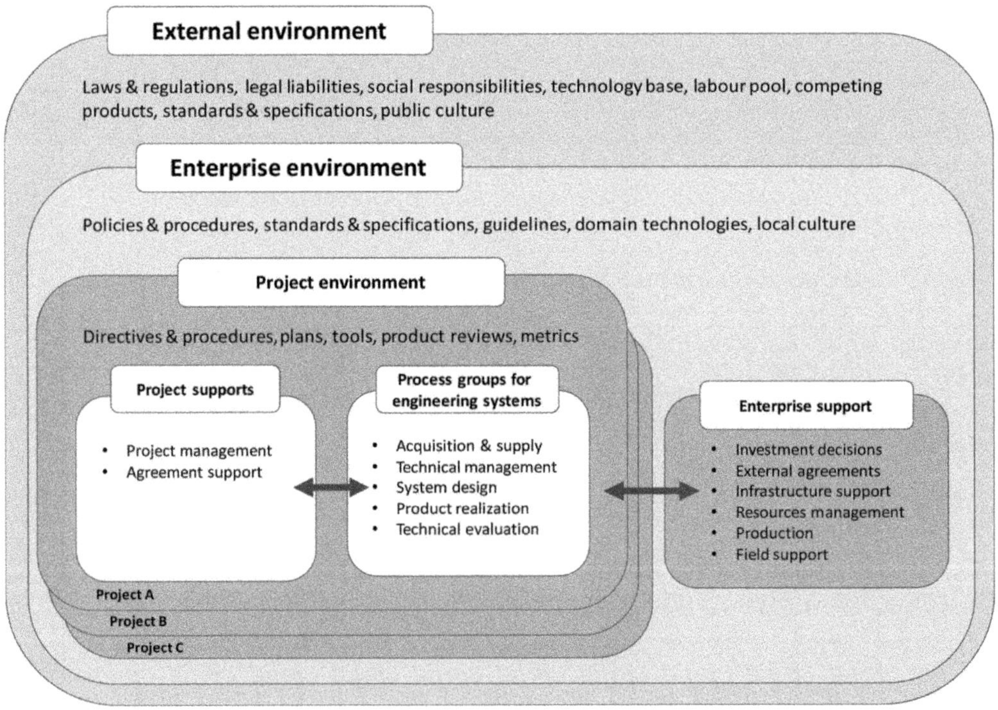

FIGURE 2.14
Generic process approach and structure. *EIA 632 & MAP Système.*

legislated requirements of society. These processes lead to the creation of a full set of requirements that address the desired capabilities within the bounds of performance, *environment*, external interfaces, and design constraints. Fig. 2.14 summarizes the generic process approach and structure of these systems according to EIA 632 & MAP *Système*.

Ervin Laszlo (1972) summarizes the open system property of holism (or *systemic state*) as a property of the system elements and how they are related in the system structure that leads them to create a cohesive whole. Open systems can persist when the relationships between the elements reach a balance which remains stable within its environment. Laszlo describes three kinds of system response to *environmental disturbance*.

- *Adaptive self-regulation*—Systems will tend to return to their previous state in response to external stimulus.
- *Adaptive self-organization*—Some systems not only return to a previous state, but also reorganize to create new stable states which are more resistant to change.
- *Holonic*—Systems displaying characteristics one and two will tend to develop increasingly complex (hierarchical) structures.

The observed behavior of a system in its environment leads to the fundamental property of *emergence*. At some point the nature of the relationships between elements within and across boundaries in *a hierarchy* of systems may lead to behavior which is difficult to understand or predict. This system *complexity* can only be dealt with by considering the systems as a collective whole.

2.4.5 Complexity

Complexity is a major issue. As system elements are added, the complexity of system interaction grows in a nonlinear fashion. Furthermore, conflicting or missing interface standards can make it hard to define data exchanges across system element interfaces. Characteristics of a complex system is illustrated in Fig. 2.15. Complexity levels and management will be discussed in chapter 3.0 (Fig. 2.16).

2.4.6 Emergence

This topic is closely related to the complexity.

Emergence is a consequence of the fundamental system concepts of holism and interaction. System wholes have behaviors and properties arising from the organization of their elements and their relationships, which only become apparent when the system is placed in different environments.

Questions that arise from this definition include: What kinds of systems exhibit different kinds of emergence and under what conditions? Can

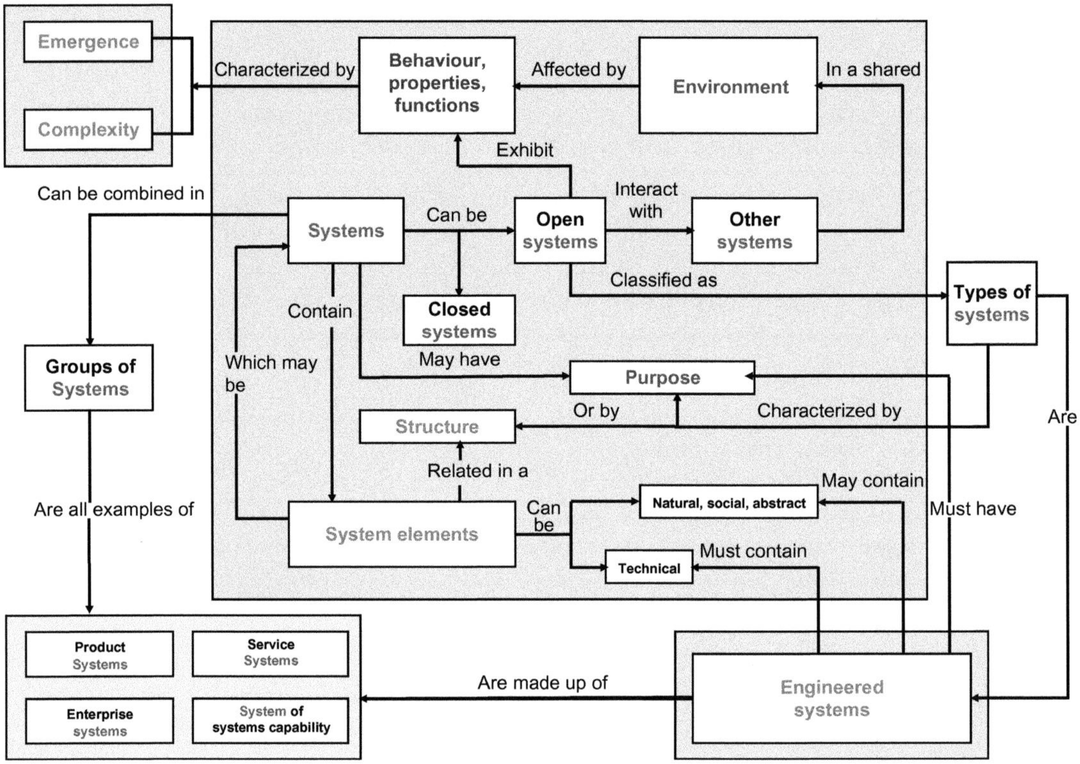

FIGURE 2.15
System fundamentals and engineered systems (SEBoK v.01).

emergence be predicted, and is it beneficial or detrimental to a system? How do we deal with emergence in the development and use of engineered systems? Can it be planned for? How?

There are many varied and occasionally conflicting views on emergence. Emergent system behavior can be viewed as a consequence of the interactions and relationships between system elements rather than the behavior of individual elements. It emerges from a combination of the behavior and properties of the system elements and the systems structure or allowable interactions between the elements, and may be triggered or influenced by a stimulus from the systems environment.

Emergence is common in nature. The pungent gas ammonia results from the chemical combination of two odorless gases: hydrogen and nitrogen. As individual parts, feathers, beaks, wings, and gullets do not have the ability to overcome gravity. Properly connected in a bird, however, they create the

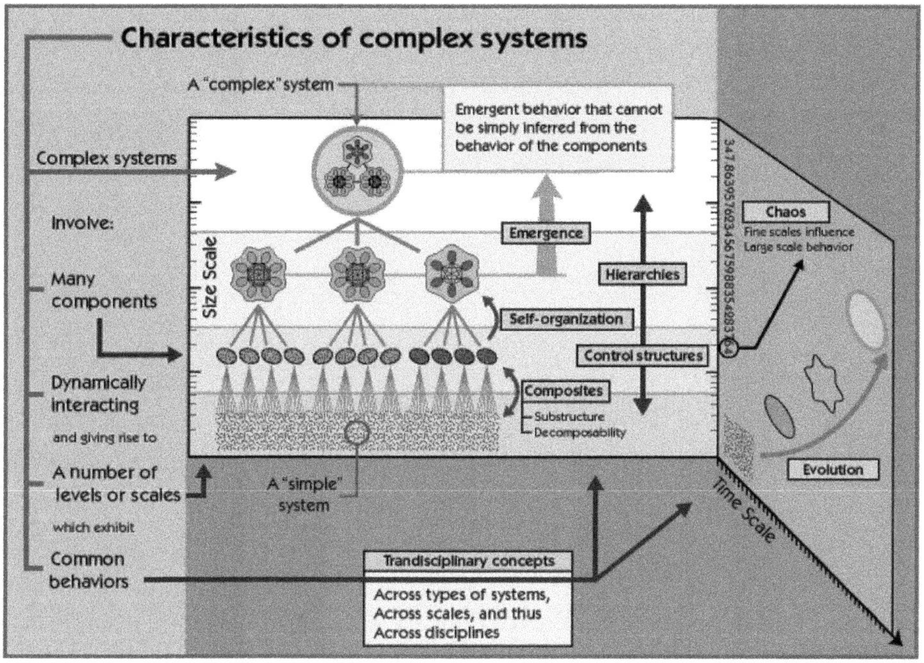

FIGURE 2.16
Characteristics of a complex system. From: http://www.necsi.edu/visual/systems.html.

emergent behavior of flight. What we refer to as "self-awareness" results from the combined effect of the interconnected and interacting neurons that make up the brain.

We can observe a number of levels of outcome which arise from interaction between elements in an engineered system context. At a simple level, some system outcomes or attributes have a fairly simple and well defined mapping to their elements; for example, center of gravity or top speed of a vehicle result from a combination of element properties and how they are combined. Other behaviors can be associated with these simple outcomes, but their value emerges in complex and less predictable ways across a system. The single lap performance of a vehicle around a track is related to center of gravity and speed; however, it is also affected by driver skill, external conditions, component ware, etc. Getting the "best" performance from a vehicle can only be achieved by a combination of good design and feedback from real laps under race conditions.

There are also outcomes which are less tangible and which come as a surprise to both system developers and users. How does lap time translate into a winning motor racing team? Why is a sports car more desirable to many than other vehicles with performances that are as good or better?

Emergence can always be observed at the highest level of system. However, the extent that the systems elements themselves can be considered as systems, they also exhibit emergence.

Three types of emergence can be distinguished: "simple," "weak," and "strong."

Simple emergence is generated by the combination of element properties and relationships and occurs in noncomplex or "ordered" systems. To achieve the emergent property of "controlled flight" we cannot consider only the wings, or the control system, or the propulsion system. All three must be considered, as well as the way these three are interconnected-with each other, as well as with all the other parts of the aircraft. Simple emergence is the only type of emergence that can be predicted. This view of emergence is also referred to as *synergy*. The safety system of Fig. 2.2A illustrates a simple emergence.

Weak emergence as expected emergence which is desired (or at least allowed for) in the system structure. However, since weak emergence is a product of a complex system, the actual level of emergence cannot be predicted just from knowledge of the characteristics of the individual system components. The personnel motivation of Fig. 2.2B illustrates a weak emergence.

The term *strong emergence* is used to describe unexpected emergence; that is, emergence not observed until the system is simulated or tested or, more alarmingly, until the system encounters in operation a situation that was not anticipated during design and development.

Strong emergence may be evident in failures or shutdowns. For example, the US–Canada Blackout of 2003 as described by the US–Canada Power System Outage Task Force (2004) was a case of cascading shutdown that resulted from the design of the system. Even though there was no equipment failure the shutdown was systemic. This example shows that emergent properties are not always beneficial.

2.5 WHAT IS SYSTEM ENGINEERING?

2.5.1 Background

The term systems engineering dates back to Bell Telephone Laboratories in the early 1940s. Fagen [1978] traces the concepts of systems engineering within Bell Labs back to early 1900s and describes major applications of systems engineering during World War II. Hall [1962] asserts that the first attempt to teach systems engineering as we know it today came in 1950 at MIT by Mr. Gilman, Director of Systems Engineering at Bell.

Defense, spatial, aeronautics, automotive, railway, and nuclear industries are the pioneers in using the system engineering in their projects.

Process manufacturing uses system engineering in an implicit way in the design and implementation of the

- Project management,
- Management systems frameworks such as HSEMS, PSM, AIM, etc.
- Determination and management of the safety-critical elements (SCEs)
- Safety integrity level (SIL) assessment and verification activities
- Reliability, availability, maintain ability, and safety,
- Management of change,
- Ergonomics and human factors,
- Brainstorming sessions such as HAZard OPeratbility (HAZOP), HAZard IDentification (HAZID), job hazard analysis (JHA), etc.

Understanding the system engineering scope and train the relevant personnel for system engineer skills and competency are crucial to obtain the optimal performance of the process manufacturing industries.

2.5.2 System Engineering Derivative Disciplines

Today, many related fields may be considered tightly coupled to systems engineering. These areas have contributed to the development of systems engineering as a distinct entity.

2.5.3 Cognitive Systems Engineering

Cognitive systems engineering (CSE) is a specific approach to the description and analysis of human–machine systems or socio-technical systems. The three main themes of CSE are how humans cope with complexity, how work is accomplished by the use of artifacts, and how human–machine systems and socio-technical systems can be described as joint cognitive systems (JCS). CSE has since its beginning become a recognized scientific discipline, sometimes also referred to as cognitive engineering. The concept of a JCS has in particular become widely used as a way of understanding how complex socio-technical systems can be described with varying degrees of resolution. The more than 20 years of experience with CSE has been described extensively.

2.5.4 Configuration Management

Like systems engineering, configuration management as practiced in the manufacturing and process industries is a broad systems-level practice. The field parallels the tasking of systems engineering; where systems engineering deals with requirements development, allocation to development items and verification, configuration management deals with requirements capture, traceability to the development item, and audit of development item to ensure that it has achieved the desired functionality that systems engineering and/or test and verification engineering have proven out through objective testing.

2.5.5 Control Engineering

Control engineering and its design and implementation of control systems, used extensively in nearly every industry, is a large subfield of systems engineering. The cruise control on an automobile and distillation column control system are two examples. Control systems theory is an active field of applied mathematics involving the investigation of solution spaces and the development of new methods for the analysis of the control process.

2.5.6 Software Engineering

From its beginnings, software engineering has helped shape modern systems engineering practice. The techniques used in the handling of complexes of large software-intensive systems have had a major effect on the shaping and reshaping of the tools, methods, and processes of SE.

2.5.7 Industrial Engineering

Industrial engineering is a branch of engineering that concerns the development, improvement, implementation and evaluation of integrated systems of people, money, knowledge, information, equipment, energy, material, and process. Industrial engineering draws upon the principles and methods of engineering analysis and synthesis, as well as mathematical, physical, and social sciences together with the principles and methods of engineering analysis and design to specify, predict, and evaluate results obtained from such systems.

2.5.8 Interface Design

Interface design and its specification are concerned with assuring that the pieces of a system connect and interoperate with other parts of the system and with external systems as necessary. Interface design also includes assuring that system interfaces be able to accept new features, including mechanical, electrical, and logical interfaces, including reserved wires, plug-space, command codes, and bits in communication protocols. This is known as extensibility. Human–computer interaction or human–machine interface is another aspect of interface design and is a critical aspect of modern systems engineering. Systems engineering principles are applied in the design of network protocols for local-area networks and wide-area networks.

2.5.9 Mechatronic Engineering

Mechatronic engineering, like systems engineering, is a multidisciplinary field of engineering that uses dynamical systems modeling to express tangible constructs. In that regard it is almost indistinguishable from systems engineering, but what sets it apart is the focus on smaller details rather than larger generalizations and relationships. As such, both fields are distinguished by the scope of their projects rather than the methodology of their practice.

2.5.10 Operations Research

Operations research supports systems engineering. The tools of operations research are used in systems analysis, decision making, and trade studies. Several schools teach SE courses within the operations research or industrial engineering department, highlighting the role systems engineering plays in complex projects. Operations research, briefly, is concerned with the optimization of a process under multiple constraints.

2.5.11 Performance Engineering

Performance engineering is the discipline of ensuring a system meets customer expectations for performance throughout its life. Performance is usually defined as the speed with which a certain operation is executed, or the capability of executing a number of such operations in a unit of time. Performance may be degraded when an operations queue to execute is throttled by limited system capacity. For example, the performance of a packet-switched network is characterized by the end-to-end packet transit delay, or the number of packets switched in an hour. The design of high-performance systems uses analytical or simulation modeling, whereas the delivery of high-performance implementation involves thorough performance testing. Performance engineering relies heavily on statistics, queueing theory, and probability theory for its tools and processes.

2.5.12 Program Management and Project Management

Program management (or program management) has many similarities with systems engineering, but has broader-based origins than the engineering ones of systems engineering. Project management is also closely related to both program management and systems engineering.

2.5.13 Proposal Engineering

Proposal engineering is the application of scientific and mathematical principles to design, construct, and operate a cost-effective proposal development system. Basically, proposal engineering uses the "systems engineering process" to create a cost-effective proposal and increase the odds of a successful proposal.

2.5.14 Reliability Engineering

Reliability engineering is the discipline of ensuring a system meets customer expectations for reliability throughout its life; i.e., it does not fail more frequently than expected. Reliability engineering applies to all aspects of the system. It is closely associated with maintainability, availability (dependability or RAMS preferred by some), and logistics engineering. Reliability engineering is always a critical component of safety engineering, as in failure modes and effects analysis (FMEA) and hazard fault tree analysis, and of security engineering.

2.5.15 Risk Management

Risk management, the practice of assessing and dealing with risk is one of the interdisciplinary parts of systems engineering. In development, acquisition, or operational activities, the inclusion of risk in trade-off with cost, schedule, and performance features, involves the iterative complex configuration management of traceability and evaluation to the scheduling and requirements management across domains and for the system lifecycle that requires the interdisciplinary technical approach of systems engineering.

2.5.16 Safety Engineering

The techniques of safety engineering may be applied by nonspecialist engineers in designing complex systems to minimize the probability of safety-critical failures. The "System Safety Engineering" function helps to identify "safety hazards" in emerging designs and may assist with techniques to "mitigate" the effects of (potentially) hazardous conditions that cannot be designed out of systems.

2.5.17 Security Engineering

Security engineering can be viewed as an interdisciplinary field that integrates the community of practice for control systems design, reliability, safety, and systems engineering. It may involve such subspecialties as authentication of system users, system targets, and others: people, objects, and processes.

2.5.18 Scheduling

Scheduling is one of the systems engineering support tools as a practice and item in assessing interdisciplinary concerns under configuration management. In particular the direct relationship of resources, performance features, and risk to duration of a task, or the dependency links among tasks and impacts across the system lifecycle are systems engineering concerns.

2.5.19 System Engineering Scope of Work

There is not community consensus on a single definition for "systems engineering"; the term has different connotations in different domains. However, one of the more commonly recognized definitions in the field is that of the International Council on Systems Engineering (INCOSE).

An interdisciplinary approach and means to enable the realization of successful systems. It focuses on defining *customer needs* and *required functionality* early in the development cycle, documenting requirements, then proceeding with design synthesis and system validation while considering the complete problem:

- Operations
- Performance
- Test
- Manufacturing
- Cost and schedule
- Training and support
- Disposal

This is the definition that is frequently referenced throughout the SEBoK. SE is also the application of a traditional engineering and holistic systems thinking that works in combination with domain engineering, human sciences, management, and commercial disciplines in order to support the engineering of one or more systems-of-interest (SoI). SE may also be considered an interdisciplinary approach, governing the complete technical and managerial effort that is required to transform a set of customer needs, expectations, and constraints into a solution and to continue to support that solution throughout its lifecycle. Each of these perspectives is valuable for gaining a holistic view of SE as shown in (Fig. 2.17).

Systems engineering is concerned with the overall process of defining, developing, operating, maintaining, and ultimately *replacing quality systems*. Where

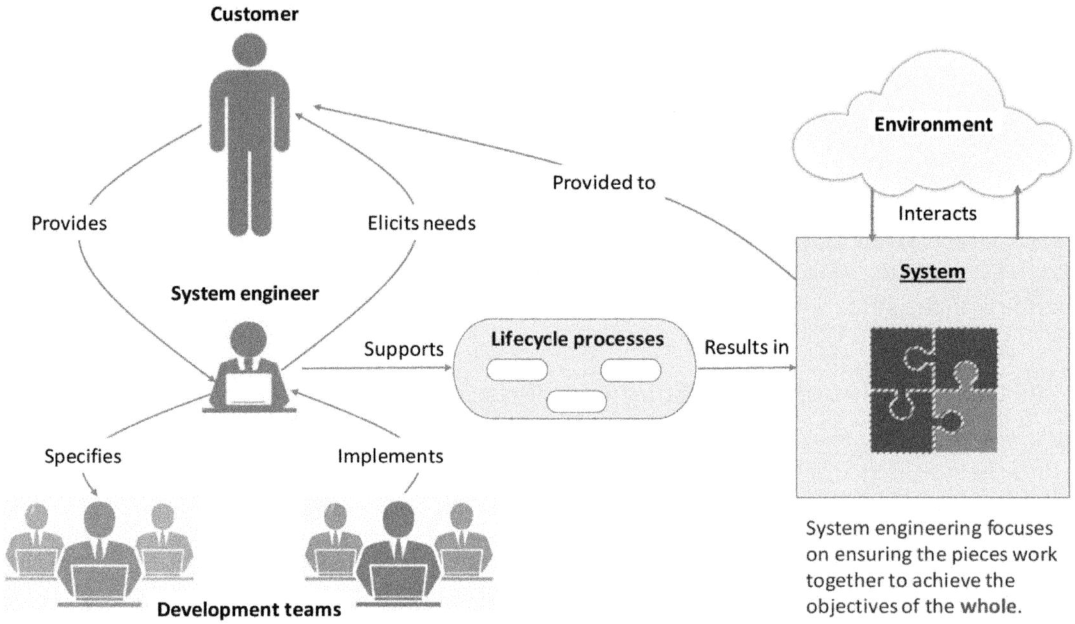

FIGURE 2.17
Key elements of systems engineering (SEBoK original v 1.7).

other engineering disciplines concentrate on the specifics of a system (electronics, mechanics, ergonometrics, aerodynamics, software, etc.), systems thinking allows the systems engineer to focus on the *integration* of these aspects as a coherent and effective system. Systems engineers bring a particular perspective to the engineering process (requirements definition, top-level functional designs, project management, lifecycle cost analysis, ...) that serves to organize and coordinate other engineering activities.

Fig. 2.18 shows that the systems engineering integrates all the disciplines and specialty groups into a team effort forming a structured development process that proceeds from concept to production to operation. Systems engineering considers both the business and the technical needs of all customers with the goal of providing a quality product that meets the user needs.

Systems engineering integrates all the disciplines and specialty groups into a team effort forming a structured development process that proceeds from concept to production to operation. Systems engineering considers both the business and the technical needs of all customers with the goal of providing a quality product that meets the user needs (INCOSE, 2012).

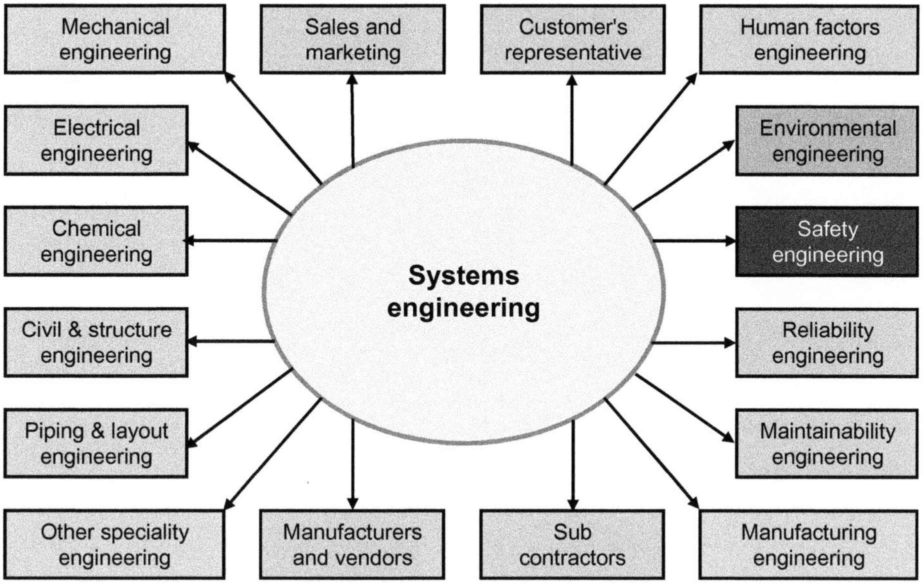

FIGURE 2.18
System engineering interfaces with the specialist disciplines.

One way to understand the motivation behind systems engineering is to see it as a method, or practice, to identify and improve common rules that exist within a wide variety of systems. Keeping this in mind, the principles of systems engineering which are holism, emergent behavior, boundary, etc. can be applied to any system, complex or otherwise, provided systems thinking is employed at all levels.

An analysis by the INCOSE Systems Engineering center of excellence (SECOE) indicates that optimal effort spent on systems engineering is about 15–20% of the total project effort. At the same time, studies have shown that systems engineering essentially leads to reduction in costs among other benefits. However, no quantitative survey at a larger scale encompassing a wide variety of industries has been conducted until recently. Such studies are underway to determine the effectiveness and quantify the benefits of systems engineering.

Systems engineering encourages the use of modeling and simulation *to validate assumptions* or theories on systems and the *interactions* within them.

Use of methods that allow early detection of possible failures, in safety engineering, are integrated into the design process. At the same time, decisions made at the beginning of a project whose consequences are not clearly understood can have enormous implications later in the life of a system, and it is the task of the modern systems engineer to explore these issues and make

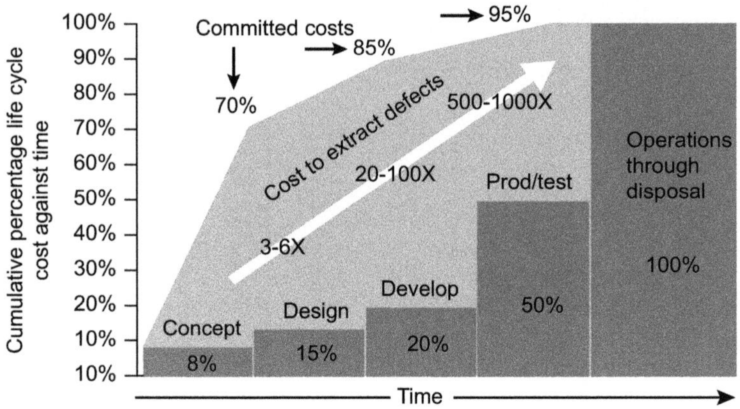

Program phase completed	Relative cost
Design	1
Unit test	5
Integration test	16
System test	40
Acceptence test	110

FIGURE 2.19

Committed lifecycle cost against time + relative error cost found, by phase. *From project manager's guide to systems engineering measurement for project success.*

critical decisions. No method guarantees today's decisions will still be valid when a system goes into service years or decades after first conceived. However, there are techniques that support the process of systems engineering.

Industry data associated with the relative cost of delaying the correction of an error or making a change is shown in Fig. 2.19. The cost of making the change/correction escalates the longer you delay. If it would take one unit to make the change/correction in the design phase, then it would take 40 units to make the same change/correction after the system test phase.

2.5.20 System-of-Interest

The systems in a process manufacturing project are man-made, created and utilized to provide products or services in defined environments for the benefit of users and other stakeholders. These systems may be configured with one or more of the following system elements: hardware, software, data, humans, processes (e.g., processes for providing service to users), procedures (e.g., operator instructions), facilities, materials, and naturally occurring entities. As viewed by the user, they are thought of as products or services.

The perception and definition of a particular system, its architecture and its system elements depend on a stakeholder's interests and responsibilities. One stakeholder's system-of-interest can be viewed as *a system element* in another stakeholder's *system-of-interest*. Furthermore, a system-of-interest can be viewed as being part of the environment for another stakeholder's system-of-interest.

The following are the key points regarding the characteristics of the SoI:

1. defined boundaries encapsulate meaningful needs and practical solutions;
2. there is a hierarchical or other relationship between system elements;
3. an entity at any level in the system-of-interest can be viewed as a system;
4. a system comprises an integrated, defined set of subordinate system elements;
5. humans can be viewed as both users external to a system and as system elements (i.e., operators) within a system;
6. a system can be viewed in isolation as an entity, i.e., a product; or as a collection of functions capable of interacting with its surrounding environment, i.e., a set of services.

2.5.21 Enabling Systems

Throughout the lifecycle of a system-of-interest, essential services are required from systems that are not directly a part of the operational environment of

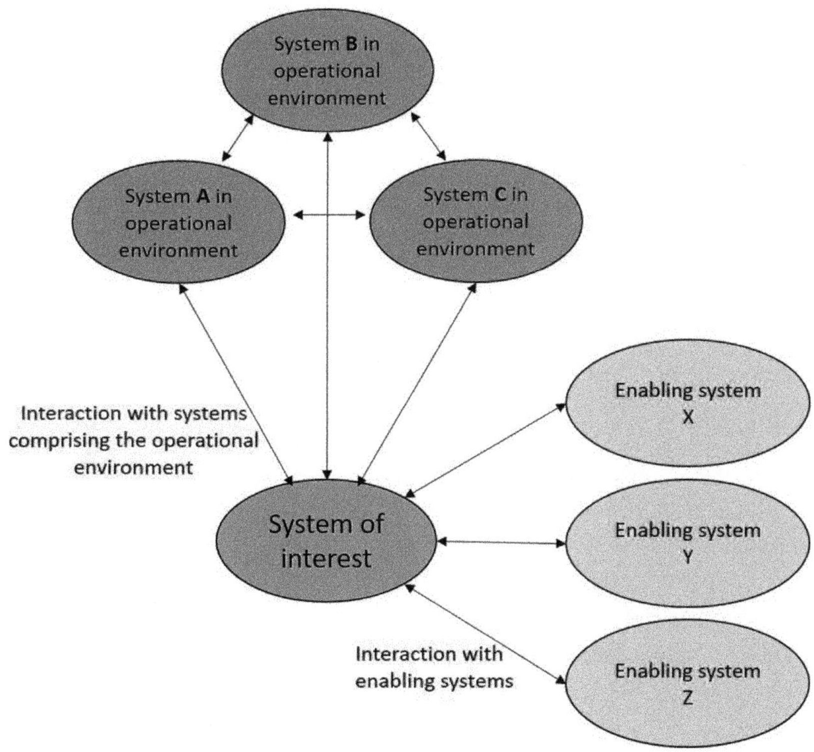

FIGURE 2.20
System-of-interest, its operational environment and enabling systems.

the system-of-interest, e.g., mass-production system, training system, maintenance system. Each of these systems enables a part, e.g., a stage of the lifecycle of the system-of-interest to be conducted. Termed "enabling systems," they facilitate progression of the system-of-interest through its lifecycle.

The relationship between the services delivered to the operational environment by the system-of-interest and the services delivered by the enabling systems to the system-of-interest are shown in Fig. 2.20. Enabling systems can be seen to contribute indirectly to the services provided by the system-of-interest. The interrelationships between the system-of-interest and the enabling systems can be bi-directional or a one-way relationship. In addition to interacting with enabling systems the system-of-interest may also interact with other systems in the operating environment, shown as systems A, B, and C. Requirements for interfaces with enabling systems and other systems in the operational environment will need to be included in the requirements for the system-of-interest.

During a stage in the system lifecycle the relevant enabling systems and the system-of-interest are considered together. Since they are interdependent, they can also be viewed as a system. When a suitable enabling system does not already exist the project that is responsible for the system-of-interest can also be directly responsible for creating and using the enabling system. Creating the enabling systems can be viewed as a separate project and subsequently another system-of-interest.

Further elaboration of these concepts can be found in the ISO/IEC/IEEE TR 24748 guides, on the application of lifecycle processes.

2.5.22 System Boundary

The relationships among a system-of-interest and its enabling systems are defined by boundaries. Such a system boundary establishes which systems are or are not under direct control of a project. Systems inside a boundary are under direct project control. While the project cannot control those systems external to the boundary it could need to influence those systems in order to reduce the impact on the system-of-interest or to make them fit system-of-interest requirements. When systems external to the boundary cannot be influenced, they become design or other lifecycle process constraints with respect to the systems inside the boundary.

Enabling systems can be either inside or outside the boundary for which a project is responsible. Whether a particular enabling system is inside or outside the boundary the enabling system is of interest to a project because of interface and scheduling constraints. Fig. 2.21 illustrates a boundary for a system-of-interest with some enabling systems inside and others outside the span of control of the project but all within the span of interest.

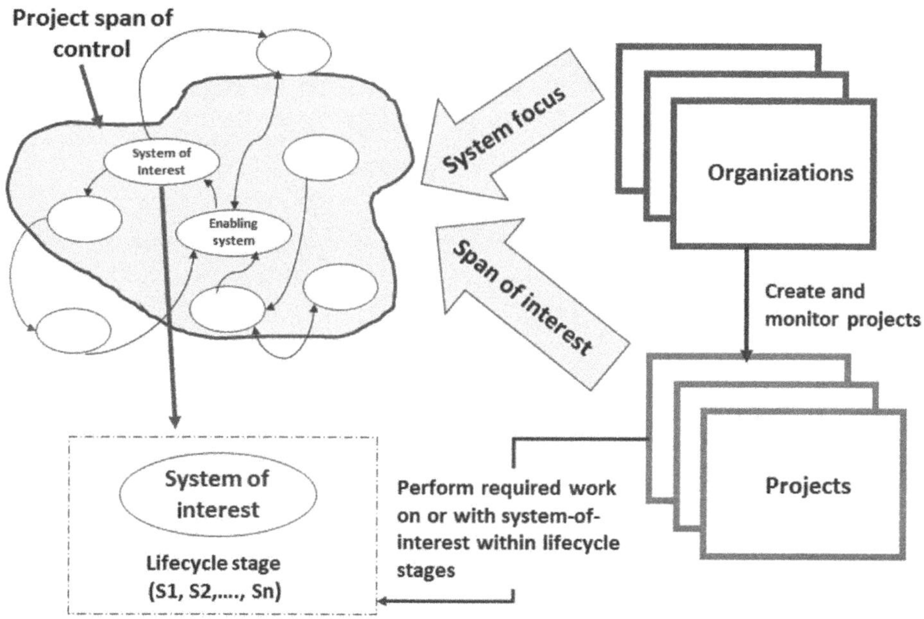

FIGURE 2.21
Systems boundaries ISO 15288–2002.

System boundaries also establish interfaces where requirements have to be defined for environmental, physical, functional, or data connectivity between those systems inside the boundary and those systems external to the boundary.

2.5.23 System Structure

The system is composed of a set of interacting system elements, each of which can be implemented to fulfill its respective specified requirements and accomplish a defined objective. Every system is delineated by its spatial and temporal boundaries, surrounded and influenced by its *environment*, described by its structure and purpose and expressed in its functioning.

Responsibility for the implementation of any system element may therefore be delegated to another party through an agreement.

The organized, purposeful structure of a system consists of interrelated and interdependent elements (components, entities, factors, members, parts, etc.). These elements continually influence one another (directly or indirectly) to maintain their activity and the existence of the system, to achieve the goal of the system.

If failure of an element leads to the failure of whole system, then it is considered as a *Critical Element*. For example, if our heart stops we will die but if

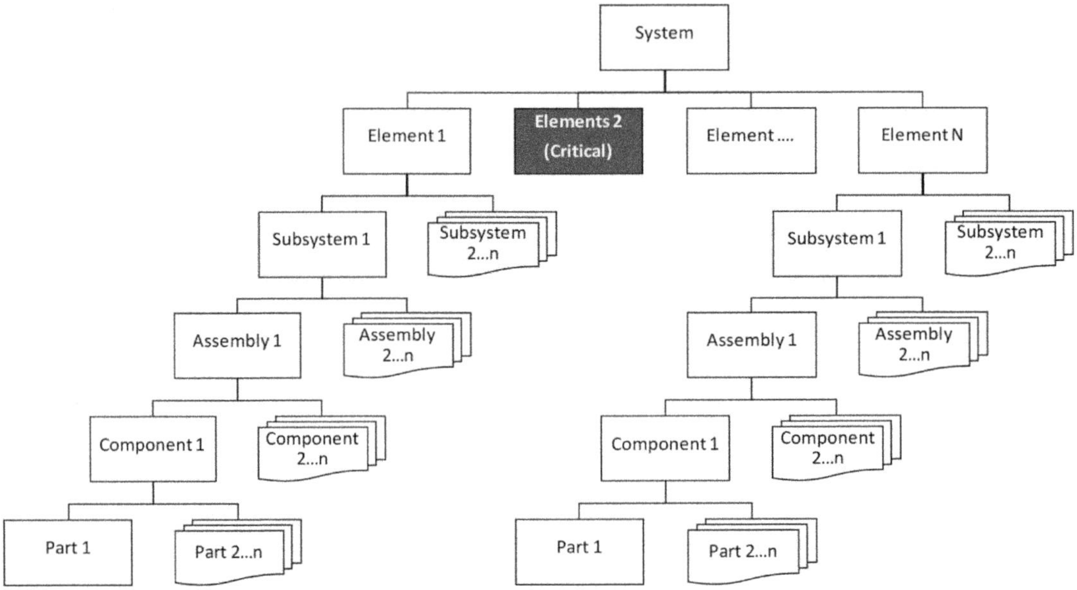

FIGURE 2.22
Hierarchy of system elements.

one of our figures is cut, we will not die immediately and can survive without a figure after treatment. Critical elements should be determined systematically and highlighted as shown in Fig. 2.22.

An element can be critical in terms of performance, cost, safety, environmental, or any other acceptance criteria that we may set.

The relationship between the system and its complete set of system elements can typically be represented in a hierarchy for the simplest of SoI. For more complex SoI a prospective system element may itself need to be considered as a system (that in turn is comprised of system elements) before a complete set of system elements can be defined with confidence. In this manner the appropriate system lifecycle processes are applied recursively to a system-of-interest to resolve its structure to the point where understandable and manageable system elements can be implemented (made, bought, or reused).

There are an increasing number of systems that, from one or more aspects, are not hierarchical, such as networks and other distributed systems which is called "system of systems (SoS)."

2.5.24 System of Systems Characteristics and Types

SoS are characterized by managerial and operational independence of the constituent systems, which in many cases were developed and continue to support originally identified users concurrently with users of the SoS. In other contexts, each constituent system itself is a SoI; its existence often predates the SoS, while its characteristics were originally engineered to meet the needs of their initial users. As constituents of the SoS, their consideration is expanded to encompass the larger needs of the SoS. This implies added complexity particularly when the systems continue to evolve independently of the SoS. The constituent systems also typically retain their original stakeholders and governance mechanisms, which limits alternatives to address the needs of the SoS.

SoS have been characterized into four types based on the governance relationships between the constituent systems and the SoS (Table 2.2). The strongest governance relations apply to directed SoS, where the SoS organization has authority over the constituent systems despite the fact that the constituent systems may not have originally been engineered to support the SoS. Somewhat less control is afforded for acknowledged SoS, where allocated authority between the constituent systems and the systems of systems has an impact on application of some of the systems engineering processes. In collaborative SoS, which lack SoS authorities, application of systems engineering depends on cooperation among the constituent systems. Virtual SoS are largely self-organizing and offer much more limited opportunity for systems engineering of the SoS.

Emergence is a key characteristic of SoS—the unanticipated effects at the systems of systems level attributed to the complex interaction dynamics of the constituent systems. In SoS, constituent systems are intentionally considered in their combination, so as to obtain and analyze outcomes not possible to obtain with the systems alone. The complexity of the constituent systems

Table 2.2 System of System Types

Virtual	• Lack a central management authority • Lack of centrally agreed upon purpose • Emerging behaviors that rely upon relatively invisible mechanisms to maintain it
Collaborative	• Component systems interact voluntarily to fulfill agreed upon purposes • Collectively decide how to interoperate, enforcing and maintaining standards
Acknowledged	• Recognized objectives, a designated manager, and resources for the SoS • Constituent systems retain their independent ownership, management, and resources
Directed	• Integrated SoS built and managed to fulfill specific purposes • Centrally managed and evolved • Component systems maintain ability to operate independently • Normal operation mode is subordinated to central purpose

and the fact they may have been designed without regard to their role in the SoS, can result in new, unexpected behaviors. Identifying and addressing unanticipated emergent results is a particular challenge in engineering SoS.

2.5.25 Requirement Engineering

Technical processes enable systems engineers to coordinate the interactions between engineering specialists, other engineering disciplines, system stakeholders and operators, and manufacturing. They also address conformance with the expectations and legislated requirements of society. These processes lead to the creation of a sufficient set of requirements and resulting system solutions that address the desired capabilities within the bounds of performance, environment, external interfaces, and design constraints. Without the technical processes the risk of project failure would be unacceptably high. As illustrated in Fig. 2.24 the technical processes begin with the development of needs and requirements (Ryan, 2013).

Needs—Per the Oxford English Dictionary a need is a thing that is wanted or required. For a system, needs are often capabilities or things that are lacking but wanted or desired by one or more stakeholders. These can be viewed in at least three contexts in which SE is performed:

1. projects with customers internal to the enterprise that is doing the engineering,
2. development under an agreement with an external entity, and
3. entrepreneurial product development in anticipation of future sales.

Requirements—Requirements are formal structured statements that can be verified and validated. There may be more than one requirement defined for each need.

Needs and requirements exist at a number of levels or layers. There is an enterprise view, a business management view, a business operations view, a systems view, and system element views. A system may comprise a number of elements including products, people, and processes. Together these elements enable a needed capability for the organization. At each level, needs are documented and transformed through analysis into requirements. The resulting requirements will be documented, agreed-to, baselined and will be put under configuration management. Once a set of requirements has been documented, agreed-to, and baselined at one layer they will low down to the next layer as shown in Fig. 2.23. In all cases, for each layer shown in Fig. 2.23, the set of requirements can be traced back to the requirements at the previous layer from which they were either decomposed or derived. This process continues for the next layer of system elements.

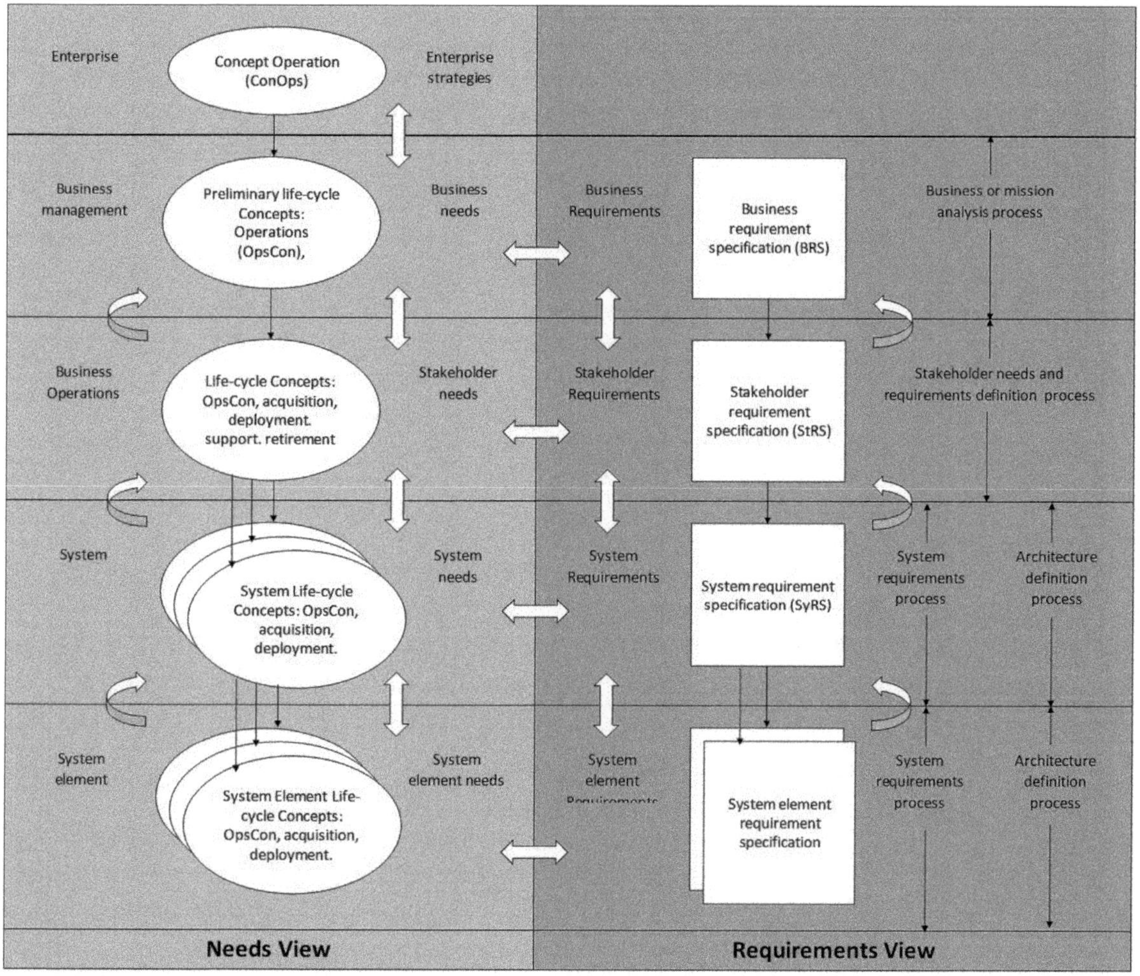

FIGURE 2.23

Transformation of needs into requirements (Ryan, 2013). *Note: One underlying principle illustrated by Fig. 4.1 is that when a decision is made to satisfy a need, that need gives rise to a corresponding requirement or set of requirements.*

ISO/IEC/IEEE 15288 includes 14 technical processes, the roles of the first four of which are illustrated in Fig. 2.23:

- *Business or mission analysis process*—Requirements definition begins with the business vision of the organization or enterprise, the concept of operations (ConOps), and other organization strategic goals and objectives from which business management define business needs (aka mission needs). These needs are supported by preliminary lifecycle concepts—acquisition concept, deployment concept, operational concept (OpsCon), support concept, and retirement concept—see

Section 4.1.2.2 for a detailed description of the roles of the ConOps and the OpsCon. Business needs are then elaborated and formalized into business requirements, which are often captured in a business requirements specification.

- *Stakeholder needs and requirements definition process*—Using the enterprise-level ConOps from the acquiring enterprise and the system-level preliminary OpsCon from the development enterprise as guidance, requirements engineers lead stakeholders from business operations through a structured process to elicit stakeholder needs (in the form of a refined system-level OpsCon and other lifecycle concepts). Stakeholder needs are then transformed by requirements engineers into a formal set of stakeholder requirements, which are often captured in a stakeholder requirements specification (StRS).
- *System requirements definition process*—The stakeholder requirements in the StRS are then transformed by requirements engineers into system requirements, which are often contained in a system requirements specification.
- *Architecture definition process*—Alternative system architectures are defined and one is selected.
- *Design definition process*—System elements are defined in sufficient detail to enable implementation consistent with the selected system architecture.
- *System analysis process*—Mathematical analysis, modeling, and simulation are used to support the other technical processes.
- *Implementation process*—System elements are realized to satisfy system requirements, architecture, and design.
- *Integration process*—System elements are combined into a realized system.
- *Verification process*—Evidence is provided that the system, the system elements, and the work products in the lifecycle meet the specified requirements.
- *Transition process*—The system moves into operations in a planned, orderly manner.
- *Validation process*—Evidence is provided that the system, the system elements, and the work products in the lifecycle will achieve their intended use in the intended operational environment.
- *Operation process*—The system is used.
- *Maintenance process*—The system is sustained during operations.
- *Disposal process*—The system or system elements are deactivated, disassembled, and removed from operations.

"Verification" and "validation" are very ambiguous unless preceded by a modifier that clearly indicates what concept the term is referring to. When using these terms, it should be clear as to which concept is being referred to: requirement verification or requirement validation; design verification or design validation; system verification or system validation.

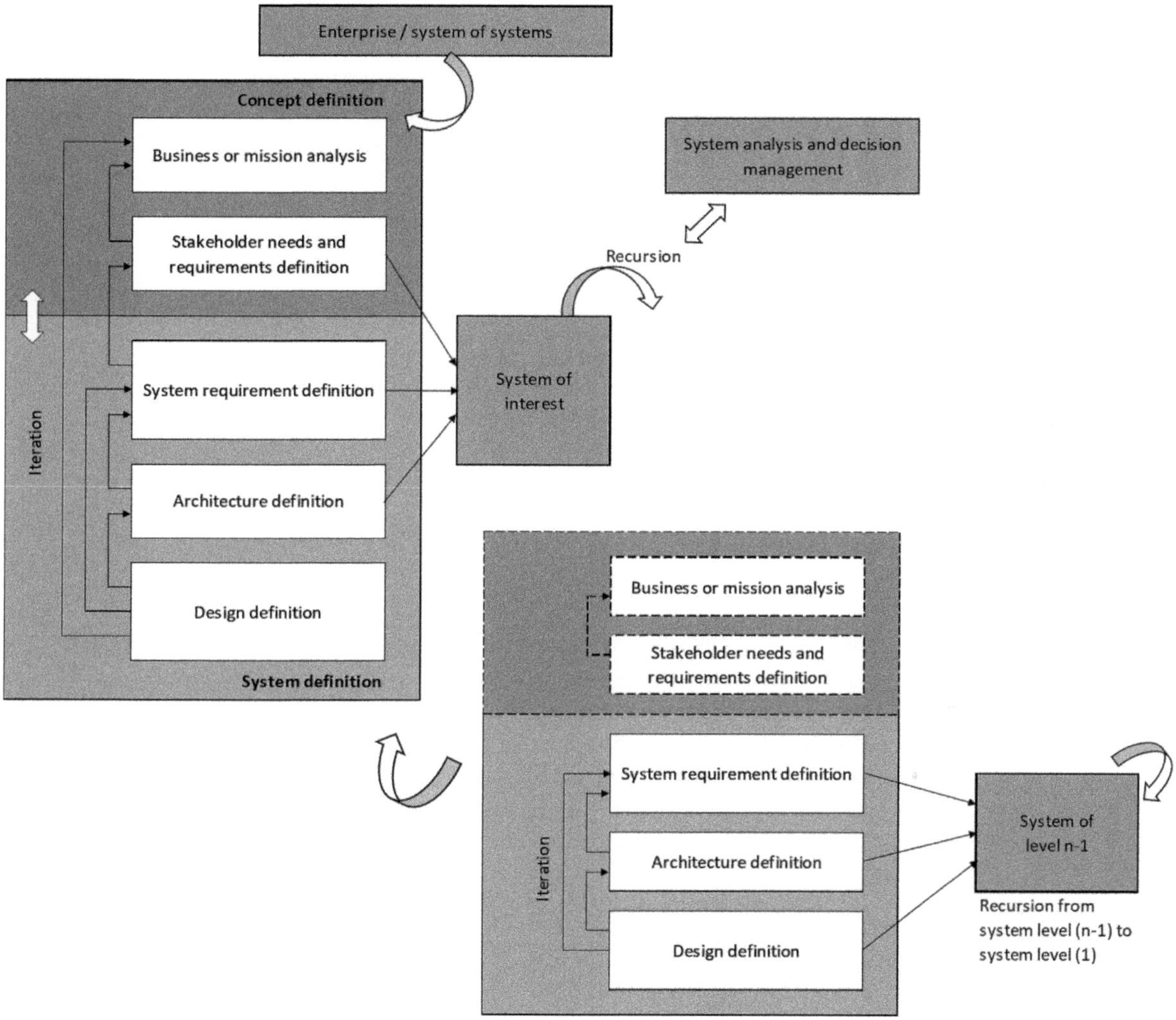

FIGURE 2.24

Iteration and recursion. *INCOSE Handbook 2015—Reprinted with permission from Garry Roedler. All other rights reserved.*

Fig. 2.24 shows an illustration of iteration and recursion of the processes. Iteration is the repeated application of and interaction between two or more processes at a given level in the system structure or hierarchy. Iteration is needed to accommodate stakeholder decisions and evolving understanding, account for architectural decisions/constraints, and resolve trades for affordability, adaptability, feasibility, resilience, etc. Although the figure only shows a subset of the lifecycle technical processes, there can be iteration between any of the processes. For example, there is often iteration between system requirements definition and architecture definition. In this case, there is a concurrent application of the processes with iteration between them, where the evolving system requirements help to shape the architecture through identified constraints and functional and quality requirements. The architecture trades, in turn, may identify requirements that are not feasible, driving further requirements analysis with trades that change some requirements. Likewise, the design definition could identify the need to reconsider decisions and trades in the requirements definition or architecture definition processes. Any of these can invoke additional application of the system analysis and decision management processes.

Recursion is the repeated application of and interaction of processes at successive levels in the system structure. The technical processes are expected to be recursively applied for each successive level of the system structure until the level is reached where the decision is made to make, buy, or reuse a system element. During the recursive application of the processes the outputs at one level become inputs for the next successive level (below for system definition, above for system realization).

Rules for requirement statements and sets of requirements are as follows:

Precision

R1—Use definite article "the" rather than the indefinite article "a."
R2—Use the active voice with the actor clearly identified.
R3—Make the subject of the requirement appropriate to the layer in which the requirement exists.
R4—Only use terms defined in the glossary.
R5—Quantify requirements precisely—avoid imprecise quantifier that provide vague quantification, such as "some," "any," "several," "many," "a lot of," "a few," "approximately," "almost always," "nearly," "about," "close to," "almost," "approximate," "significant," "flexible," "expandable," "typical," "sufficient," "adequate," "appropriate," "efficient," "effective," "proficient," "reasonable."
R6—Use appropriate units, with tolerances or limits, when stating quantities—explicitly state units for all numbers.
R7—Avoid the use of adverbs—words that end in -ly" Such as "usually," "approximately," "sufficiently," "typically."
R8—Avoid the use of vague adjectives such as "ancillary," "relevant," "routine," "common," "generic," and "customary."
R9—Avoid escape clauses such as "so far as is possible," "as little as possible," "as much as possible," "if it should prove necessary," "where possible," and "if practicable."
R10—Avoid open-ended clauses such as "including but not limited to," "etc." and "and so on."

Concision

R11—Avoid superfluous infinitives such as ".. be designed to . . .," ". . . be able to . . ..," ". . .be capable of . . .."
R12—Use a separate clause for each condition or qualification.

Nonambiguity

R13—Use correct grammar.
R14—Use correct spelling.
R15—Use correct punctuation.
R16—Use the logical construct "X AND Y" instead of "both X AND Y."
R17—Avoid the use of "X and/or Y."
R18—Avoid the use of the oblique ("/") symbol except in units, i.e., km/h

Singularity

R19—Write a single, simple, single-thought, affirmative, declarative sentence, conditioned, and qualified by relevant subclauses.
R20—Avoid combinators such as "and," "or," "then," "unless," "but," "/," "as well as," "but also," "however," "whether," "meanwhile," "whereas," "on the other hand," and "otherwise."
R21—Use an agreed typographical device to indicate the use of propositional combinators for expressing a logical condition within a requirement.
R22—Avoid phrases that indicate the purpose of the requirement.
R23—Avoid parentheses and brackets containing subordinate text.
R24—Enumerate sets of entities as explicit requirements instead of using generalizations.
R25—When a requirement is related to complex behavior, refer to the supporting diagram or model rather than a complex textual description

Completeness

R26—Avoid the use of pronouns.
R27—Avoid using headings to support explanation of subordinate requirements.

Realism

R28—Avoid using unachievable absolutes such as 100% reliability or 100% availability.

Conditions

R29—State applicability conditions explicitly instead of leaving applicability to be inferred from the context.
R30—Express the propositional nature of a condition explicitly instead of giving lists of conditions.

Uniqueness

R31—Classify the requirement according to the aspects of the problem or system it addresses.
R32—Express each requirement once and only once.

Abstraction

R33—Avoid stating a solution—focus on the problem "what" rather than the solution "how."

Quantifiers

R34—Use "each" instead of "all," "any" or "both" when universal quantification is intended.
R35—Define the range of acceptable values associated with quantities.
R36—Provide specific measurable performance targets.
R37—Define temporal dependencies explicitly instead of using indefinite temporal keywords.

Uniformity of language

R38—Define a consistent set of terms to be used in requirement statements in a glossary—avoid the use of synonyms.
R39—If acronyms are used in requirement statements, use a consistent set of acronyms.
R40—Avoid the use of abbreviations in requirement statements.
R41—Use a project-wide style guide.

Modularity

R42—Group mutually dependent requirements together.
R43—Group related requirements together.
R44—Conform to a defined structure or template.

2.5.26 Modeling and Simulation

Stakeholders of the SE lifecycle have used models and simulations for some time both to check their own thinking and to communicate their concepts to others. The benefit is twofold:

1. models and simulations confirm the need for the systems and the anticipated system behaviors before proceeding with the development of an actual system, and
2. models and simulations present a clear, coherent design to those who will develop, test, deploy, and evolve the system, thereby maximizing productivity and minimizing error.

The ability to detect limitations and incompatibilities via system models and simulations early in a project helps avoid higher project cost and schedule overruns later in a project, especially during system operation. The value of modeling and simulation increases with the size, be it physical or complexity, of the system or SoS under development.

Early in the SE lifecycle the objective of modeling and simulation is to obtain information about the system before significant resources are committed to its design, development, construction, verification, or operation. To that end, modeling and simulation helps generate data in the domain of the analyst or reviewer, not available from existing sources, in a manner that is affordable and timely to support decision making. An adequate, accurate, and timely model and simulation informs stakeholders of the implications of their preferences, provides perspective for evaluating alternatives, and builds confidence in the capabilities the system will provide. They also help the development, deployment, and operational staffs comprehend the design requirements, appreciate imposed limits from technology and management, and ensure an adequate degree of sustainability. Finally, adequate, accurate, and timely models and simulations help the organization and its suppliers provide the necessary and sufficient personnel, methods, tools, and infrastructure for system realization.

The long-term benefits of modeling and simulation are commensurate with the gap between the extent, variety, and ambiguity of the problem and the competencies of downstream staffing. A relatively simple model of an intended system may be sufficient for a highly competent staff, whereas a much more elaborate simulation may be necessary for a less competent staff, especially one faced with producing a novel, large scale system that is capable of autonomously coping with unpredictable mission situations. Ultimately, the benefit of modeling and simulation is proportional to the stakeholders' perception of the timeliness, trustworthiness, and ease of use and maintenance of the model or simulation. Consequently, the planned resources anticipated to be spent in development, verification, validation, accreditation, operation, and maintenance of the model must be consistent with the expected value of the information obtained through use of the model.

Types of models: "The image of the world around us, which we carry in our head, is just a model" (Forrester, 1961). Most systems start as a mental model that is elaborated and translated in several stages to form a final model or simulation product. A model may be a mental image of selected concepts, and relationships between them, that can be translated to sketches, textual specifications, graphics/images, mock ups, scale models, proto- types, or emulations. Often, separate models are prepared for distinct viewpoints, such as functional, performance, reliability, survivability, operational availability, and cost.

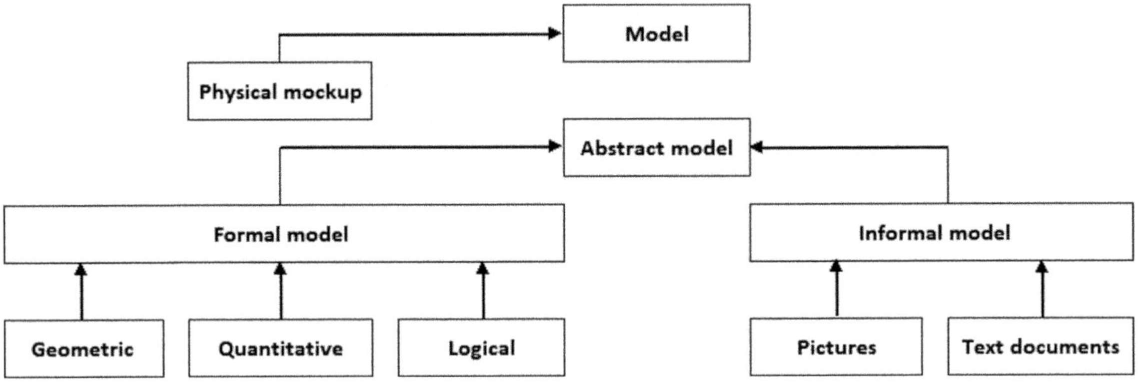

FIGURE 2.25
Sample model taxonomy. *Reprinted with permission from Sandy Friedenthal. All other rights reserved.*

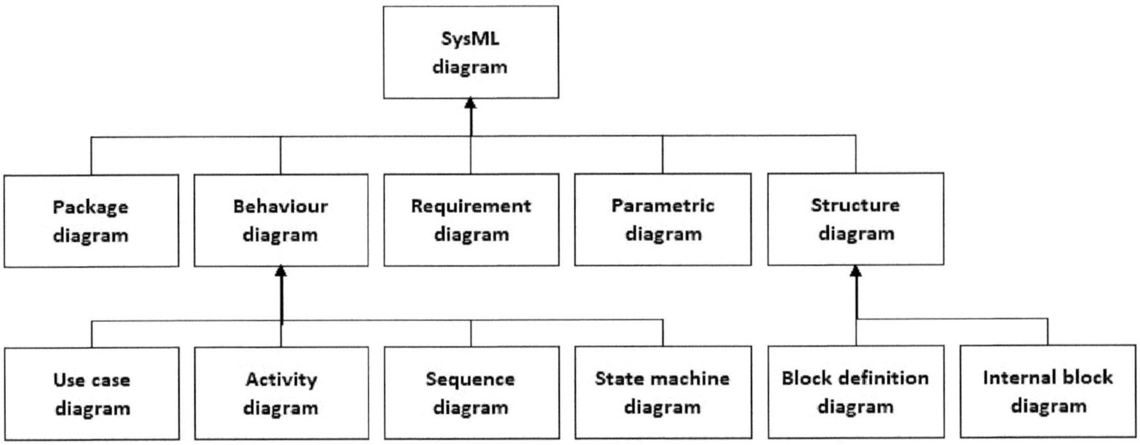

FIGURE 2.26
SysML diagram types. *Reprinted with permission from Sandy Friedenthal. All other rights reserved.*

It is useful to classify models to assist in selecting the right kind of model for the intended purpose and scope. Models can be classified in many different ways. The model taxonomy shown in Fig. 2.25 is one such taxonomy and provides a useful classification for one instance as an illustration and not necessarily providing an exhaustive set of model classes; other classes may exist:

- *Physical mock ups*—A model that represents an actual system, such as a model airplane or wind tunnel model, or a more abstract representation, such as a model that is often represented using a computer.

- *Abstract models*—An abstract model can have many different expressions to represent a system, entity, phenomena, or process, which vary in degrees of formalism. Therefore the initial classification of models that distinguishes between informal models and formal models is noted, with this guidance focusing on formal models.

The SysML™ from the object management group (OMG) has emerged as an important modeling language for systems. Summary descriptions of the SysML diagram types shown in Fig. 2.26 are as follows:

- A *package diagram* (pkg) is used to organize the model into packages that contain other model elements. This facilitates model navigation and reuse, as well as access and change control.
- A *requirements diagram* (req) captures text-based requirements. Having requirements within the model enables fine-grained traceability from requirement to requirement and between requirements and design, analysis, and verification elements in the model.
- System structure is represented using block diagrams:
  - A *block definition diagram* (bdd) describes the system hierarchy and classifications of system elements.
  - An *internal block diagram* (ibd) depicts the internal structure of a system in terms of how its parts are interconnected using ports and connectors, describing how the parts within the system are interconnected.
- Behavior is captured in use case, activity, sequence, and state machine diagrams:
  - A *use case diagram* (uc) provides a high-level description of the system functionality in terms of how users and external systems use the system to achieve their goals.
  - An *activity diagrams* (act) represents the transformation of inputs to outputs through a controlled sequence of actions.
  - A *sequence diagram* (sd) represents interaction in terms of the time-ordered exchange of messages between collaborating parts of a system.
  - A *state machine diagram* (stm) describes the states of a system or its parts; the transitions between the states; the actions that occur within states or upon transition, entry, or exit; and the events that trigger transitions.
- A *parametric diagram* (par) represents constraints on system property values as necessary to support detailed engineering analysis. These constraints may include performance, reliability, and mass properties,

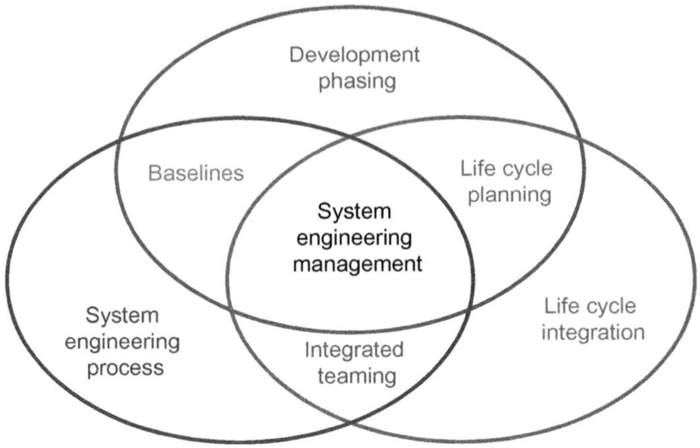

FIGURE 2.27
Three activities of systems engineering management.

among others. SysML can be integrated with other engineering analysis models and tools to execute the analysis.

SysML includes an allocation relationship to represent allocation of functions to elements, allocation of logical to physical elements, and other types of allocations. SysML is a general purpose modeling language that is intended to support many different model-based methods, such as structured analysis methods and object oriented methods. A particular method may require only a subset of the diagrams. For example, a simplified functional analysis method may only require activity diagrams augmented by bdds, ibds, and perhaps requirements diagrams.

For general information on SysML, along with links to tool vendors, articles, and books, see the official OMG SysML website at http://www.omgsysml.org.

2.5.27 System Engineering Management

As illustrated by Fig. 2.27, systems engineering management is accomplished by integrating three major activities:

- *Development phasing* that controls the design process and provides baselines that coordinate design efforts,
- *A systems engineering process* that provides a structure for solving design problems and tracking requirements flow through the design effort, and

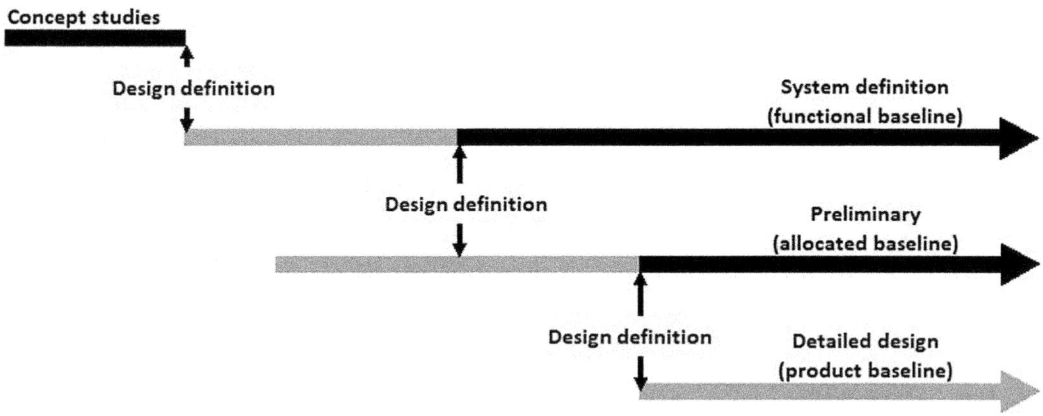

FIGURE 2.28
Development phasing.

- *Lifecycle integration* that involves customers in the design process and ensures that the system developed is viable throughout its life.

Each one of these activities is necessary to achieve proper management of a development effort. *Development phasing* has two major purposes:

- it controls the design effort and
- it is the major connection between the technical management and the overall acquisition efforts.

It controls the design effort by developing design baselines that govern each level of development. It interfaces with acquisition management by providing key events in the development process, where design viability can be assessed as shown in (Fig. 2.28). The viability of the baselines developed is a major input for acquisition management milestone (MS) decisions. As a result the timing and coordination between technical development phasing and the acquisition schedule is critical to maintain a healthy acquisition program.

The *systems engineering process* is the heart of systems engineering management. Its purpose is to provide a structured but flexible process that transforms requirements into specifications, architectures, and configuration baselines. The discipline of this process provides the control and traceability to develop solutions that meet customer needs. The systems engineering process may be repeated one or more times during any phase of the development process. Fig. 2.29 summarizes the system engineering process.

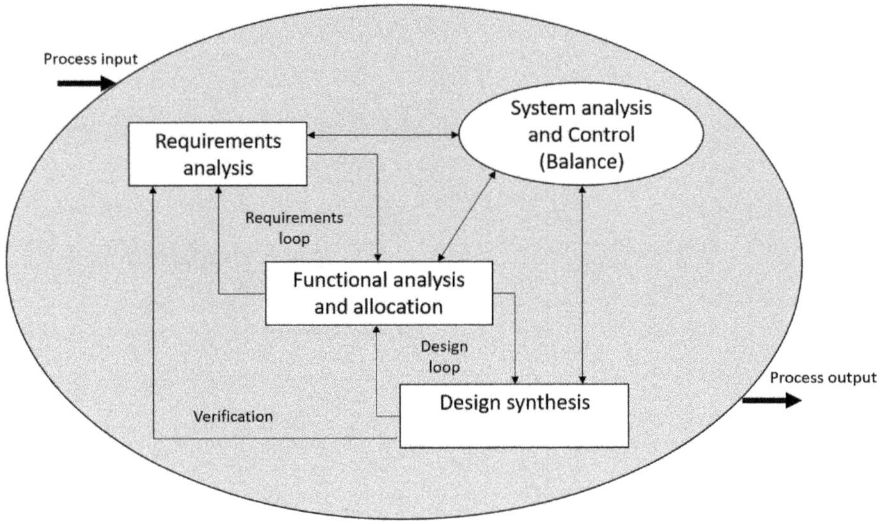

FIGURE 2.29
System engineering process.

Lifecycle integration is necessary to ensure that the design solution is viable throughout the life of the system. It includes the planning associated with product and process development, as well as the integration of multiple functional concerns into the design and engineering process. In this manner, product cycle-times can be reduced, and the need for redesign and rework substantially reduced.

Verification, supporting activities, and training are relevant to the all stages of lifecycle, as shown in Fig. 2.30.

2.5.28 Risk and Opportunity Management

Most projects are executed in an environment of uncertainty. Uncertainty influences the ability of the project team to achieve the project objectives. Uncertainty includes events that could harm the project (threats) and those that could help the project (opportunities). Well-established techniques exist for managing threats. There is some debate whether the same techniques are applicable to recognizing opportunities. In an optimal situation, opportunities are maximized at the same time as threats are minimized, resulting in the best chance to meet project objectives. The Øresund Bridge case shown in Fig. 2.31 illustrates this; the man-made Peberholm Island was created from the materials dredged from the Strait to meet environmental requirements and is now a sanctuary for a rare species of tern.

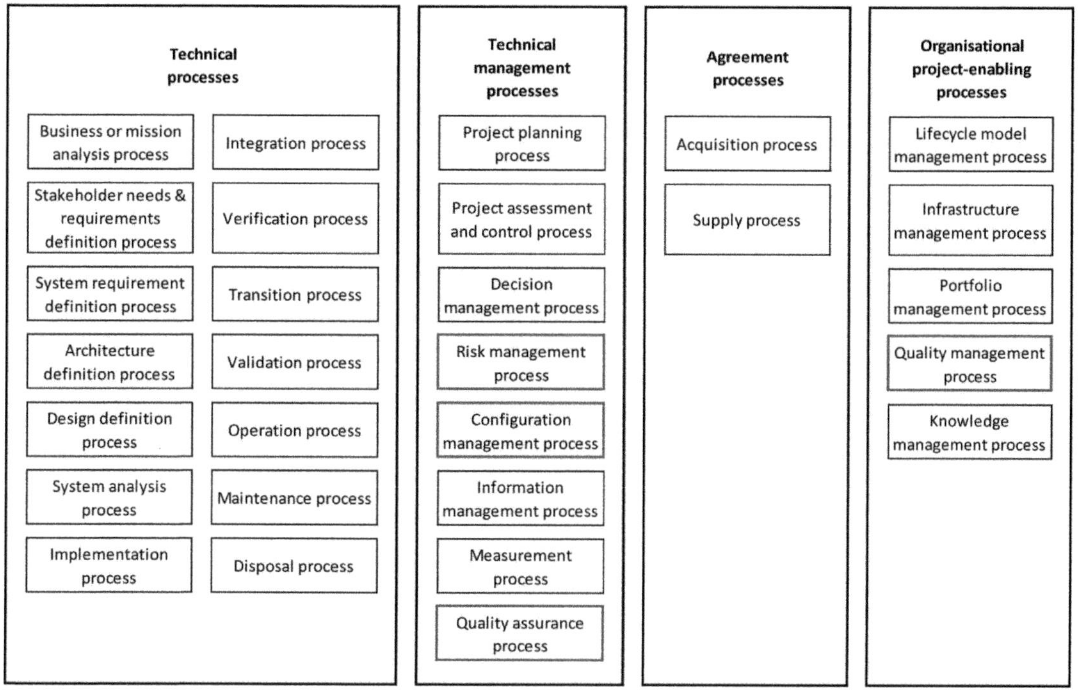

FIGURE 2.30

System lifecycle processes. Fig. 1.1. System lifecycle processes per ISO/IEC/IEEE 15288. *This figure is excerpted from ISO/IEC/IEEE 15288:2015, Fig. 4 on page 17 with permission from the ANSI on behalf of the ISO.*

FIGURE 2.31

Oresund Bridge is a combined railway and motorway bridge across the Øresund strait between Sweden and Denmark.

The main reason for constructing the island was to have a crossover point between the tunnel and the bridge.

The tunnel was built since a bridge spanning the entire link between Malmö and Copenhagen would have interfered with obstacle-free zones around Kastrup Airport. Another reason was to provide an opportunity for large ships to pass the Öresund without worrying about the height of the bridge. To make the Øresund Bridge higher would interfere with the freight train traffic.

The other main reason was that the Øresund had to be dug deeper in order to not decrease the flow of the water because of the bridge. The material was used to create this island. The area was so shallow, and the amount of material so large, that it was chosen to make the island as long as 4 km.

The original designs for a bridge across the Øresund required extensive use of Saltholm as a stepping-stone for the bridge. They required either the construction of both a motorway and railway across Saltholm or an expansion of the existing island to serve the same purpose.

Both of these options were ultimately abandoned in order to protect the island's ecology; Danish and Swedish politicians decided instead to construct an artificial island immediately south of Saltholm to achieve that goal. The location was chosen in the belief that it would ensure the freest flow of water through the sound, a key element in the debate about whether or not the bridge should be constructed.

Peberholm is protected by strict laws. Only biologists are allowed one annual visit to the areas of the island outside of the railroad and highway to which the island owes its creation. The island is part of Natura 2000 area 142, comprising Saltholm and surrounding waters with a total area of 7218 ha. The same area is also under two birdprotections as habitats H110 and H126.

Risks are events that if they occur can jeopardize the successful completion of the project. Risks should be identified and assessed for probability of occurrence and impact on the project.

The likelihood that an undesirable event will occur often is expressed as a probability. The consequence of the event is expressed in terms that depend on the nature of the event (e.g., lost investment, inadequate performance). The combination of low likelihood and low undesirable consequences gives low risk, while high risk is produced by high likelihood and highly undesirable consequences.

By changing the adjective from *undesirable* to *desirable* the noun changes from *risk* to *opportunity*, but the diagram remains the same. As suggested by the shading, most projects experience a comparatively small number of high risk or high opportunity events.

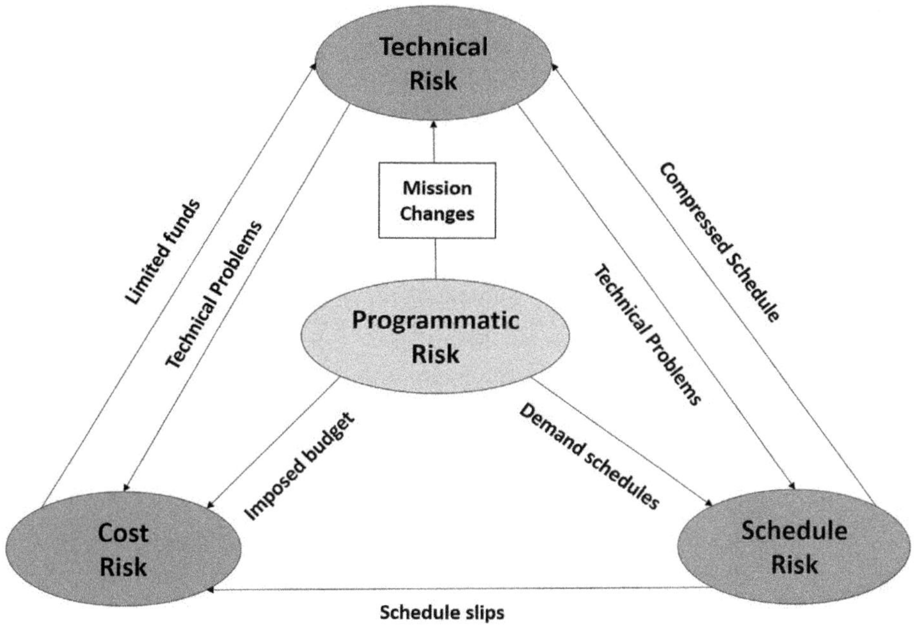

FIGURE 2.32
Typical relationship among the risk categories.

Risk pervades the lifecycle of systems. The system may be intended for technical accomplishments near the limits of the state of the art, creating technical risk. System development may be rushed to deploy the system as soon as possible to meet an imminent threat, leading to schedule risk. All systems are funding-limited so that cost risk is present. Risk can be introduced by external constraints or can develop from within the project, since technical risk can create schedule risk, which in turn can create cost risk.

There is no alternative to the presence of risk in system development. The only way to remove risk is to set technical goals very low, to stretch the schedule, and to supply unlimited funds. None of these events happen in the real world. No realistic project can be planned without risk. The challenge is to define the system and the project which best meet overall requirements, which allow for risk, and which achieve the highest chances of project success.

Fig. 2.32 illustrates the major interactions between the four risk categories: *technical, cost, schedule,* and *programmatic*. The arrow names indicate typical risk relationships; others certainly are possible.

Technical risk is the possibility that a technical requirement of the system may not be achieved in the system lifecycle. Technical risk exists if the system may fail to achieve performance requirements; to meet operability, producibility, testability, integration requirements; or to meet environmental protection requirements. A potential failure to meet any requirement which can be expressed in technical terms is a source of technical risk.

Cost risk is the possibility that available budget will be exceeded. Cost risk exists if the project must devote more resources than planned to achieve technical requirements; if the project must add resources to support slipped schedules due to any reason; if changes must be made to the number of items to be produced; or, if changes occur in the enterprise or national economy. Cost risk can be predicted at the total project level or for a system element. The collective effects of element-level cost risks can produce cost risk for the total project.

Schedule risk is the possibility that the project will fail to meet scheduled milestones. Schedule risk exists if there is inadequate allowance for acquisition delays. Schedule risk exists if difficulty is experienced in achieving scheduled technical accomplishments, such as the development of software. Schedule risk can be incurred at the total project level for milestones such as deployment of the first system element. The cascading effects of element-level schedule risks can produce schedule risk for the total project.

Programmatic risk is produced by events which are beyond the control of the project manager. These events often are produced by decisions made by personnel at higher levels of authority. Programmatic risks can be produced by reductions in project priority, by delays in receiving authorization to proceed with a project, by reduced or delayed funding, by changes in enterprise or national objectives, etc. Programmatic risk can be a source of risk in any of the other three risk categories.

Risk mitigation activities conform to the risk handling options. There are some steps that can be taken to avoid unnecessary risks.

Requirements scrubbing—Requirements that significantly complicate the system can be scrutinized to ensure that they deliver value equivalent to their investment. Find alternative solutions that deliver the same or comparable capability.

Selection of most promising options—In most situations several options are available, and a trade study can include project risk as a criterion when selecting the most promising alternative.

Staffing and team building—Projects accomplish work through people. Attention to training, teamwork, and employee morale can help avoid risks introduced by human errors.

For high-risk technical tasks, risk avoidance is insufficient and can be supplemented by the following approaches:

- Early procurement
- Initiation of parallel developments
- Implementation of extensive analysis and testing
- Contingency planning

The high-risk technical tasks generally imply high schedule and cost risks. Cost and schedule are impacted adversely if technical difficulties arise and the tasks are not achieved as planned. Schedule risk is controlled by early procurement of long-lead items and provisions for parallel-path developments. However, these activities also result in increased early costs. Testing and analysis can provide useful data in support of key decision points. Finally, contingency planning involves weighing alternative risk mitigation options.

In China the authorities built the short Maglev train line in Shanghai as a proof-of-concept. In spite of the high investment, this represented lower risk to the project than attempting a longer line with an unproven technology. The results collected from this project are inspiring others to consider Maglev alternatives for greater distances.

2.5.29 Technical Performance Measures

Technical measurement is the set of measurement activities used to provide the supplier and/or acquirer insight into progress in the definition and development of the technical solution and the associated risks and issues. This insight helps project management make better decisions throughout the lifecycle to increase the probability of delivering a technical solution that meets both the specified requirements and the mission needs. The insight is also used in trade-off decisions when performance exceeds the threshold. Technical measurement is planned early in the lifecycle and then performed with increasing levels of fidelity as the technical solution is developed. Fig. 2.33 the relationship between the various types of technical measures which are defined as follows.

Measures of effectiveness (MOEs): The "operational" measures of success that are closely related to the achievement of the mission or operational objective being evaluated, in the intended operational environment under a specified set of conditions; i.e., how well the solution achieves the intended purpose.

Measures of performance (MOP): The measures that characterize physical or functional attributes relating to the system operation, measured, or estimated under specified testing and/or operational environment conditions.

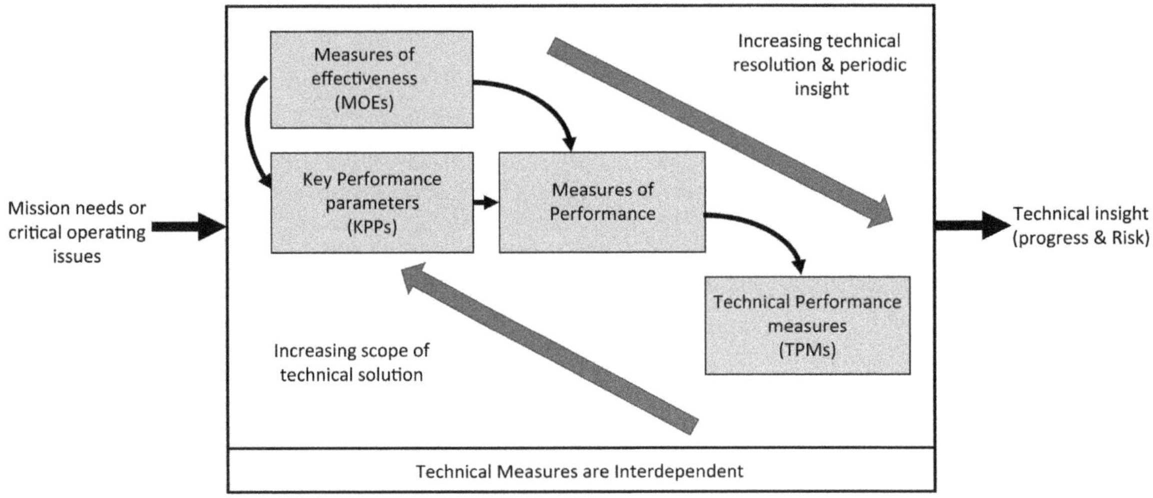

FIGURE 2.33
Relationship of the technical measures.

Technical performance measures (TPMs): TPMs measure attributes of a system element to determine how well a system or system element is satisfying or expected to satisfy a technical requirement or goal.

Key performance parameters (KPPs): A critical subset of the performance parameters representing those capabilities and characteristics so significant that failure to meet the threshold value of performance can be cause for the concept or system selected to be reevaluated or the project to be reassessed or terminated.

It is important to understand and plan for the interaction of the measurement activities with both the technical and management processes of the project. Fig. 2.33 illustrates the key interactions of these processes with the measurement activities. The technical processes will aid planning and provide essential insight into risk/opportunity and quality management of the technical solution through the lifecycle.

The management processes require information to enable good and timely decisions. The information needs of these management processes are considered together with the technical requirements and KPPs of the technical processes to form the inputs to the selection and definition of measures.

The definitions of the measures determine the data collection requirements (also known as base measures). As the data are collected, it needs to be processed, analyzed, and made accessible to management for review and to

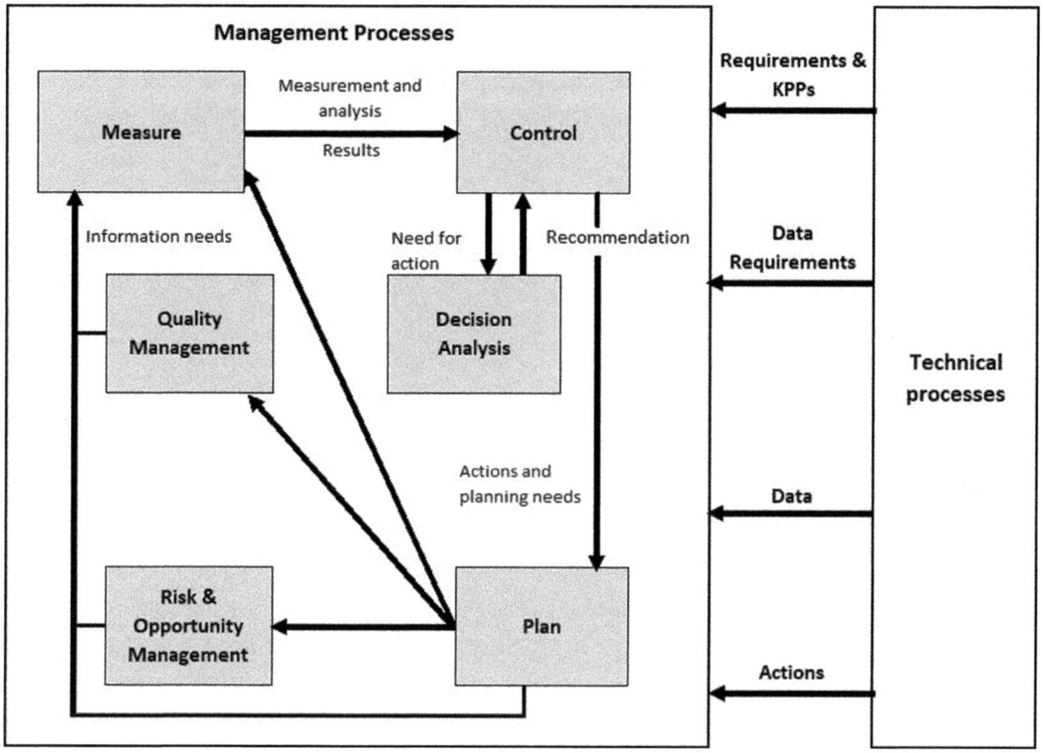

FIGURE 2.34
Integration of technical measurement with project processes.

determine whether actions are warranted. This integration needs to be defined in the plan, including timing, milestones, responsibilities, etc. The project team needs to ensure that the data to be collected are cost-effective to collect versus the insight provided and can be collected as the process is executed.

The determination of the fit of data collection to the processes should consider the measurement specifications, the acquirer and supplier processes, and decision points. Responsibility for tracking the technical measures throughout the project must be assigned. Typically, this is assigned to an organization, IPT, or individual with overall system oversight (Fig. 2.34).

Failure to achieve the technical measures (especially TPMs) often results in significant cost/schedule impacts to projects and may result in penalties or project cancellation. Where the measure indicates, margin exists for the technical requirement being tracked, there may be an opportunity to use the margin in a technical trade to reduce the overall system risk. Therefore these

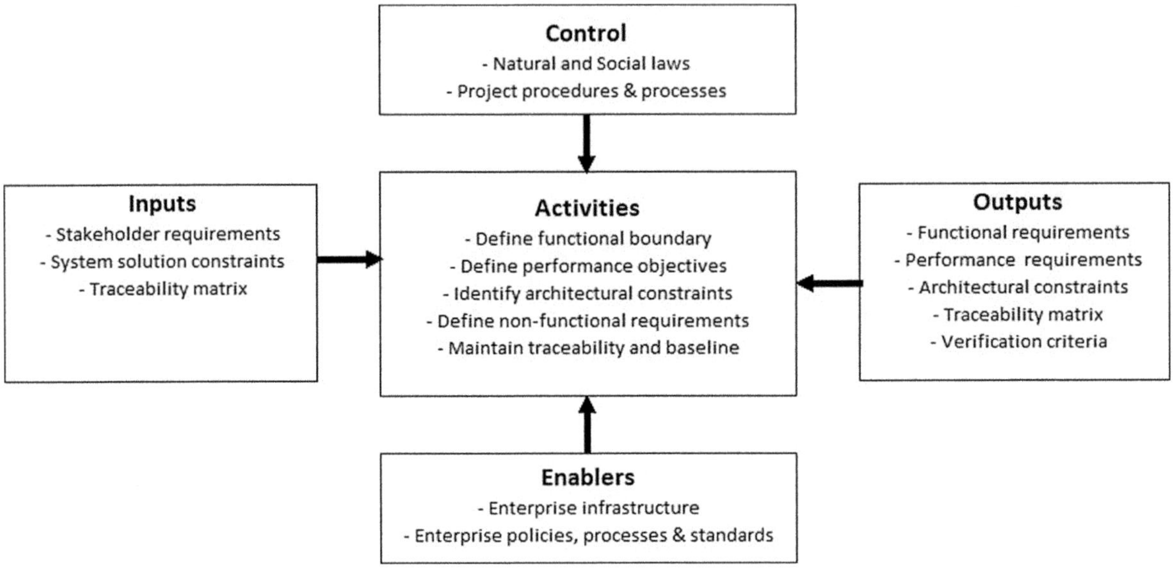

FIGURE 2.35
Context diagram for the requirements analysis process.

measures are always candidates for incorporation into the project's risk and opportunity management program.

Interaction with the risk and opportunity management process includes providing the list of TPM parameters for risk assessment, providing the TPM status/profile charts for risk analysis, and receiving the risk "watch list" from the risk and opportunity management process, for potential additional TPM parameters.

A guidance entitled "Technical Measurement" published by INCOSE provides the detailed information on implementing technical measurement on a project.

2.5.30 Define Other Nonfunctional Requirements

The nonfunctional requirements such as availability, supportability, security, and training are not directly related to the primary capability provided by the system-of-interest. Addressing these requirements from the earliest stages is a good way to ensure that they are not forgotten and that they are satisfied. Fig. 2.35 summarises the Context diagram for the requirements analysis process.

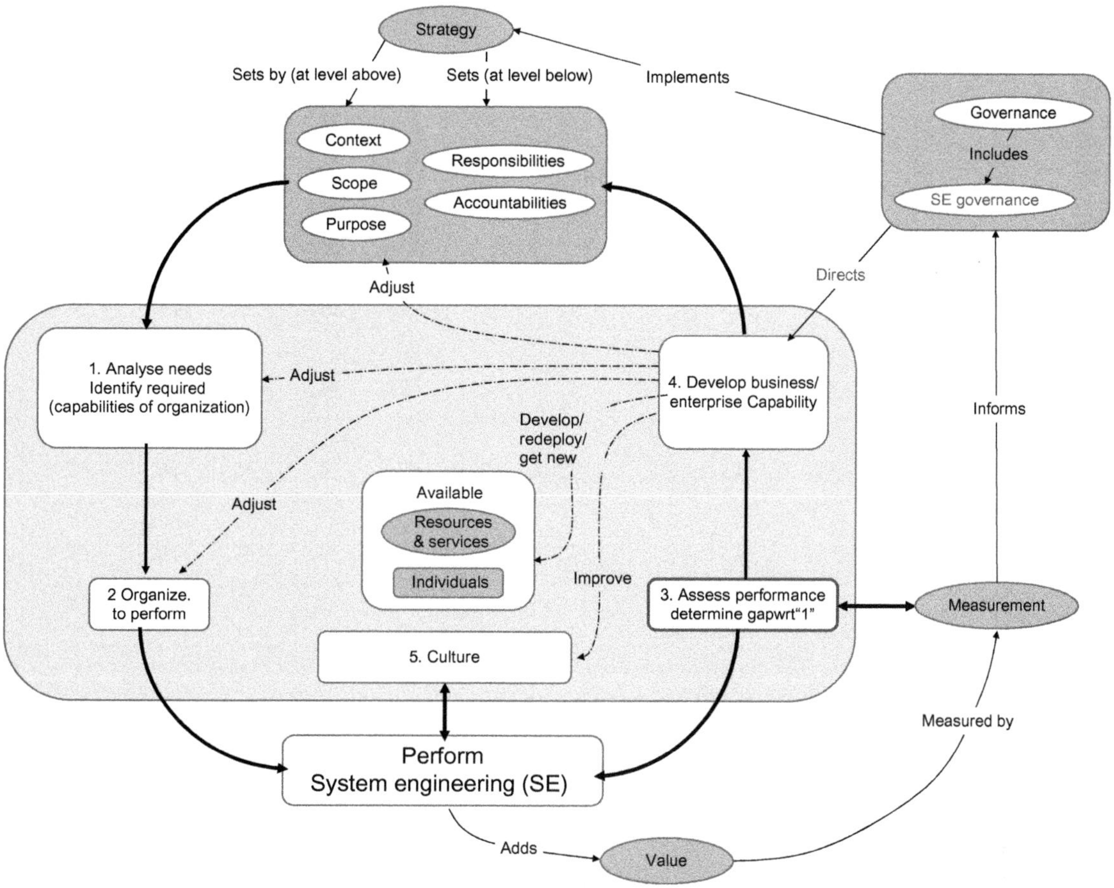

FIGURE 2.36

Assessing systems engineering performance in business or enterprise: part of closed loop governance (SEBoK Original, V.1.7).

The Øresund Bridge case which is covered in section 2.5.28 is a good example of the impact of nonfunctional requirements. In this case, the constraints on the construction practices have been established to avoid the negative environmental impacts.

2.5.31 Performance Assessment Measures

Typical measures for assessing SE performance of an enterprise include the following:

- Effectiveness of SE process
- Ability to mobilize the right resources at the right time for a new project or new project phase

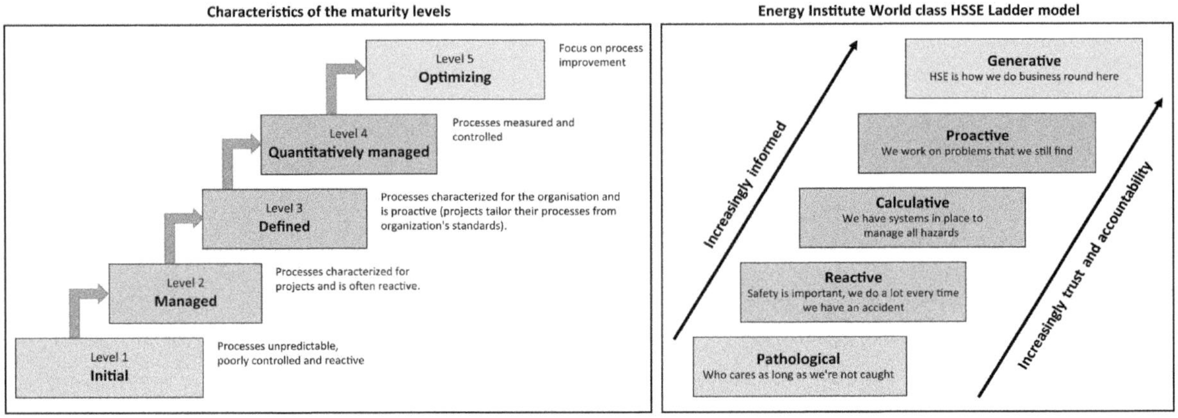

FIGURE 2.37
Application of maturity model for HSSE. From: Energy Institute Heart & Mind model.

- Quality of SE process outputs
- Timeliness of SE process outputs
- SE added value to project
- System added value to end users
- SE added value to organization
- Organization's SE capability development
- Individuals' SE competence development
- Resource utilization, current, and forecast
- Productivity of systems engineers
- Deployment and consistent usage of tools and methods

Since collecting data and analyzing it takes effort that is often significant, measurement is best done when its purpose is clear and is part of an overall strategy. The "goal, question, metric" paradigm should be applied, in which measurement data are collected to answer specific questions, the answer to which helps achieve a goal, such as decreasing the cost of creating a system architecture or increasing the value of a system to a particular stakeholder. Fig. 2.36 shows one way in which appropriate measures inform enterprise level governance and drive an improvement cycle such as the Six Sigma DMAIC (define, measure, analyze, improve, control) model.

Process is a "*set of interrelated or interacting activities which transforms inputs into outputs*." The SEI CMMI Capability Maturity Model provides a structured way for businesses and enterprises to assess their SE processes. In the CMMI a process area is a cluster of related practices in an area that, when implemented collectively, satisfies a set of goals considered important for making improvement in that area. There are CMMI models for acquisition, for development, and for services. CMMI defines how to assess individual process areas against capability levels on a scale from 0 to 3, and overall organizational maturity on a scale from 1 to 5 as shown on the Fig. 2.37.

2.5.32 Affordability/Cost-Effectiveness/Lifecycle Cost Analysis

As stated in Blanchard and Fabrycky (1990) and Fabrycky and Blanchard (1991),

Many systems are planned, designed, produced, and operated with little initial concern for affordability and the total cost of the system over its intended lifecycle. . .The technical [aspects are] usually considered first, with the economic [aspects] deferred until later.

An affordability design model must be able to provide the ability to effectively manage and evolve systems over long lifecycles. The derived requirements we will focus on in this section are as follows:

- Design perspective that assumes the system will change based on environmental influences, new uses, and disabling system causing performance deterioration

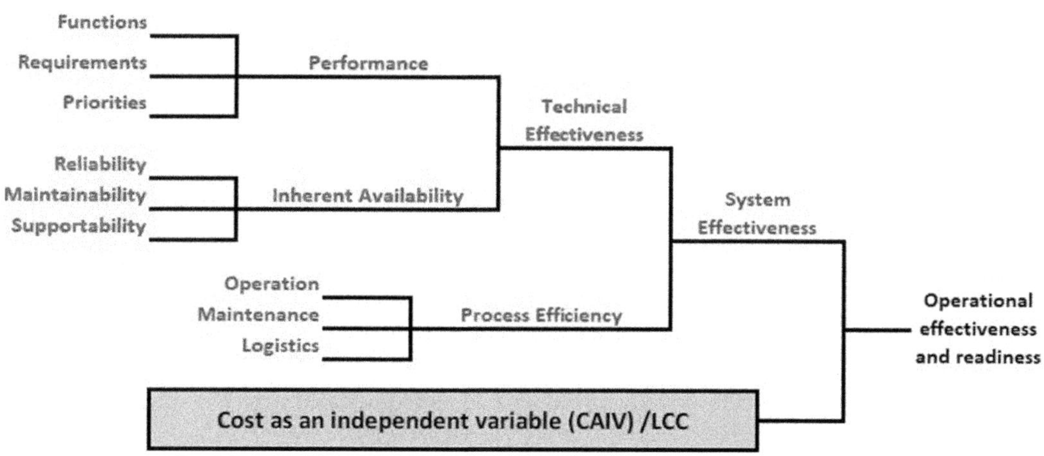

FIGURE 2.38

System operational effectiveness. *Derived from Bobinis, J. and T. Herald. 2012. "An Enterprise Framework for Operationally Effective System of Systems Design." Journal of Enterprise Architecture. 8(2), May 2012. Available at: https://www.mendling.com/publications/JEA12-2.pdf. Fig. 4. Reprinted with permission from Joseph Bobinis. All other rights reserved.*

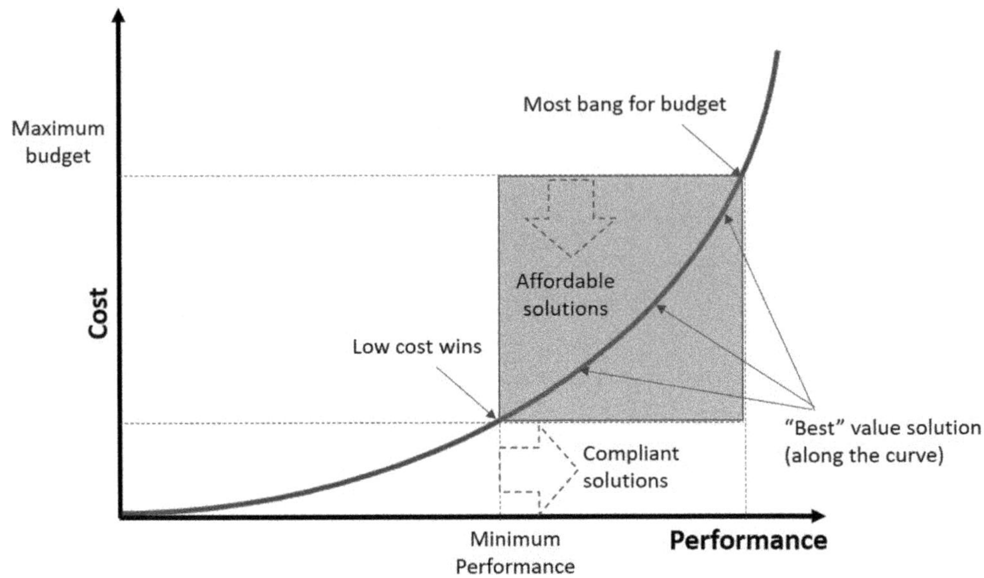

FIGURE 2.39
Cost versus performance. *Reprinted with permission from Joseph Bobinis. All other rights reserved.*

- Causes of system and system element lifecycle differences (technology management)
- Feedback functions and measurement of system behaviors with processes to address emergence (lifecycle control systems)
- A method to inductively translate system behaviors into actionable engineering processes (adaptive engineering)

One of the major assumptions for measuring the affordability of competing systems is that given two systems, which produce similar output capabilities, it will be the *nonfunctional* attributes of those systems that differentiate system value to its stakeholders. The affordability model is concerned with operational attributes of systems that determine their value and effectiveness over time, typically expressed as the system's "ilities" or specialty engineering as they are called in this handbook.

These attributes are properties of the system as a whole and as such represent the salient features of the system and are measures of the ability of the system to deliver the capabilities it was designed for over time. "System integration, and its derivatives across the lifecycle, requires additional discipline and a long-term perspective during the SE and design phase. This approach includes explicit consideration of issues such as system reliability, maintainability and supportability to address activities pertaining to system operation, maintenance, and logistics. There is also a need to address real-world realities pertaining to

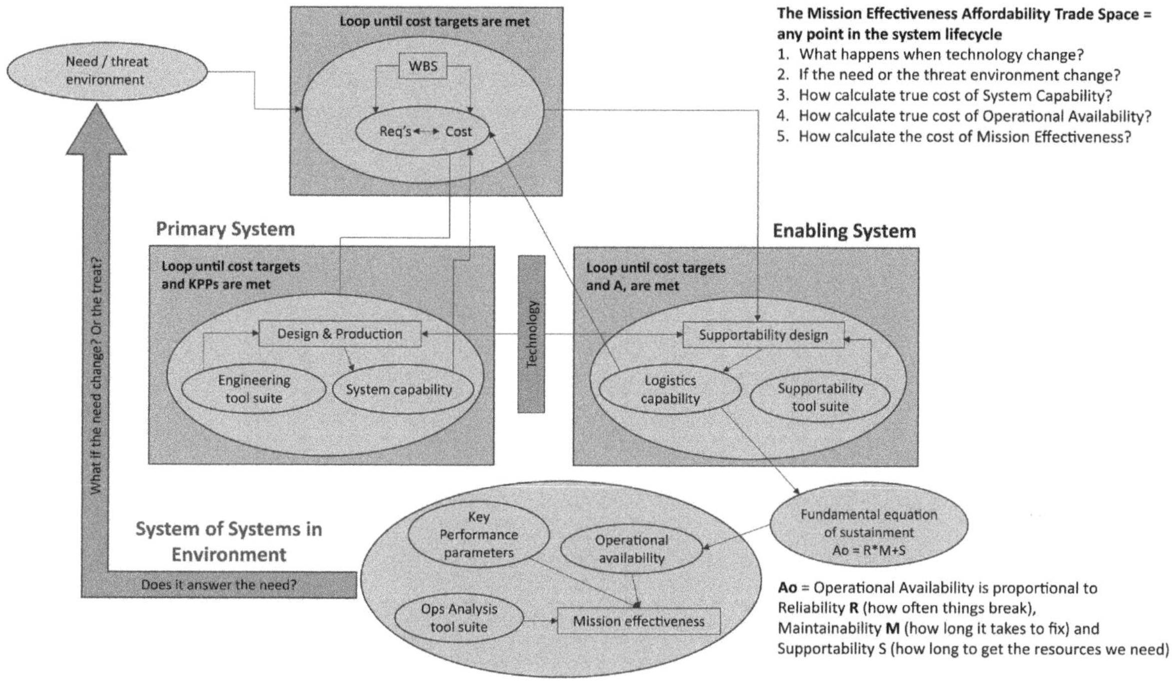

FIGURE 2.40

Affordability cost analysis framework. *Derived from Bobinis, J. and T. Herald. 2012. "An Enterprise Framework for Operationally Effective System of Systems Design." Journal of Enterprise Architecture. 8(2), May 2012. Available at: https://www.mendling.com/publications/JEA12-2.pdf. Reprinted with permission from Joseph Bobinis. All other rights reserved.*

changing requirements and customer expectations, changing technologies, and evolving standards and regulations" (Gallios and Verma, n.d.). Fig. 2.38 summarizes the requirements for the system operational effectiveness.

When one considers the entire SoI, both the primary and enabling systems should be treated as an SoS. In the example in Fig. 2.39 the mission-effectiveness affordability trade space brings together primary and enabling systems into an SoS. Note that the SoS is treated as a closed-loop system where requirements are modified as the mission needs evolve. These iterations allow for technology insertion as the design is updated. Design-to-cost targets are set for the primary and enabling systems, which ensure that affordability throughout the system lifecycle is considered, even as the system evolves.

Affordability measurements that feedback into the system assessment are KPPs and operational availability, which ensures that the mission can be accomplished over time (e.g., those KPPs are met across the system lifecycle; see Fig. 2.40).

Table 2.3 System Engineering Leading Indicator Overview

Leading Indicator	Insight Provided	Phases/Stages									
		p1	p2	p3	p4	p5	S1	S2	S3	S4	S5
Requirements trends	Rate of maturity of the system definition against the plan. Additionally, characterizes the stability and completeness of the system requirements that could potentially impact design, production, operational utility, or support	•	•	•	•	•	•	•	•	•	•
System definition change backlog trend	Change request backlog which, when excessive, could have adverse impact on the technical, cost, and schedule baselines			•	•	•		•	•	•	
Interface trends	Interface specification closure against plan. Lack of timely closure could pose adverse impact to system architecture, design, implementation, and/or V&V any of which could pose technical, cost, and schedule impact	•	•	•	•	•	•	•	•	•	
Requirements validation trends	Progress against plan in assuring that the customer requirements are valid and properly understood. Adverse trends would pose impacts to system design activity with corresponding impacts to technical, cost, and schedule baselines and customer satisfaction	•	•	•	•	•	•	•	•	•	
Requirements verification trends	Progress against plan in verifying that the design meets the specified requirements. Adverse trends would indicate inadequate design and rework that could impact technical, cost and schedule baselines. Also, potential adverse operational effectiveness of the system	•	•	•	•	•	•	•	•	•	•
Work product approval trends	Adequacy of internal processes for the work being performed and also the adequacy of the document review process, both internal and external to the organization. High reject count would suggest poor quality work or a poor document review process each of which could have adverse cost, schedule and customer satisfaction impact	•	•	•	•	•	•	•	•	•	
Review action closure trends	Responsiveness of the organization in closing postreview actions. Adverse trends could forecast potential technical, cost and schedule baseline issues.	•	•	•	•	•	•	•	•	•	•
Risk exposure trends	Effectiveness of risk management process in managing / mitigating technical, cost & schedule risks. An effective risk handing process will lower risk exposure trends.	•	•	•	•	•	•	•	•	•	•
Risk treatment trends	Effectiveness of the SE organization in implementing risk mitigation activities. If the SE organization is not retiring risk in	•	•	•	•	•	•	•	•	•	•

	a timely manner, additional resources can be allocated before additional problems are created.										
Technology maturity trends	Risk associated with incorporation of new technology or failure to refresh dated technology. Adoption of immature technology could introduce significant risk during development while failure to refresh dates technology could have operational effectiveness/customer satisfaction impact.		•	•	•	•		•	•	•	
Technical measurement trends	Progress towards meeting the MOEs/MOPs/KPPs and Technical Performance Measures (TPMs). Lack of timely closure is an indicator of performance deficiencies in the product design and/or project team's performance.			•				•			
Systems engineering staffing & skills trends	Quantity and quality of SE personnel assigned, the skill and seniority mix, and the time phasing of their application throughout the project lifecycle.	•	•	•	•	•	•	•	•	•	•
Process compliance trends	Quality and consistency of the project defined SE process as documented in SEP/systems engineering management plan (SEMP). Poor/inconsistent SE processes and/or failure to adhere to SEP/SEMP, increase project risk.	•	•	•	•	•	•	•	•	•	•
Facility and equipment availability trends	Availability of non-personnel resources (infrastructure, capital assets, etc.) needed throughout the project lifecycle.	•	•	•	•	•	•	•	•	•	•
Defect/error trends	Progress towards the creation of a product or the delivery of a service that meets the quality expectations of its recipient. Understanding the proportion of defects being found and opportunities for finding defects at each stage of the development process of a product or the execution of a service.	•	•	•	•	•	•	•	•	•	•
System affordability trends	Progress towards a system that is affordable for the stakeholders. Understanding the balance between performance, cost, and schedule and the associated confidence or risk.	•	•	•	•	•	•	•	•	•	•
Architecture trends	Maturity of an organization with regards to implementation and deployment of an architecture process that is based on an accept set of industry standards and guidelines.	•	•	•			•	•			
Schedule and cost pressure	Impact of schedule and cost challenges on carrying out a project	•	•	•	•	•	•	•	•	•	•

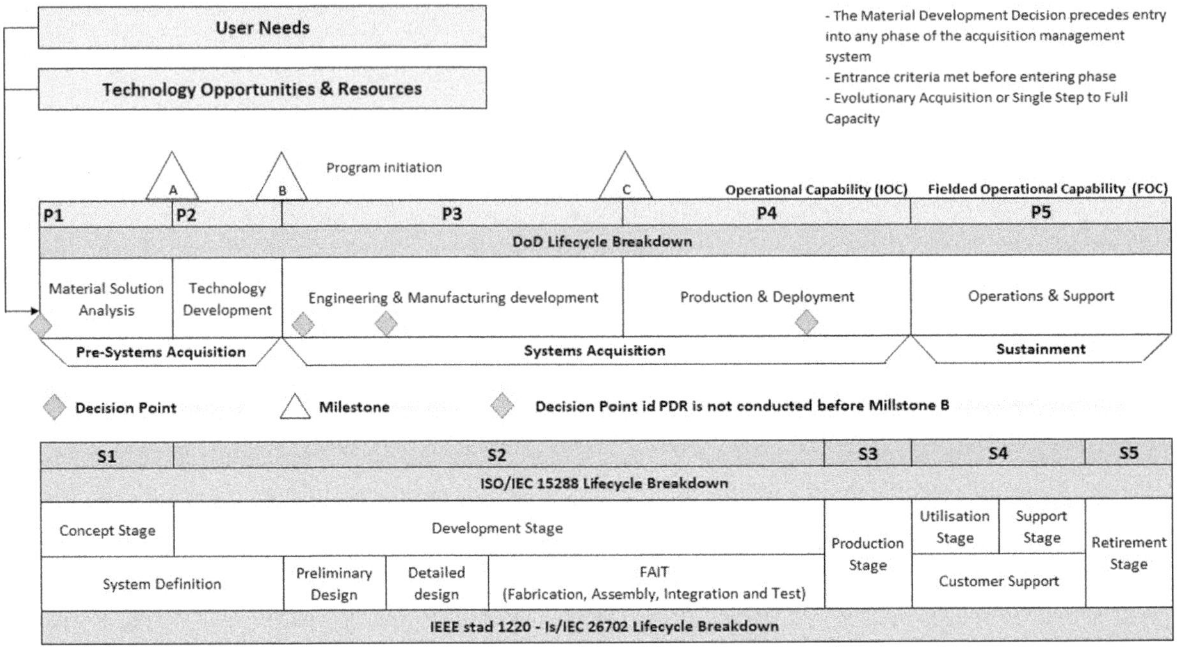

FIGURE 2.41
Definition of the lifecycle breakdown of the phases and stages in Table 2.3.

2.5.33 System Engineering Leading Indicators

A leading indicator is a measure for evaluating the effectiveness of how a specific activity is applied on a project in a manner that provides information about impacts that are likely to affect the system performance objectives. A leading indicator may be an individual measure, or collection of measures and associated analysis that are predictive of future systems engineering performance before the system is fully realized. Systems engineering performance itself could be an indicator of future project execution and system performance. Leading indicators aid leadership in delivering value to customers and end users, while assisting in taking interventions and actions to avoid rework and wasted effort.

Subsequent to the June 2004 workshop the "SE Leading Indicators Action Team" was formed under the auspices of LAI, comprised of engineering measurement experts from industry, government, and academia, involving a collaborative partnership with INCOSE. To date, 18 SE leading indicators have been developed, as summarized in Table 2.3.

As demonstrated in Table 2.3, most of the leading indicators are trend measures that have broad applicability across both US department of defense (DoD) and commercial lifecycle phases/stages. The DoD lifecycle phases are

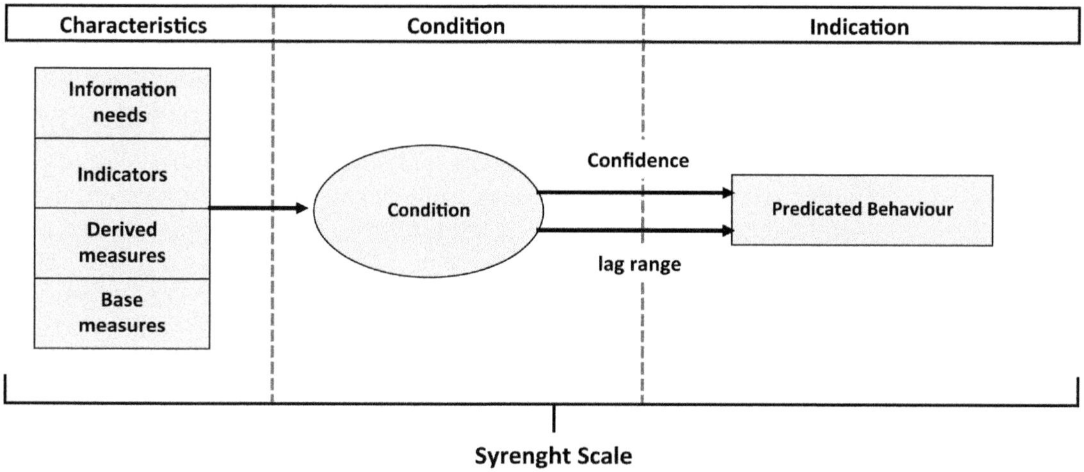

FIGURE 2.42
Relationship of the characteristic and probability of an indicators.

denoted as P1 through P5 in and the industry standard lifecycle stages are denoted as S1 through S5 in Table 2.3.

Fig. 2.41 defines the columns of Table 2.3.

Leading indicators support the effective management of systems engineering by providing visibility into expected project performance and potential future states. Visibility into the future state of a project has not traditionally been part of a measurement process. Additionally, without the use of leading indicators, it is difficult for leadership to establish the likelihood of delivering a complex system within the project constraints such as scope, schedule, quality, and budget.

These leading indicators have been specifically selected to provide insight into key systems engineering activities across the phases of a project.

The further validation efforts will be monitored by the core team, in collaboration with the participating collaboration organizations. Based on results of project use, leading indicators will be adjusted as required. Additionally, recommendations will be developed regarding which leading indicators are most effective for particular types of projects.

The ongoing maintenance of this guidance will be facilitated collaboratively by a leadership team from INCOSE, MIT, and PSM. This leadership team will meet on a regular basis, leveraging existing functions such as the INCOSE International Workshop, PSM Users Group Conference, and LAI Annual Conference in order to involve as many of the collaborative partners as possible. The ongoing work will include workshops with the contributors and interested parties to examine results of the indicators and opportunities for improvement. Consideration will be given to establishing a website to facilitate ongoing communication of contributors and users.

A leading indicator is composed of characteristics, a condition and a predicted behavior. The characteristics and condition are analyzed on a periodic or as-needed basis. The leading indicator predicts behavior within a given confidence and within an accepted time range into the future. Based on accuracy within an organization a leading indicator is given a strength value that represents the belief an organization has in the indicator. As organizational experience proves the value of the leading indicator, the organization would increase the strength value.

For example, an invalidated leading indicator that is used for the first time might predict that there is a 100% probability that the project will fail in 3 months. However, managers would not rely heavily on this leading indicator because it has a very low strength, in this case 0. Fig. 2.42 illustrates the relationship of the characteristic and probability of an indicator.

Within regard to base measures an important assumption and implementation consideration is the additional associated attributes are also recorded with the base measurement values. These attributes are necessary to aid in categorization and analysis. Attributes are needed to convert the base measures into derived measures. This additional insight makes indicator data much more relevant to the information users. Without such data, users would find it difficult, if not impossible to appropriately interpret the indicator or to investigate, and potentially, take appropriate corrective action.

These categories should be selected by relevance to the organizational. As examples, some very useful attribute categories to consider include ISO/IEC 15288, IEEE 1220, and ISO/IEC 26702 process milestone, stage, or phases, disposition action (opened, started, approved, incorporated, rejected, stopped, closed, overdue), maturity states (planned, interim, actual), priority levels (critical, high, medium, low), cause (error, customer request, external), impact level (high, medium, low), classification type, and dates & times of associated events. This list of attributes is not exhaustive. It is simply a list of those attributes widely used in the existing leading indicators.

For example, "Requirements" must be converted to "% Requirements Approved" ("Approved" being a disposition action) for the user to get a sense of the system definition progress and maturity. Further, if at a monthly status review the "% Requirements Approved" is observed to be trending lower than established thresholds, then appropriate action can be taken. Ideally, additional associated attributes are also recorded, including but not limited to the process phases,

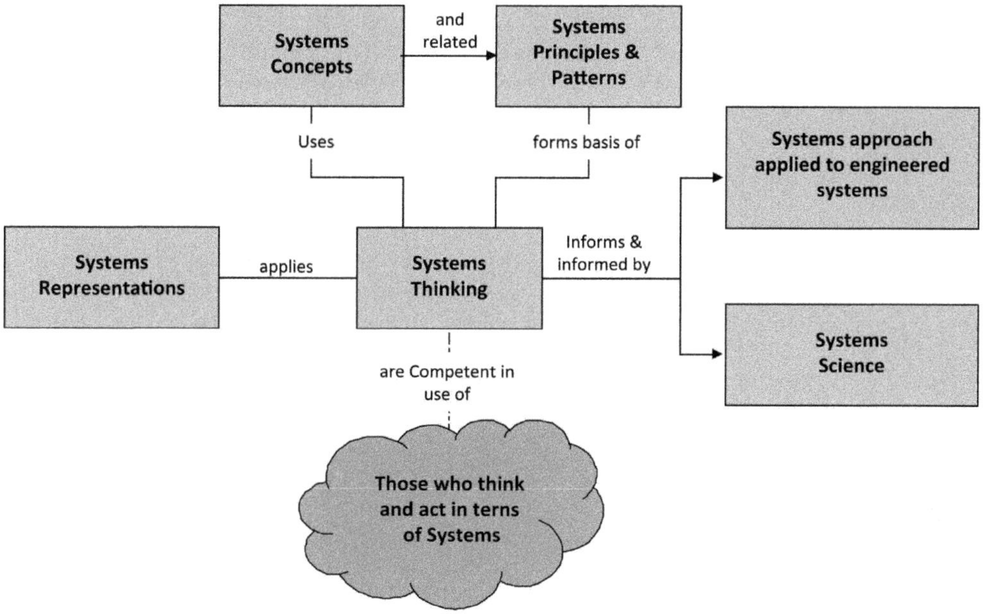

FIGURE 2.43
Systems thinking in the SEBoK (SEBoK Original).

disposition action, maturity states, priority levels, cause, impact level, classification type, and dates & times associate with the requirement approval events.

INCOSE document entitled "System Engineering Leading Indicators guide" provides the description of the leading indicators. Each description provides a rationale that justifies the value of the indicator, describes the decision insight it offers, and specifies how to measure and calculate it. In addition, there are sample graphics intended to illustrate the use of the indicator.

2.6 SYSTEM THINKING

2.6.1 Background

Systems thinking is the *process of understanding* how those things which may be regarded as systems influence one another within a complete entity, or larger system. In nature, systems thinking examples include ecosystems in which various elements such as air, water, movement, plants, and animals work together to survive or perish. In organizations, systems consist of people, structures, and processes that work together to make an organization "healthy" or "unhealthy."

Systems thinking has roots in the general systems theory that was advanced by Ludwig von Bertalanffy in the 1940s and furthered by Ross Ashby in the 1950s. The field was further developed by Jay Forrester and members of the Society for Organizational Learning at MIT which culminated in the popular book The Fifth Discipline by Peter Senge which defined systems thinking as the capstone for true organizational learning. Fig. 2.43 summarizes the inputs and outputs of systems thinking according to the SEBoK model.

Systems thinking has been defined as an approach to problem solving, by viewing "problems" as parts of an overall system, rather than reacting to specific parts, outcomes or events, and thereby potentially contributing to further development of unintended consequences. Systems thinking is not one thing but a set of habits or practices within a framework that is based on the belief that the component parts of a system can best be understood in the context of relationships with each other and with other systems, rather than in isolation. Systems thinking focuses on cyclical rather than linear cause and effect.

In systems science, it is argued that the only way to fully understand why a problem or element occurs and persists is to understand the parts in relation to the whole. Standing in contrast to Descartes's scientific reductionism and philosophical analysis, it proposes to view systems in a holistic manner. Consistent with systems philosophy, systems thinking concerns an understanding of a system by examining the linkages and interactions between the elements that compose the entirety of the system.

Systems science thinking attempts to illustrate how small catalytic events that are separated by distance and time can be the cause of significant changes in complex systems. Acknowledging that an improvement in one area of a system can adversely affect another area of the system, it promotes organizational communication at all levels in order to avoid the silo effect. Systems thinking techniques may be used to study any kind of system—*physical, biological, social, scientific, engineered, human,* or *conceptual.*

The systems thinking approach incorporates several tenets:

- *Interdependence* of objects and their attributes—independent elements can never constitute a system
- *Holism*—emergent properties not possible to detect by analysis should be possible to define by a holistic approach
- *Goal seeking*—systemic interaction must result in some goal or final state
- *Inputs and outputs*—in a closed system inputs are determined once and constant; in an open system additional inputs are admitted from the environment

- *Transformation of inputs into outputs*—this is the process by which the goals are obtained
- *Entropy*—the amount of disorder or randomness present in any system
- *Regulation*—a method of feedback is necessary for the system to operate predictably
- *Hierarchy*—complex wholes are made up of smaller subsystems
- *Differentiation*—specialized units perform specialized functions
- *Equifinality*—alternative ways of attaining the same objectives (convergence)
- *Multifinality*—attaining alternative objectives from the same inputs (divergence)

A treatise on systems thinking ought to address many issues including:

- Encapsulation of a system in space and/or in time
- Active and passive systems (or structures)
- Transformation by an activity system of inputs into outputs
- Persistent and transient systems
- Evolution, the effects of time passing, the life histories of systems and their parts
- Design and designers

2.6.2 System Thinking in Practice

Systems thinking in practice involves stepping back from messy situations of complexity, change, and uncertainty, and clarifying key interrelationships and perspectives on the situation. It further requires engaging with multiple often contrasting perspectives amongst stakeholders involved with and affected by the situation so as to best direct responsible joined-up thinking with action to bring about morally justifiable improvements.

As [Literature item 4] suggests, to achieve an effective system thinking the following steps should be performed:

- *Identify the stakeholders*. Identify who contributes or delivers resources to the system and who benefits, i.e., system actors (including staff and service users), system designers, system decision makers, system influencers.
- *Consider system purposes*. Consider the common or superordinate purpose(s) that defines the system as a whole, considering customer needs. Study how parts of the system contribute to this purpose, including any conflicts or tension between parts of the system, or with the superordinate system purpose(s).
- *Explore the system and its boundary*. Model the system, its interactions and an agreed boundary, for the purpose, question or problem in mind (concerning investigation, assessment, design, etc.). Continually adapt

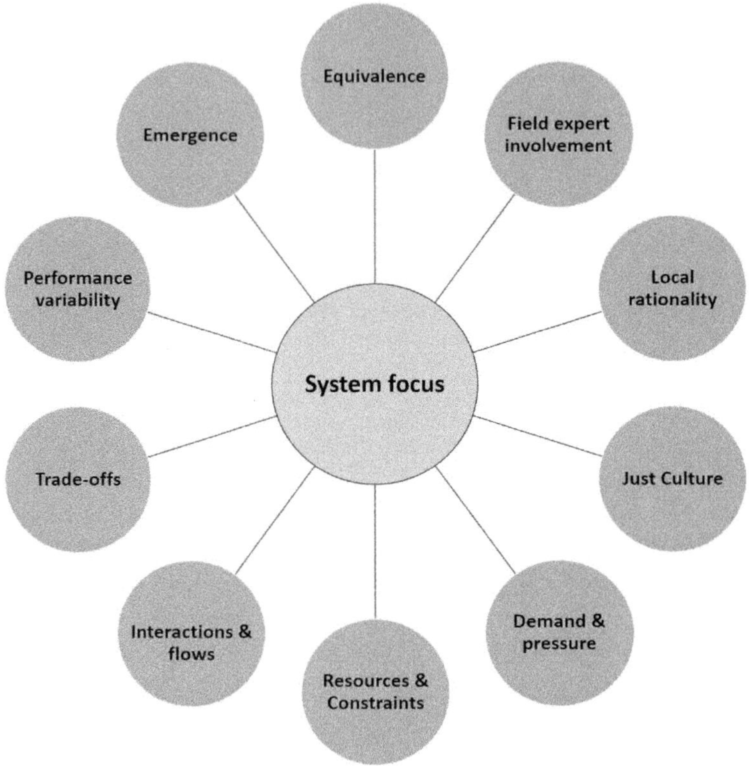

FIGURE 2.44
System thinking principles. Cited in [Literature item 4.]

this as you get data, exploring the differences between the system-as-imagined and the system-as-found.

- *Study system behavior and system conditions*. Consider how changes to one part of the system affect other parts. Bear in mind that decisions meant to improve one aspect can make system performance worse.

Then the 10 principles of Fig. 2.44 should be applied to understand the interactions between the parts of the system (human, social, technical, information, political, economic, and organizational) in light of system goals and implement the appropriate solutions.

There are concepts, theories, and methods to help do this, but they are often not used in practice. We therefore continue to rely on outdated ways of thinking in our attempts to understand and influence how socio-technical systems work. The 10 principles shown in Fig. 2.15 provide a practical "systems thinking" framework to make sense of—and improve—the system performance and safety.

Principle 1. Field expert involvement

The people who do the work are the specialists in their work and are critical for system improvement. To understand work-as-done and improve how things really work, involve those who do the work.

Principle 2. Local rationality

People do things that make sense to them given their goals, understanding of the situation and focus of attention at that time. Work needs to be understood from the local perspectives of those doing the work.

Principle 3. Just culture

People usually set out to do their best and achieve a good outcome. Adopt a mindset of openness, trust, and fairness. Understand actions in context, and adopt systems language that is nonjudgmental and nonblaming.

Principle 4. Demand and pressure

Demands and pressures relating to efficiency and capacity have a fundamental effect on performance. Performance needs to be understood in terms of demand on the system and the resulting pressures.

Principle 5. Resources & constraints

Success depends on adequate resources and appropriate constraints. Consider the adequacy of staffing, information, competency, equipment, procedures, and other resources, and the appropriateness of rules and other constraints.

Principle 6. Interactions & flows

Work progresses in flows of interrelated and interacting activities. Understand system performance in the context of the flows of activities and functions, as well as the interactions that comprise these flows.

Principle 7. Trade-offs

People have to apply trade-offs in order to resolve goal conflicts and to cope with the complexity of the system and the uncertainty of the environment. Consider how people make trade-offs from their point of view and try to understand how they balance efficiency and thoroughness in light of system conditions.

Principle 8. Performance variability

Continual adjustments are necessary to cope with variability in demands and conditions. Performance of the same task or activity will vary. Understand the variability of system conditions and behavior. Identify wanted and unwanted variability in light of the system's need and tolerance for variability.

Principle 9. Emergence

System behavior in complex systems is often emergent; it cannot be reduced to the behavior of components and is often not as expected. Consider how systems operate and interact in ways that were not expected or planned for during design and implementation.

Principle 10. Equivalence

Success and failure come from the same source—ordinary work. Focus not only on failure, but also how everyday performance varies, and how the system anticipates, recognizes, and responds to developments and events.

The principles do not operate in isolation; they interrelate and interact in different ways, in different situations. This is illustrated in the following scenario.

2.6.3 Case Study—Practical System Thinking for Alarm Management

Imagine an engineering control and monitoring position. There is variability in the way that alarms are handled, and some important alarms are occasionally missed. This must be understood in the context of the overall control and supervisory system. Since success and failure come from the same source—everyday work—it is necessary to understand the system and day-to-day work in a range of conditions over time (*Principle 10: Equivalence*). This can only be understood with the engineers and technicians who do the work (*Principle 1: Field experts*). They will view their work from their own (multiple) perspectives, in light of their experience and knowledge, their goals at their focus of attention, and how they make sense of the work (*Principle 2: Local rationality*).

In particular, it is necessary to understand how performance varies over time and in different situations (*Principle 8: Performance variability*). For this, we must understand demand over time (e.g., the number, pattern, and predictability of alarms) and the pressure that this creates in the system (time pressure; pressure for resources) (*Principle 4: Demand and pressure*). Through observation and discussion, it is possible to understand the adequacy of resources (e.g., alarm displays, competency, staffing, procedures), and the effect of constraints and controls (e.g., alarm system design) (*Principle 5: Resources and constraints*) on interactions and the end-to-end flow of work (*Principle 6: Interactions and flow*)—from demand (alarm) to resolution in the field.

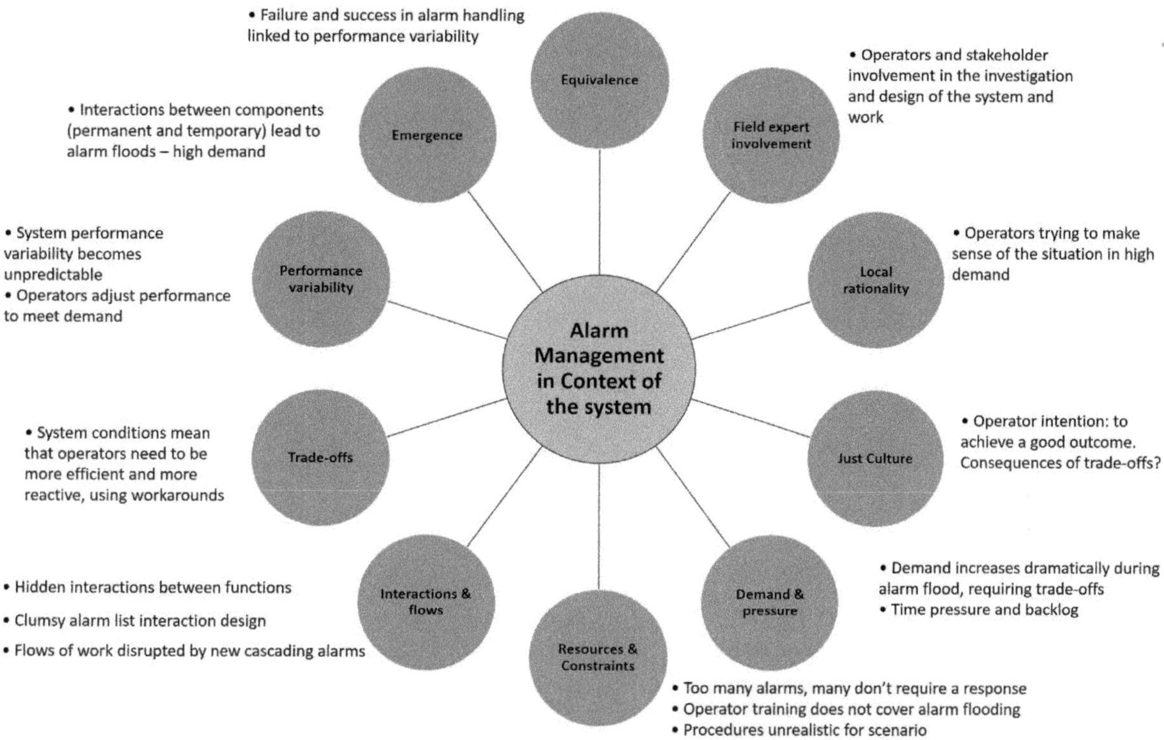

FIGURE 2.45
Application of the system thinking principles for the alarm management system. From: System Thinking for Safety: Ten Principles, A white paper for Moving towards Safety II http://www.skybrary.aero/bookshelf/books/2882.pdf.

It will likely become apparent that engineers must make trade-offs (*Principle 7: Trade-offs*) when handling alarms. Under high pressure, with limited resources and particular constraints, performance must adapt. In the case of alarms handling, engineers may need to be more reactive (tactical or opportunistic), trading off thoroughness for efficiency as the focus shifts toward short-term goals.

Through system methods, observation, discussion, and data review, it may become apparent that the alarm flooding emerges from particular patterns of interactions and performance variability in the system at the time (*Principle 9: Emergence*), and cannot be traced to individuals or components. While the alarm floods may be relatively unpredictable, the resources, constraints, and demand are system levers that can be pulled to enable the system to be more resilient—anticipating, recognizing, and responding to developments and events. Fig. 2.45 summarizes the application of system thinking principles for the alarm management system.

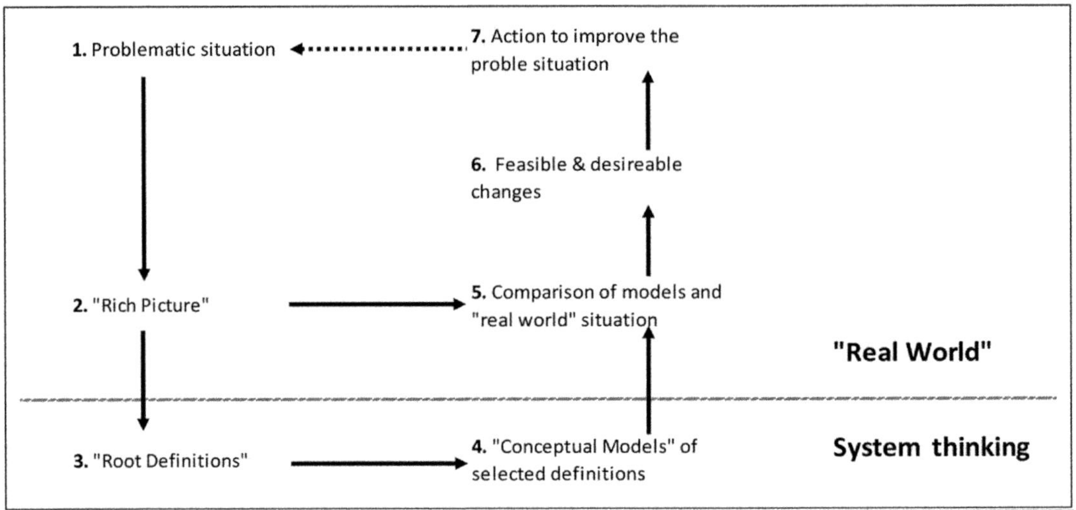

Stage 1 **The "problem situation" unstructured**
The remit of intervention is identified, providing the site sometimes referred to as "the mess"

Stage 2 **The problem-situation expressed**
Ordinarily carried out through a *rich picture*; a tool used for brainstorming ideas of the problem situation usually with the active collaboration of those involved with the system. Designed to generate the main *issues* (e.g. conflicts) and *tasks* (e.g. communication links, social or institutional norms and roles) of the system.

Stage 3 **Root definitions of relevant systems**
Systems deemed relevant to the problem situation are first conceptually identified through isolation of "problem themes", *named* and provided with concise *root definitions*; formulated around fundamental systems questions encapsulated in the mnemonic **CATWOE**:

Customers/clients : beneficiaries of "T"
Actors/agents: those who would do "T"; "experts"
Transformation ("T"): the purpose of the system
Weltanschauung (worldview): value-informed view which makes 'T' meaningful
Owner(s): decision makers who control conditions of the system
Environment: "constraining" elements outside the system

Stage 4 **Conceptual modelling**
To encapsulate the key activities which the system must undertake in order to fulfil the requirements of the root definition. System models are expressions of ideal forms of organised activities. Derived from the root definition rather than rich picture. Used for pro-active interrogative and analytical work, rather than a contrivance at representing 'reality'.

Stage 5: **Comparative analysis**
Conceptual constructs (models) compared with the "realities of the mess". Carried out both *monologically*, comparing conceptual models with the "rich picture" formulated in Stage 2, and/or, more preferably, *dialogically*, in a process of re-presenting models back to those involved with the system in order gain critical feedback.

Stage 6: **Debate changes**
The critique emerging from Stage 5 is used as the basis for a debate amongst those involved with the system, concerning the *desirability* and *feasibility* of future changes

Stage 7: **Action to improve the situation**
Implementing changes agreed upon during the Stage 6 debate or reverting back to Stage 3.

FIGURE 2.46
Stages of soft systems methodology. *Adapted from Reynolds, Martin (2011). Critical thinking and systems thinking: towards a critical literacy for systems thinking in practice. In: Horvath , Christopher P. and Forte, James M. eds. Critical Thinking. New York, USA: Nova Science Publishers, pp. 37–68. http://oro.open.ac.uk/30464/1/Reynolds-CTchapterFINAL.pdf.*

2.6.4 Hard and Soft Systems Thinking

Hard systems views of the world are characterized by the ability to define purpose, goals, and missions that can be addressed via engineering methodologies in an attempt to, in some sense, "optimize" a solution.

In hard system approaches the problems may be complex and difficult, but they are known and can be fully expressed by the investigator. Such problems can be solved by selecting from the best available solutions (possibly with some modification or integration to create an optimum solution). In this context the term "systems" is used to describe real world things; a solution system is selected, created, and then deployed to solve the problem.

Soft systems views of the world are characterized by complex, problematical, and often mysterious phenomena for which concrete goals cannot be established and which require learning in order to make improvement. Such systems are not limited to the social and political arenas and also exist within and amongst enterprises where complex, often ill-defined patterns of behavior are observed that are limiting the enterprise's ability to improve.

Soft system approaches reject the idea of a single problem and consider *problematic* situations in which different people will perceive different issues depending upon their own viewpoint and experience. These problematic situations are not solved, but managed through interventions which seek to reduce "discomfort" among the participants. The term system is used to describe systems of ideas, conceptual systems which guide our understanding of the situation, or help in the selection of intervention strategies.

The success of SSM might be attributed to the detailed development of methodological procedures for addressing complex human-based problem situations. These are commonly represented as an iterative learning cycle with seven stages of enquiry (Fig. 2.46) although Checkland himself more recently describes it in terms of two parallel streams of enquiry.

The three ideas of "problem vs. problematic situation," "solution vs. discomfort reduction," and "the system vs. systems understanding" encapsulate the differences between hard and soft approaches.

2.6.5 Critical Systems Thinking

The development of a range of hard and soft methods naturally leads to the question of which method to apply in what circumstances. Critical systems thinking (CST), or *critical management science*, attempts to deal with this question.

The word *critical* is used in two ways. Firstly, critical thinking considers the limits of knowledge and investigates the limits and assumptions of hard and

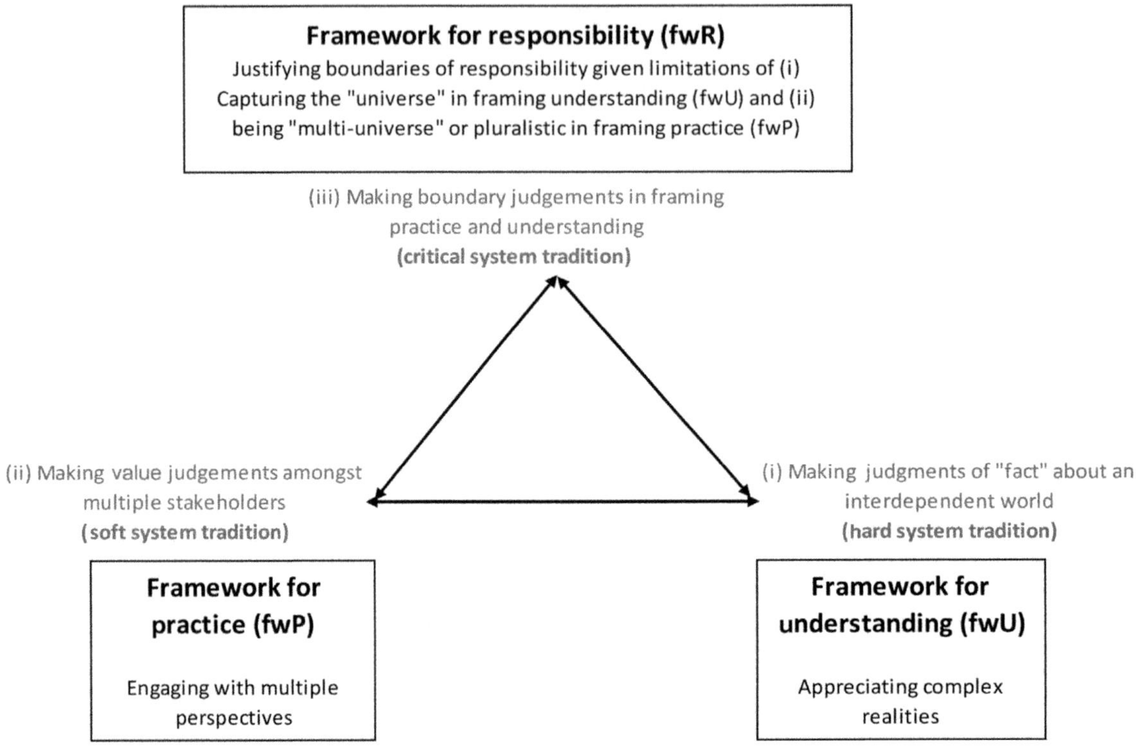

FIGURE 2.47
Critical systems framework illustrating systems thinking in practice activities. Reynolds, Martin (2011). Critical thinking and systems thinking: towards a critical literacy for systems thinking in practice. In: Horvath, Christopher P. and Forte, James M. eds. Critical Thinking. New York, USA: Nova Science Publishers, pp. 37—68.

soft systems, as discussed in the above sections. The second aspect of critical thinking considers the ethical, political, and coercive dimension and the role of system thinking in society.

Systems thinking in practice can be achieved by the following three generalized purposeful orientations:

- *Making sense of, or simplifying* (in *understanding*), relationships between different entities associated with a complex situation. The prime intention is not to get some thorough comprehensive knowledge of situations, but rather to acquire a better appreciation of wider dynamics in order to improve the situation.
- *Surfacing and engaging* (through *practice*) contrasting perspectives associated with complex situations. The prime intention here is not to embrace all perspectives on a predetermined problem so as to solve the problem, but rather to allow for possibilities in reshaping a problem-situation for improved possibilities of resolution.

- *Exploring and reconciling* (with *responsibility*) ethical issues and power relations, both expressions of boundary issues associated with inevitable partial understandings.

The criticality of systems thinking in practice can be expressed in terms of promoting continual and meaningful conversation. The "conversation" works at two levels:

- One is an expression of boundary reflection, a conversation between our *conceptual constructs of real world realities*—constructs called "systems"—and the *actual realities* being addressed.
- The other is an expression of boundary discourse, a conversation between people involved with and affected by the systems used to construct and engage with reality.

Whereas boundary reflection is a conversation influenced by conventional CST, boundary discourse is influenced also by traditions of social learning. Both conversations constitute what might be referred to as boundary critique, a triadic interplay between judgements of "fact," value judgements, and boundary judgements, underpinning systems thinking in practice.

Boundary critique can be described in terms of activities underpinning a framework of systems thinking in practice; constituting what has been referred to as an overall critical systems framework as shown in Fig. 2.47. The framework is supported by three (sub)frameworks, respectively—framework for understanding (fwU), framework for practice (fwP), and a framework for responsibility (fwR). The activities of boundary critique involve continual revising of boundary judgements (systems thinking) with judgements of 'fact' (observing) and value judgements (evaluating).

In developing this into a broader heuristic for systems thinking in practice, three complementary entities can be added: firstly, real world *contexts* of change and uncertainty associated with a fwU; secondly, people or *practitioners* involved with making change associated with a fwP; and thirdly, the ideas and concepts—including systems—as *tools* for effecting change associated with a framework for responsibility.

What is critical in systems thinking in practice is

1. an appreciation that complex realities, despite good intentions with complexity sciences, can never be holistically comprehended;
2. an acknowledgment that any perspective on a situation is laden with values that inhibit any sense of neutral engagement; and
3. an awareness of the limitations of systems design in the light of (1) and (2).

Table 2.4 Critical-Heuristic Categories

Categories				Dimensions of Intentionality	
1	Client? Purpose?	(role)			
2	Measure of	(concerns)	Sources of		The purposeful system of
3	Improvement?	(problems)	Motivation		concern (or context of
4	Decision maker?	(role)			application)
5	Components?	(concerns)	Sources of	Those involved	On which depends the
6	Environment?	(problems)	Control		Meaning of "*improvement*"
7	Planner?	(role)	Sources of		
8	Expertise?	(concerns)	Expertise		
9	Guarantor?	(problems)			
10	Witness?	(role)	Sources of		
11	Emancipation?	(concerns)	Legitimization	Those affected	
12	Worldviews?	(problems)			

Adapted from Ulrich (1983, p. 258, 1993, p. 595, 1996, p. 43).

2.7 EMERGENCE OF BOUNDARY CRITIQUE

Churchman's characterization of purposeful systems dealt initially with only those *involved* in the systems design. Nine conditions that must be fulfilled for a system (S) to demonstrate purposefulness were identified (derived from the philosophy of Immanuel Kant). The conditions are reproduced in summary below (adapted from Churchman, 1971, p. 43)

1. S is teleological
2. S has a measure of performance
3. There is a client whose interests are served by S
4. S has teleological components which coproduce the measure of performance of S
5. S has an environment (both social and ecological)
6. S has a decision maker who can produce changes in the measure of performance of S's components and hence changes in the measure of performance of S
7. S has a designer who influences the decision maker
8. The designer aims to maximize S's value to the client
9. There is a built in guarantee that the purpose of S defined by the designer's notion of the measure of performance can be achieved and secured

Ulrich (1983) distinguished between those *involved* in a system's design and those *affected* by a systems design so as to define the latter role more concisely for social systems planning. The category of those affected by, but not

Table 2.5 Typical Critical Thinking Questions

Who	– benefits from this?	– have you also heard discuss this?
	– is this harmful to?	– would be the best person to consult?
	– makes decisions about this?	– will be the key people in this?
	– is most directly affected?	– deserves recognition for this?
What	– are the strength/weaknesses?	– is the best/worst case scenario?
	– is another perspective?	– is most/least important?
	– is another alternative?	– can we do to make a positive change?
	– would be a counter-argument?	– is getting in the way of our action?
Where	– would we see this in real world?	– can we get more information?
	– are there similar concepts/situations?	– do we go for help with this?
	– is there the most need for this?	– will this idea take us?
	– in the world would this be a problem?	– are the areas for improvement?
When	– is this acceptable/unacceptable?	– will we know we've succeeded?
	– would this benefit our society?	– has this played a part in our history?
	– would this case a problem?	– can we expect this to change?
	– is the best time to take action?	– should we ask for help with this?
Why	– is this a problem /challenge?	– should people know about this?
	– is it relevant to me /others?	– has it been this way for so long?
	– is this the best/worst scenario?	– have we allowed this to happen?
	– are people influenced by this?	– is there a need for this today?
How	– is this similar to __________?	– does this benefit us/others?
	– does this disrupt things?	– does this harm us/others?
	– do we know the truth about this?	– does we see this in the future?
	– will we approach this safely?	– can we change this for our good?

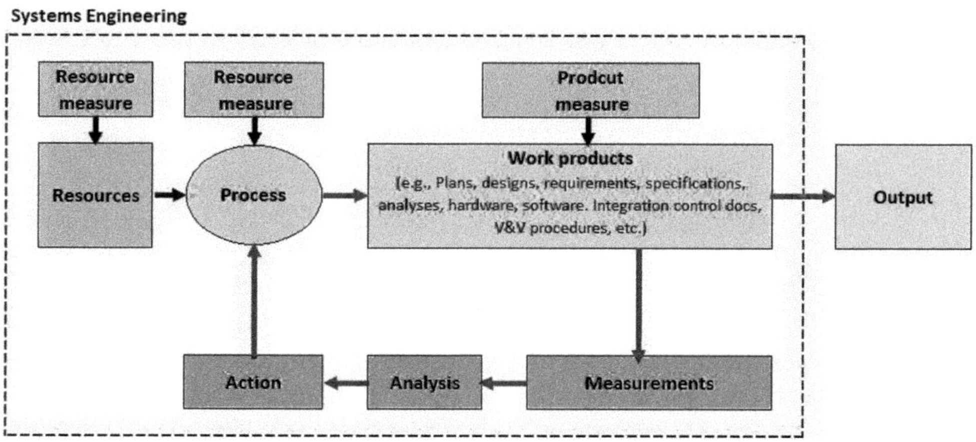

FIGURE 2.48
Systems engineering measurement as a feedback control system.

involved in, are the *witness*. The role specific concerns of the witness are conceptualized as those of *emancipation*; liberation from oppressive material conditions and false consciousness.

The "key problem" category represents the possibilities of a conflict in *world-views*—"different visions of what social reality and human life in it ought to be"—between the involved and the affected. Consequently the "source of influence" for this category group is defined as the *source of legitimization*. Table 2.4 summarizes the twelve Ulrich "critical-heuristic categories."

At more practical level, Table 2.5 is a simple checklist the day-to-day critical thinking questions.

2.8 SYSTEMS ENGINEERING COMPETENCIES FRAMEWORK

The competency of system engineers is the key element in success of a system engineering project. The required competency and evaluation methods should be defined clearly to select the right system engineers and measure their performance to identify the gaps and plan for their training programs. Fig. 2.48 illustrates the system engineering measurement as a feedback control system.

There are many ongoing and lively debates about specific aspects of systems engineering. It is beyond the scope of this framework to provide definitive answers or a single route to applying systems engineering. As a general principle, systems engineering has been defined as the overall discipline, which has specific roles contained within it, e.g., system designer, capability engineer, architect, enterprise architect, requirements engineer, and operational analyst.

Given the complex nature of organizations and the variation in approach and role definition it may be that different terminology and/or a different hierarchy of roles is appropriate.

Systems engineering is a broad discipline that interacts with all other engineering disciplines and as such can be deployed in a variety of ways. To support this the framework can be tailored to make it relevant and appropriate to the specific use. The terminology for the different levels of competency may be relabeled as needed (e.g., to remove any reference to specific roles).

The range of competencies encompassed by the systems engineering framework is very large and it *is not expected that an individual will be "expert" in more than a few of these competencies*. It is important therefore that this framework be used as the common starting point for tailoring the description of systems engineering relevant to an organization and individual. It is expected that an organization will have a set of roles, each with a profile against these competencies (or a tailored subset), with different levels of competence needed. These roles may well include requirements for expertise in other engineering skills and domain specific knowledge/experience. An important check for the enterprise will be to ensure that the roles are balanced (expertise not diluted and all key competencies covered) and the means of communication and integration of the roles understood—so that the "team" is appropriately competent in systems engineering.

INCOSE UKAB Working Group published two documents in which a comprehensive generic systems engineering competency framework and evaluation methods are explained. This competency framework follows the recommendations of the systems engineering standards such as:

- International Standards Organisation ISO15288;
- Capability Maturity Model Integration;
- EIA731;
- INCOSE Systems Engineering Body of Knowledge and Handbook;
- NASA Handbook;
- IEE/BCS Safety Competency Guidelines.

INCOSE Systems Engineering competency framework can be applied in the context of an application, a project, an organization, an individual, and/or a training program or training development. This framework cannot be specific for all these potential (and different) situations, and hence a degree of tailoring is necessary. However, it is expected that to get a more holistic and balanced view as an individual, these competencies would be combined with other complimentary competency and capability frameworks (e.g., process, process safety engineering, and program management) and other relevant skills and behaviors tailored to their particular role or area in which they operate.

As a stand-alone document, this framework focuses on the competencies and skills that when linked with processes, organizations, and infrastructure capabilities create successful systems engineering applications. Therefore depending upon circumstances, subsets of this framework may be selected and used independently. When using the indicators for levels of competency the indicators accumulate—i.e., to reach practitioner level all the indicators for the levels below apply.

For an organization the framework needs to be combined with the terminology, processes, and job roles that exist when the framework is applied. Therefore the framework can be used to:

- Identify what systems engineering competencies are embedded in each of the different roles in the organization;

- Redefine the roles to ensure suitable systems engineering competencies are included;
- Define new roles that apply systems engineering competencies that are not currently implicit in the organization—or to make more explicit the need to exploit these competencies;
- Guide changes to the processes, organization, etc. (its capability) to exploit systems engineering competencies;
- Review the status/strength of systems engineering, and prioritize organization improvements and/or training programs;
- Provide a better understanding of systems engineering by explaining what it is people do when undertaking systems engineering;
- Support enterprise profiling;
- Support team profiling;
- Support recruitment;

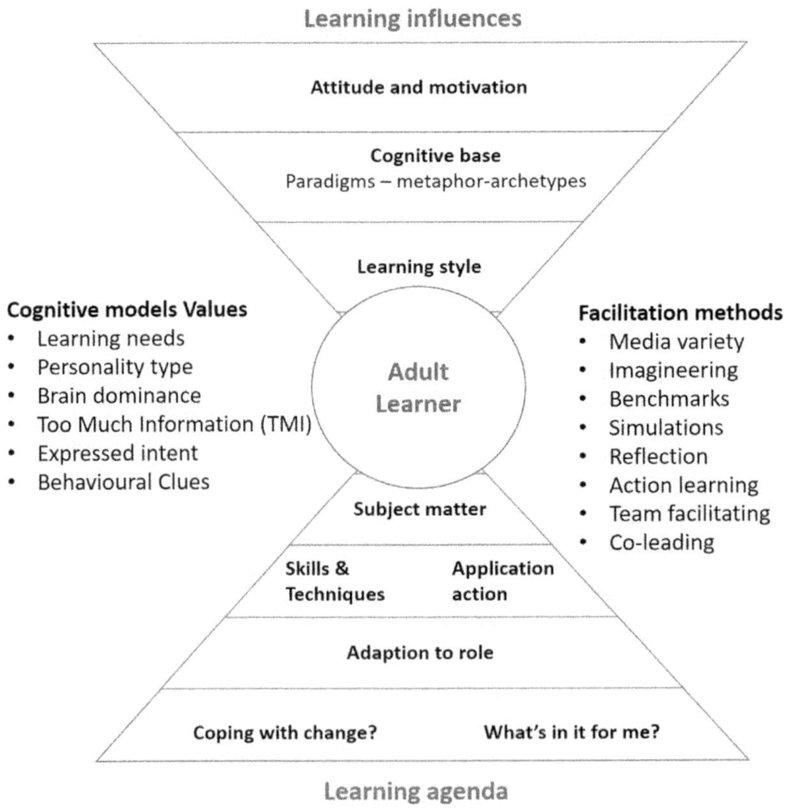

FIGURE 2.49
Adult learner model.

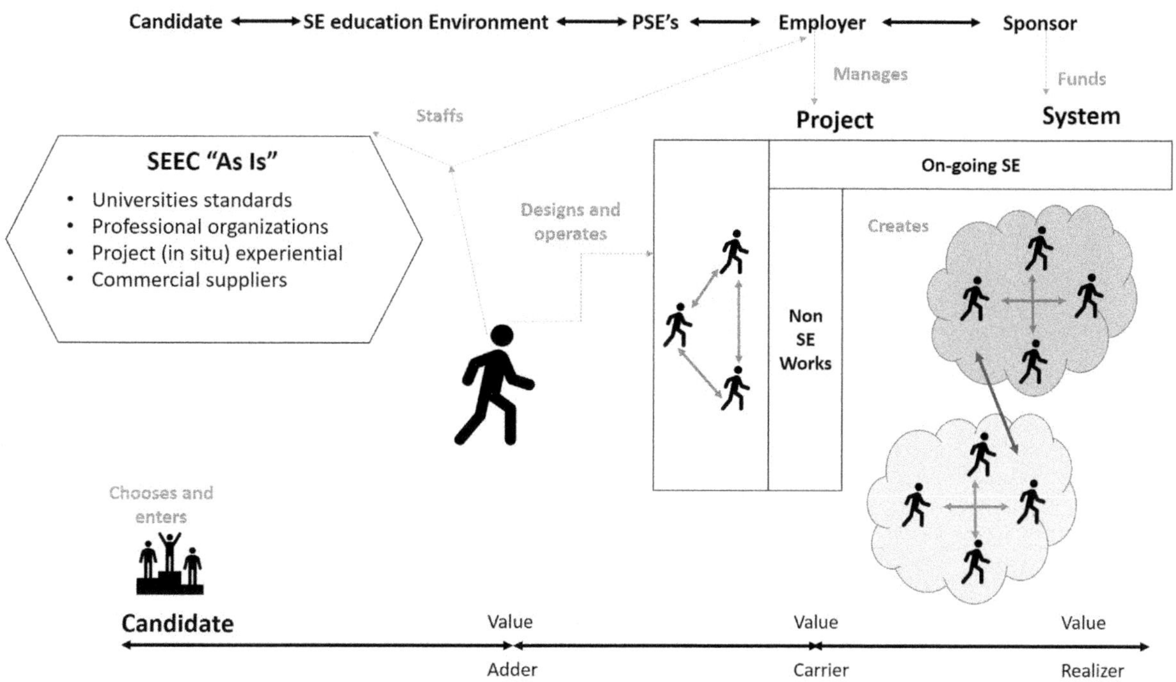

FIGURE 2.50
Context of a systems engineering education community (SEEC).

- Measure and benchmark the state of systems engineering in the organization;
- Identify gaps in skill base.

This competency definition is nonprescriptive, and so organizations need to consider the size and complexity of their products and services when using and tailoring this definition to roles in their organization. Therefore comparison between different roles or different organizations needs to consider the specific complexity of the domain, to ensure comparisons are "like with like."

2.8.1 Systemized Systems Engineering Education Community

The population of professional system engineers (PSEs) is far from homogeneous. However, some invariants are expected:

- *First,* PSEs are primarily adults, most having earned one or more college degrees but not necessarily in engineering disciplines.

- *Secondly*, each PSE will tend to be a generalist, have a tolerance for ambiguity, and an ability to exhibit both divergent and convergent thinking styles.
- *Thirdly*, the PSE population, although representing perhaps only about 4% of the spectrum of personality types, will still represent distinctly different learning styles.

Consequently, an SEEC must offer multifaceted learning environments even for the same subject matter. Adult learner model shown in Fig. 2.49, several factors must be brought together and harmonized if the SEEC is to serve the SE community. Learning to learn is especially important during in situ sessions of direct project work. As the PSE's level of awareness increases and he/she "learns to learn."

The SEEC must span the needs not only of PSEs on projects but also of PSEs who act as employer managers and PSEs who act as principals in the SEEC provider organizations. The more capable PSEs should be able to contribute effectively in any of these roles. The SEEC must also provide opportunities for engineers of other persuasions to gain competencies to work with PSEs.

An SEEC in context as shown in Fig. 2.50. The SEEC consists of at least five types of providers:

- Universities and other academic centers,
- Commercial suppliers of SE learning,
- SE relevant standards bodies,
- Professional organizations,
- In situ or on-the-job-training (OJT) situations on real projects.

These five types of providers are necessitated by the level of understanding of the foundations, tools, and skills required by PSEs, the span of domain knowledge required by PSEs and the spectrum of learning styles exhibited by PSEs. Currently, this five part montage exhibits gaps and overlaps in subject matter, learning environment offered and learning style supported.

Although academic centers are striving to create more realistic learning environments, we believe that structural and cultural limits in academia will preclude universities and related laboratories from being sufficiently responsive.

The reader should note that, in the interest of readability, Fig. 2.1 does not illustrate the fact that the project (in situ) experiential learning can be congruent with the project operations on the right side of the figure. All projects may be used as SE learning environments unless project considerations make this imprudent. In Fig. 2.50 the SEEC is located within the hexagon on the left. All else is SEEC context. The context elements are as follows:

SEEC "As Is": Within this context the SEEC adds value to PSE participants by increasing each PSEs competencies (repertoire of capabilities and proficiency level in each).

The value-added to a PSE is considered latent until the PSE produces artifacts. Only when a competency is applied can the value-added be realized. Consequently, value-added is an emergent property rather than a guaranteed effect of a resource. As will be discussed later in this paper, this fact significantly influences the envisioned SEEC planning and control system.

Candidates: The SEEC processes candidates to produce PSEs. A small percentage of candidates will experience the SEEC for the first time but the majority of candidates will be recycling through the SEEC multiple times as they incrementally augment their respective capabilities. One mode of recycle is the real plane where the practitioner physically returns for more experiences. The more prevalent mode of recycle, however, is the reflection plane wherein a practitioner imagines, recalls, and reaches deeper understanding [Schon] without returning to the original learning location. The more advanced PSE may transit an SEEC, in real or in reflection, more than a thousand times during a 40 year career.

Practitioners of SE (PSEs): As part of the overall project, PSEs, as employees or subcontractors, collaboratively design the SE project (perhaps documenting the scheme as a SEMP). Then PSEs operate the SE project by forming relationships with one another and with other people who are in non-SE roles on the project. The degree to which PSEs are effective in a project setting is modulated by three factors:

- Competencies of individual PSEs.
- Collaborativeness (the emergent competencies of the set of PSEs on a project) and immunity to situations which tend to disturb and degrade individual and workgroup enthusiasm.
- SE project work climate, method, and tooling. The project work climate and overall culture is generally created by the employer and influenced by the system project's sponsor. The methods and tooling may not match the SE challenge. Mismatch can affect productivity and innovation by a factor of 10 or more for both the PSEs and other project members [Livingston].

Project: The project is where time, resources, competencies, goals, status reviews, and activities are mixed together and results emerge in terms of quality, cycle time, cost, and learning. For purposes of this paper a project is viewed in three parts: a front-end part consisting mostly of SE work, an ongoing SE part that encompasses development, deployment, and adaptation, and a third part that consists of the non-SE work such as component

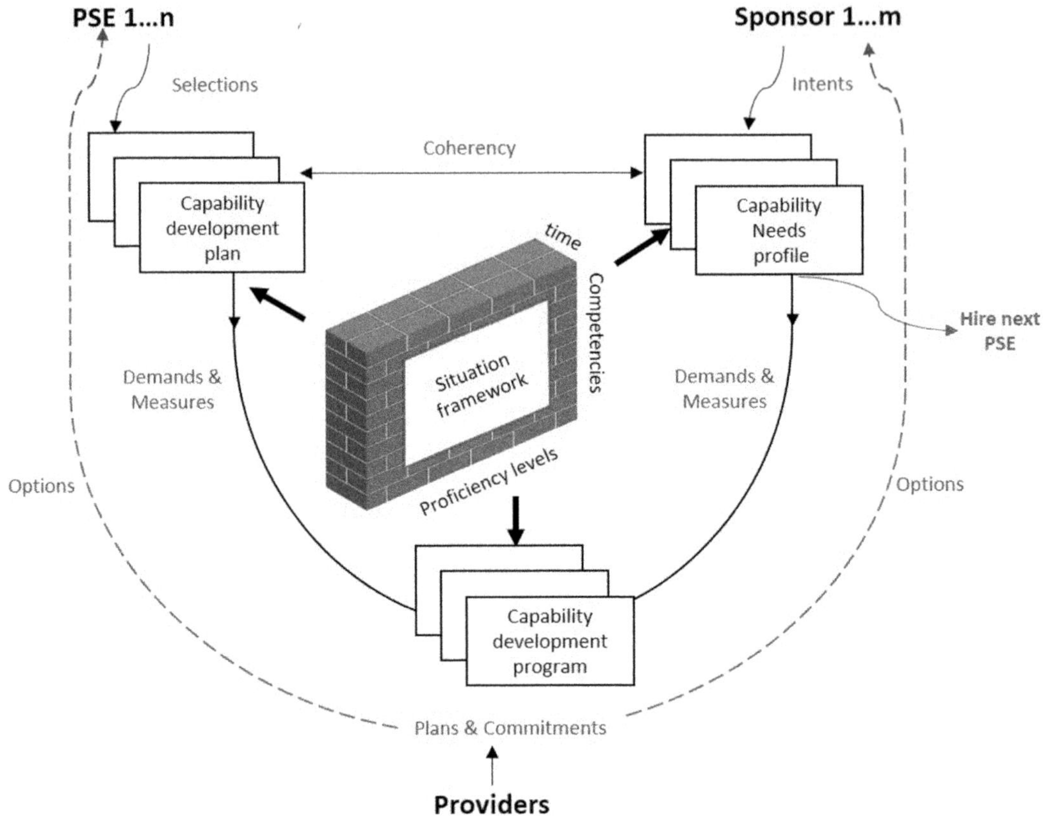

FIGURE 2.51
A unifying framework.

engineering, integration, test, training, deployment, activation, and adaptation. For clarity, project management and other administrative tasks are not shown. Although projects vary widely, heuristics regarding successful projects indicate that the front-end part consumes about 5% of the project budget and the ongoing part consumes another 5–10%.

Employer/contractor: The sponsor typically relies on an employer/contractor to manage the project that creates the system. The sponsor selects the employer who convincingly proposes to accomplish the project "better, faster, cheaper" than competitors.

Ideally, sponsors and employers share the risks inherent in creating the system and share the rewards that flow from the value exhibited by the system when operational.

Sponsor: The sponsor funds the system, including its development. The sponsor and related beneficiaries realize value from the system after it is delivered, operated and adapted to the operational situation [Ring 98].

SE artifacts: SE artifacts (aka work products) are the results of SE work. The SE artifacts inform PSEs and the non-SE performers on the overall project who then produce the balance of the deliverable artifacts that complete the project. Collateral outputs of a project embellish the SEBoK.

Yet other outputs of the project are by-products that may be benign or may be toxic waste and must be purged from the body of knowledge.

2.8.2 Unified Framework

This suggests that a three axis competency, proficiency, time (CPT) coordinate system will provide an appropriate framework in which the participants can locate themselves and estimate the probable effects of transition choices. Such a framework and its use is illustrated in Fig. 2.51.

2.8.3 System Engineering Abilities

Systems engineering ability comprises four key elements:

- Competencies;
- Supporting techniques;
- Basic skills and behaviors;
- Domain knowledge.

2.8.4 Competencies

The *competencies* are grouped into three themes; systems thinking, holistic lifecycle view, and systems engineering management.

Systems thinking contains the underpinning systems concepts and the system/super-system skills including the enterprise and technology environment. It can be breakdown into:

- Systems concepts
- Super-system capability issues
- Enterprise and technology environment

Holistic lifecycle view contains all the skills associated with the systems lifecycle from need identification, requirements through to operation and ultimately disposal. System engineer should be capable to determine and manage stakeholder requirements. The system design encompasses the followings:

- Architectural design
- Concept generation

Table 2.6 Supporting Techniques for System Engineering

Category	Supporting Techniques	Specific Techniques
Analysis and design	Operational analysis	
	Behavioral analysis	Event simulation transaction analysis
	Logical analysis	
	Physical analysis	N2 partitioning DSM axiomatic design
	Functional analysis	Functional decomposition
	Structured methods	Yourdon quality function deployment—QFD SSADM agile methods OOAD
	Decision analysis and resolution	Trade studies
	Failure analysis	FMECA, FTA, FMEA
	Lean design	
	Management of margins	
	Six sigma design	Statistical analysis
Systems thinking	System definition	SSM seven samurai
Management	Estimating	COCOMO, COSYSMO
	Budgeting	Earned value management
	Scheduling	Material requirements planning (MRP) Manufacturing resource plaNning (MRP II)
	Planning	Network analysis schedule analysis critical path analysis
	Change management	
	Configuration management	
	Progress monitoring	Earned value management critical parameter management
	Technical risk and opportunity management	PESTEL, SWOT, Delphi Technique
	Technology planning	TRL, SRL, DML
Specialist	Human factors	Hierarchical task analysis
	Availability reliability	Reliability availability maintainability (RAM) analysis
	Maintainability analysis	RAM analysis
	Reliability analysis	RAM analysis
	Testability analysis	
	Safety analysis	FMECA, FMEA, HAZOPS
	Security analysis	
Modeling and simulation	Mathematical modeling	
	Graphical modeling	
	Physical modeling	
	Synthetic Environments	

- Design for . . .
- Functional analysis
- Interface management
- Maintaining design integrity

Table 2.7 Basic Skills and Behaviors for System Engineering

Basic Skills and Behavior	Specific Techniques
Abstract thinking	Ability to see multiple perspectives, ability to see the big picture
Knowing when to ask	Asking for advice, engaging an expert, peer review, requesting training
Knowing when to stop	Pareto, 80:20 rule, decision making skills
Creativity	Lateral thinking, brainstorming, TRIZ, six thinking hats
Objectivity	Reference of policy, baselining, viewpoint analysis
Problem solving	TQM tools (cause/effect, force field, pareto, etc.), SWOT analysis, PESTEL analysis, decision trees, logical reasoning
Developing others	Coaching, mentoring, training
Two way communicating	Listening skills, verbal & nonverbal communication, body language, writing skills, presentation skills
Negotiating	Win-win, bartering, diplomacy, cultural awareness, stakeholder management, management of expectations
Team working	Belbin team roles, Meyers-Briggs Type Indicator, TQM tools (cause/effect, force field, pareto, etc.)
Decision making	Risk/benefit analysis, pareto analysis, pair-wise comparison, decision trees, force field analysis, six thinking hats

Note: *The Pareto Principle claims that only a "vital few" peapods produce the majority of peas.*

- Modeling and simulation
- Select preferred solution
- System robustness
- Systems integration & verification
- Validation
- Transition to operation

Systems engineering management deals with the skills of choosing the appropriate lifecycle and the planning, monitoring, and control of the systems engineering process. It can be breakdown into the followings:

- Concurrent engineering
- Enterprise integration
- Integration of specialisms
- Lifecycle process definition
- Planning, monitoring, and controlling

The distinguishing feature of systems engineering is its interdisciplinary nature. All these competencies may be present in single discipline individuals, for example, software systems engineers. However, to be a "Systems Engineer" requires the definition and integration of a system solution that comprises a number of discipline areas, for example mechanics, electronics, software, including specialist disciplines such as human factors and electromagnetic compatibility.

2.8.5 Supporting Techniques

Supporting techniques for system engineering are summarized in Table 2.6. This list is not exhaustive.

2.8.6 Basic Skills and Behaviors

Basic skills and behaviors for system engineering are summarized in Table 2.7. This list is not exhaustive.

The "levels of cognition" to achieve these skills in order from least complex to most complex according to the Bloom's taxonomy (Bloom, 1984) are as follows:

- *Remember*: Recall or recognize terms, definitions, facts, ideas, materials, patterns, sequences, methods, principles, etc.
- *Understand*: Read and understand descriptions, communications, reports, tables, diagrams, directions, regulations, etc.
- *Apply*: Know when and how to use ideas, procedures, methods, formulas, principles, theories, etc.
- *Analyze*: Breakdown information into its constituent parts and recognize their relationship to one another and how they are organized; identify sublevel factors or salient data from a complex scenario.
- *Evaluate*: Make judgments about the value of proposed ideas, solutions, etc., by comparing the proposal to specific criteria or standards.
- *Create*: Put parts or elements together in such a way as to reveal a pattern or structure not clearly there before; identify which data or information from a complex set is appropriate to examine further or from which supported conclusions can be drawn.

Collaboration and individual characteristics includes competencies that describe important, universal communications skills, and personal characteristics. The most important for systems engineer's competencies selected by the MITRE Institute's Leadership and Management Model are as follows:

- Building trust
- Building a successful team
- Communicating with impact
- Persuasiveness and influence
- Facilitating, managing, and championing change
- High quality standards
- Results orientation

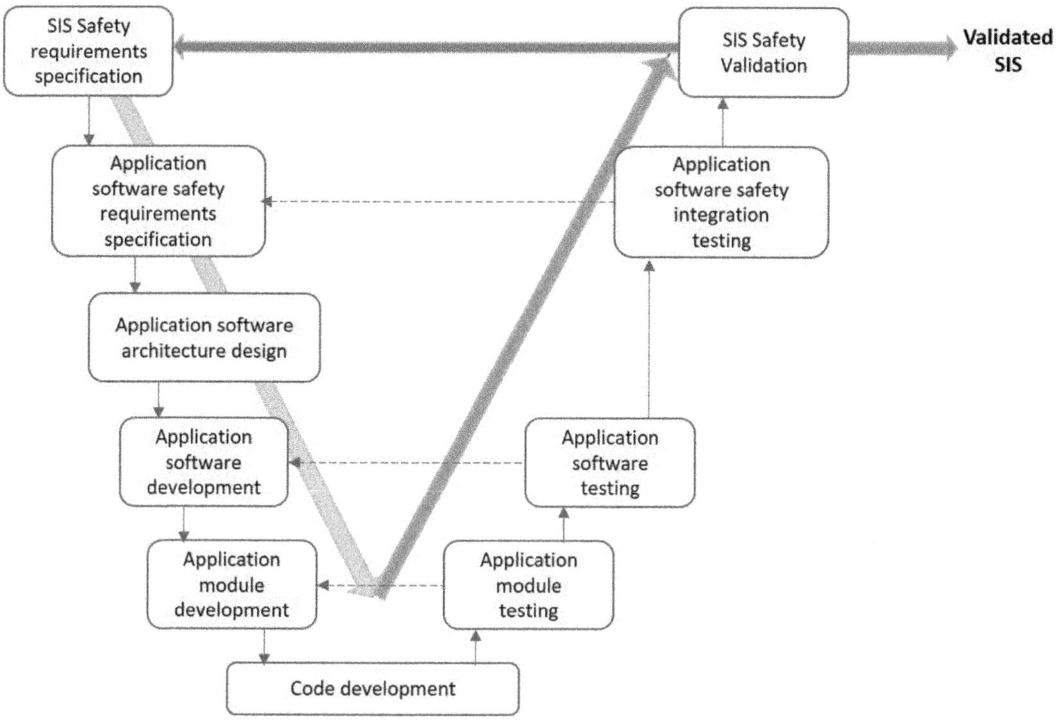

FIGURE 2.52
V-model for verification and validation of safety instrumented-based systems (from 61511–1, figure 12).

- Adaptability
- Integrity

2.8.7 Domain Knowledge

Domain knowledge will vary from industry to industry. Domain knowledge acknowledges that industrial context, the specific commercial environment, and types of supply chain have a big impact on the systems engineering being conducted and that this will be specific to particular industrial fields. It is therefore difficult to produce a generic set of competencies for domain knowledge and will be left to the enterprise implementing these competencies to define what domain knowledge is required.

Systems engineers generally become systems engineers by two mechanisms:

1. They may start their career in a deep technical specialty and broaden their system view over the years as they take on more project responsibilities;
2. They may be trained in systems engineering through either academic or corporate related education. In either case, on the job work experience is a critical requirement for becoming a good systems engineer.

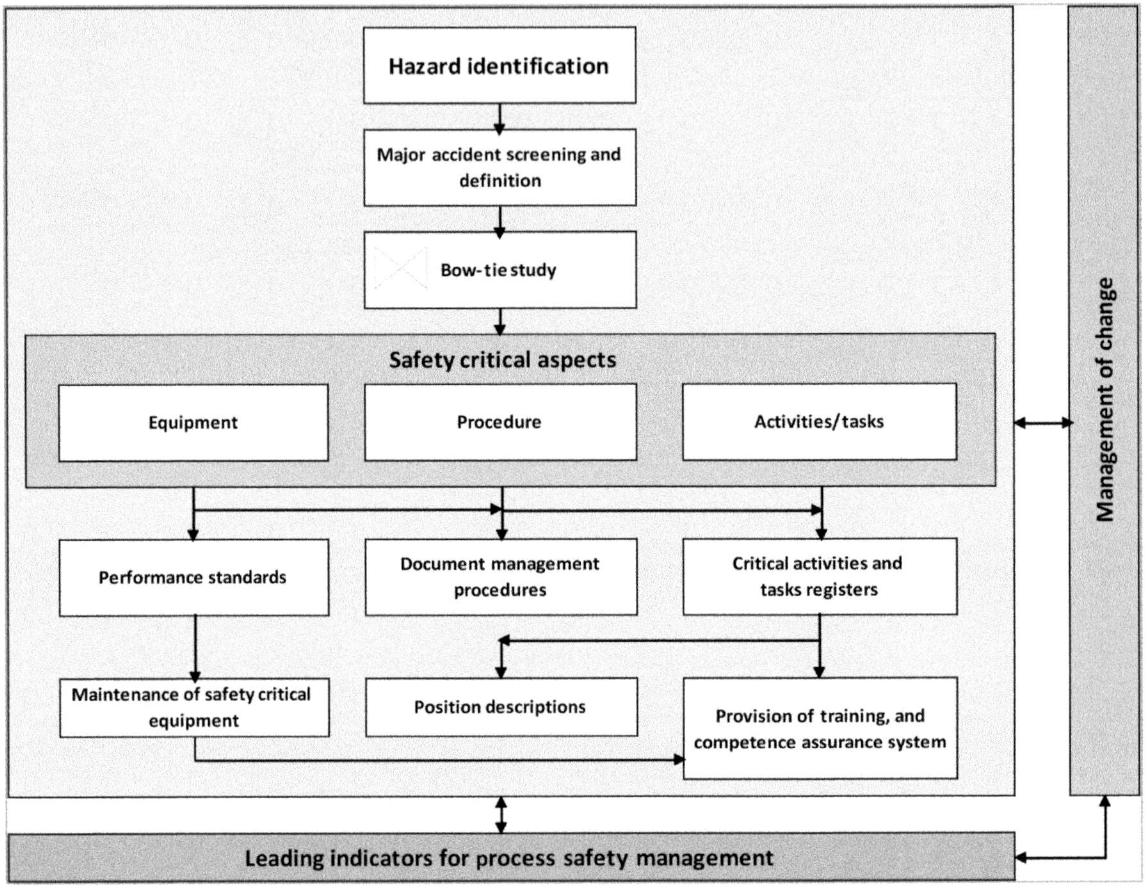

FIGURE 2.53
Safety critical system management. *From: Bow-Tie diagrams in downstream hazard identification and risk assessment.*

There is no specific discipline or competency framework for the system engineering in process manufacturing industry. Very often the specialists informally board their system view over the years upon their assigned tasks. It is crucial that process manufacturing industry tailor a specific system engineering competency model to:

- improve performance of the project, HSE process safety mangers and technical safety engineers.
- involve effectively the other managers and specialists in the reverent system engineering processes.

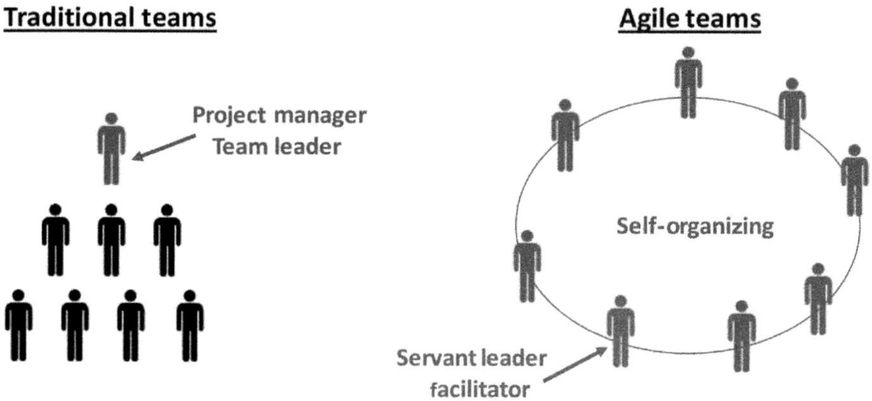

FIGURE 2.54
Brainstorming activities such as HAZOP are inspired by the Agile Scrum method.

Process manufacturing uses system engineering mainly in project, quality, HSE, PSM, and the other management systems in an implicit way. System engineering is applied in an explicit manner in two process manufacturing projects:

1. SIL assessment in compliance with IEC 61508 and IEC 61511. Fig. 2.52 illustrates the V model for Verification and Validation of Safety Instrumented based Systems.
2. Determination and management of the SCEs in compliance with the energy institute guidelines for the management of SCEs. SCEs are the regulatory requirements of the UK Offshore Regulation for Fire & Explosion Emergency Response (PFEER) which is on force since 1995.

   Today, the major oil & gas operators apply the SCEs for both onshore and offshore facilities. The authors of this book believe that this approach can be very beneficial for any process manufacturing project. Wherever, the chemical reactors are involved the SCEs should be specified case by case.
3. Performance standards and verification schemes—in compliance with PFEER, duty holder shall prepare the performance standards and verification schemes for the safety-critical systems. Fig. 2.53 illustrates how Bow-tie analysis is used in the Safety Critical System Management.

   Requirement engineering techniques should be applied to clarify the lifecycle goals of the element and its subsystems and specify the required functionalities, reliability/availability, survivability, and dependency of the SCEs.

   Verification schemes defines how the performance standards lifecycle requirements should be "managed" through the quality control/assurance or risk-based inspection/preventive maintenance routines.

Table 2.8 INCOSE Competency Levels for the System Engineers

Competency Level	Possible Contributory Types of Evidence
Awareness	• Learning and development • Tell me about it (overview)
Supervised practitioner	• Certified education • Tell me about it (can explain and understands why) • Experience of doing (on a training course or as part of a team)
Practitioner	• Experience of doing—relevant and recent (last 5 years) • Objective evidence
Expert	• Experience of doing—relevant and recent (last 5 years) • Objective evidence • Peer references/assessment

4. Brainstorming sessions such as HAZOP, HAZID, JHA, etc. share the following general practices and principles of the scrum agile method:
   - The project team understands, respects, works, and behaves within a defined SE process;
   - The project is executed as fast as possible with minimum down time or staff diversion during the project and the critical path is managed;
   - All key players are physically or electronically collocated, and "notebooks" are considered team property available to all.
   - Baseline management and change control are achieved by formal, oral agreements based on "make a promise—keep a promise" discipline. Participants hold each other accountable.
   - Opportunity exploration and risk reduction are accomplished by expert consultation and rapid model verification coupled with close customer collaboration; software development is done in a rapid development environment while hardware is developed in a multidisciplined model shop; and
   - A culture of constructive confrontation pervades the project organization. The team takes ownership for success; it is never "someone else's responsibility." Fig. 2.54 compares the self-organized configuration of the scrum method and conventional management hierarchy.

Recognition of the system engineering skills is the key to a successful implementation of any management system. Therefore not only the health, safety, environment and quality (HSEQ) and PSM managers but also the other personnel at managerial and operational levels should be trained for system thinking approach.

2.8.8 Competency Levels

The INCOSE competency tables describe the required competencies and explains why they matter. Then they provide the required competencies in

Table 2.9 Project Management and System Engineering Competencies

Project Management Competencies		Common Competencies		System Engineering Competencies	
PM 1.0	Project conceptualization	C 1.0	NASA internal and external environment	SE 1.0	System design
PM 1.1	Project proposal	C 1.1	Agency structure, mission, and internal goals	SE 1.1	Stakeholder expectation definition
PM 1.2	Requirements development and management	C 1.2	NASA procedures and guidelines	SE 1.2	Technical requirements definition
PM 1.3	Acquisition management	C1.3	External relationships	SE 1.3	Logical decomposition
PM 1.4	Project planning	C2.0	Human capital management	SE 1.4	Design solution definition
PM 1.5	Cost-estimating	C2.1	Staffing and performance	SE 2.0	Product realization
PM 1.6	Risk management	C2.2	Team dynamics and management	SE 2.1	Product implementation
PM.1.7	Earned value management	C3.0	Security, safety, and mission assurance	SE 2.2	Product integration
PM 2.0	Resource management	C3.1	Security, safety, and mission assurance	SE 2.3	Product verification
PM 2.1	Budget and full cost management	C3.2	Workplace safety	SE 2.4	Product validation
PM 2.2	Capital management	C 3.3	Safety and mission assurance	SE 2.5	Product transition
PM 3.0	Project closeout	C 4.0	Professional and leadership development	SE 3.0	Technical management
PM 3.1	Systems engineering	C 4.1	Mentoring and coaching	SE 3.1	Technical planning
PM 3.2	Contract management	C 4.2	Communication	SE 3.2	Requirements management
PM 4.0	Project closeout	C 4.3	Leadership	SE 3.3	Interface management
PM 4.1	Stakeholder management	C 4.4	Ethics	SE 3.4	Technical risk management
PM 4.2	Technology transfer and commercialization	C 5.0	Knowledge management	SE 3.5	Configuration management
PM 5.0	Program control and evaluation	C 5.1	Knowledge capture and transfer	SE 3.6	Technical data management
PM 5.1	Tracking/trending of project performance	C 5.2	Knowledge sharing	SE 3.7	Technical assessment
PM 5.2	Project control			SE 3.8	Technical decision analysis
PM 5.3	Project review and evaluation				

terms of the levels of comprehension and experience defined by "Awareness" through to "Expert". Table 2.8 summarizes the INCOSE Competency Levels for the System Engineers.

Awareness: The person is able to understand the key issues and their implications. They are able to ask relevant and constructive questions on the subject. This level is aimed at enterprise roles that interface with systems engineering and therefore require an understanding of the systems engineering role within the enterprise.

Supervised practitioner: The person displays an understanding of the subject but requires guidance and supervision. This level defines those engineers who are "in-training" or are inexperienced in that particular competency.

Practitioner: The person displays detailed knowledge of the subject and is capable of providing guidance and advice to others.

Expert: The person displays extensive and substantial practical experience and applied knowledge of the subject.

Effectiveness indicators of knowledge and experience given in the tables of INCOSE Systems Engineering Competencies Framework (Document No.: INCOSE-TP-2010–003) and are designed to be entry level requirements, i.e., an individual must satisfy all the effective indicators for a particular level to be considered competent at that level. The time-lapse involved since a particular effectiveness indicator was last met should be taken into consideration.

Guide to competency evaluation is found in Annex-A od Systems Engineering Competencies Framework (Document No.: INCOSE-TP-2006–002).

2.8.9 Overlaps Between the Project Management and System Engineering Competencies

There are some overlaps between the project management and system engineering competencies. Following the recommendations of the Rogers Commission after the 1986 *Challenger* accident, NASA created the academy to develop an agency-wide professional development program for project management. This academy has developed a "project management and systems engineering competency framework" which offers a detailed description of the skills, behaviors, actions, and experiences that demonstrate proficiency in each competency at four career levels, ranging from project team members to managers of programs or very large projects. Table 2.9 summarizes the framework competency areas which includes five project management competency areas, three systems engineering competency areas, and five competency areas common to both disciplines.

References

1. von Bertalanffy, L. (1901–72), Wikipedia: https://en.wikipedia.org/wiki/Ludwig_von_Bertalanffy.
2. Wiener, N. (1894–1964), Wikipedia: https://en.wikipedia.org/wiki/Norbert_Wiener.
3. Ashby, R. (1903–72), Wikipedia: https://en.wikipedia.org/wiki/W._Ross_Ashby.

4. Holland, J.H. Wikipedia: https://en.wikipedia.org/wiki/John_Henry_Holland.
5. Gell-Mann, M. Wikipedia: https://en.wikipedia.org/wiki/Murray_Gell-Mann.
6. Kahneman, D. Wikipedia: https://en.wikipedia.org/wiki/Daniel_Kahneman.
7. Tversky, A. Wikipedia: https://en.wikipedia.org/wiki/Amos_Tversky.
8. Prospect theory, Wikipedia: https://en.wikipedia.org/wiki/Prospect_theory.
9. Laszlo, E., 1972. Wikipedia: https://en.wikipedia.org/wiki/Ervin_L%C3%A1szl%C3%B3.
10. System Philosophy, Wikipedia: https://en.wikipedia.org/wiki/Systems_philosophy.
11. Forrester, J.W. Wikipedia: https://en.wikipedia.org/wiki/Jay_Wright_Forrester.
12. Mental Model, Wikipedia: https://en.wikipedia.org/wiki/Mental_model.
13. Fabrycky, W. Wikipedia: https://en.wikipedia.org/wiki/Wolt_Fabrycky.
14. Blanchard, B.S. Wikipedia: https://en.wikipedia.org/wiki/Benjamin_S._Blanchard.
15. B.S. Blanchard, W.J. Fabrycky, Systems Engineering and Analysis, Vol. 4, Prentice Hall, Englewood Cliffs, NJ, 1990.
16. W.J. Fabrycky, B.S. Blanchard, Life-Cycle Cost and Economic Analysis, Prentice Hall, Englewood Cliffs, NJ, 1991.
17. Checkland, P. Wikipedia: https://en.wikipedia.org/wiki/Peter_Checkland.
18. Soft Systems Methodology, Wikipedia: https://en.wikipedia.org/wiki/Soft_systems_methodology.
19. Blog: https://problemlosning.wordpress.com/tag/checkland-1981/.
20. P.B. Checkland, Systems Thinking, Systems Practice, John Wiley & Sons Ltd., 1981, 1998. ISBN 0-471-98606-2
21. Churchman, C.W. Wikipedia: https://en.wikipedia.org/wiki/C._West_Churchman.
22. Churchman, C.W. Champion of the Systems Approach1 Nicholas Berente University of Georgia (https://papers.ssrn.com/sol3/papers.cfm?abstract_id=2411176).
23. C.W. Churchman, The Design of Inquiring Systems, Basic Concepts of Systems and Organizations, Basic Books., New York, 1971.
24. Bloom, B. Wikipedia: https://en.wikipedia.org/wiki/Benjamin_Bloom.
25. Bloom's 2 Sigma Problem, Wikipedia: https://en.wikipedia.org/wiki/Bloom%27s_2_Sigma_Problem.
26. Ryan, M.J., 2013. An Improved Taxonomy for Major Needs and Requirements Artefacts. INCOSE International Symposium IS2013.
27. Sandeep Verma, Understanding Integrity Requirements for Government Contracts in India: A Quick Roundup of Select Legal and Regulatory Developments, Department of Science & Technology, 2013, 22. September
28. J. Bobinis, T. Herald, An Enterprise Framework for Operationally Effective System of Systems Design, Journal of Enterprise Architecture 8 (2) (2012). Available at: https://www.mendling.com/publications/JEA12-2.pdf
29. Friedenthal, S., Moore, A., Steiner, R., 2014. A Practical Guide to SysML, 3rd Ed. eBook ISBN: 9780128008003, Paperback ISBN: 9780128002025 (https://www.elsevier.com/books/a-practical-guide-to-sysml/friedenthal/978-0-12-800202-5).
30. W. Ulrich, Critical Heuristics of Social Planning: A New Approach to Practical Philosophy. Bern: Haupt. Paperback reprint edition, Wiley, Chichester, 1983, 1994. (Note: the 1994 edition is still in print)
31. A Brief Introduction to Critical Systems Heuristics (CSH) Werner Ulrich, University of Fribourg, Switzerland, and The Open University, United Kingdom http://projects.kmi.open.ac.uk/ecosensus/publications/ulrich_csh_intro.pdf.

Literature

1. Wastell, D., 2012 March. Systems Thinking: an introductory Essay
2. From Project Manager's Guide to Systems Engineering Measurement for Project Success
3. Systems Engineering Fundamentals
4. Systems Thinking for Safety: Ten Principles A White Paper Moving towards Safety-II
5. Critical Thinking and Systems Thinking: Towards A Critical Literacy For Systems Thinking in Practice
6. Concept of Operations (ConOps) of A Systems Engineering Education Community (SEEC)
7. Project Management and Systems, Engineering Competency Framework
8. The Taxonomy of Systems Engineering Competency for the New Millennium
9. Systems Engineering at the Undergraduate Level: Boom or Bust?
10. Assessing the Competencies Of Systems Engineers
11. INCOSE Guide to Competency Evaluation, Document No.: INCOSE-TP-2006-002 , Version/Revision: 2, January 2010
12. Ackoff, R.L., A Systemic View Of Transformational Leadership
13. Ackoff, R., From Data to Wisdom
14. Ackoff, R., A Systemicviewof transformationalleadership
15. A Note on Systems Science
16. Pyster, A., Olwell, D.H. (eds). 2013. "Principles of Systems Thinking." The Guide to the Systems
17. Checkland, P. B. and Scholes, Soft Systems Methodology in Action, Wiley & Sons: Chichester, 1990.
18. Guide for Writing Requirements, Document No.: INCOSE-TP-2010-006-02, Version/Revision: 2, July 1, 2015
19. Systems Engineering Leading Indicators Guide
20. Roedler, G.J., Martin, L., Jones, C., December 27, 2005. Technical Measurement, A Collaborative Project of PSM, INCOSE, and Industry. US Army, Version 1.0, INCOSE-TP-2003-020-01
21. Defining Explanation in Probabilistic Systems
22. Explanation in Probabilistic Systems: Is It Feasible? Will It Work?
23. Multilevel System Analysis - An Introduction to Systems Thinking
24. System of Systems Management: A Network Management Approach
25. Systems Engineering for DUMmIES
26. Critical Thinking and Systems Thinking: Towards A Critical Literacy for Systems Thinking in Practice
27. Establishing Profiles for Systems Engineering Standards: A Great Help for Companies to Manage Their Processes
28. Project Management and Systems Engineering Handbook, NASA MSFC-HDBK-3173, Revision B Effective Date: October 16, 2012
29. NASA Systems Engineering Handbook
30. Notions de Système et D'ingénierie de Système, Alain FAISANDIER, 2014, ISBN: 979-10-91699-04-4
31. NASA Systems Engineering Handbook, NASA/SP-2007-6105, Rev1

32. Michael Hanratty, J., Lightsey, R.H., Larson, A.G., Open Systems and the Systems Engineering Process
33. Systems Engineering Guide for Systems of Systems, August 2008
34. Systems Engineering Standards: A Summary
35. Systems Engineering for Software Intensive Projects Using Agile Methods
36. Establishing Profiles for Systems Engineering Standards: A Great Help for Companies to Manage Their Processes
37. Seven Systems Engineering Myths and the Corresponding Realities
38. The Relationship Between Systems Engineering and Innovation
39. Service Engineering Life-cycles
40. Lean Enablers for Systems Engineering
41. Quantifying the Benefit of Introducing Systems Engineering Processes – Myth or Reality?
42. A Systemic Failure Modes: A Model for Perrow's Normal Accidents in Complex, Safety Critical Systems
43. An Overview of the Soft Systems Methodology
44. Soft Systems Methodology
45. Systems Thinking: An Introductory Essay
46. A Summary of the Principles of Hierarchy Theory

Blog

1. Systems Thinking World
2. Visualizing Complex Systems Science (CSS)
3. Heart & Mind Program
4. A Summary of the Principles of Hierarchy Theory
5. Analytic vs. Systemic Approaches

Handbook

1. INCOSE Handbook-2002
2. INCOSE Handbook-2006
3. INCOSE Handbook-2012
4. INCOSE Handbook-2015
5. Guide to the Systems Engineering Body of Knowledge (SEBoK)
   - SEBoK v. 1.7
   - SEBoK v. 1.6
   - SEBoK v. 1.5.1
   - SEBoK v. 1.5.
   - SEBoK v. 1.4
   - SEBoK v. 1.3.1
   - SEBoK v. 1.3
   - SEBoK v. 1.2
   - SEBoK v. 1.1
   - SEBoK v. 1.0

Standards

1. ISO/TR 12489:2013: Petroleum, Petrochemical and Natural Gas Industries—Reliability Modelling and Calculation of Safety Systems
2. EIA 632 & MAP Système: Processes for Engineering a System
3. ISO/IEC 15228: Systems and Software Engineering – System Life Cycle Processes

Guidance

1. Guidelines for the Management of Safety Critical Elements, EI Publication, ISBN: 9780852934623, 2nd edition, April 2007
2. Guidelines for the Identification and Management of Environmentally Critical Elements, EI Publications, ISBN: 978-0-85293-632-0, 1st edition, October 2012

CHAPTER 3

Fundamentals of the Complexity

3.1 WHAT IS COMPLEXITY?

According to the INCOSE Complex Systems Working Group White Paper entitled "A Complexity Primer for Systems Engineers", there is no absolute definition of what complexity means; the only consensus among researchers is that there is no agreement about the specific definition of complexity. However, a characterization of what is complex is possible.

Complexity is generally used to characterize something with many parts where those parts interact with each other in multiple ways. The study of these complex linkages at various scales is the main goal of complex systems theory.

Complexity describes the behavior of a *system* or *model* whose components interact in multiple ways and follow local rules, meaning there is no reasonable higher instruction to define the various possible interactions.

The stem of the word *complexity*, i.e., *complex*, is composed of the Latin words *com* (meaning: "together") and *plex* (meaning: woven). This is best contrasted with *complicated* where *plic* (meaning: folded) refers to many layers. A *complex* system is thereby characterized by its interdependencies, whereas a *complicated* system is characterized by its layers.

Neil Johnson states that "even among scientists, there is no unique definition of complexity—and the scientific notion has traditionally been conveyed using particular examples..." Ultimately he adopts the definition of "complexity science" as "*the study of the phenomena which emerge from a collection of interacting objects.*"

In ordinary language, we often call something complex when we can't fully understand its structure or behavior: it is uncertain, unpredictable, complicated, or just plain difficult. Sillitto (2009) described the inability of a

A Systems Approach to Managing the Complexities of Process Industries.
DOI: http://dx.doi.org/10.1016/B978-0-12-804213-7.00003-7

human mind to grasp the whole of a complex problem and predict the outcome as *subjective complexity*.

Sillitto's *objective complexity* describes technical or system characteristics that lead to the subjective complexity or difficulty. As systems engineers, we have the ability to modify these characteristics; they are also the ones most frequently addressed by complex systems science.

The standard systems engineering process breaks down a problem into parts, recursively, until the parts are simple enough that we understand them and can design solutions; we then re-assemble the parts to form the whole solution. The approach works well for systems whose parts interact in fixed ways (also known as "complicated" systems—an example might be a car), even when there are many interacting parts and the systems may have unpredictable behavior.

For example, this hypothetical "system" shown in Fig. 3.1 is composed of 12 items (of some kind) that interact with each other. By dividing these items into four smaller assemblages, the total number of interactions has been reduced from 66 to 18, and the system is now much less complex. That being said, it is still very complex.

Other systems, however, present significant problems when analyzed in piecewise fashion. Systems such as transportation networks have autonomous parts whose interactions lead to emergent *self-organized patterns* of behavior. In these systems, defined here as "complex" systems, the emergent properties that we really care about are not understandable from the perspective of the parts in isolation. It is especially for these systems that we are providing guidance to additional tools and techniques created in complex systems sciences and recommend their use in systems engineering of complex systems.

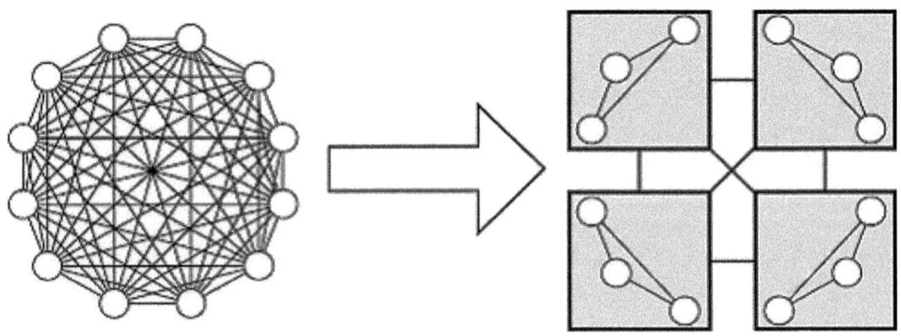

FIGURE 3.1

Reducing complexity. *From http://testerstories.com/2011/11/the-ubiquity-of-ubiquitousness/*

From NASA: The Illusion of Complexity: Neuroactivity And Complex Behaviour

Nature gives us a lot of examples of complex behavior emerging from the interaction of many individuals. The members of an ant colony are capable of a wide variety of complex behaviors: finding and transporting food, constructing elaborate underground complexes of tunnels and chambers, and defending their territory from invaders. Such activity would seem to involve a great deal of planning, memory, and coordination. But there is no head ant that draws up a blueprint for the colony and gives directions to the worker ants. Instead, each ant follows a very simple set of rules, based on cues from its environment and from the activity of ants nearby. For example, how do ants know to walk in a single file to pick up food, and then return in single file to the nest? They don't. They know three simple following rules which is illustrated in Fig. 3.2:

- First, if you come upon something that smells like food—pick it up!
- Second, when you pick up a piece of food, release a chemical signal. . .a pheromone.
- Third, if you come across a pheromone trail left by another ant. . .follow it!

Let's see what happens when a group advanced follow these three simple rules. The first ant wanders about in its environment and picks up a piece of food. She releases a chemical signal and begins to leave a trail. As the ant continues to wander, the other ants stumble upon her trail. They follow this trail and are led to the food. Some of them may follow the trail the wrong way but that's okay. Once the first ant reaches the nest, she drops off the

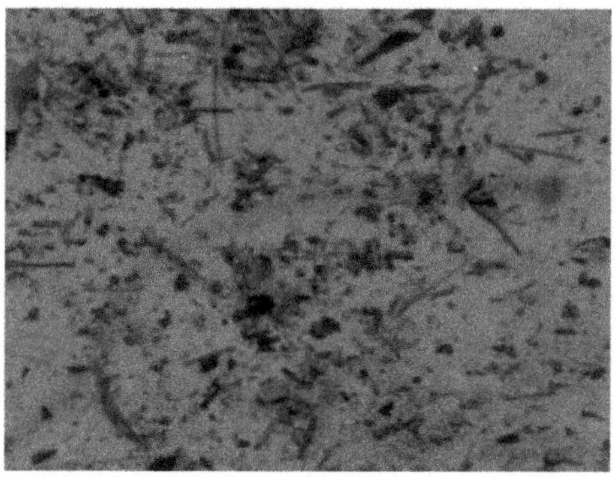

FIGURE 3.2

Illusion of the complex ant food picking process. *From NASA:-The Illusion Of Complexity - Neuroactivity And Complex Behaviour.*

morsel of food and when she exits the nest again, she will come upon her own pheromone trail and follow it.

The ant doesn't remember where the food is. She doesn't even know that she's following her own trail. She is just following rule number 3; "always follow the trail." Meanwhile, more and more ants have come upon the original trail and followed it to the food. As they follow the first dance path back, they strengthen and refine the chemical trail. Soon the ants will appear to march in single file to the food and back.

There is no plan, no memory, no need for a leader to point the way. Three simple rules and a few environmental clues are all that is needed to orchestrate this remarkably efficient process.

The overall result is that when one ant finds a good (i.e., short) path from the colony to a food source, other ants are more likely to follow that path, and positive feedback eventually leads to all the ants following a single path.

The ant colony optimization algorithm (ACO) is a probabilistic technique for solving computational problems mimic the ant behavior to find the shortest path.

Another example of the natural self-organized system is illustrated in Fig. 3.3. A fish swarm is a multiagent, self-organizing dynamic structure, where the control is distributed and each agent has limited information. In general, empirical observations indicate effective group coordination, with an absence of individual leaders ("alpha" fish). The collective behavior is often altered in the presence of a predator.

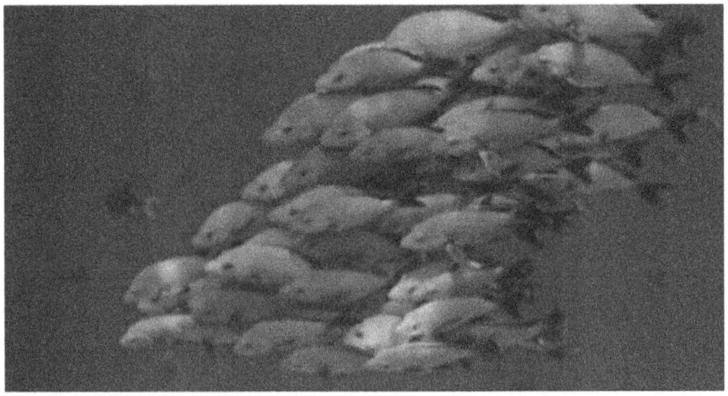

FIGURE 3.3
Self-organized fish swarm systems.

FIGURE 3.4

Self-organisation in the crystallisation process. *From https://en.wikipedia.org/wiki/Self-organization.*

Fig. 3.4 illustrates the self-organization in micron-sized Nb3O7(OH) cubes during a hydrothermal treatment at 200°C. Initially, amorphous cubes gradually transform into ordered 3D meshes of crystalline nanowires as summarized in the model later. Human examples of self-organized behavior can be observed in economy, psychology, social sciences, and brain theory, to mention a few disciplines. To give further examples, population growth, the spread of diseases, the formation of public opinion, and the formation of clouds can be modeled as a self-organizing process.

All these phenomena share the fact that the participating entities establish a structure or function without requiring central coordination. Instead, entities interact directly with each other and continuously react to changes in their environment. Such systems are often flexible, adaptive, and scalable. Although individual entities are often unreliable, the overall system shows a high level of resilience.

When work people should deal with nonrepeatable types of production and services, or find using the defined processes unmanageable, or there are the uncertain requirements combined with the risk of unpredictable technology implementation the plan-driven approaches such as those described by the PMBOK Guide which consider the underlying mechanisms are well-understood are inefficient. In these cases, the work depends on knowledge creation and collaboration. The project managers should create the sufficient commitment to grow a self-organizing team.

When requirements are uncertain and technology risks are high, adding too many people to the situation makes things worse because as Fig. 3.4 shows the communication pathways increase as a square of team size. Grouping people by specialty also makes things worse. Grouping people by architectural components (a.k.a. component teams) makes things worse eventually.

Scrum addresses uncertain requirements and technology risks by grouping people from multiple disciplines into one team (ideally in one team room) to maximize communication bandwidth, visibility, and trust. Fig. 3.5 compares the traditional hierarchical versus agile self-organized teams.

The most successful approach to this problem has been the creation of fully cross-functional "feature teams," able to operate at all layers of the architecture in order to deliver customer-centric features. In a large system this requires learning new skills.

As teams focus on learning—rather than short-term microefficiencies—they can help create a learning organization.

Scrum is the leading agile development methodology used by Fortune 500 companies around the world. It is an example of design and implementation of the self-organized systems. Scrum is an agile framework for completing complex projects.

Scrum originally was formalized for software development projects, but it works well for any complex, innovative scope of work.

The brainstorming HAZID, HAZOP, SIL, JHA, etc. are also the scrum-type self-organized verification and learning activities.

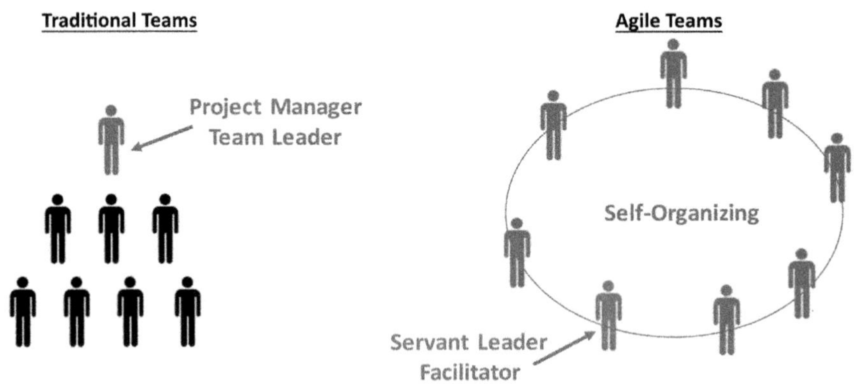

FIGURE 3.5
Traditional versus agile teams.

When Jeff Sutherland created the scrum process in 1993, he borrowed the term "scrum" from an analogy put forth in a 1986 study by Takeuchi and Nonaka, published in the Harvard Business Review. In that study, Takeuchi and Nonaka compare high-performing, cross-functional teams to the scrum formation used by Rugby teams.

3.1.1 Sources and Factors of Complexity

There are generally rules which can be invoked to explain the origin of complexity in a given system.

The source of disorganized complexity is the large number of parts in the system of interest, and the lack of correlation between elements in the system.

In the case of self-organizing living systems, usefully organized complexity comes from beneficially mutated organisms being selected to survive by their environment for their differential reproductive ability or at least success over inanimate matter or less organized complex organisms. See, e.g., Robert Ulanowicz's treatment of ecosystems.[8]

Complexity of an object or system is a relative property. For instance, for many functions (problems), such a computational complexity as time of computation is smaller when multitape Turing machines are used than when Turing machines with one tape are used. Random access machines allow one to even more decrease time complexity (Greenlaw and Hoover 1998, p. 226), while inductive Turing machines can decrease even the complexity class of a function, language, or set (Burgin, 2005). This shows that tools of activity can be an important factor of complexity.

3.1.2 Disorganized Complexity Versus Organized complexity

One of the problems in addressing complexity issues has been formalizing the intuitive conceptual distinction between the large number of variances in relationships extant in random collections, and the sometimes large, but smaller, number of relationships between elements in systems where constraints (related to correlation of otherwise independent elements) simultaneously reduce the variations from element independence and create distinguishable regimes of more uniform, or correlated, relationships, or interactions.

Weaver perceived and addressed this problem, in at least a preliminary way, in drawing a distinction between "disorganized complexity" and "organized complexity."

In Weaver's view, disorganized complexity results from the particular system having a very large number of parts, say millions of parts, or many more.

Though the interactions of the parts in a "disorganized complexity" situation can be seen as largely random, the properties of the system as a whole can be understood by using probability and statistical methods.

A prime example of disorganized complexity is a gas in a container, with the gas molecules as the parts. Some would suggest that a system of disorganized complexity may be compared with the (relative) simplicity of planetary orbits—the latter can be predicted by applying Newton's laws of motion. Of course, most real-world systems, including planetary orbits, eventually become theoretically unpredictable even using Newtonian dynamics, as discovered by modern chaos theory.

Organized complexity, in Weaver's view, resides in nothing else than the nonrandom, or correlated, interaction between the parts. These correlated relationships create a differentiated structure that can, as a system, interact with other systems. The coordinated system manifests properties not carried or dictated by individual parts. The organized aspect of this form of complexity vis-à-vis to other systems than the subject system can be said to "emerge," without any "guiding hand."

The number of parts does not have to be very large for a particular system to have emergent properties. A system of organized complexity may be understood in its properties (behavior among the properties) through modeling and simulation, particularly modeling and simulation with computers. An example of organized complexity is a city neighborhood as a living mechanism, with the neighborhood people among the system's parts.

3.1.3 Complexity Topics

Complex behavior: The behavior of a complex system is often said to be due to emergence and self-organization. Chaos theory has investigated the sensitivity of systems to variations in initial conditions as one cause of complex behavior.

Complex mechanisms: Recent developments around artificial life, evolutionary computation, and genetic algorithms have led to an increasing emphasis on complexity and complex adaptive systems.

Complex simulations: In social science, the study of the emergence of macroproperties from the microproperties is also known as macro-micro view in sociology. The topic is commonly recognized as social complexity that is often related to the use of computer simulation in social science, i.e., computational sociology.

Complex systems: Systems theory has long been concerned with the study of complex systems (in recent times, complexity theory and complex systems have also been used as names of the field). These systems are present in the research of a variety of disciplines, including biology, economics, and technology. Recently, complexity has become a natural domain of interest of real-world sociocognitive systems and emerging systemics research. Complex systems tend to be high-dimensional, nonlinear, and difficult to model. In specific circumstances, they may exhibit low-dimensional behavior.

Complexity in data: In information theory, algorithmic information theory is concerned with the complexity of strings of data.

3.1.4 Fields of Complexity in Enterprises

Complexity appears in the following fields:

- *Product portfolio*: Number and type of products, features, and services sold to the customer base (e.g., SKUs)
- *Markets/segments*: Number and type of market segments in the respective countries/regions served
- *Customer portfolio*: Number and variety of all customers served (including customer accounts and ship-to customers)
- *Material/components*: Variety of raw materials, components, and services used in value generation
- *IT systems*: Variety and type of systems and applications
- *Organization*: Company organization, structures, and governance
- *Processes*: Number and type of processes in the enterprise
- *Production/supply chain*: Production plants and network, manufacturing processes, asset base, and distribution
- *Technologies*: R&D platforms and project portfolio and product technologies involved

Complexity in enterprises is driven by:

- *Market volatility*: Changing market conditions like raw material supply and sales volumes drive business process complexity
- *Fragmented customer* demands drive product portfolio and feature complexity
- *Globalization* drives complexity of served markets and company locations
- *Mergers and acquisitions* drive complexity in all fields
- *Silo-oriented cultures* drive complexity in organization, IT systems, and business processes
- *Increasing customer pressure* drives complexity in product portfolio and features

3.2 CHARACTERISTICS OF COMPLEXITY

Complexity is not limited to the interaction of the physical parts of an asset. The interaction of people, organizations, and the environment which surround the physical system creates the most important complexity features.

Complexity results from not only the numerous parts but also the:

1. diversity,
2. connectivity,
3. interactivity, and
4. adaptivity of a system and its environment.

Constant change makes it difficult to define stable goals for a project or system. Technical systems that worked well in the past to solve an environmental problem become obsolete quickly. Intricate networks of evolving cause-and-effect relationships lead to subtle bugs and *surprising dynamics*. Unintended consequences can overwhelm or even negate the intended consequences of actions.

When systems are complex, their structures cannot be described at a single level or with a single view; multiscale descriptions are needed to understand complex systems. Their *emergent behavior*, derived from the relationships among their elements and with the environment, via internal and external feedback loops, gives rise to observed patterns that may not be understood or predicted. Describing the behavior of a system as a response function may require an unobtainable amount of information. It is often impossible to predict future configurations, structures, or behaviors of a complex system, given finite resources.

A complex system may have *multiple stable states* (meaning each state is actually metastable), transient states, or even no lasting stable states, exhibiting continuous evolution. Perturbations in the system may result in recovery to the former state but may also lead to transitions to another state and consequent radical changes of properties. In addition, details seen at the fine scales can influence large-scale behavior. Dynamics of different parts and patterns cannot be reproduced using simple averages.

Complex systems are perpetually *generating novelty*, many key variables are opaque, boundaries are indeterminate, and weak ties can have a disproportional effect on system behavior. Duality is common: tension between large and small, distributed and central, agile and planned calls for perpetual seeking of balance. In short, complex systems are different.

Complex adaptive systems are extremely hard to predict in great detail. This means their future is fundamentally *uncertain*. In the next activities of this

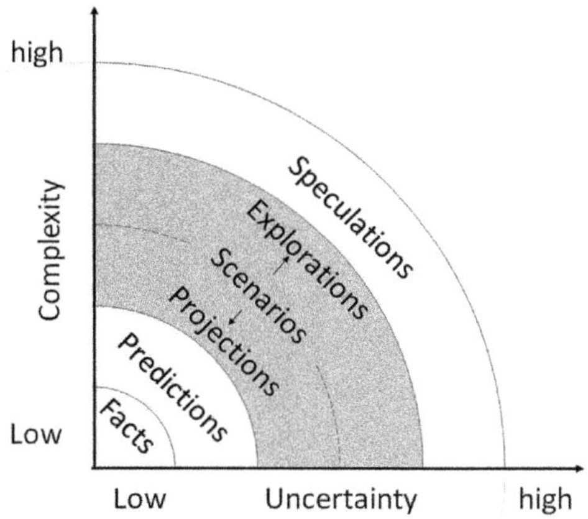

FIGURE 3.6
Complexity versus uncertainty of a context.

week we will introduce the notion of uncertainty, how it differs from risk, and how this relates to complexity. (Fig. 3.6) relates the complexity versus uncertainty of a context.

Although complexity can present challenges, *complexity is often inherent and may even be a necessary or desirable attribute of a solution system.* Systems that have been engineered to rule out any but deterministic behaviors are necessarily limited by the prescribed behaviors, and do not extend well into unplanned environments. In contrast, complex systems can be engineered to have sufficient adaptability to operate well in a changing environment, responding to change in appropriate and effective ways. A complex system provides a welcome kind of variety that can help provide control of different dimensions and enables the system to adapt to environmental change. To provide new capabilities or graceful degradation, a complex system can adapt by reorganizing its structure, responses, or patterns of parts.

3.3 IDENTIFYING THE RIGHT LEVEL OF COMPLEXITY

Science and engineering help humans to make successful systems because they provide ways to understand, predict, and control technology, in order to create desirable effects on the world. All science involves abstraction of the complexity of the world into approaches and models that use *simplifying assumptions.* This allows us to generalize from one complex situation to another. The best engineering methods take advantage of the simplicity in

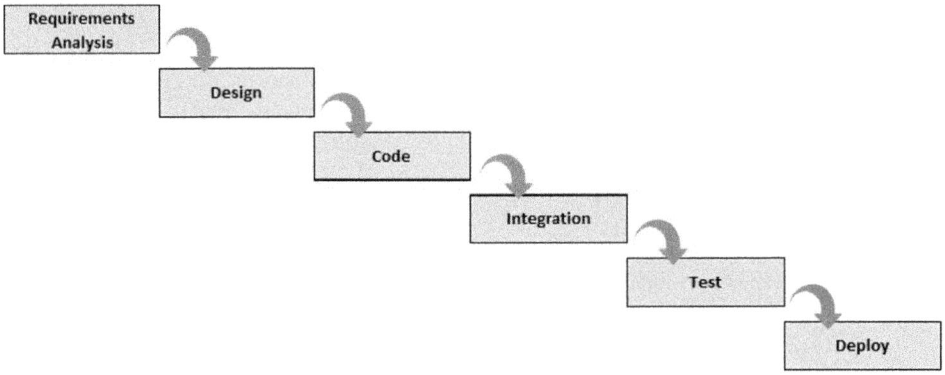

FIGURE 3.7
"Waterfall" development.

the models without diverging so far from reality that behavior can no longer be predicted and controlled.

As a system's diversity, connectivity, interactivity, or adaptivity increases, the risk associated with using simpler methods and simplifying assumptions also increases, and more advanced techniques may be needed. Tools and techniques apply differently to systems on a spectrum of increasing complexity.

At the less complex end, the waterfall model for top-down sequential design applies well. The development sequence shown in Fig. 3.7 is applied in most of the industrial projects. However, the effective implementation of this approach becomes more difficult for the mega projects.

At the more complex end, tools such as agent-based models for model-based systems engineering can be used to understand and address complex, dynamic systems design challenges. (Fig. 3.8) provides an example of the agent-based models.

Techniques at the lower end of the spectrum tend to be easier to learn, and simpler and faster to apply, because they make simplifying assumptions that ignore some of the complexity. The practitioner must apply judgment to utilize a mixture of tools along this spectrum that satisfies Einstein's razor: *make things as simple as possible, but not simpler.*

It is rarely possible to fully assess in advance what complexity within a project must be addressed and what can be assumed away. Because complex systems perpetually generate novelty, systems engineers will often have to adapt their approach to unfolding conditions and use flexible tools. Systems

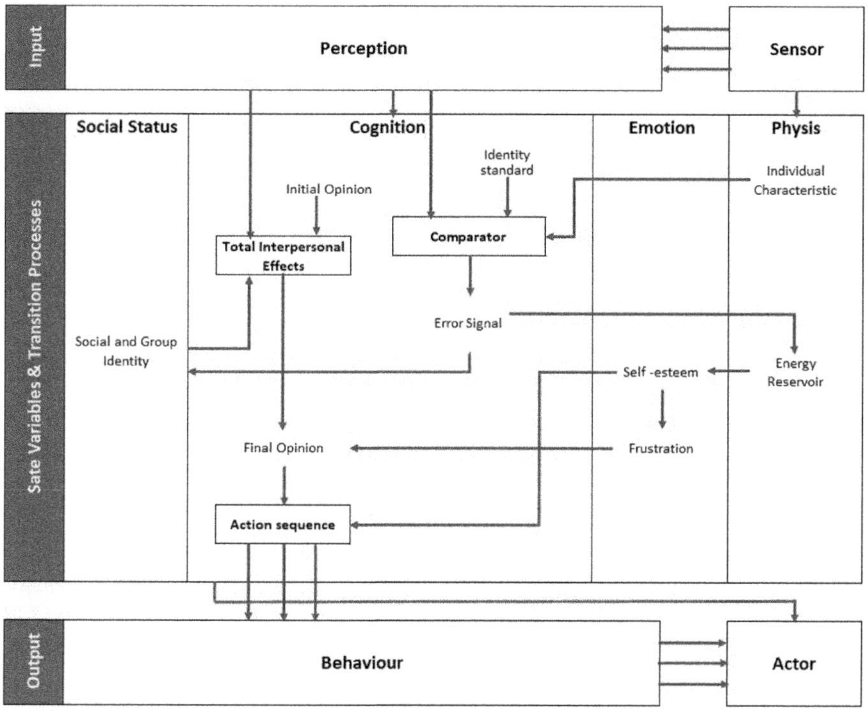

FIGURE 3.8

Example of the agent-based models. Note: Physis is a Greek theological, philosophical, and scientific term usually translated into English as "nature".

engineers should also recognize that the complexity of a system, as manifested in many diverse types of parts and relationships, provides a welcome kind of variety that can help facilitate control of different dimensions and enables the system to adapt to environmental change.

3.4 CYNEFIN COMPLEXITY FRAMEWORK

Scholar David J. Snowden used the word "Cynefin" to describe a framework he developed in 1999, to categorize the complexity level of a given "context." Cynefin, pronounced kuh-nev-in, is a Welsh word that signifies the multiple factors in our environment and our experience that influence us in ways we can never understand.

So, the framework not only tells us how to approach a set of different situations, but also the characteristics also explain enough to help us recognize

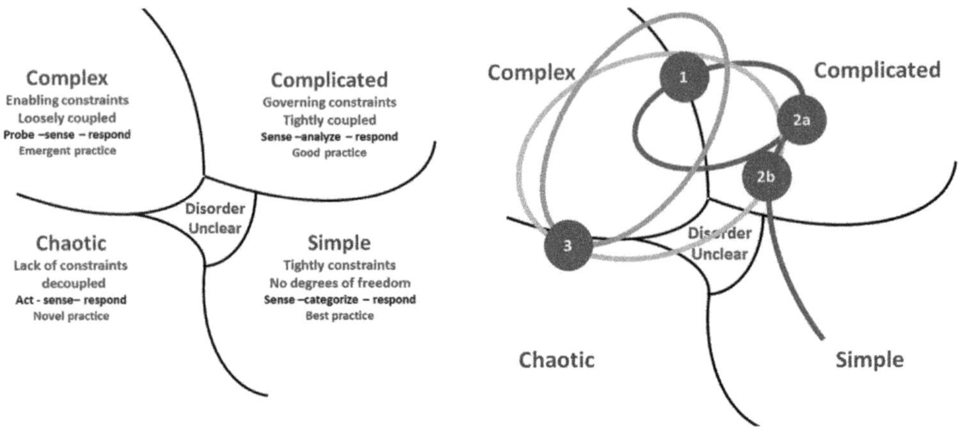

FIGURE 3.9
Cynefin framework and trigger points.

the situation in which we currently reside. You can have great solutions, but if they are applied in the incorrect context, they will be worthless or worse, harmful.

The Cynefin framework (Fig. 3.9) categorize the contexts into five domains of "obvious, complicated, complex, chaotic, and disorder" defined by cause-and-effect relationships. Each of these domains has a specific decision-making approach that helps to make better sense of the situation, and choose the most appropriate way forward.

Fig. 3.9b illustrates the main dynamic movement between domains and domains themselves as follows:

- *Blue*: The natural cadence between complex and complicated in which you experiment with the tightening of constraints. If that produces repeatable results, i.e., a consistent relation between cause and effect, then you continue. You shift from exploration in the complex to exploiting opportunities in the complicated. Then if you don't get repetition of results, you shift back to the *safe-to-fail* experimental mode that characterizes complexity. That dynamic has a natural cadence linked to organization type, market conditions, etc.
- *Green*: Where you have long-term stability and little dissent then (and only then) do you shift into the best practice domain of the obvious. Getting back from that domain to any other is very difficult as process conservative will always creep in. Inertia is the main enemy of innovation in most organizations so you need to be very very careful before making this shift.

- *Yellow*: It is the shallow dive into chaos, the need to break up existing constraints to allow something novel to emerge. If you don't keep up the natural cadence of the blue dynamic, then this becomes a means of recover. If the market suddenly shifts, then it may be forced on you. Better to take a small hit now than a catastrophic failure, but attention needs to be paid to make sure it is a shallow dive, not a plummeting shift into the abyss.
- *Purple*: It is the *grazing dynamic*. Here the position never stabilizes enough for full exploitation. This is the domain of the perpetual beta (or even the perpetual alpha) in software development but it is also a strategic decision. To maintain this dynamic needs a strong focus on granularity and linkages to allow rapid combination and recombination. The dynamic is also at the heart of how you scale success in a complex system so expect a few posts on that in the near future. It is a dynamic that can be forced on you, but also one that you can choose to adopt to gain strategic advantage but the architecture of systems and organizational structure needs to be radically different from the dominant paradigm.

As you shift from the complex to complicated you have to decide if the constraint shift from *enabling to governing* is really sustainable. If it takes a lot of effort or compromise, then you may be in a grazing dynamic rather than a natural cadence. Equally you may not need to keep spinning around the grazing dynamic you may be able to stabilize the whole, or some aspect and get sustainable exploitation. So in software where SCRUM really focuses on complex to complicated transitions in software code you might realize that the number of iterations is getting to the point where you really need to start again or open up the architecture. In product marketing you may realize that the overall utility of your product is getting difficult to sustain and you are getting leakage of formerly loyal customers. If that is an ordinary change, then you just shift to complex, but it may indicate an overall shift in the market that will force you into grazing.

When you are at full exploitation you reach a key point at which you have to decide if you can shift to best practice or not. There may still be dissent, not everyone buys in but enforcement can be justified. For example, if we have all agreed to drive on the left not the right, then there comes a point (and its pretty soon) where driving on the right is no longer a lovable eccentricity but something dangerous to life and limb. I've shown this as two points A and B. The reason for that is once you make the choice to shift to best practice you need to closely monitor for gaming or work arounds, not to punish them but as an indication that you may have made the wrong call. So you need a double take on this one.

Finally if we have done a radical reset because the old stable patterns have started to fail, and we have started to pull ourselves out again then there is a key decision. Was the radical reset caused by our own complacency? If so we go back to the natural cadence, if it was caused by a radical shift in the external world then we need to move to grazing for a period to ensure we are resilient. Grazing is all about being inefficient enough to be rapidly adaptive or agile.

It can be extremely difficult to identify when you're in a "disorder" situation of the Cynefin framework. In disorder domain, it isn't clear which of the other four domains is dominant, and people generally rely on decision-making techniques that are known and comfortable. The primary goal in this situation is to gather more information, so that you can move into a known domain and then take the appropriate action.

Effective information gathering is the most basic perspective-widening tool an effective leader requires. Good quality information marks out the context in which the leader operates, creates the information patterns from which ideas emerge, and provides the criteria by which ideas are screened and assessed. Effective leaders gather two main types of information:

- Background data
- Task-related data

Leaders gather background data to build their view of the world in which they operate. This information is made up of the countless facts, trends, and opinions that they encounter and the observations they make on a daily basis. The higher the quality of background data they gather and the more effectively they prioritize it, the more accurate their view of the world will be, and the better their judgment and "common sense."

By contrast with the steady, slow gathering of background data, task-related information is gathered for a specific purpose.

Cynefin has parallels with the strategy pyramid. In this the strategy pyramid has been extended from its more common trichotomy of operational, tactical, and strategic decisions by including a fourth category of instinctive or recognition-primed decisions. Strategic decisions set a broad direction, a framework in which more detailed tactical and operational decisions may be taken.

In delivering operational decisions, many much smaller decisions have to be taken. These are the instinctive, recognition-primed ones. Simon (1960) noted that strategic decisions tend to be associated with unstructured, unfamiliar problems. Indeed, strategic decisions often have to be taken in the face of such a myriad of ill-perceived issues, uncertainties, and ill-defined objectives that Ackoff (1974) dubbed such situations *messes*.

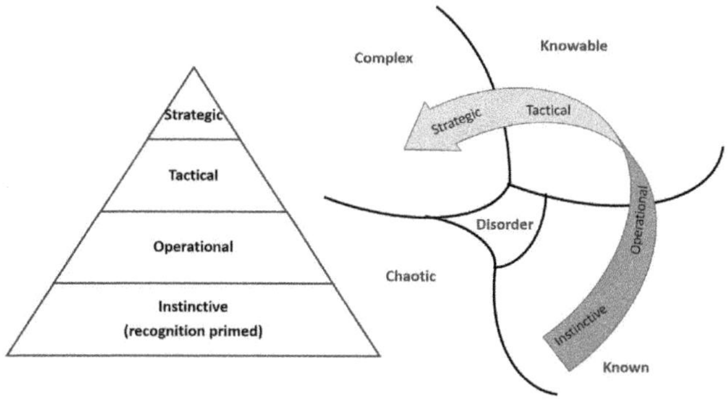

FIGURE 3.10
Relationship between the perspectives offered by the strategy pyramid and Cynefin.

There is a clear alignment of the context of strategic decision-making and the complex and even chaotic spaces of Cynefin. Tactical, operational, and instinctive decision contexts have increasing familiarity and structure, and occur with increasing frequency. Again the alignment with Cynefin is clear. Jacques (1989) distinguished four domains of activity, and hence decision-making, within organizations: the *corporate strategic, general, operational,* and *hands-on work*. French et al. (2009) relate these directly to the strategic, tactical, operational, and instinctive categories in the extended strategy pyramid, and hence they also relate to Cynefin as in the curved arrow in Fig. 3.10.

Therefore, at early stage of any PSM projects, it is necessary to explore context much more than sometimes we do if we are to adopt an appropriate form of statistical or decision analysis. Cynefin seems to offer a very catalytic and supportive framework for achieving this. It encourages us to seek an appropriate balance between qualitative sense-making and quantitative analysis. I hope that this paper stimulates further discussion on this.

Table 3.1 summarizes the Cynefin framework for decision-making context characteristics and the leader's job to manage them.

Like quality and safety, the complexity management is the "business of all." Because every day, no matter what our roles and responsibilities are we encounter situations that require proper decision-making and choices.

The authors of this book aim to attract the attention of top managers and business system designers to the fact that practical complexity assessment and management methods and tools are available and can be embedded in the existing procedures and job descriptions to assure a systematic context complexity consideration by all personnel, contractors, and third parties.

Table 3.1 The Cynefin Framework for Decision-Making

	Nature	Response	Prepare
Chaos	Unknowable unknowns	Act-sense-respond	Establish heuristics and parables
	Temporary state—no time	Speed of authoritative response vital	Human sensor networks
	No evidence of any constraint	Follow and enforce heuristics	Crews as crisis management teams
	High turbulence no patterns	Focus on constraints not solution	Simulation games for key managers
	Old certainties no longer apply	Use the opportunity to innovate	Multiperspective dissent feedback
Complex	Unknown unknowns	Probe-sense-respond	Create time and space for reflection
	Messily coherent, patterns discernible	Monitor safe-to-fail experiments	Human sensor networks operational
	Partial changing constraints	All contradiction within heuristics	SNS and the tile to create networks
	Flex within stabilities	Flex constraints to manage emergence	Scenario planning (inc. micro)
	Evidence supports contradiction	Agility key to amplification/dampening	Build and monitor "requisite diversity"
Complicated	Knows unknowns	Sense-analyze-respond	Right people and process, right time
	Ordered, predictable, forecastable	Determine experts or process to resolve	Process engineering with feedback
	Constraints evident and enforceable	Manage and enforce process	Diversity of experts in network
	Stable within constraints	Monitor effectiveness of constraints	Sound analytical practice
	Evidence susceptible of analysis	Focus on exploitation not exploration	Stand aside but stay in touch
Simple	Known knowns	Sense-categorize-respond	Watch for outliers
	Familiar, certain, well-worn pathways	Ensure sound process in place	Usable process
	Constraints self-evident to all	Monitor for noncompliance and deviance	Right support people for key staff
	Stable within universal constraints	Test for complacency	Automate, but no automata
	Self-evident solutions	Protect some pet mavericks	Anonymous appeal/whistle blowers

3.5 HOW COMPLEX SYSTEMS FAIL?

Both hard and soft complex systems fail when the interfaces between the critical parts are

- interrupted or
- partially damaged or
- develop an unpredictable dynamic behavior.

During the design phase the following requirements shall be defined and engineered for the safety critical systems:

- *Goal*—What the system does
- *Scope*—What are the boundary limits to which the Performance Standard applies
- *Functionality*—What the system must do and the criteria it must achieve
- *Availability/reliability*—How often it will work when required
- *Survivability*—The extent is it required to function after a hazardous event has occurred
- *Dependency/interactions*—Affected by or effects on other critical systems
- *Constructability*—Which measures must to be considered during construction
- *Accessibility*—Which measures should be considered to ease accessibility if inspection and maintenance is required
- *Maintainability*—Which type of maintenance routines, procedures, and permit to work is required
- *Operability*—How operations are performed (manually, remote control, automatically, etc.) and which procedures are critical
- *Procurement criticality rating and option*—What are the safety, operational, and financial consequences if the system under study fails? What are the specific design factors and manufacturing complexity?
- *Decommissioning*—How the obsolete systems should be dismantled and their environment cleaned-up.

Then during the operation phase the duty holder shall clearly explain:

- How these items will be assured (written *scheme of examination*).
- What means are to be adopted for verifying the compliance of the system of interest with the standards set for its performance (*verification scheme*).

All these requirements should be specific, measurable, achievable, realistic and timely (SMART) to assure the integrity of the individual parts as well as their interfaces with the other parts.

For the simple systems the clear cause and effect relationships is evident to everyone. The right answer exists and can be verified by the response of "Yes or No" to a well-prepared checklist.

When the requirement is more complex and depends on the more than just one assumption/fact/result of study/code of practice then the simple and straightforward "textual rational" of a checklist is not sufficient to clarify all the rationale behind of the requirement in a traceable and auditable manner. The management of change of this type of requirements is complex and difficult.

To overcome this difficulty a multilayer traceability model should be applied (see Fig. 3.11). In the multilayer traceability model, three types of inputs are distinguished:

- *Systems requirements* from the code, standards, guidelines, and project documents like philosophies and specifications.
- *Assumptions*
- *Domain knowledge* from the specific knowledge of project, specific studies, or experiments and lesson learnt from other projects.

"Textual rationales" are derived from a combination of these inputs to satisfy the requirements of client (end user) which is described in the contract.

"Assumption" is the Achille heel of any traceability model. The major accidents occur when at least one of the basic assumption of design become invalid during operation.

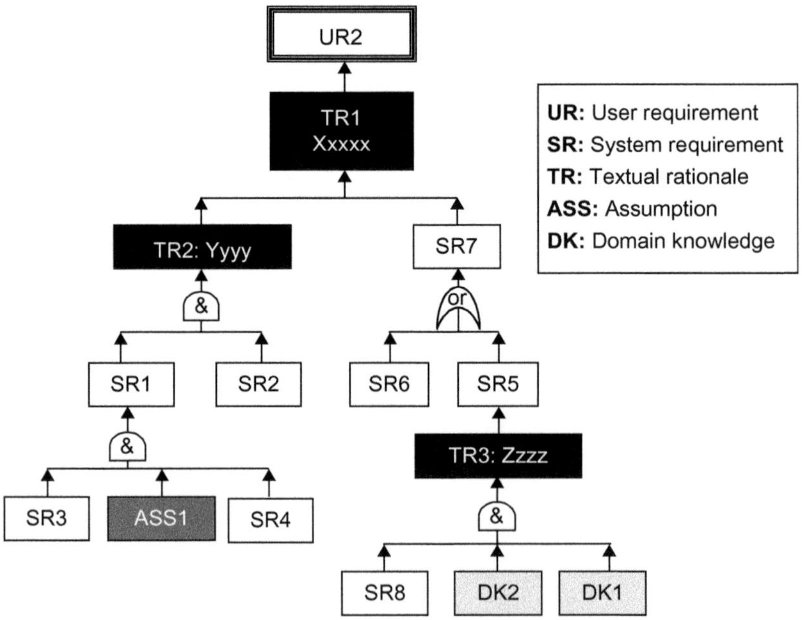

FIGURE 3.11
Multilayer rich traceability.

Table 3.2 Rational for setting the performance standard for survivability requirements at compression area

Step	Source	Type	Statement
1	Contract	UR1	HSEMS of project shall be in compliance with Company HSE Plan Doc. (Client-00aa)
2	Company HSE Plan, Doc. (Client-00aa)	User TR1	The safety barriers shall be designed to limit the probability of accident in fire and explosion hazard areas to expand to the installation as a whole. The criterion is that the frequency of immediate loss of safety barrier protecting persons or safety systems shall not be more than 1/10,000 year
3	Design HSE Plan Doc. (Design-HSE-xxx)	Designer TR1	Requirement from Hazard & Effect Management Process: Fire & explosion study (Doc. Design-HSE-yyy) shall be performed to identify the foreseeable scenarios and quantify their frequency and consequences
4	Design HSE philosophy Doc (Design-HSE-002)	SR1	The impairment criteria of API 2218 for fire impairment shall be applied
5	Design HSE philosophy Doc (Design-HSE-002)	ASS1	The safety barriers will be designed for one fire at the time
6	*API 2218*	DK1	Impairment criteria: • Exposure to 8 kW/m^2 thermal radiation from jet or pool fires
7	*API 2218*	DK2	If unprotected equipment is exposed to the jet fire (300 kW/m^2) it will be failed with 5 min
8	*Fire and explosion study (Doc. Design-HSE-003)*	DK3	Duration of medium jet fires at compressors pipework is about 25 min. This fire could impinge the compressor scrubbers
9	*Performance standard for the passive fire protection (PFP) at compression area. Section: Survivability (Doc. Design-HSE-PS-zzz)*		
9.1	*Option 1*	SR2	The equipment and associated pipework of compression area shall be passively protected to resist a jet fire impingement and/or thermal radiation for at least 30 min
9.2	*Option 2*	SR3	The welded pipeworks shall be used at compression area and number of flanges and instrumentation shall be minimized Flange guards should be applied where safe direction jet fire is practical

UR, *user requirement;* TR, *textual rational;* SR, *system requirement;* ASS, *assumption;* DK, *domain knowledge.*

To understand how the multilayer rich traceability is used, consider the performance standard of the survivability requirement at compression area as an example. Table 3.2 summarizes the rationale behind this requirement. Fig. 3.12 illustrates this rational in a user-friendly pictorial way.

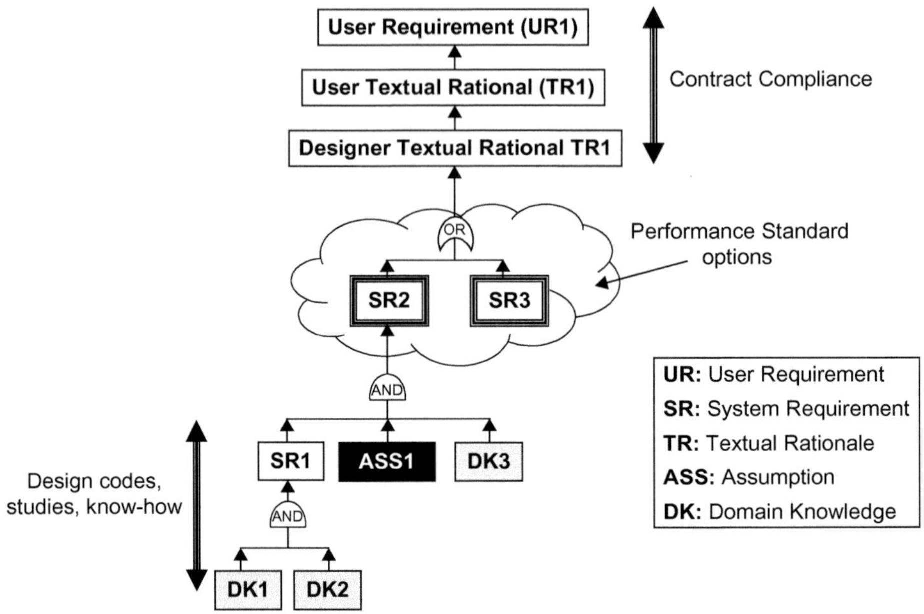

FIGURE 3.12
Multilayer rich traceability for the example of the survivability requirement at compression area.

If, for example, project meets the difficulties in budget to provide the required passive fire protection then they refer to the multilayer traceability model of the requirement to find which part of rational can be reworked and asserted with client without compromising the safety integrity level of the plant.

Project may attract the attention of client on the fact that the impairment criteria suggested by API 2218 is very tight.

Assuming that there is enough available space the same level of safety and asset integrity could be achieved if they increase the impairment criteria from 8 to 10 kW/m^2 and manage the risk of jet fire impingement and thermal radiation with a better layout and longer safety distances.

If the new option is approved then the content of Table 3.2 and configuration of traceability model will be changed accordingly.

The traceability model serves as the mirror of the requirements and their complexity level. The rationale behind any change or decision should be reversible and could be corrected/adjusted at any time. An efficient traceability model provides an objective support for:

- knowing what is a current requirement
- knowing why a requirement was changed

- documentation to form the basis of testing
- understanding where requirements have been built into the system
- forming the basis for ongoing system documentation

With a clear traceability model the audit activities get much more comprehensive. This will lead to a better communication and ease the decision-making process for the proposed options and changes.

In conclusion the hard and soft complex systems can survive by two different mechanisms:

1. A rigorous traceable and auditable structure which can assure the integrity of the interfaces of the system of interest such as a clock, a compressor, or entire chemical plant. The configuration management, risk-based inspection, and preventive maintenance are the techniques in place to detect the discrepancies between the design requirements and actual conditions. The goal of these standards and method is to detect the deviations on time and take the proper actions before the critical failures happen.
2. A viable extent of the self-organization, which gives the required flexibility to transform and adapt system to the postchange situation. This approach is more efficient for the complexity management of the organizational and human systems. The disciplinary and interdisciplinary checks, brainstorming and peer reviews, scheduled internal and external audits are in place to detect the critical issues on time and remove the burden of responsibility from one individual or a discipline.

   The wireless sensors network, adaptive control systems, and fuzzy logic are example of technical self-organized systems.

The flexible and sometimes loose interface between the parts is essential characteristics of a self-organized system. Self-organized criticality (SOC) should be considered to manage the complexity of this type of systems.

In physics, *SOC* is a property of dynamical systems that have a critical point as an attractor. Their macroscopic behavior thus displays the spatial and/or temporal scale-invariance characteristic of the critical point of a phase transition, but without the need to tune control parameters to a precise value, because the system, effectively, tunes itself as it evolves toward criticality.

SOC is typically observed in slowly driven nonequilibrium systems with extended degrees of freedom and a high level of nonlinearity.

The idea of SOC is commonly illustrated conceptually with avalanches in a pile of grains. The grains are dropped onto a pile one by one, and the pile ultimately reaches a stationary "critical" state in which its slope fluctuates about a constant angle of repose, with each new grain being capable of inducing an avalanche on any of the relevant size scales.

3.6 RESILIENCE ENGINEERING

In the early 2000s, resilience engineering was proposed as an alternative (or as a complement) to the conventional view of safety. This led to early discussions about resilience versus robustness, resilience versus brittleness, etc. The focus of resilience engineering is "resilient performance," rather resilience as a property (or quality) or resilience in a "X versus Y" dichotomy.

Whereas conventional risk management approaches are based on hindsight and emphasize error tabulation and calculation of failure probabilities, resilience engineering looks for ways to enhance the ability at all levels of organizations to create processes that are robust yet flexible, to monitor and revise risk models, and to use resources proactively in the face of disruptions or ongoing production and economic pressures.

In resilience engineering "failures" do not stand for a breakdown or malfunctioning of normal system functions, but rather represent the converse of the adaptations necessary to cope with the real-world complexity. Individuals and organizations must always adjust their performance to the current conditions; and because resources and time are finite it is inevitable that such adjustments are approximate. Success has been ascribed to the ability of groups, individuals, and organizations to anticipate the changing shape of risk before damage occurs; failure is simply the temporary or permanent absence of that.

Resilience or more accurately, the ability to perform in a resilient manner is not about avoiding failures and breakdowns, i.e., it is not just the opposite of a lack of safety. In fact, things that go wrong happen in (more or less) the same way as things that go right. This has later been elaborated by Erik Hollnagel in *The ETTO Principle* and in *Safety-I and Safety-II*.

Safety has traditionally been defined as a condition where the number of adverse outcomes was as low as possible (Safety-I). From a Safety-I perspective, the purpose of safety management is to make sure that the number of accidents and incidents is kept as low as possible, or as low as is reasonably practicable. This means that safety management must start from the manifestations of the absence of safety and that—paradoxically—safety is measured by counting the number of cases where it fails rather than by the number of cases where it succeeds. This unavoidably leads to a reactive approach based on responding to what goes wrong or what is identified as a risk—as something that could go wrong. Focusing on what goes right, rather than on what goes wrong, changes the definition of safety from *avoiding that something goes wrong* to *ensuring that everything goes right.*

More precisely, Safety-II is the ability to succeed under varying conditions, so that the number of intended and acceptable outcomes is as high as possible. From a Safety-II perspective, the purpose of safety management is to ensure that as much as possible goes right, in the sense that everyday work achieves its objectives. This means that safety is managed by what it achieves (successes, things that go right), and that likewise it is measured by counting the number of cases where things go right. In order to do this, safety management cannot only be reactive, it must also be proactive. But it must be proactive with regard to how actions succeed, to everyday acceptable performance, rather than with regard to how they can fail, as traditional risk analysis does.

As technologies change and people adapt to take advantage of those technologies, we can contrast successful and unsuccessful cases of managing the complexity that results—whether the system is human, biological, or engineered. In the successful cases multiple interacting and interdependent groups continuously readjust to each other as they cope with, and as they exploit, changing capabilities, demands, contexts, and constraints. These systems exist in layered networks with extensive and sometimes hidden interdependencies, yet they are able to adapt in the face of variation. But much more importantly, they are able to sustain adaptability as those forms and sources of variation change. In shorthand, these are complex adaptive networks which demonstrate sustained adaptability. Sustained adaptability refers to the ability to continue to adapt to changing environments, stakeholders, demands, contexts, and constraints (in effect, to adapt how the system in question adapts).

On the other hand, much less successful complex adaptive networks are more common: initially successful adaptations unwind over time and suffer sudden performance collapses—tactics become stale, groups work at cross purposes, actions seem unable to keep pace with change, and cascades from surprising events. The failures to sustain adaptability are vivid when we witness an ecosystem degrade in a tragedy of the commons, a market unravel, or a dramatic failure occur in safety-critical operations.

Resilience engineering asks three questions about this contrast:

1. What governance or architectural characteristics explain the difference between networks that produce sustained adaptability and those that fail to sustain adaptability?
2. What design principles and techniques would allow one engineer a network that can produce sustained adaptability?

3. How would one know if one succeeded (how can one confidently assess whether a system has the ability to sustain adaptability over time)?

Web services exemplify the above contrast as periods of rapid florescence occur but eventually stall, leading some organizations to the edge of collapse while others experience new periods of florescence. The participants in these experiences oscillate between extreme confidence about future decisions and a sense of precariousness and vulnerability to unanticipated surprises. Can the experiences of the organizations delivering or dependent on web services provide the insights to answer these three questions? Can the provisional answers to these questions help network web services learn how to sustain adaptability even as their past successes produce new upheavals?

3.7 IMPROVISATION THINKING

Improvisation is the process of devising a solution to a requirement by making-do, despite absence of resources that might be expected to produce a solution. In a technical context, this can mean adapting a device for some use other than that which it was designed for, or building a device from unusual components in an ad hoc fashion. Improvisation as a context of performing arts is a very spontaneous performance without specific preparation. The skills of improvisation can apply to many different faculties, across all artistic, scientific, physical, cognitive, academic, and nonacademic disciplines.

Improvisation in engineering is to solve a problem with the tools and materials immediately at hand. Examples of such improvisation were the re-engineering of carbon dioxide scrubbers with the materials on hand during the Apollo 13 space mission, or the use of a knife in place of a screwdriver to turn a screw.

Engineering improvisations may be needed because of emergencies, embargo, obsolescence of a product and the loss of manufacturer support, or just a lack of funding appropriate for a better solution. Users of motor vehicles in parts of Africa develop improvised solutions where it is not feasible to obtain manufacturer-approved spare parts.

The popular television program MacGyver used as its gimmick a hero who could solve almost any problem with jury rigged devices from everyday materials, a Swiss Army knife, and some duct tape.

The skills of improvisation can apply to many different abilities or forms of communication and expression across all artistic, scientific, physical, cognitive, academic, and nonacademic disciplines. For example, improvisation can make a significant contribution in music, dance, cooking, presenting a speech, sales, personal or romantic relationships, sports, flower arranging, martial arts, psychotherapy, and much more.

Techniques of improvisation are widely used in training for performing arts or entertainment; e.g., music, theater, and dance. To *extemporize* or ad lib is basically the same as improvising. Colloquial terms such as "let's play it by the ear," "take it as it comes," and "make it up as we go along" are all used to describe "improvisation."

The simple act of speaking requires a good deal of improvisation because the mind is addressing its own thought and creating its unrehearsed delivery in words, sounds, and gestures, forming unpredictable statements that feed back into the thought process (the performer as listener), creating an enriched process that is not unlike instantaneous composition (with a given set or repertoire of elements).

Where the improvisation is intended to solve a problem on a temporary basis, the "proper" solution being unavailable at the time, it may be known as a "stop-gap." This applies to the field of engineering. Another improvisational, group problem-solving technique being used in organizations of all kinds is *brainstorming*, in which any and all ideas that a group member may have are permitted and encouraged to be expressed, regardless of actual practicality. As in all improvisation, the process of brainstorming opens up the minds of the people involved to new, unexpected, and possibly useful ideas. The colloquial term for this is "thinking outside the box."

Improvisation as a concept has traditionally been associated with handling of exceptions. As resilience engineering emphasizes the constantly changing environment, it could be argued that all operational activities should be considered from a nonroutine perspective. A similar point is made by Weick and Sutcliffe (2007) in terms of their principle of *Sensitivity to Operations*, which emphasize that any function must be implemented in a specific context, and that even minor contextual differences may imply some degree of adaptation. The distinction between regular, irregular, and unexampled events may therefore be somewhat misleading, creating an impression that the lion's share of dangerous events can be handled as nonsingular instantiations of categories of events, in a repeatable manner. A distinct focus on improvisation however, renders plain and intelligible the need for constant awareness and adaptation.

Theories of resilience implies the use of "provoking" concepts and terms, advocating a certain degree of "letting go" that jeopardize traditional assumptions of staying in control. Highlighting improvisation will not bring more comfort in that sense, but contribute to a more realistic view on the challenges of managing complex systems.

From the definitions of *improvisation*, only one is easily mapped to the right (ordered) side of Cynefin in terms of a high-speed analytical process. The others may be interpreted as a kind of sensitivity to changing circumstances and emergent problems, emphasizing the significance of both *probing* and *acting on* the system.

The improvisation is about interplay of action and interpretation, and about interchangeably looking forward and backward in order to make sense of what is happening. This maps neatly to the very idea of Cynefin dynamics; to use different frames of interpretation, according to circumstances. Improvisation could thus be perceived as the "engine" that actually drives movements between (Cynefin) domains, and thus makes resilience something substantially different from economically motivated substitution of rational analysis in advance. Resilience Engineering in terms of Cynefin dynamics, refers to the right side of Fig. 3.13 as follows:

- Detecting a possible unexpected variation, and stabilizing it to avoid harm ("Exploration" and "Just-in-time-transfer")
- "Bouncing back" and mitigating surprising variations that can develop into dangerous sequences ("Divergence-Convergence")
- "Bouncing back" on severe circumstances that threatens the system as a whole, into a degraded mode of operation ("Swarming" or "Imposition").

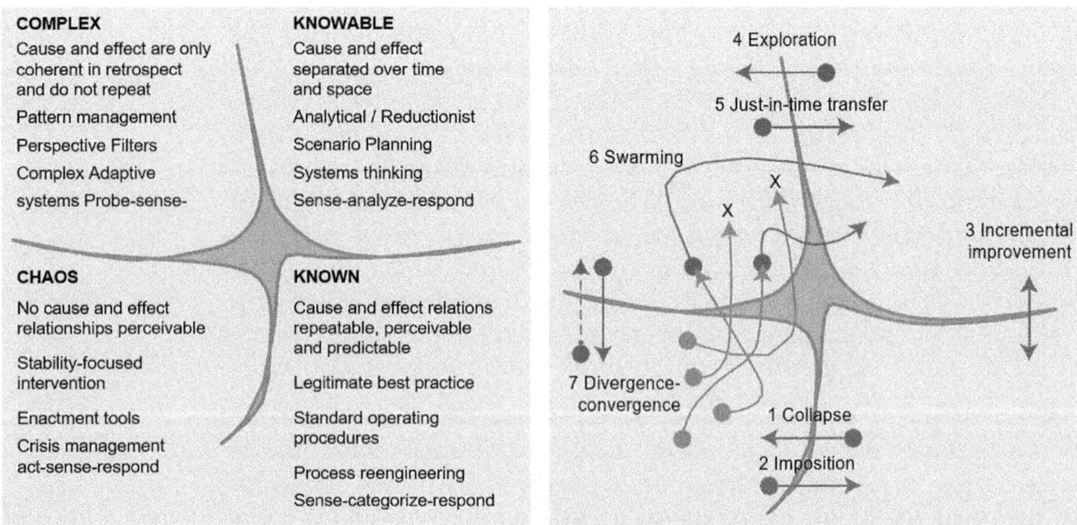

FIGURE 3.13
The *Cynefin* framework. Excerpts from Kurtz and Snowden (2003).

3.8 EFFICIENCY-THOROUGHNESS TRADE-OFF

In the daily activities, at work or at leisure, people (and organizations) routinely make a choice between being effective and being thorough, since it rarely is possible to be both at the same time. For example:

- If demands to productivity or performance are high, thoroughness is reduced until the productivity goals are met.
- If demands to safety are high, efficiency is reduced until the safety goals are met.

The essence of this balance or trade-off between efficiency and thoroughness is described by *Erik Hollnagel* as the efficiency-thoroughness trade-off (ETTO) principle.

Accident investigation and risk assessment have for decades focused on the "equipment failure" and human factor, particularly "human error." Countless books and papers have been written about how to identify, classify, eliminate, prevent, and compensate for it. This bias toward the study of performance failures leads to a neglect of normal or "error-free" performance and the assumption that as failures and successes have different origins there is little to be gained from studying them together.

Erik Hollnagel believes this assumption is false and that safety cannot be attained only by eliminating risks and failures. The ETTO principle looks at the common trait of people at work to adjust what they do to match the conditions—to what has happened, to what happens, and to what may happen. In practice, usually sacrificing thoroughness for efficiency. While in some cases the adjustments may lead to adverse outcomes, these are due to the very same processes that produce successes, rather than to errors and malfunctions. The ETTO principle removes the need for specialized theories and models of failure and "human error" and offers a viable basis for effective and just approaches to both reactive and proactive safety management.

The ETTO principle refers to the fact that people (and organizations) as part of their activities frequently—or always—have to make a trade-off between the resources (primarily time and effort) they spend on preparing to do something and the resources (primarily time and effort) they spend on doing it. The trade-off may favor thoroughness over efficiency if safety and quality are the dominant concerns, and efficiency over thoroughness if throughput and output are the dominant concerns. It follows from the ETTO principle that it is never possible to maximize efficiency and thoroughness at the same time. Nor can an activity expect to succeed, if there is not a minimum of either.

Efficiency means that the level of investment or amount of resources used or needed to achieve a stated goal or objective are kept as low as possible. The resources may be expressed in terms of time, materials, money, psychological effort (workload), physical effort (fatigue), manpower (number of people), etc. The appropriate level or amount is determined by the subjective evaluation of what is sufficient to achieve the goal, i.e., good enough to be acceptable by whatever stop rule is applied as well as by external requirements and demands. For individuals, the decision about how much effort to spend is usually not conscious, but rather a result of habit, social norms, and established practice. For organizations, it is more likely to be the result of a direct consideration—although this choice in itself will also be subject to the ETTO principle.

Thoroughness means that an activity is carried out only if the individual or organization is confident that the necessary and sufficient conditions for it exist so that the activity will achieve its objective and not create any unwanted side effects. These conditions comprise time, information, materials, energy, competence, tools, etc. More formally, thoroughness means that the preconditions for an activity are in place, that the execution conditions can be ensured, and that the outcome(s) will be the intended one(s).

The ETTO fallacy is that people are required to be both efficient and thorough at the same time—or rather to be thorough when with hindsight it was wrong to be efficient!

3.9 SPECIFIC METHODS TO ADDRESS ENVIRONMENTAL AND SYSTEM COMPLEXITY

The principles and approaches described above are important points of departure for systems engineers facing complexity in their SE activities. Fig. 3.14 illustrates the Map of Complexity Science. It is also useful to describe how these concepts apply to particular aspects of the systems engineer's work—aspects that are fundamental regardless of the nature of the system being developed or the problem being addressed. Systems engineers' toolkits should include a wide range of methods and processes to address environmental and system complexity in appropriate and useful ways.

Diagnosis is the key step. The systems engineer must identify the kind and extent of complexity that bears on the problem set. As we have seen, complexity can exist in the problem being addressed, in its environment or context, or in the system under consideration for providing a solution to the problem. The diagnoses made will allow the systems engineer to tailor his/her approaches to key aspects of the systems engineering process: requirements elicitation, trade studies, the selection of a development process life

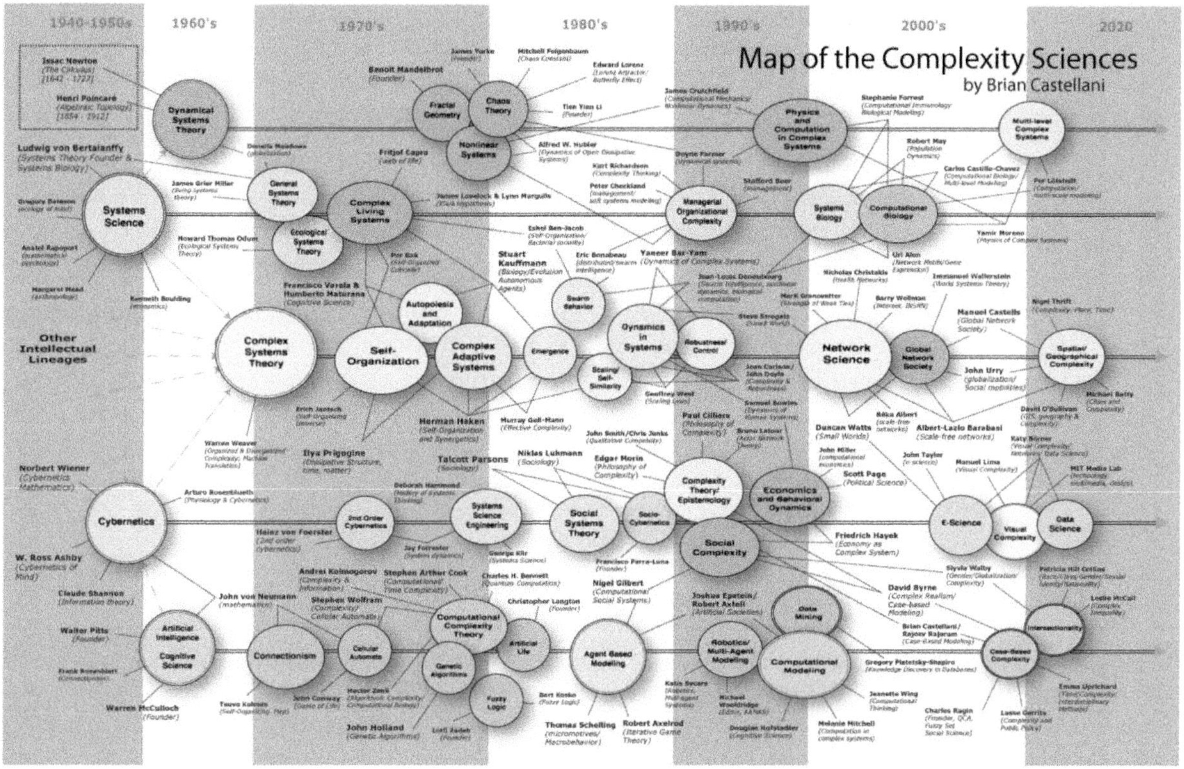

FIGURE 3.14

Map of complexity science. From (http://www.scimaps.org/detailMap/index/map_of_complexity_sc_154).

cycle, solution architecting, system decomposition and subsystem integration, test and evaluation activities, and others.

In addition, the diagnosis will allow the systems engineer to consider whether there may be mechanisms for shifting complexity to a more desirable region of the problem space. There may be choices available or investments that can be made to allow the decoupling of aspects of the system or of the system to its environment. Likewise, there may also be options for shaping the feedbacks within and across problem-environment-solution elements, allowing the complexity of the situation to be harnessed via the leveraging of beneficial adaptation and self-organization.

The tables below show examples of the kinds of choices that can improve the success of engineering complex systems when complexity of different types and from different sources exists. These lists are by no means exhaustive; indeed, astute readers will note that some entries in the tables below are blank.

These tables are intended to convey the types of approaches and departures from traditional SE methods that may be required when dealing with complexity in problem, solution, and environment contexts. The authors expect that this table will be augmented and evolved in future updates to this primer.

These approaches are offered to help systems engineers ensure that their processes are appropriate for dealing with the dynamics, uncertainty, and behaviors that can arise when significant complexity exists in the problem, the solution, or the environment. Experience suggests that when SE activities do not account for these factors when they exist, projects fail and problems go unaddressed.

Our goal in this section is to provide systems engineers with techniques for better recognizing complexity and its consequences as it pertains to their activities, and for expanding the envelope of the degree and types of complexity that can be dealt with. Without doubt, there is additional work to do to expand our understanding of these phenomena—and to provide practical tools to allow the systems engineering community to leverage and better apply emerging insights. The concepts and approaches in this section, however, should provide a useful starting point (Fig. 3.14).

In addition to new approaches to SE activities, different approaches to systems modeling and simulations are also needed to deal with complexity. INCOSE Complex Systems Working Group suggests a matrix showing modeling methods, their benefits, and conditions for which each method is and is not appropriate. A subset of applicable modeling methods have been extracted from this matrix and organized into categories relevant to systems engineering: Analyze, diagnose, model, and synthesize. The tools in each column generally range from the simplest at the top to the most complex at the bottom. The resulting matrix, shown in Table 3.3, provides a starting point for systems engineers seeking complexity-appropriate modeling approaches to systems engineering. In this context, Table 3.4 summarises the Candidate approaches to address complexity in problem context or environment and Table 3.5 summarises the Candidate approaches to address system/solution complexity.

3.10 COMPLEXITY THINKING: GUIDING PRINCIPLES

Complex systems engineering requires both a shift in thinking and an expanded set of tools and techniques. In this section, we summarize the shift in thinking needed to acknowledge and embrace complexity within systems engineering. Several principles are provided below that encourage systems engineers to think differently about how to engage with complexity.

1. *Think like a gardener, not a watchmaker.* Consider the complexity of the environment and the solution, and think about evolving a living solution to the problem rather than constructing a system from scratch.

Table 3.3 INCOSE Selected Modeling Methods for Complex Systems

Analyze	Diagnose	Model	Synthesize
Data mining	Algorithmic complexity	Uncertainty modeling	Design structure matrix
Splines	Monte Carlo methods	Virtual immersive modeling	Architectural frameworks
Fuzzy logic	Thermodynamic depth	Functional/behavioral models	Simulated annealing
Neural networks	Fractal dimension	Feedback control models	Artificial immune system
Classification and regression trees	Information theory	Dissipative systems	Particle swarm optimization
Kernel machines	Statistical complexity	Game theory	Genetic algorithms
Nonlinear time series analysis	Graph theory	Cellular automata	Multiagent systems
Markov chains	Functional information	System dynamics	Adaptive networks
Power law statistics	Multiscale complexity	Dynamical systems	
Social network analysis		Network models	
		Agent-based models Multiscale models	

2. *Combine courage with humility.* It takes courage to acknowledge complexity, relinquish control, encourage variety, and explore unmapped territory. It takes humility to accept irreducible uncertainty, to be skeptical of existing knowledge, and to be open to learning from failure. A combination of courage and humility enables the complex systems engineer to risk genuine innovation and learn fast from iterative prototyping of solutions in context.
3. *Take an adaptive stance.* Systems engineers should mimic how living systems cope with complexity by identifying and creating variation, selecting the best versions, and amplify the fit of the selected versions. This means, e.g., to think "influence" and "intervention" rather than "control" and "design." Designing or evolving a complex system requires recognition that the designer may not ever be able to control or even understand the system completely.
4. *Use free order.* In architecting and designing solutions, build in "order for free" using self-organization, presuming it has been modeled and can be limited to desired effects. This in particular applies when the system being designed must be resilient.

Table 3.4 Candidate Approaches to Address Complexity in Problem Context or Environment

	Requirements Elicitation and Derivation	Trade Studies	Solution Architecture and Design	Development Process
Complexity in the environment (general)	Use multiple methods for requirements elicitation Elicit requirements from multiple perspectives and at multiple levels of aggregation Emphasize capture of system objectives and desired outcomes rather than thousands of detailed requirements	Emphasize robustness over local efficiency and performance	Include both positive and negative feedback mechanisms to provide mechanisms to compensate for the effects of higher-than-linear positive feedback and runaway system behavior	Employ soft systems methodologies to surface the nature of the problem space, its internal structure and information flows, and produce simple representations, e.g. "rich pictures" to communicate these.
Intricate and evolving/self-organizing interactions with the environment	Include requirements for the system to provide adaptive local control, rather than strong, deterministic control	Trade end-to-end system performance and behavior against problem space complexity. Think hierarchy rather than flat networks.	Early implantation (or at least prototyping) of external interfaces	Early deployment of system functionality with feedback to developers
Environment susceptible to "black swan" events (unlikely, unpredictable, high-consequence events) and/or recursive complexity	Use power laws rather than Gaussian distributions to characterize phenomena in requirements and sell-off criteria Focus requirements elicitation on resiliency, robustness and adaptiveness vice optimizing to particular assumptions	Make resilience a key trade-space attribute and use trades to identify aspects of the problem space that will drive the system architecture	Design for resilience to "beyond-design-envelope" events to provide robustness and timely recovery to a minimally functional state	Resilience analysis. Enterprise development: study how enterprises or societies survive catastrophes
Complexity in the problem/ mission	Emphasize identification of constraints as well as requirements Capture scenarios and mission threads in preference to large numbers of requirements	Use scalability and agility as criteria in appropriate trade studies	Use solution elements which are adaptable and/or reconfigurable Design to achieve scenarios rather than detailed requirements Satisfice at the system level rather than satisfy detailed requirements	Use Agile, evolutionary SE processes instead of Waterfall SE processes Define multilayer processes and their interface points

Continued

Table 3.4 Candidate Approaches to Address Complexity in Problem Context or Environment *Continued*

	Requirements Elicitation and Derivation	Trade Studies	Solution Architecture and Design	Development Process
Complexity in stakeholder relationships	Use multiple scales (or a balanced scorecard approach) instead of a single utility function to determine "goodness" or fitness for use	Seek stakeholder buy in to trade studies	Use modeling and simulation to enable stakeholders to experience (rather than just be briefed about) interactions of solution elements and the environment	Employing a multi-methodological approach, i.e., soft systems methodologies plus SE plus boundary critique, identify stakeholders and achieve buy-in.
Complexity in interactions between different mission elements	Capture scenarios and mission threads in preference to large numbers of detailed requirements Develop understanding and means of controlling nonlinearities, disruptive events	Make minimizing interactions/ interdependences between complex systems elements a key trade-off attribute. Model with system dynamics.	Use exploratory modeling and simulation to assess ability of solution elements to address mission elements under wide range of conditions and assumptions. Use system dynamics to link economics and system changes.	Establish and maintain an interdependency database between constituent systems that is used for all major design decisions

Table 3.5 Candidate Approaches to Address System/Solution Complexity

	Requirements Elicitation and Derivation	Trade Studies	Solution Architecture and Design	Development Process
Complexity in system design and development (general)	Use multiscale modeling (linking macro- and microlevel models), including exploratory analysis and agent-based modeling, and experimentation: • To generate insight into the implications of derived requirements • As the basis for trade studies and to inform trade-off decisions		Emphasize selection of robust and adaptive elements and structures over optimizing to meet current requirements	Use SoSE methodologies to synchronize constituent systems; incentivize collaboration. Ensure prototyping and experimentation are used.
Emergent properties or behaviors in solution system	Maximize description of emergent properties in scenarios and mission definition.	Employ modeling and experimentation to ensure relevant effects of trades are explored at different levels of aggregation	Built-in feedback mechanisms to enable the system and system elements to adapt to the environment and each other in effect ways.	Conduct development activities always within context of the whole Employ collaborative development processes so that information about

Continued

Table 3.5 Candidate Approaches to Address System/Solution Complexity *Continued*

	Requirements Elicitation and Derivation	Trade Studies	Solution Architecture and Design	Development Process
	Elicit requirements from multiple perspectives and at multiple levels of aggregation; ensure requirements and constraints at all relevant levels are understood. Understand the real value of predictability at different levels; encourage the definition of requirements at higher "essential value" levels		Acknowledge the limits to the value of decomposition-based methods; emergence is a collective phenomenon that requires aggregation—emergence will not be observed until the system is considered as a whole	design decisions are visible throughout the project Prototyping and holistic testing are critical to explore and check for the manifestation of emergent behavior
Complexity in system deployment and operation	Employ soft systems methodologies to surface the nature of the deployed solution, and its internal structure and information flows; produce simple representations, e.g., "rich pictures" to communicate these. Use problem definition methods from an evolutionary SE or SoSE methodology.	Trade criteria need to value cost and ease of training and logistical support over acquisition cost. Model system evolution with genetic algorithms.	Use self-organizing and self-repairing elements when possible. Model the cost of change, the benefits, and the balance.	Employ soft systems methodologies to surface issues, engage stakeholders, identify approaches to improve the deployed system, and to achieve stakeholder buy-in to the solution. Use an evolutionary SE or SoSE methodology. Identify utility and cost of using and modifying legacy systems.
Complexity in system evolution and support	Focus on capabilities, not requirements	Use resilience and robustness as criteria in trade studies	Ensure that the most important system elements are composable (i.e., capable of being reconfigured adaptively with other elements in the future to satisfy emergent operational needs not previously envisioned) and have clear and accessible interfaces.	Emphasize understanding the problem, and at each stage of this iterative process, intervene to add an incremental capability and watch carefully over time to see whether there is improvement If not, try something else.

5. *Identify and use patterns.* Patterns are exhibited by complex systems, can be observed and understood, and are a key mechanism in the engineering of complex systems. Patterns are the primary means of dealing specifically with emergence and side effects—i.e., the means of inducing desired emergence and side effects, and the means of avoiding undesired emergence and side effects.
6. *Zoom in and zoom out.* Because complex systems cannot be understood at a single scale of analysis, systems engineers must develop the habit of looking at their project at many different scales, by iteratively zooming in and zooming out. Can problems be solved more elegantly by addressing them at a higher or lower hierarchical level? The complex systems engineer must be especially open to solutions that arise from the bottom-up through self-organization, rather than only seeking to impose order from the top-down.
7. *See through new eyes.* A complex situation often looks very different from the perspectives of the variety of stakeholders. By empathizing with these multiple perspectives, a systems engineer can sometimes find creative ways to solve several problems at once.
8. *Collaborate.* Collaboration includes information sharing, active listening, establishment of trust to enable candid dialogue, and making decisions transparent. A collaborative mindset can lead to deeper stakeholder engagement practices that may include crowdfunding and crowdsourcing, to enable co-creation and co-evolutionary systems design.
9. *Achieve balance.* Optimization is often counterproductive within a complex system. Either the whole is suboptimized when a part is optimized, or an optimized whole becomes rigid, unable to flex with changing conditions. Instead of optimizing, complex systems engineers should seek balance among competing tensions within the project. Systems engineers can leverage integrative thinking to generate improved solutions and avoid binary either/or trade-offs. The goal is a system that would continue to meet the need even if a number of current conditions change.
10. *Learn from problems.* In a changing context, with an evolving system, where elements are densely interconnected, problems and opportunities will continually emerge. Moreover, they will emerge in surprising ways, due to phase transitions, cascading failures, fat tailed distributions, and "black swan" (Taleb, 2007) events. A traditional approach to risk management and mitigation should be augmented by a complexity mindset that balances risk management with exploiting opportunity and expects and learns from error.
11. *Metacognition.* Metacognition, or reflecting on how one reflects, helps to identify bias, make useful patterns of thinking more frequent, and improve understanding of a complex situation.

12. *Focus on desired regions of outcome space rather than specifying detailed outcomes.* Instead of zeroing in on an exact solution, focus on what range of solutions will have the desired effects, and design to keep out of forbidden ranges.
13. *Understand what motivates autonomous agents.* Changing rewards will shape collective behavior. Implement incentives that will move the system toward a more desired state.
14. *Maintain adaptive feedback loops.* Adaptive systems correct for output variations via a feedback mechanism. Over time, feedback loops can either hit the limit of their control space, or may be removed in the interest of maintaining stability. To maintain robustness, periodically revisit feedback and ensure that adaptation can still occur.
15. *Integrate problems.* Focus on the relationships among problems rather than addressing each problem separately. This allows fewer solutions that take care of multiple problems in an integrative fashion.

Literature

1. A Complexity Primer for Systems Engineers
2. What is a Complex System?
3. Cynefin, Statistics and Decision Analysis
4. Snowden, D.J., Boone, M.E. (Cynefin Definition Excerpts) A Leader's Framework for Decision Making. Copyright © 2007 Harvard Business School Publishing Corporation. All Rights Reserved. www.hbrreprints.org, Reprint R0711C
5. Get a Grip on Sense-Making and Exploration Dealing with Complexity through Serious Play
6. How Complex Systems Fail
7. White Paper: Foresight Review of Resilience Engineering
8. Resilience, Adaptation And Improvisation—Increasing Resilience By Organising For Successful Improvisation
9. Complexity of System Configuration Management
10. Cracking the Complexity Code
11. PMI White Paper, Complexity and Change: Birds of a Feather?, by Michel Thiry, PhD, PMP, PMI Fellow
12. The New Dynamics of Strategy: Sense-Making in a Complex and Complicated World
13. Mastering Complexity – Capture the Hidden Opportunity
14. Decision-Support and Complexity in Decision Making
15. The Process of Solving Complex Problems
16. The Decision-Making Process in a Complex Situation

Blog

1. The Ubiquity of Ubiquitousness
2. Resilience Engineering

3. The Cynefin Framework and Emotional Intelligence
4. Simon Powers Blog on the Cynefin Review
   Part 1: The Sense Making Framework
   Part 2: Dynamics
   Part 3: The Complex Domain
   Part 4: The Chaos Domain
   Part 5: The Complicated Domain
   Part 6: The Obvious Domain
   Part 7: Finding Your Place on the Framework
   What is the Agile Mindset?
   A Game for Determining Next Steps Towards a Desired Outcome.
5. Cognitive Edge – David Swendon Blog on Cynefin Framework
   Triggering Attention in Cynefin
   Cynefin Dynamics
6. Complexity Thinking

Video

1. The Illusion of Complexity – Neuroactivity and Complex Behaviour
2. Cook, R., Velocity 2012: "How Complex Systems Fail"
3. Complextity Labs Multimedia (http://complexitylabs.io/multimedia/)
4. Complexity Labs on Youtube Channel (https://www.youtube.com/channel/UCutCcajxhR33k9UR-DdLsAQ)

Handbook

http://sebokwiki.org/wiki/INCOSE_Systems_Engineering_Handbook

1. INCOSE Handbook-2002
2. INCOSE Handbook-2006
3. INCOSE Handbook-2012
4. INCOSE Handbook-2015
5. Guide to the Systems Engineering Body of Knowledge (SEBoK)
   - SEBoK v. 1.7
   - SEBoK v. 1.6
   - SEBoK v. 1.5.1
   - SEBoK v. 1.5.
   - SEBoK v. 1.4
   - SEBoK v. 1.3.1
   - SEBoK v. 1.3
   - SEBoK v. 1.2
   - SEBoK v. 1.1
   - SEBoK v. 1.0

Standards

1. ISO/IEC 15228: Systems and Software Engineering – System Life Cycle Processes

Guidance

1. Project Complexity and Risk Assessment Tool

CHAPTER 4

System Engineering of the Complex Megaprojects

World energy needs are predicted to be 50% higher in 2030 than in 2007. On the other hand, the advanced modern technology enables for implementing the larger plants operating under much more challenging environmental conditions. As a result, an increasing number of energy megaprojects are being considered by governments and enterprises.

Megaprojects are defined by their size, complexity, long duration, and high demands on resources and technology. The practice and outcomes of megaprojects have shown an alarming rate of failure, with complexity aspects having been recognized as one of the main causes of these failures. Contractors in megaprojects reply to overall and to task complexity by breaking it down to functional departments, to social complexity by trust and commitment, and to cultural complexity by sense-making processes.

The design and construction processes of a megaproject should be carefully planned and controlled through implementation. A structured, systematic approach should be applied to control and manage the three critical factors of cost, time, and quality. The robust system engineering techniques should be applied to manage modularization, interface management, and the risks associated with the likely significant degree of change in the environment and associated scope of over the extended timescales.

Complexity is recognized as one of the main idiosyncratic attributes of megaprojects and, at the same time, a cause of failure in energy megaprojects. Complexity is the most unknown and pathless attribute of megaprojects that needed to be addressed to manage the uncertainty associated with the cost estimation and changing scope and could improve construction productivity hence making the project more cost effective and therefore possibly more beneficial (profitable) for constructors.

Many works on project complexity have been carried out on the complexity management of the megaprojects in the recent years. But the efforts to date seem to have generated more confusion than precision, as complexity and project complexity have been interpreted in many different ways.

A Systems Approach to Managing the Complexities of Process Industries.
DOI: http://dx.doi.org/10.1016/B978-0-12-804213-7.00004-9

The majority of the studies on project complexity have been mostly devoted to the conceptual aspects of project complexity and their accuracy and practice applicability are challenged by the following three issues:

1. The indicators contributing to project complexity are not fully identified and have not been organized in a standard categorization, or taxonomy;
2. The proposed methods mainly neglected the process of consistency checking and consensus building, resulting in imprecise final results;
3. The development of objective and numerical rating criteria for all project complexity indicators has been entirely omitted from those.

Megaproject managerial performance could be improved by modifying the work practices in order to handle megaprojects' characteristics. This can be possible through the adequate system integration. The advances in industrial Internet of the things (IIOT) give us the possibility of testing these methods quickly and safely on the virtual/augmented reality simulation of the construction site.

The project complexity assessment (PCA) tool should combine the mathematical group decision-making (GDM) method, consistency-checking, and consensus-building processes.

It is crucial to establish a specific international standard for system engineering and complexity management of the development, implementation, and operation of the megaprojects. This standard could be similar to ISO 15288 covering the megaprojects complexity taxonomies which are summarized in Fig. 4.1.

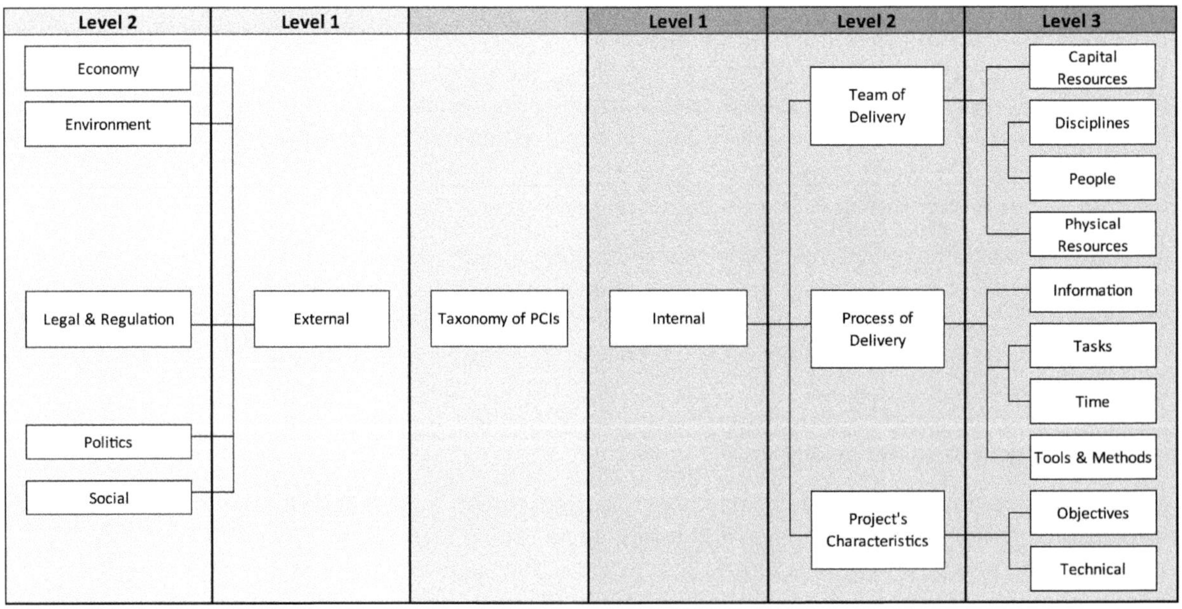

FIGURE 4.1
Taxonomy of project complexity of megaprojects.

INCOSE has also published a document entitled "Guide for the Application of Systems Engineering in Large Infrastructure Projects" which tends to provide the practical insights for the system engineering and project management of the megaprojects.

4.1 MEGAPROJECT DEFINITION

The "megaprojects" are the large-scale engineering projects with the five following attributes:

1. "High" capital cost—at least 1 billion USD,
2. "Long" duration but program urgency—average development duration 6.5 years with a construction period of 4 years,
3. Technologically and logistically demanding,
4. Requiring multidisciplinary inputs from many organizations, and
5. Leading to a "virtual enterprise" for the execution of the project.

One characteristic of megaprojects is the lack of scope and cost certainty. This has led to the development of a number of procurement approaches that allocate the risk of this uncertainty in different ways, e.g., engineer procure construct (EPC), design and build (D&B), private finance initiative (PFI), public private partnership (PPP), and alliancing and build own operate (BOO). Fig. 4.2 illustrates that the contract serves as an interface between enterprises.

Megaprojects attract a lot of public attention because of substantial impacts on communities, environment, and budgets. Megaprojects can also be defined as "initiatives that are physical, very expensive, and public."

The megaprojects are realized by the multiorganizational enterprises (MOEs) which are characterized by the followings features:

1. singularity
2. complexity
3. goal-orientation (technical, financial, time)
4. the nature and number of project owners

For many types of infrastructure project, such as high-rise office blocks and residential buildings, the construction process is relatively efficient as a result of having been performed many, many times, such that we can consider each project being an incremental step in a long, continuously improving development process. However, because megaprojects have a degree of *uniqueness* it is difficult to standardize and therefore to optimize the construction process itself.

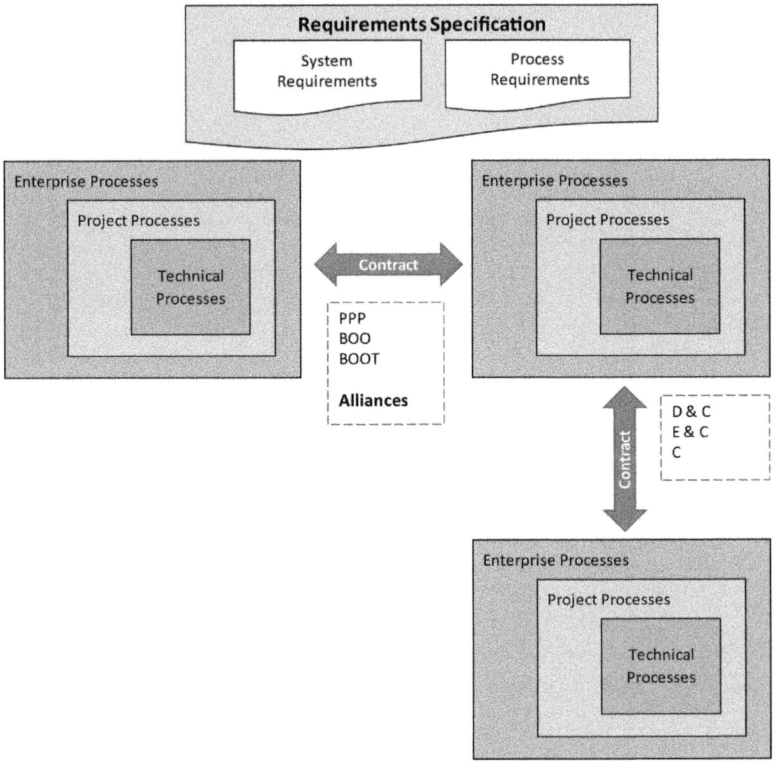

FIGURE 4.2
The contract as an interface between enterprises.

4.2 MEGAPROJECTS IN OIL AND GAS INDUSTRY

As reserves of oil are inevitably depleted, oil companies are faced with the challenge of finding and exploiting new sources. Driven by the need to keep their reserve replacement ratio, the amount of proven reserves added to a company's base relative to the amount of oil produced each year, at or above 100%, oil companies are increasingly investing in fields that are very challenging, such as in the Arctic Circle or at ever deeper sites at sea.

The investment and continuing expenditure needed to develop and produce from these fields puts them in the realm of "megaprojects," costing 5 billion USD or more. They require huge supply chains, needing expensive shipping, massive transport networks, and thousands of workers.

Around the world, the giant oil companies of the United States and Europe are putting the brakes on a decade-long spending spree focused on finding and developing offshore oil fields in ever-tougher environments. This is a major shift in an industry that has been focused for years on chasing production growth at any cost. Oil majors like Total and Chevron are now prepared to ditch or delay costly megaprojects, putting off investment decisions or reducing exposures that they deem too risky.

A report by global assurance, tax, transaction, and advisory services specialists, EY, entitled "Spotlight on oil and gas megaprojects," found that the major causes of *these delays* and *added costs* are *nontechnical factors* such as *people, organization,* and *governance,* with other related causes being *management processes* and *contracting* and *procurement strategies.*

This report also reveals that almost *two-thirds (64%)* of multibillion-dollar, technically and operationally demanding megaprojects continue to exceed budgets, with *three quarters (73%)* missing project schedule deadlines. On average, current project estimated completion costs were *59% above* the initial estimate.

Because these projects are so big, *complex,* and inaccessible, needing highly skilled workers, service costs are also rising, with increases of up to *8%* a year since 2011.

Among the many challenges are project complexity and completing projects on time and on budget. For example, Chevron has created a dedicated unit to manage expenses and oversee contractors.

The EY report identified several management areas in which oil companies are failing, including poor procurement of contractors, poor contractor management, and ineffective project management. The challenge is to identify the areas of poor performance and put in place management practices that will lead to cost and time reductions and better quality. One way is to place *greater reliance and trust* in the *suppliers* of technical solutions.

Far from taking control away from the operator or main EPC, this has the benefit of freeing them to focus on their strengths in project construction, while relying on solution vendors and other suppliers to handle the detailed technical integration and commissioning of systems.

Cost control is another factor, as soaring costs are outpacing foreseeable rises in energy prices. Rising labor costs are also significant, driven by fierce competition for *skilled workers and increasing productivity.*

In other words, the conventional management systems cannot respond to the requirements of unconventional megaprojects.

A successful staffing model for the megaprojects is a challenge too. It takes about 4 years to move from entry level to journey level, while petrochemical construction projects, e.g., typically take between 18 and 24 months. In that scenario, a new hire who is partly trained at the start could likely move to an experienced freelancer by the end of the project.

The best staffing model is to hire and train as many local people as we can to work their way through the construction phase. By the time the construction phase is over, they will be the experienced freelancers or close to it and able to help with maintenance of these new facilities and future capital projects.

The web-based and virtual/augmented reality training programs of the learning process of the entry level personnel can be accelerated.

4.3 EXAMPLES OF MEGAPROJECTS FAILURES

Some of the major megaprojects failure histories are as follows:

4.3.1 Kashagan-Kazakhstan

A casebook example is Kashagan, a giant Caspian oil and gas field discovered off Kazakhstan in the early 1990s and developed by international oil companies. A highly challenging project, it is years behind schedule, plagued by cost overruns—and now halted by leaks from *corroded pipelines*.

The investment per barrel in Kashagan was almost *five times* higher than a typical Iraqi or Saudi field, Mr. Terzian said, and companies have become wary of the danger of finding themselves trapped in excessively high-cost projects if oil and gas prices should fall.

The massive oil and gas field in the Caspian Sea has become the most expensive oil project in the world and is perhaps the most technically daunting. Originally discovered in 2000, the project was billed as the largest oil discovery in decades. It is the fifth largest field in the world by reserves, and the largest outside of the Middle East.

Kashagan is probably the leading example of a megaproject in development—while also serving as a stark reminder of the difficulties involved in such work. The Kazakh field is an undisputed giant, with reserves of 9–13 billion barrels of oil, and started up briefly in 2013, before shutting down as a result of pipeline corrosion.

Despite the alluring bounty, the project has turned into a nightmare for the consortium of companies involved, which includes Eni, ExxonMobil, Royal Dutch Shell, Total, and the Kazakh government-owned oil company.

Dubbed by The Economist as "cash all gone," Kashagan has been plagued with delays, cost overruns, and engineering difficulties. The consortium has had to build artificial islands, work through winter ice conditions, and drill at great depths and pressure. An official with the consortium has called

Kashagan "one of the largest and most complex industrial projects currently being developed anywhere in the world."

But perhaps the most challenging is the "sourness" of the gas the companies are pulling out of the seabed. Production finally began in 2013, after an 8-year delay. However, just a few weeks later, operations were quickly shut down because the pipelines suffered from leaks.

The problem is toxic hydrogen sulfide gas, which is corroding the pipelines. Fixing the pipes could tack on an additional *3.6 billion USD* (or more) to the project, with production not set to resume until 2017.

The project's cost has been tipped to pass the *130 billion USD mark*, while participants in the development bicker and the Kazakh government demands progress.

Kashagan was made possible by the opening up of Kazakhstan—and other former Soviet Union states—following the end of the USSR. But the hunt for new frontiers has moved on, for instance, with companies bullish about the capacity for Brazil's pre-salt to yield new giant finds. The resources are there, but the complexity of developing the Libra field offshore Brazil is a challenge, particularly when combined with governmental requirements and restrictions.

4.3.2 Gorgon LNG Project, Australia

Although oil remains the world's pre-eminent energy source, gas supply and demand has grown rapidly in recent years. Helping drive this have been major gas discoveries, for instance, in offshore Australia, which have also come at a high price. Chevron's Gorgon LNG project is believed to now (April 2014) carry a price tag of *US$54 billion*, up from the original estimate of *US$37 billion* made in 2009.

4.3.3 Kearl Oil Sands, Canada

One of the largest and most expensive oil sands projects, Kearl has been a logistical challenge to say the least. It holds an estimated *4.6 billion barrels* of oil. The open-pit mining operation is supposed to produce 345,000 barrels per day, but has experienced equipment malfunctions and setbacks, and production has run well low of what Imperial Oil, the project's lead operator, had planned. The facility had to be temporarily shut down in late 2014 because of failed equipment.

To build the operation, equipment had to be shipped from Asia to the United States, and then broken up so that they could be transported on highways through Montana and Idaho. The equipment then had to be broken up into even smaller loads because of public opposition to the convoys of trucks moving through those states. Delays and cost overruns have bedeviled Imperial, but it is pushing on with an expansion of Kearl, despite low oil prices. The project has now cost more than *20 billion USD*.

Preliminary cost estimates suggested that a phased mining and bitumen-separation development, with no associated upgrading facilities would involve capital spending between *4.5 billion USD and 6.5 billion USD* (CDN, 2005 $). Approximately 40% of this expenditure was expected to occur for the first train and 60% for the two subsequent trains.

4.4 MEGAPROJECTS PROBLEMS AND THEIR CAUSES

Megaprojects' poor results can be contributed to a variety of reasons. Several different researchers have determined different causes of poor performance. The causes of poor performance are:

- Lack of realism in initial cost estimates
- Underestimation of the length and cost of the delays
- Contingencies are too low
- Changes in specifications and designs are not adequately taken into account
- Underestimated changes in currencies exchange rates
- Underestimated geological risk undervalued quantity and price changes
- Underestimated expropriation costs and safety and environmental demands
- High risk as a result of technological innovation that is translated into cost increase

The Construction Industry Institute CII summarizes the general reasons of large-scale project problems in the following areas: Front end planning, design, procurement, startup phase, human resources issues, organization structure, project processes, and project control (CII 1987).

Throughout the construction management literature, numerous case studies have been conducted to determine the causes of the poor performance. The causes of poor performance are summed up in the following list:

Planning phase causes of poor performance:

- Incomplete design especially with innovative designs
- Incomplete execution requirements
- Nonrealistic planning in terms of cost and time (compressed schedule and competitive low prices)
- Underestimation of the project's complexity (optimism in ability to manage such complexity)

- Underestimation of the size and material requirements (optimism in ability to supply and manage resources with such size)
- Underevaluated risks (low contingencies for technical, operational, and business risks)
- Difficult long-term planning that doesn't solve uncertainty issues in the long run
- Inefficient governmental procedures and regulations, environmental and other time-consuming effects

Execution phase causes of poor performance:

- Variations and mistakes due to inadequate planning, incomplete execution requirements, and ambiguous design documents
- Poor project culture leading to productivity loss parties' adversarial relationships and disputes
- Inadequate project organization that is insufficient for the size and complexity of the project
- Poor communication and team work
- Inefficient owner decision-making structure caused by governmental intervention, paralyzed public/private inadequate partnership operation
- Poor coordination and integration of work crews; inexperienced personnel in critical positions

Several causes of poor performance obtained from the literature review, i.e., underestimation of size and complexity, nonrealistic planning, inadequate project organization, inefficient structure, and poor integration, highlight the fact that the conventional management practices are insufficient to handle megaprojects' characteristics. The following section analyzes how megaprojects characteristics impose managerial challenges that are difficult to handle leading to poor performance in terms of cost and time.

4.5 SYSTEM ENGINEERING FOR THE MEGAPROJECTS

4.5.1 What Is the Product of the Magaprojects?

The "product system" is the system to be produced by megaproject at the conceptual design stage to fulfill the stakeholders requirements. At this stage the system architecture is a graphical description of the system (often supported by text and data) that shows the complete system at the highest level and, as a minimum, includes:

1. The total system
2. External interfaces
3. Next lower level systems/subsystems and their interfaces

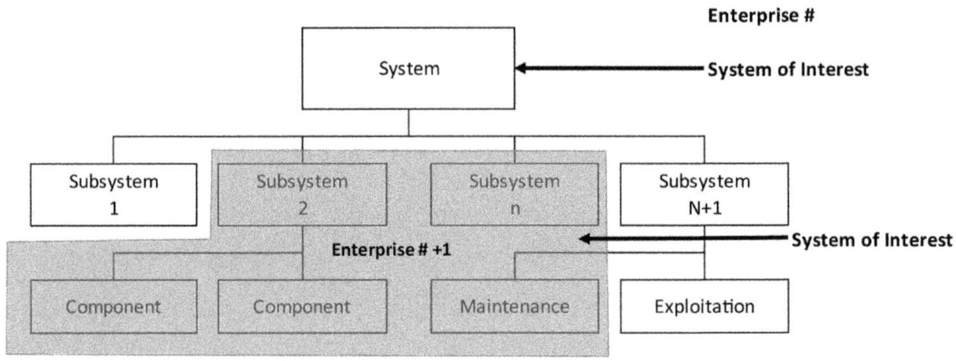

FIGURE 4.3
The systems of interest is relative to the enterprise.

Once the system architecture defining the design solution is sufficiently developed, the lower level systems, subsystems, and components can be progressively developed as shown in Fig. 4.3. The logical decomposition of the system is called the system breakdown structure (SBS) and needs to consider:

1. How the system is going to be procured, i.e., the elements to be packaged into separate contracts and agreements;
2. How the system will be designed, built, and integrated—the disciplines involved (e.g., process, civil, mechanical, electrical, and software engineering) and the order of manufacture or construction and proving and bringing into service;
3. The arrangement of the functions and management of critical interfaces.

The product system is expected to achieve specific functions which have certain performance, reliability, availability, maintainability, and safety (PRAMS) characteristics. At the start of the construction stage, as well as the specification of the product system, we also have a defined budget (the estimated cost) and time scale (program or schedule). This defines each side of the megaprojects' project management triangle for cost, time, and quality, as illustrated in Fig. 4.4.

The objective of our megaprojects is to deliver the system [achieve the agreed solution and meet all requirements (as progressively changed under control through the project lifecycle)] while maintaining a "balanced PM triangle," in the sense of balancing the weights assigned to the three sides of the triangle as agreed between all the parties involved.

The megaproject has a lifecycle with discrete phases (separate delivery and introduction into service of a version of the system) and stages (separate

FIGURE 4.4
The project management triangle.

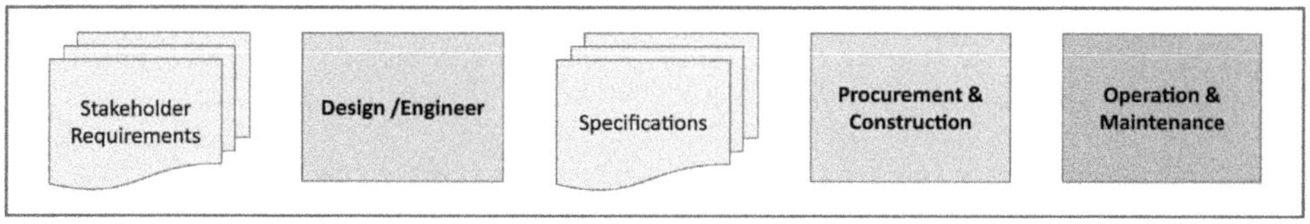

FIGURE 4.5
High-level view of a traditional megaproject lifecycle.

increments of the lifecycle for each phase). A megaproject consists of two main stages as shown in Fig. 4.5:

- the development of an engineering solution in the form of a detailed design and
- the delivery of the design solution through procuring contractors to build or construct the solution.

For the manufacturing industry, the deliverable is most often numerous items of the same product, such as the same airplane, car, radar, or mobile phone, and the effort that goes into developing and testing the detailed design before entering mass production is usually many times the cost of a single item. Design is also a technically complex stage, with leading edge technology from different disciplines required to meet very challenging specifications, intense interaction with multiple users, and so on. Consequently, many of the SE processes are either directly concerned with this design stage or relate to the management of the technical complexity when addressing other project aspects.

This high-level lifecycle is decomposed in a work breakdown structure (WBS) that must relate to both the SBS and the way the work is to be divided between different organizations, the organizational breakdown structure

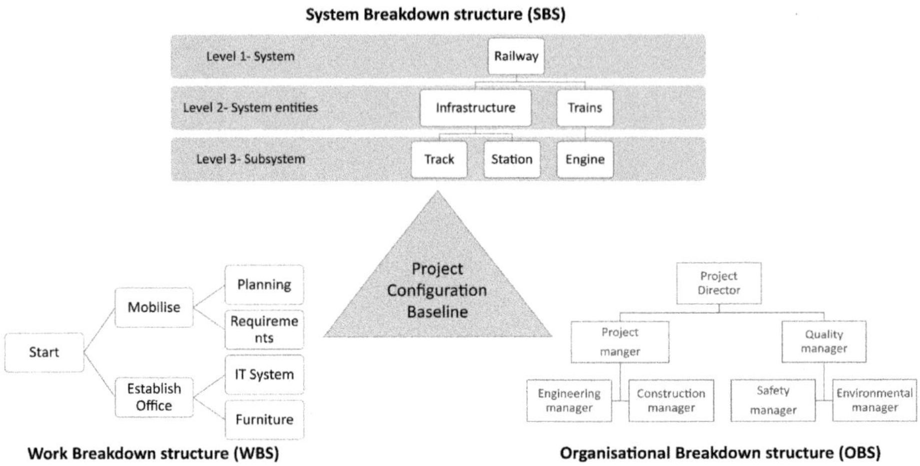

FIGURE 4.6
Relationship between system, work, and organizational breakdown structures.

(OBS), as it will define who and how each part of the system is going to be successfully completed. Fig. 4.6 illustrates the relationship between System, Work & Organizational Breakdown Structures.

Using a framework based on an interrelated WBS, OBS, and SBS the time and cost of each work package, each contributing organization, and each element of the system can be seen, hence making it relatively straightforward to monitor and, if necessary, make adjustments so as to maintain a balanced PM triangle. The project should be baselined (captured) using the WBS, OBS, and SBS at appropriate milestones (e.g., at the end of each lifecycle stage) so that the megaprojects' time, cost, and quality are synchronized to a consistently defined scope. Changes to the project from one baseline to the next are controlled using the *configuration management* SE practice.

In the case of a megaproject, such as a major highway or railway it is not always possible to develop or test a prototype; the first item is the only item. The technology involved may not develop as quickly as, e.g., in electronics; the design is, to a large extent, circumscribed by standards and codes; the technical solutions will be similar to those of recent projects; and the cost of the design will be a fraction of that of the implementation, which in this case is the construction stage. All process manufacturing plants lie in the same category of these major construction megaprojects. The 3D model of plant and not integrated 2D and 3D models and simulations of the critical technical issues substitute the prototype for these projects.

Modularization is crucial for cost optimization of the process manufacturing megaprojects. The modules are unique and should be fabricated according to the client specifications. Very often, the fabrication of the module starts when the engineering phase is not finished yet. This makes project very vulnerable to any changes or misalignment or lack of timely communications of the multiple vendors, EPC, and duty holder.

Due to these factors, the design processes for megaprojects should be well established and reasonably optimized to address any complexity which lies more in issues of interfaces, procurement, and constructability, given the specific political, commercial, and local conditions, than in the design of a certain type of bridge, the particular alignment of a railway or highway, or the type of process manufacturing plant.

The progress in virtual reality (VR) and IIOT enables to develop a "digital twin" of the project and safely experience it at real size. Fig. 4.7 illustrates how the engineers can have a better understanding of the systems that they design or operate using the digital twins. We will cover this subject in details in Chapter 5 of this book.

The procurement and construction stage, on the other hand, is usually rather unique and quite specific. Each project faces new "on-the-ground" conditions and a host of challenging decisions relating to temporary works, enabling works, land use for equipment and material storage, site and service access without disturbing existing operations, allocation of heavy equipment, access to utility services, management of multiple work

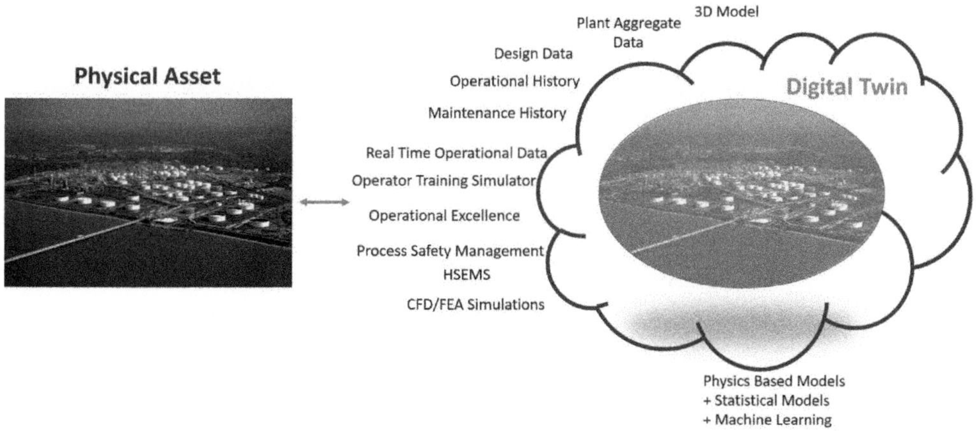

FIGURE 4.7
A Digital Twin continuously collects sensor data on the asset and applies advanced analytics and self-learning artificial intelligence to gain unique insights about its performance and operation.

sites, workforce accommodation at remote work sites, sequencing of work, the interaction of numerous contractors on the same site, and many more.

Cloud and fog computing enables the digital transformation of the businesses in which the big data generated by the management systems are effectively managed.

4.5.2 Modularization

Modular design, or "modularity in design," is a design approach that subdivides a system into smaller parts called modules or skids that can be independently created and then used in different systems. A modular system can be characterized by functional partitioning into discrete scalable, reusable modules, rigorous use of well-defined modular interfaces, and making use of industry standards for interfaces.

Besides reduction in cost (due to less customization and shorter learning time), and flexibility in design, modularity offers other benefits such as augmentation (adding new solution by merely plugging in a new module), and exclusion.

Examples of modular systems are cars, computers, solar panels and wind turbines, elevators, and modular buildings. Earlier examples include looms, railroad signaling systems, telephone exchanges, pipe organs, and electric power distribution systems.

Computers use modularity to overcome changing customer demands and to make the manufacturing process more adaptive to change. Modular design is an attempt to combine the advantages of *standardization* (high volume normally equals low manufacturing costs) with those of *customization*. A downside to modularity (and this depends on the extent of modularity) is that low-quality modular systems are not optimized for performance. This is usually due to the cost of putting up interfaces between modules.

In recent decades process industry and in particular oil and gas sector strive to improve the delivery of major projects by adopting a modular approach successfully used by other industries.

This section will discuss modular plant design and how it differs from conventional stick-built design, including cost and project schedules. Modularization in this section refers to the use of shop assemblies, subassemblies, and full-scale modular packages.

It's no secret that cost and time overruns are commonplace in the oil and gas industry. In fact, most current projects incur one or both problems, and spiraling project-development costs may soon put severe strain on the ability of companies to earn adequate returns. Yet there is an alternative: the industry could reduce costs and accelerate project delivery times by more aggressively implementing advanced modularization and standardization approaches such as those already used in industries that range from autos and electronics to satellite manufacture.

Many oil companies think they are standardizing already. Yet most are actually only scratching the surface and, as a result, missing a major opportunity to gain a competitive edge. To truly capture the benefits, companies need an approach that takes the best of what was developed for these other industries and tailor it to large capital projects. Called "modular standardization," this approach entails dividing a plant into modules that are standardized and then reused multiple times.

Studies on modularization of the oil and gas companies show that:

- Direct project costs can be reduced by as much as *15%*;
- Project delivery can be speeded up by as much as *20%*;
- The constraints of transport of the modules push the design boundaries for a more optimized elevation, congestion, and accessibility of the units.

Because common equipment is used, safety performance is improved by increasing the sharing of information about failure modes. And the effectiveness of technical personnel is raised because they can stop "reinventing the wheel" and instead work on more challenging and value-creating projects. In fact, modular standardization is already being successfully used by other industries that work at *large scale* and with high *complexity*, such as industrial turbine manufacturing and steel-mill construction.

To be able to move to modular standardization, oil and gas companies need to make changes in two main areas:

- The first is *project design*, where they must adopt modular architecture and reuse standardized modules across multiple major oil and gas plants.
- The second is *organizational*: most oil and gas companies come from a tradition of building stand-alone projects designed to specific geological conditions. Going forward, they must make the reuse of existing modules the norm in their organizations.

What is the most effective way for an oil and gas company to move to a modular architecture? For common projects, it needs to break the work up into corresponding modules and then determine the appropriate level of standardization for each. Clear guidelines, including a sound business case, direct which modules or submodules can be standardized and which must be customized. In the end, each will have its own standardization strategy, ranging from identical design to a set of discrete options to fully customized.

In our experience, it's possible to standardize at least one-third of all modules or submodules for even the most complex equipment. For some types of projects (e.g., those with little variation), it's possible to standardize up to two-thirds. The frequency of use, complexity, and nature of the module will determine the appropriate level of standardization; decisions must be driven by a clear business case on a module-by-module basis. At the same time, the benefits from modularization increase with the scale of a company's portfolio: the more units, the greater the impact on a company's bottom line.

There are a few smaller plants where complete plant-level standardization is possible—e.g., the standardized "monotower" unmanned platforms used for gas production in the North Sea. For a floating production, storage, and offloading vessel, the oil and gas processing module can use a standard design template, but it must be scaled to the oil and gas flow characteristics of the particular well. However, the power and compression module can be standardized because the same design and equipment can be reused on many vessels. For larger plants, such as offshore platforms or liquefied natural gas (LNG) plants, the focus moves to replicating modules that make up the plant, such as helicopter-landing pads on platforms or compressor trains in LNG plants.

While it is believed that modularization and standardization can provide major benefits to the oil and gas industries, two issues have proven to be stumbling blocks:

- First, companies must overcome the natural reflex among many project managers: to think that their projects are *unique* and therefore resistant to common approaches.
- Second, companies often fail to *convince* project-design engineers that standardization brings benefits that more than compensate for limited design choice.

Organization changes in four areas can help resolve these issues:

1. *In engineering and design activities*, companies that have successfully embraced modular standardization employ common design specifications and guidelines for each project type (e.g., refinery or production platform). Typically, such organizations have a library of modules built with cross-functional input (engineering, commercial, and procurement) and use design software that provides access to approved modules and equipment lists covered by supplier purchasing agreements.
2. *In project management*, these companies broaden stage-gate-review criteria to include plant- or module-design reuse, and to minimize design changes. Some companies are developing metrics to track reuse and accelerate adoption.

3. *The procurement* group should be involved from an early stage, making decisions alongside the engineering department. Successful companies involve internal procurement personnel in project decisions from the concept-assessment stage, and also tap the expertise of suppliers from the beginning.
4. Finally, successful companies make changes to *staffing and individual incentives* such as staffing project teams centrally to ensure the same team members work together on subsequent related projects. They designate technical staff as "module owners" who are responsible for making sure their modules are used across the portfolio. These companies develop a culture that promotes and values standardization through the use of common training and capability-building programs, and incentivize engineers to work across different projects and reuse modules.

By pushing aggressively in these areas of modular project design and in organizational changes that can help advance standardization, oil and gas companies can make substantial progress toward reining in project time and cost overruns. It's an opportunity they must grasp.

4.5.3 Measuring Successful Delivery of the Megaprojects

To measure our success, we need to start out with an agreed definition of the system with both the stakeholders that are to accept the product of the megaproject and the parties responsible for its delivery. This definition is called the *required system build configuration (RSBC)*. Fig. 4.8 illustrates how the SBC

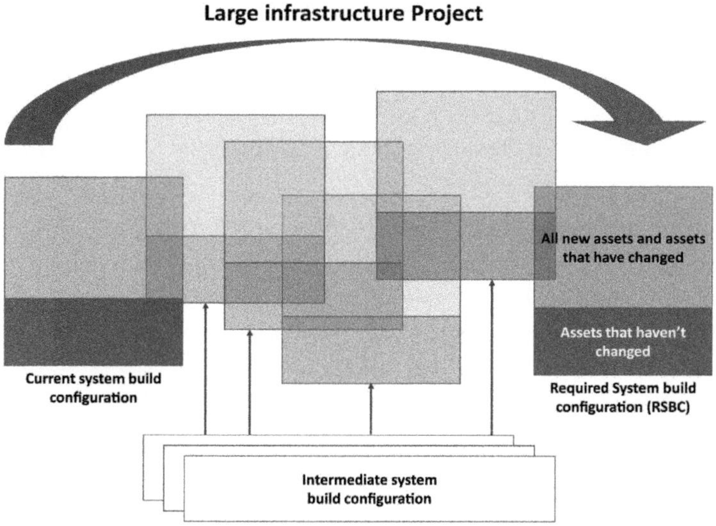

FIGURE 4.8
How the SBC evolves through the project lifecycle.

evolves through the Project Lifecycle. The success of the megaprojects will, in part, be measured by demonstrating achievement of (validating) each item of the RSBC (i.e., each configuration item). The extent to which the RSBC is delivered in compliance with specifications (such as regulations, design codes, and standards) and how well the resulting system is verified, validated, and documented will determine the quality of the megaproject.

Other measures of success will be associated with the balance of the project management triangle; how much it cost and how long it took. If we control what is actually being built, in the form of the system build configuration (SBC), throughout the project lifecycle, then the metrics associated with changes to the system, its cost, and its timescales should be captured in the process so as to provide the relevant metrics against which the successful balance of the project management triangle is judged.

Thus, the SBC needs to be closely monitored and managed. Designs often change, for instance, due to unknown field conditions which become apparent during the construction stage. In addition, stakeholders often change their requirements as their expectations vary over time, they better understand the detail of the solution or some significant external factors emerge. These changes will result in an update of the RSBC. The actual definition of the SBC therefore needs to be a living document, with progress and modifications captured throughout the project lifecycle. The final SBC will be validated against the revised RSBC, and the project as a whole against the stakeholders' goals and objectives.

4.5.4 Managing Change/Configuration Control

The constructor will start with a design, but inevitably changes will be required due to constructability, efficiency, fit, unforeseen circumstances, or a number of other reasons. Controlling change is one of the most important processes during construction. A poorly run change management process can cause significant cost overruns and schedule delays.

Configuration control of the process design is essential and critical. Standards such as ISO 10007 and ANSI/EIA-649 define configuration management processes. However, these standards were established for common industrial circumstances and sometimes need to be tailored to meet the needs of large, complex infrastructure projects. Many infrastructure processes include development of control systems or other systems containing significant software. Processes that address the particular characteristics of software should not be overlooked.

Change management requirements will vary with the procurer and the selected project approach. For example, highly regulated projects (e.g., in the nuclear industry) may require design documents to accurately reflect the facility as construction progresses. Other projects may allow the design documents to be updated to reflect the as-built facility after construction completes. In addition, deciding what must be maintained under configuration control will also greatly influence the cost of any change.

A fundamental concept of change control is that the design, facility, and support documentation (e.g., operations, maintenance, and training manuals) must be fully aligned not later than the time that the system is transitioned into service. Thus, change processes for the procurer, project manager, engineer, constructor, and O&M organizations must align at this point of integration. Coordination between all affected parties is essential to success, especially to prevent this process from becoming a bottleneck restricting progress. The structure and nature of these processes will depend on contract strategies, and on where the technical authority lies.

A change management process will have the following characteristics:

1. A set of documented processes for configuration identification, control, status accounting, and audit. Ad hoc methods should not be used.
2. Identifies items and documentation that are required to be maintained under configuration control.
3. Requires identification of all affected configuration-controlled items, requirements and supporting assurance evidence (e.g., calculations), associated design and procurement documents, and support documentation. (Items and documentation that are not required to be controlled need not be identified.)
4. Requires change documentation to be packaged and reviewed for completeness and for impact to the project baselines.
5. Requires a sequence for implementation and integration with other change packages.
6. Requires clear approval authorities to be established.
7. Ensures alignment among design, facility, and support documentation.
8. Identifies and records construction evidence required to document as-built conditions.

Refer to the configuration management standards for a more complete list.

Tightly coupled design and construction processes require more rigid configuration management processes to ensure changes are correctly and fully implemented. In such circumstances, construction is often accomplished at risk that something completed may have to be changed. For example, when developing a large complex of integrated facilities, design and construction might be coincident. If a change is necessary, the impact to previously acquired and constructed items needs to be determined and carefully managed to ensure configuration control is not lost. Thus, the risk for change needs to be considered when sequencing and scheduling construction activities.

4.5.5 System Engineering Process Verification and Validation

SE practice dictates that criteria for validating that requirements are met are specified early in the project lifecycle and, where practical, at the same time as the requirements themselves are defined. Assurance (by verification) that the processes are being followed, and that the intention of each stage has been achieved in the subsequent stage of the lifecycle, is also important.

Processes are measured to ensure they are producing the expected results within time and cost constraints. There are a number of tools used for this purpose, including assessments, audits, and measuring process factors that provide leading indicators of performance.

Assessments are self-performed investigations designed to examine compliance against requirements, or to determine how well a process meets the intended objectives. Results identify areas for improvement. These are effective tools to discover where process designs or personnel are not performing as intended, and provide warning of potential noncompliance and other project issues.

Audits are performed by those independent of doing the work (e.g., by quality assurance staff, the procurer, or the regulator) for the purpose of checking process results against requirements. These audits are not controlled by the contractor but need to be included in their budgets. Like assessments, audits reveal process or implementation weaknesses or noncompliances that need to be addressed to ensure project success.

The most important approach to process verification is the measurement of factors contributing to success. Each process can be considered an equation where the final result is based on the contributing factors. These factors can be supplier quality, weather, specific construction methods, tooling arrangements, process controls, or other factors that influence output quality. Each process could have hundreds of such factors, but not all factors are important. Identifying these factors will require collecting data. Six sigma methods may be useful in assessing these factors. Once found, these factors will provide leading indicators useful to point to problems that can be addressed before they adversely affect quality.

It is important to also validate that we are achieving the results expected (or required) from the construction process. Evidence of requirements satisfaction tends to be associated with the measurement of how well the product of the project is meeting the defined quality, time and cost targets assuming an optimized construction process. The quality target for the product will be measured using established SE methods (e.g., analysis, inspection, demonstration, factory, and on-site testing).

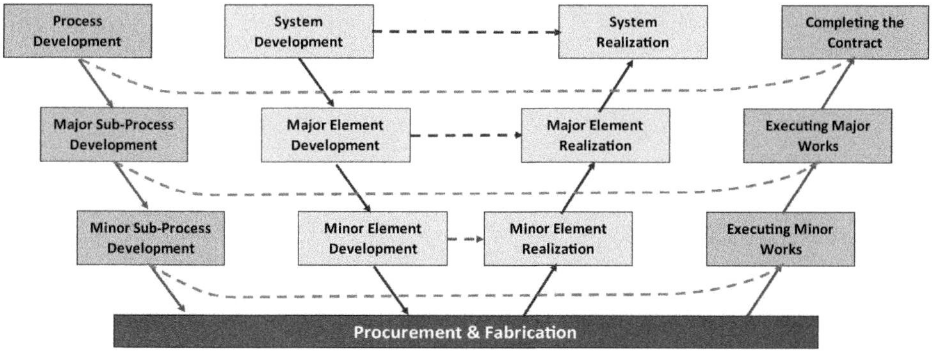

FIGURE 4.9
The modified "V" model.

For a megaproject many systems will be integrated on-site during the construction stage as the product system of the project is progressively built. It is important that all the systems are proved prior to integration (some off-site) and then again as part of the integrated system.

A recent approach to verification and validation that is particularly applicable to megaprojects is that of progressive assurance, as used on railway projects.

The modified "V" model in Fig. 4.9 shows the duplication of the legs to explicitly show the simultaneous development of the product and the process. The direction of the arrows indicates the idealized flow of information and/or material, in reality there will always be some corrective feedback between stages.

4.5.6 Defining and Allocating the Hand-over Responsibilities

A common model used within infrastructure projects is the "V" Model. A version updated by the Dutch Ministry of Public Works as illustrated in Fig. 4.10.

The situation today is that there is frequently a misunderstanding of the responsibilities within the model. Acquirers often underestimate their responsibilities vis-à-vis stakeholders and the interfaces with them at system level. The contractors are often thought to be doing a good job for the acquirer. On the other hand, contractors often have a poor understanding about the goals of the acquirer on a political and environmental level.

The hand-over responsibilities defined by the V-model are bidirectional. At each system level, the hand-over responsibilities are shifting from the acquirer to the contractor. The acquirer is always the main party related to system responsibilities. Within megaprojects the political and environmental impact can be very significant. The acquirer must ensure that the contract is

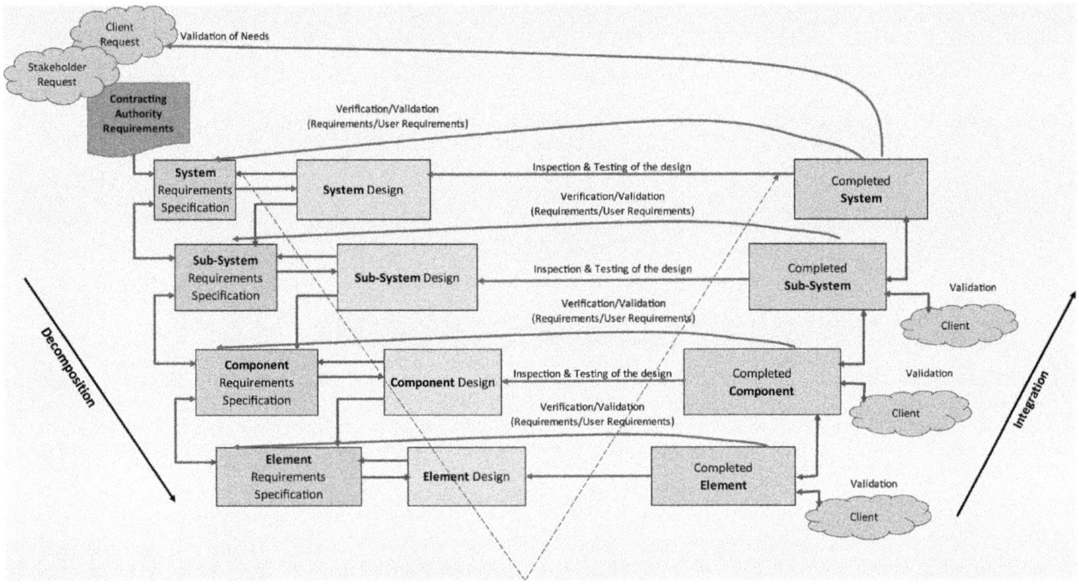

FIGURE 4.10
The updated 'V' Model for the infrastructure projects.

complete and clear. In other words, the acquirer has to verify that the contract will deliver an operational system that will meet the requirements and will meet the needs and goals of all stakeholders, which were the basis for starting the project in the first place. The acquirer is, e.g., responsible for the stakeholder analysis at the system level, and from that point of view the acquirer must set up a system assurance strategy. This is extremely important with stakeholders in the area of state and local politics.

When contracting at a certain system level, the acquirer is responsible for delivering a system specification(s), system assurance strategy and verification and validation management plan at that system level. The contractor is responsible for delivering the design (starting at system level), verification and validation plans based on the system assurance strategy, a compliant system and support for the final validation of the system itself.

Contracting at a lower system level will give the acquirer more responsibilities on design, implementation, integration, verification, and validation, and Fig. 4.10 shows a case where the acquirer provides the system design and contracts out individual subsystem D&C packages.

Fig. 4.11 shows the allocation of responsibilities in a case where the acquirer (contracting authority) has developed the system design and is contracting out D&C packages for individual subsystems (only one shown).

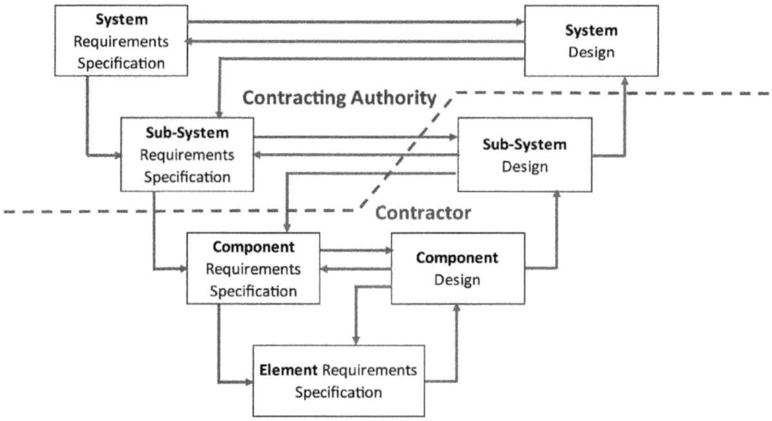

FIGURE 4.11
Allocation of responsibilities.

4.6 DEFINITION OF COMPLEXITY FOR MEGAPROJECTS

Luhmannian defines the complexity in his system theory as follows:

Complexity is the degree of manifoldness, interrelatedness, and consequential impact of a decision field.

The advantage of this definition is that it takes us to one of the core concepts of organizational science: decision-making and thus also to the topic of megaprojects. Manifoldness refers to the *differentiation* of functions in a megaproject. This could be client, designer, contractor, subcontractors, suppliers, banks, authorities, and the public or it could be the internal differentiation of the contractor's organization.

The *interrelatedness* describes the dependencies between supersystem and the different subsystems or among the latter ones. The consequential impact, finally, draws our attention to the number and importance of the causal chains or the consequential processes set in motion by any one decision. The degree of complexity is then the sum within each of the three components and overall three of them.

Every system is defined according to a size and a degree of complexity. Complexity is determined by the number of different elements in the system and the interrelatedness of these elements; size is determined by the quantity of similar items per element. Accordingly, a project diagram (Fig. 4.12) is developed to define and differentiate projects based on size and complexity dimensions. The complexity dimension is defined by two aspects, differentiation, i.e., total number of dissimilar elements (shown in different geometrical shapes) in the system, and interdependence among these elements (shown in the number of arrows).

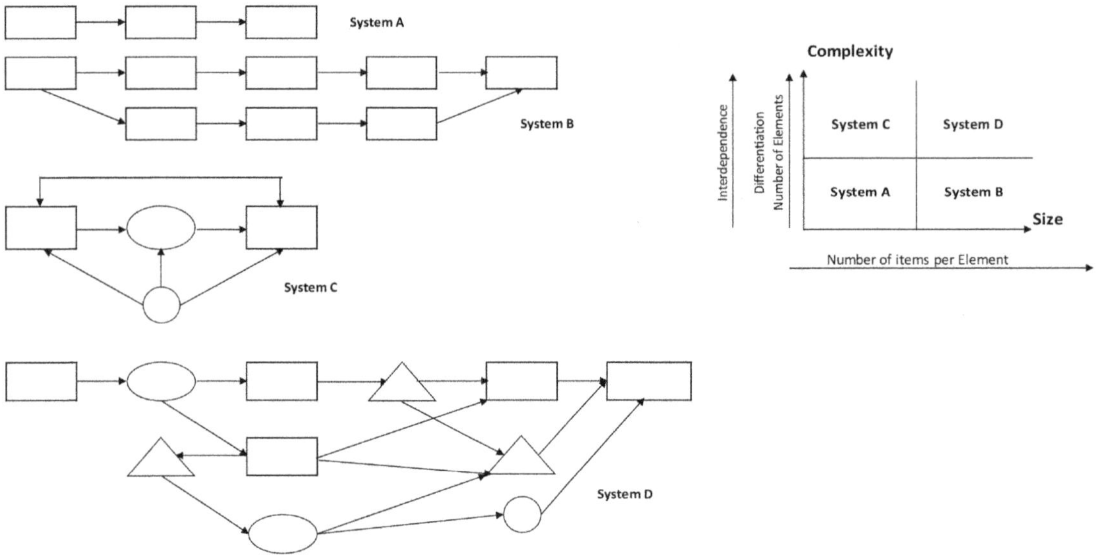

FIGURE 4.12
Size-complexity project diagram.

Project size can be explained by a multitude of aspects. In this paper, the most basic size aspects considered are the constructed area and the time frame needed to build the project. The constructed area and the construction time frame are indications of the quantity of items per element used in the project such as labor, equipment, work groups, and material used.

Gidado (1996) determines in an article on project complexity four different sources:

1. Employed resources
2. Environment
3. Level of scientific and technological knowledge required
4. Number and interaction of different parts in the work flow.

A large amount of required resources, a turbulent environment, working on the edge of technology, and innumerable possible interactions are certainly identifiable for megaprojects. So by this definition, megaprojects are complex and yet the definition does not consider all possible layers of complexity because it just is concerned with task complexity. There are four other layers that need to be considered:

1. Social complexity
2. Cultural complexity
3. Operative complexity
4. Cognitive complexity.

Overall Project Complexity: Manifoldness, interrelatedness, consequential impact of a decision field		
Task Complexity	**Social Complexity**	**Cultural Complexity**
Density of activities in a spatial and temporal frame	Number and diversity of actors	Diversity of the cultural software of the mind

FIGURE 4.13
Overall, task, social, and cultural complexity.

While task complexity can be defined as the density of activities in a given spatial and temporal frame, social complexity depends on the number and diversity of actors communicating and working with each other. Cultural complexity comprises the history, experience, and sense-making processes of the different groups that join their efforts in a megaproject and that have taken place before its start. Operative complexity denotes the degree to which organizations of the project are independent when defining their operations to achieve given goals. Cognitive complexity can be treated on the level of a person or the level of a group. Only the latter perspective is different in megaprojects and it identifies the degree to which self-reflection (and thus sense-making processes, the emergence of an identity, or even an organizational culture) is possible (Wilke, 2000).

Due to the project character, operative and cognitive complexity have no time to develop in megaprojects and they can be discarded for the topic. What needs to be added is an overall complexity that encompasses the other three (Fig. 4.13).

Megaprojects are normally build by international construction joint ventures (ICJVs) that comprise at least two parties, which again is more complex than the organization of a single contractor or a client. Design/build approaches are also common for megaprojects, at least in newly industrialized or least developed countries and this is more complex than the traditional design/bid/build approach since the contractor organized in the form of an ICJV has to coordinate the design besides his other tasks.

4.6.1 Definition of Megaproject Based on Size and Complexity

Theoretically, a megaproject is not different from any other construction project. It includes numerous parallel activities, limited resources, tight schedules, and multiple decision-making parties. Therefore, a megaproject is defined in comparison with other projects according to size and complexity.

The project diagram is further developed to set a clear differentiation among different categories of projects.

The diagram's size axis is composed of the constructed area and time frame. The complexity axis is composed of two aspects. The first aspect is the design complexity that ranges from simple design to complex and difficult design. Simple design corresponds to a limited number of design steps and specializations. Complex and difficult design corresponds to a large number of design steps and design specializations. It also corresponds to a high degree of interdependence among different design steps and among different engineering specializations. The second aspect is the organizational complexity that ranges from simple execution to complex execution. Simple execution corresponds to a small number of work groups with limited interdependence among them. Complex execution corresponds to numerous groups with substantial interdependence among them. The following project categories can be allocated using the diagram:

4.6.2 Small and Noncomplex Projects (Small Projects, SP)

Small noncomplex projects are described to have a small constructed area and a relatively short construction time frame. They are simple to design and build. They neither require complex designs nor complex managerial organizations to have a successful project performance. Several projects could fit this category such as simple roads, houses, and small residential buildings.

4.6.3 Large and Noncomplex Projects (Large Projects, LP)

Large-scale projects have a large constructed area and a relatively long construction time frame. They neither require complex designs nor the services of many engineering trades. However, they require a complex organization to integrate numerous work groups and to manage limited as well as plenty resources. Several simple yet large projects could fit this category, for instance, long roads, highways, bridges, and airport runways.

4.6.4 Small and Complex Projects (CP)

Complex projects have a relatively small constructed area as compared to large projects such as highways. The time scale would vary among different projects. They encompass complex designs since numerous specializations are required. They require complex management, integration, and coordination of different engineering trades. Examples of such projects are complex hospitals, hotels, and other specialized projects.

4.6.5 Large and Complex (Mega Projects, MP)

As their name signifies, large-scale complex projects have a large constructed area and a long time frame. They encompass complex and challenging designs since numerous design steps and specializations are required. They require complex management, integration, and coordination of different work groups. Furthermore, they require the integration of different engineering trades within each work group. They require complex resource management given the limited labor resources and enormous material quantities. Examples of such projects are power plants, large industrial plants, airports, and large transportation such as metro-railway tunnels. Fig. 4.14 categorizes complexity levels of the projects design and execution by their size and execution time.

According to the project diagram, megaprojects are defined as large-scale projects that have a long time frame and are characterized by a complex difficult design and complex organization. Their size and design complexity necessitate a complex managerial structure in order to manage and integrate different work groups, engineering specializations, and limited resources.

4.6.6 Design Complexity

Design complexity is determined by design differentiation and design interdependence.

Design Differentiation: It is explained by two aspects. The first aspect is the number of different steps taken to achieve the final product. For instance, a

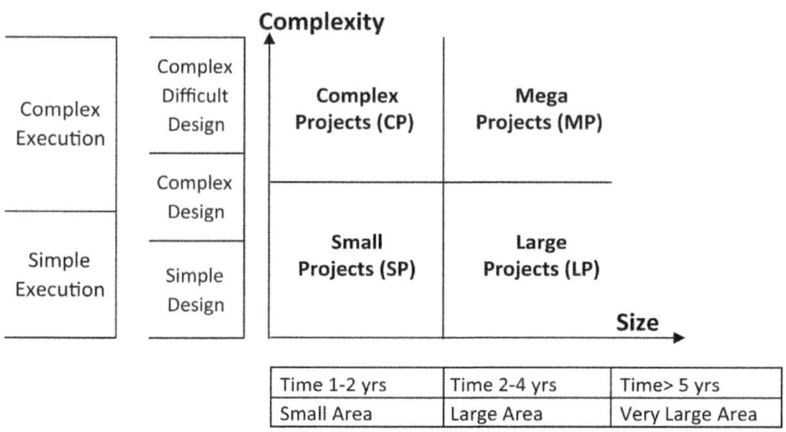

FIGURE 4.14
Complexity criteria for the and project development.

civil engineering design of a simple road would be composed of a limited number of steps. Whereas a tunnel design that includes many geotechnical, structural, environmental, and safety elements would be composed of plentiful steps. The second aspect of design differentiation is the number of different engineering specializations included in the design. For instance, designing a complex power plant would require civil, electrical, and mechanical engineers in contrast to designing a simple road that would require a limited number of specializations.

Design Interdependence: It is explained by the interrelatedness of different design elements. The elements can be different design steps that may be highly related or independent. In addition, interdependence could be explained by the relation between different engineering specializations. For instance, electrical and mechanical trades are more interdependent in industrial plants that include motor control centers, than in simple buildings that include simple electrical lighting and simple piping and ventilating systems.

4.6.7 Organizational Complexity

In the construction management context, organizational complexity is illustrated by the managerial and operational complexity. In other words, it is the complexity of integrating work crews and coordinating dissimilar engineering trades. Fig. 4.15 categorizes complexity levels of projects design and organizational by the diversity and interdependence level of their parts.

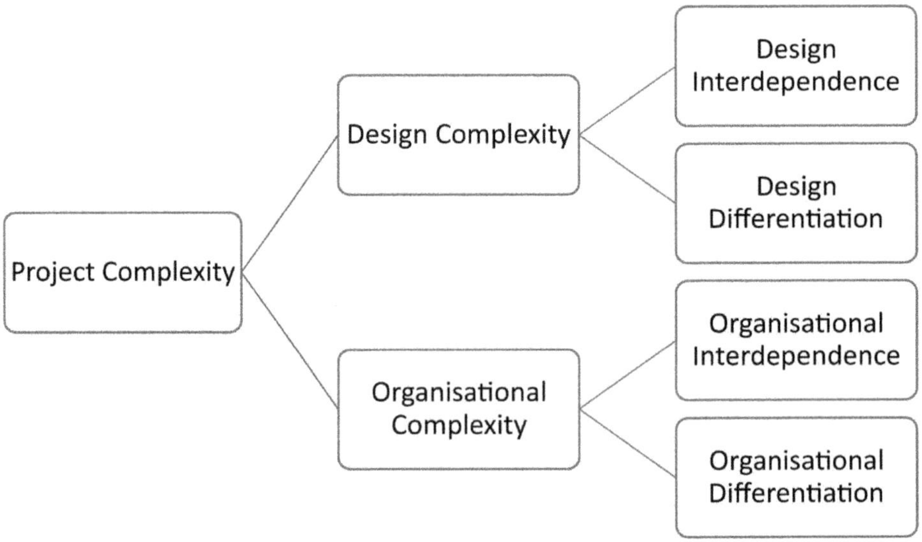

FIGURE 4.15
Types of the project complexity.

Organizational Differentiation: It is explained by the number of different entities to be managed. The entities could be different work groups and different engineering trades. As the project increases in size, more work groups are added contributing to more organizational differentiation. Also, as the project increases in complexity (namely, design complexity), more specializations are added contributing to more organizational differentiation.

Organizational Interdependence: It is explained by the interrelatedness of different entities' works. Different projects have different organizational interdependence depending on the nature of the works and the design complexity. Thus, different projects require varied efforts of integration and coordination among work groups. For instance, physically integrated electromechanical works found in electric rooms need more coordination efforts than physically distant electrical and mechanical works.

4.6.8 Development of Overall Complexity With Time

Complexity clearly is a construct that is not used by practitioners in the way defined above. In the interviews they rather describe the outcome of complexity without referring to the construct itself. In the beginning of a megaproject there is chaos, mess, disorder or there are just simply "headaches."

Complexity within the project can be reduced in many ways. Normally a functional organization is the first step, thereby transferring a part of the complexity to the departments (subsystems). In this case, the project management as supersystem gives away a good part of the control over the project. In return the coordination becomes more difficult and risk compensation a priority for project management. It must also adapt the ICJV to any changes in the environment.

Project management as well as departments then reduce complexity further through decision-making, coordinating, communicating, and learning. A decision connects parts of the project with each other in a specific way by allocating resources and choosing a technology. Coordinating allows to treat a variety of problems at the same time and then bringing them to a planned result. It is an answer to the task difficulty by specialization and the time pressure by concurrence. Communication works specifically in the areas of social and cultural complexity because such barriers can be found, discussed, and brought to a conclusion. Learning helps to standardize solutions and thus limit the search. The result is a reduction in the required resources (cf. learning curve). Of great importance in ICJVs is also the question whether the core management group has already learned to know each other on previous projects. This way cultural complexity is reduced right from the beginning.

At the end of a project, task complexity for the contractor is reduced to zero, he has fulfilled his contract and he can move on to the next project. This is a great difference between project and process complexity. The latter one might never come to an end. As social relations and cultural encounters are processes, their complexity will not be reduced to zero at the end of a project.

Fig. 4.16 summarizes the discussion above graphically by distinguishing a theoretical and a practical level of the construct "complexity." While practitioners see chaos and start working on it until the project is completed, complexity on a theoretical level is reduced to zero over time.

Going back to the theoretical level as a basis, how does complexity develop with time? Here, it is useful to differentiate between three phases of a megaproject. In the beginning everything is in disorder: the management staff does not know each other, the partner companies want to see their views implemented, the design is only in a preliminary stage, often—at least in innovative projects—the technology has not been used before. Many decisions are required, yet project knowledge is low. All tasks seem to be of highest importance, all are interlinked, and the consequences cannot easily be foreseen. This period lasts in smooth running megaprojects for several months, six is a good average. If the input from the clients is not coming forth timely, the period will be prolonged. From the outside there seems to be little progress.

The middle phase is characterized by a starting, then deepening routinizing of the work. A low level of complexity remains because of environmental influences or any other type of disorder (such as mistakes). Problems of social and cultural complexity always stay virulent.

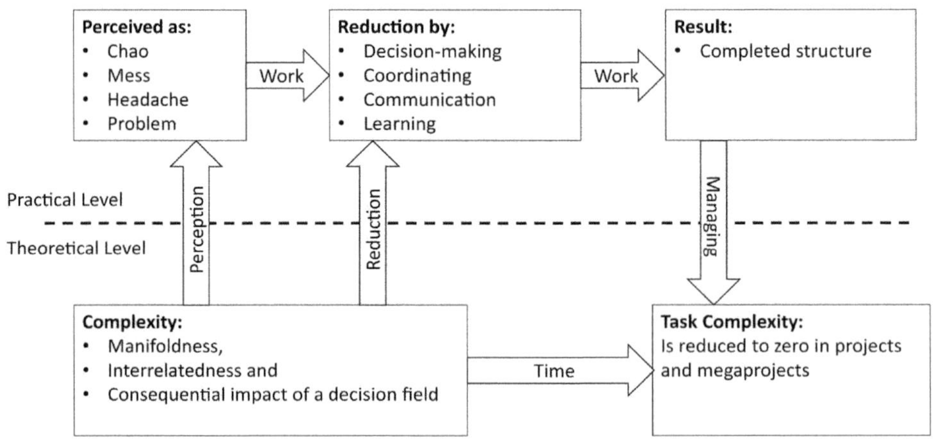

FIGURE 4.16
Complexity as practical and theoretical problem in megaprojects.

During the end phase the complexity picks up again as new tasks need to be performed. The contract must be wrapped up and the site and the offices closed down.

In Fig. 4.17 there are three examples of overall project complexity shown for different environmental conditions. The first one is for a smooth running project, the second one for a project with an important change order, and the third one shows the case of a megaproject where contractual disputes are not settled at the end of the project. The client on the Taiwan Highspeed Railway introduced a variation order with an increased earthquake loading. This is a case of environment 2. A project in Bangkok was completed in 2000 and is still not settled in the courts today (environment 3).

The complexity curves are different for the separate departments in an ICJV. The project management is faced with the overall complexity curves as shown in Fig. 4.17. The same curve for the design department is reduced much earlier to almost zero (pending the as-built status, Fig. 4.18).

The curve for administration has a very strong increase at the end because termination and sales of equipment are very different from the earlier tasks (Fig. 4.19).

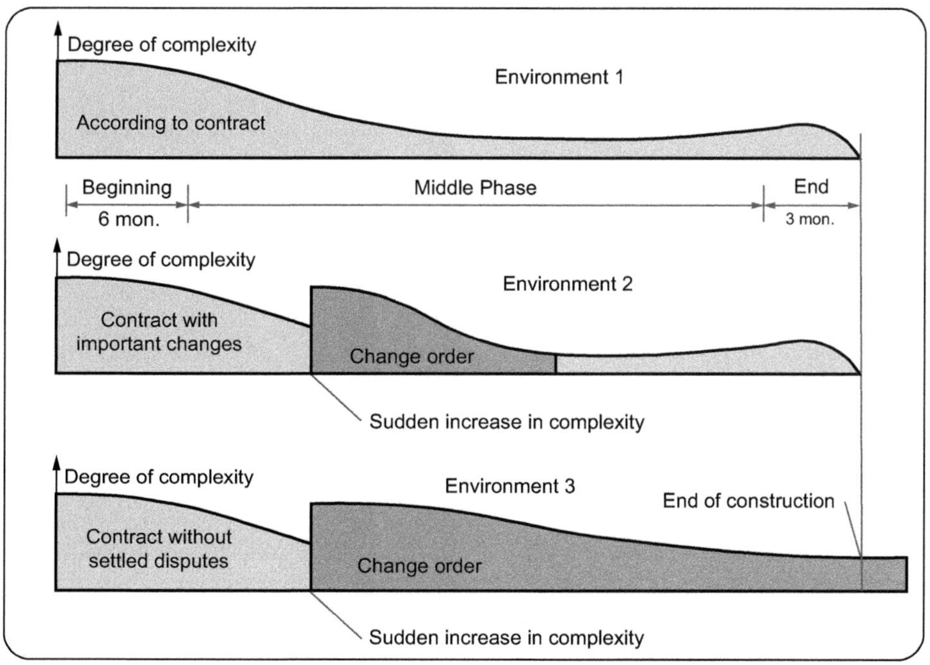

FIGURE 4.17
Overall complexity curves.

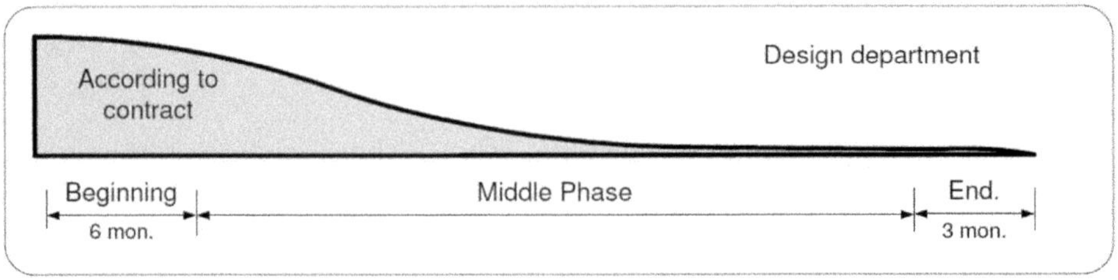

FIGURE 4.18
Complexity curve for the design department.

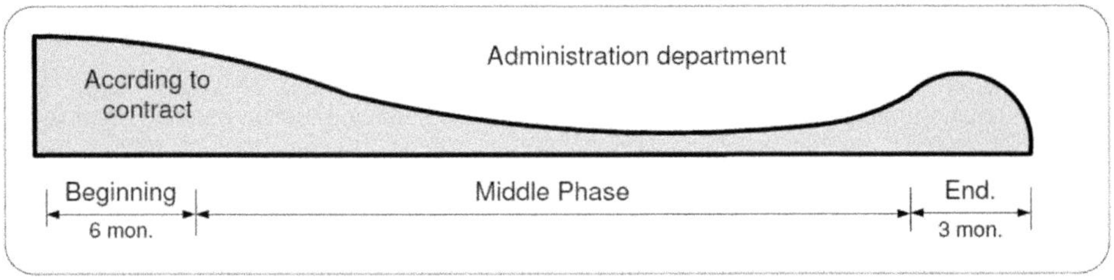

FIGURE 4.19
Complexity curve for administration department.

The figures might imply that the degree of complexity is the same for the different departments. This is not true. One of the most basic ways to deal with complexity is through departmentalization. Each department (subsystem) has to deal with a part of the complexity that is considerably smaller than the overall complexity. Each department must have competencies and resources to solve their task. This independence increases on the other hand the overall complexity for the project management.

As the discussion has demonstrated, megaprojects are more complex than normal projects, therefore they must deal with the overwhelming complexity by departmentalization. According to the task of each department, the complexity curves are different (Figs. 4.18 and 4.19). The way a contract is implemented through the client has also an impact on the complexity curve (Fig. 4.17).

4.6.9 Task Complexity

Discussions about complexity are often enough rather abstract (as above) or they do not capture the completeness of the construct. To grasp complexity, a formula will not suffice and the story of complexity is a very long one. A way out of this dilemma is a table of the task to be fulfilled on a megaproject. Yet this also has drawbacks, as neither interrelatedness nor consequential impact can be shown. They have to be guessed by the manifoldness of the topics in Table 4.1.

Table 4.1 Manifoldness of Task Complexity in Megaprojects

Area	Tasks
Organizational planning	– Organization – Organization chart – Competency matrix – Job descriptions – Contract management – Quality management – Safety management – Personnel management Purchasing – Financial accounting – Cost accounting – Communication Correspondence and filing
Design planning	– Outsourcing of design – Coordination of design – Approval procedure – Design schedule Documentation (as-built drawings)
Work preparation	– Work estimate – Controlling outsourcing – Construction methods – Scheduling – Deliveries – Planning of site installation Logistics
Site installation	– Land acquisition – Purchase of plant and equipment utilities – Offices, labor camps, canteens, lavatories Waste
Construction management	– Production processes – Quantity and quality control of materials – Quantity and quality control of subcontracts – Deployment of plant and equipment – Deployment of work force – Deviations from contract – Hand-over – Warranty

Five main areas can be distinguished for megaprojects:

1. Organizational planning
2. Design planning
3. Work preparation
4. Site installation planning
5. Construction management.

Organizational planning has much more importance than for normal projects, where standard solutions are used. The same holds true for site installation. The cost for site installation on a megaproject is around 15%, this is for a billion-dollar-project 150 million dollars and by itself larger in volume than most normal projects.

It should never be forgotten what size means in a megaproject. When looking at the tasks some examples might suffice: Personnel management means employment of several thousand staff and workers, coordination of designs means managing a worldwide network of engineers, construction methods means the use of an untested technology, labor camps mean setting up hole villages for one or two thousand people, and control of subcontracts means coordinating a hundred contracts with subcontractors from around the world.

All tasks are highly interrelated and a decision on one subject will have repercussions for many others.

4.6.10 Social Complexity

Social complexity shall again be highlighted by just one example. Especially for all decisions in the area of organizational planning, all partner companies have their own experiences and preferences. Through their employees who they send to the ICJV they try to implement these preferences. The project management includes employees from all partners who are urged to follow the wishes of their bosses. Yet on the other hand, the project management must decide on a coherent set of management procedures that assure efficiency under the terms of the contract. Moreover, the client also has through the stipulations of the contract an influence on the organization. In the end it is amazing when despite all these actors an efficient organization emerges out of the maze of ICJV politics.

The case would be different if the partners and their employees in the ICJV were committed primarily to the ICJV. Then they would have to forgo opportunism and indeed the interests would merge.

On a personal level trust is another mechanism often used to reduce complexity. By delegating work to a trusted person, the problem is solved for the superior (Girmscheid and Brockmann, 2005).

4.6.11 Cultural Complexity

Culture can be described as the software of the mind (Hofstede and Hofstede, 2005). A first programming comes about by the family (national enculturation), a second by professional training (industry culture), and a third by the mother company (organizational culture). The diversity of the cultural software causes problems.

Even more practical than in the case of social complexity are the following two examples of cultural complexity: A Korean/Taiwanese ICJV subcontracted the design for the bridges to a world-class engineering firm in the United States. By the main contract they must also employ a contractor's checking engineer and the ICJV gave the contract to a German engineering firm of highest reputation. The two firms fought for a full year to settle the design fundamentals in the design/build contract with a 4 year duration. The ICJV tried to solve the problem on the task level and despaired. With 25% of the contract time passed there existed no design and no construction. The two firms held different values and principles, big egos were also involved. At the end it became a bitter personal fight between the engineers over these values. The objectives of the contracts were lost out of sight.

A German/Thai ICJV contracted another world-class engineering firm from the United States for the design. Since most large German contractors are design/build companies, they insisted on supervising the design by giving directions. The American engineering firm was used to not being interfered with by German contractors. Again all kinds of distractions erupted because of neglect of these fundamental cultural differences.

The two examples highlight mostly the influence of different organizational cultures. Problems between national cultures are erupting in between and on top of the two explained cases.

A typical way to deal with problems of cultural complexity is through a managed process of sense-making.

4.6.12 Modularity and Complexity

4.6.12.1 Basic Definitions

The basic definitions of modularity are elegantly illustrated by Chen, C., & Crilly, N in a primer on the design and science of complex systems entitled "From modularity to emergence" (Literature 4). This primer introduces a domain-neutral framework and diagrammatic scheme for characterising the ways in which systems are modular or complex. The term 'system' is used to refer to all kinds of entities, including products, ecologies, infrastructure, animals, organisations, societies, and so on. Rather than seeing modularity and complexity as inherent attributes of systems, they are instead seen as ways in which those systems are characterised. In this section the basic system constructs and three fundamental attributes of modular system architecture, namely structural encapsulation, function-structure mapping and interfacing are introduce. These constructs and attributes encourage more precise descriptions of different aspects of complexity (e.g. emergence, self-organisation, heterarchy). This approach enables the system engineers of the megaprojects to illustrate their modularised system model for a better assessment and communication.

For the purposes of modularization, we define a "system" as a set of entities and relationships, where the relationships are connections or interactions

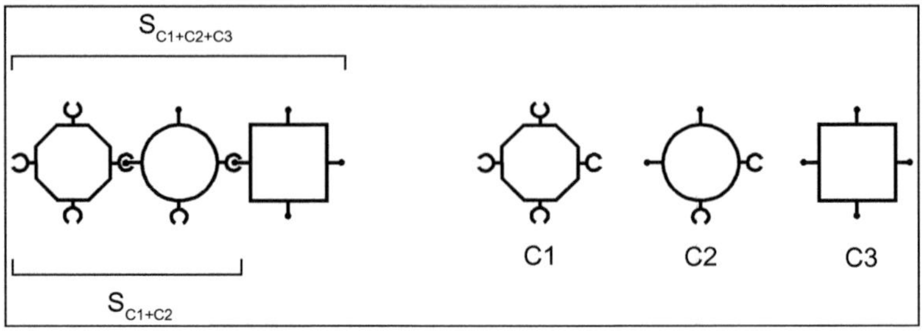

FIGURE 4.20
"*Compositional*" (part-whole) relationships.

between the entities. We call the entities in the system the "elements" of the system, which might be considered "components" or "subsystems" with respect to the system.

We use the term "characterization" to refer to any representation, model, specification, or description of an entity. Indeed, even calling an entity a "system" indicates that a certain stance is being taken toward it; the entity is being characterized as a system. The characterization might include system architecture, design specifications, functions, behavior, and so on.

In terms of the relationships between entities, we can distinguish between two formal relationships *"compositional"* (part-whole) relationships, and *"classificatory"* (subtype-type) relationships. These two relationships provide the basis for defining "levels" and "hierarchies". Fig. 4.20 illustrates the "compositional" (part-whole) relationships which is used in this section.

A composition relationship implies an entity (the "whole") that can be broken down into a set of further entities (the "parts"). The term "element" itself implies a composition relationship between the element and the system. However, different sets of a system's elements can also have part-whole relationships with each other. We use the terms "subsystem," "component," and *"supersystem"* to characterize such relationships. These are relational terms that only make sense when defined with respect to each other and with respect to a given characterization with respect to a given system "s."

- a *subsystem* of "s" is a subset of the entities and relationships in "s;"
- a *component* of "s" is an entity in s that cannot be further decomposed;
- a *supersystem* of "s" is a superset of the entities and relationships in "s."

Note that when we use the term "system," what we really mean is a system characterization; we do not make any metaphysical claims about the decomposability of physical entities. In addition to defining subsystems, components. and supersystems, with respect to a given system, we define an

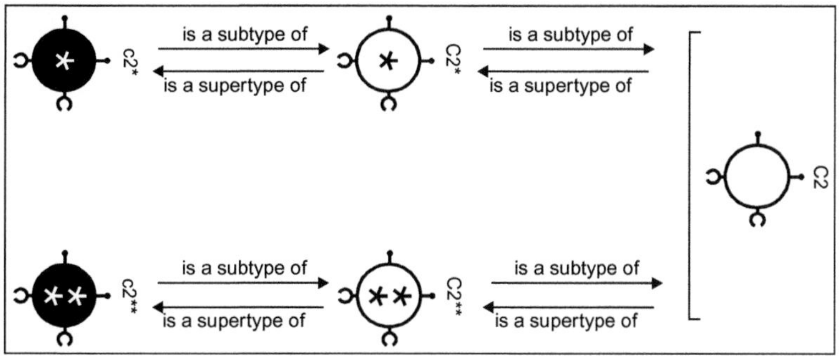

FIGURE 4.21
"*Classificatory*" (subtype-type) relationships.

"environment" of the system as a set of entities and relationships that are not in the set of entities and relationships constituting the system but that belong to a supersystem of the system. The difference between "the supersystem of s" and "the environment of s" is that the supersystem of s includes s, whereas the environment of s does not.

Entities can also be characterized at different levels of abstraction. Two elements can be seen to be different to each other at one level but the same as each other at another, more abstract level, where they belong to the same class or "type." Classificatory relationships between characterizations determine which characterizations can be treated as equivalent (see Fig. 4.21).

We define a "type" as a taxonomic group or "class" associated with a set of subtypes and instances. With respect to a given system type, S,

1. a *subtype* of S is a taxonomic group containing a subset of the entity types, entity instances, and characterizations contained in the set defined by S;
2. a *supertype* of S is a taxonomic group containing a superset of the entity types, entity instances, and characterizations contained in the set defined by S;
3. an *instance* of S is a concrete realization of S (an entity in the world) which belongs to the set of entities defined by S.

4.6.12.2 Hierarchies and Heterarchies

The terms "level" and "hierarchy" are frequently found in systems discourse. The part-whole (composition) and subtype-supertype (classification) relationships defined above give us a means of more precisely understanding these terms. Implicit in the classification relationship is the "resolution" of the characterization (also known as "granularity" or "level of abstraction"), which is the set of distinctions that can be made between the elements. Implicit in the *composition* relationship is what is known as the "scope" of the characterization, which is the set of elements involved.

In the case of complex systems characterizations, multiple hierarchies overlap a single characterization. This is what is referred to as a "heterarchy," "panarchy," or "entangled hierarchy" and can be represented by hypernetworks. Fig. 4.22 depicts a heterarchy that contrasts with the hierarchy described above.

4.6.12.3 Aspects and Mapping Relationships

As well as composition and classification relationships between different systems characterizations, there are also mapping relationships. These are used to relate characterizations of different *aspects* of the system, e.g., functions, properties, behavior, and architecture. This section considers three aspects of systems that are important in design and science: "architecture," "functions," and "properties." The pervasiveness of these three concepts is evidenced by the existence of several ontologies relating them, both in design domains and in scientific domains.

4.6.12.4 Aspects and Abstractions of Modularity

A system characterization with a straightforward compositional hierarchy describes components and subsystems as interacting (or interfacing) with each other in well-defined, well-understood ways and is said to be "modular." Although there exist many different notions of "modularity,"

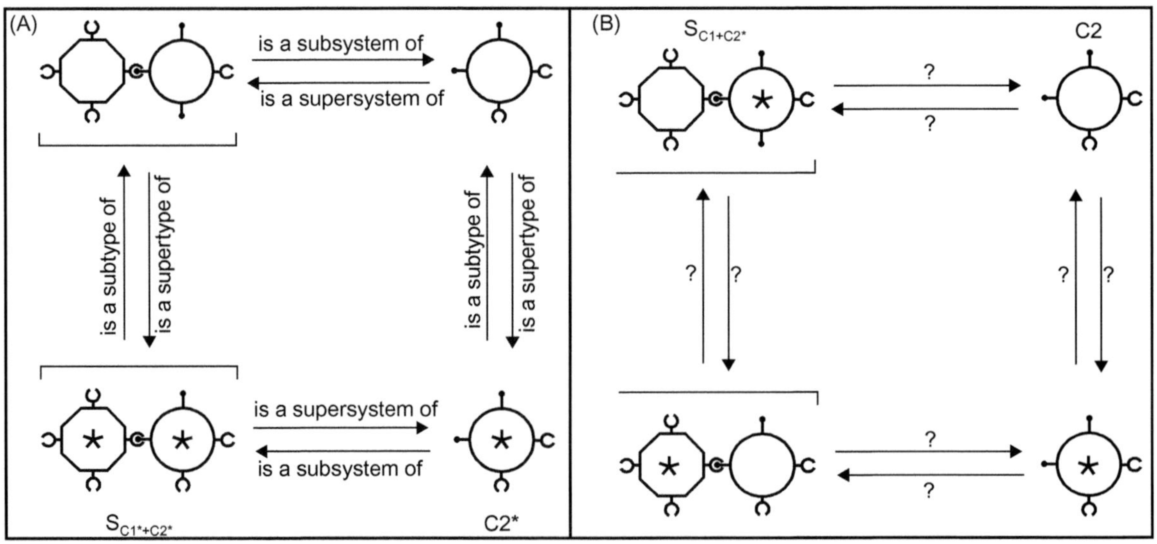

FIGURE 4.22
Schematics of the hierarchy and heterarchy relationship. (A) An example of a (clean) hierarchy. (B) An example of a heterarchy.

they can be understood and distinguished on the basis of three fundamental attributes of system architecture:

- structural encapsulation,
- function–structure mapping, and
- interfacing.

Table 4.2 illustrates how these fundamental attributes can be used to consolidate different definitions of modularity found in the literature. From these three fundamental architectural attributes, we can derive two further abstractions, function-driven encapsulation and interface compatibility.

4.6.13 Three Core Aspects of Modularity

The three fundamental attributes of system architecture that we associate with modularity are represented diagrammatically in Fig. 4.23.

A set of system elements can be collectively characterized as a "module" if all three following core aspects of modularity are satisfied:

- Structural encapsulation means that the elements can collectively be treated as a single encapsulated component;
- One-to-one function–structure mapping means that the set of elements collectively map to a particular function;
- Interfacing means that as a collective, the set of elements has well-defined interactions with other system elements.

4.6.13.1 Structural Encapsulation

We use the term "structural encapsulation" to refer to the grouping of related system elements, i.e., subsystems, into units that can then be treated as component types at some level of abstraction. Structural encapsulation also implies "interface decoupling" since it allows the relationships between a set of related system elements to be considered independently from its interactions with other system elements.

4.6.13.2 Function–Structure Mapping

We use the term "function–structure mapping" to refer to the mapping between a set of related system elements (i.e., a subsystem) and a function. This structured set of system elements can then be encapsulated into a component type because they are collectively associated with the function. We refer to such encapsulation as "function-driven" (see Section 4.6.13.3 below).

4.6.13.3 Interfacing

We define the term "interface" as an aspect of the element that allows it to interact with another element or set of other elements in the same system.

Table 4.2 Different Notions of Modularity Related to the Three Aspects of Modularity: Structural encapsulation (SE), Function–Structure Mapping (F-SM), and Interfacing (I)

Aspect of Modularity			
SE	**F-SM**	**I**	**Examples from Literature**
X			*Abstract characterization:* • A module is a physical or conceptual grouping of components (Jiao & Tseng, 1999b). • Modules contain a high number of components that have minimal dependencies upon and similarities to other components not in the module (Gershenson et al., 1999). *Network characterization:* • A subsystem is a module when the number of edges within the subsystem is much higher than the expected number of edges derived from an equivalent random network model with the same number of elements and similar distribution of links between elements with no modular structure (Newman, 2010). *In manufacturing and product design:* • The most modular architecture is one in which each functional element of the product is implemented by exactly one chunk (subassembly) and in which there are few interactions between chunks. Such a modular architecture allows a design change to be made to one subassembly without affecting the others (Ulrich & Eppinger, 1995). *In software design:* • There should be no access to, informational flow to, or inter-activity between modules (George & Leathrum, 1985). Modular programming has developed coding techniques which "(1) allow one module to be written with little knowledge of the code in another module, and (2) allow modules to be reassembled and replaced without reassembly of the whole system" (Parnas, 1972, p. 1053).
	X		*Abstract characterization* • The term "modular" refers to the minimization of the number of functions per component (Ishii et al., 1995). *In manufacturing and product design* • "Conceptual" modules perform the same functions even if they have different physical compositions (Otto & Wood, 2001).
X	X		*In manufacturing and product design* • Product modularity is defined in terms of "(1) Similarity between the physical and functional architecture of the design and (2) Minimization of incidental interactions between physical components" (Ulrich & Tung, 1991: p. 73). Therefore a modular product or subassembly has a one-to-one mapping from functional elements in the function structure to the physical components of the product (Ulrich, 1995). • A module is a set of components grouped together in a physical structure and by some characteristic based on the designer's intent (Di Marco et al., 1994; Newcomb et al., 1998). • A module is a component or group of components that can be removed from the product nondestructively as a unit, which provides a unique basic function necessary for the product to operate as desired, and modularity is the degree to which a product's architecture is composed of modules with minimal interactions between modules (Allen & Carlson-Skalak, 1998). • Modularity refers to the "building of complex product or process from smaller subsystems that can be designed independently yet function together as a whole" (Baldwin & Clark, 1997: p. 84). • Modularity requires similarity of functional interactions and suitability of inclusion of components in a module (Huang & Kusiak, 1998).

Continued

Table 4.2 Different Notions of Modularity Related to the Three Aspects of Modularity: Structural encapsulation (SE), Function–Structure Mapping (F-SM), and Interfacing (I) *Continued*

Aspect of Modularity SE	F-SM	I	Examples from Literature
			Abstract characterization • A module is a component or subsystem in a larger system that performs specific function(s) and emerges as a tightly coupled cluster of elements sharing dense intra-module interactions and sparse inter-module interactions (Sarkar et al., 2013).
		X	*In manufacturing and product design* • A module is a group of standard and interchangeable components (Galsworth, 1994). • Modular systems are those that are constructed from standardized units of dimensions for flexibility and use (Walz, 1980).
	X	X	*In manufacturing and product design* • A modular product is "a function-oriented design that can be integrated into different systems for the same functional purpose without (or with minor) modifications" (Chang & Ward, 1995 in Gershenson et al., 2003: p. 298). • Modules are groups of components that can easily be reused or remanufactured, also considering material compatibility (Sosale et al., 1997). *In software design* • Modularity refers to "tools for the user to build large programs out of pieces" (Chen, 1987, in Gershenson et al., 2003: p. 297).
X		X	*Abstract characterization* • A module is a structurally independent building block of a larger system with fairly loose connections to the rest of the system. They have well-defined interfaces which allow independent development of the module as long as the interconnections at the interfaces are retained (Holtta & Salonen, 2003). *In manufacturing and product design* • Modularity is design with subsystems "that can be assembled and tested prior to integration. . .to reduce the time and cost of manufacturing" (Carey, 1997, in Gershenson et al., 2003: p. 298). • Modularity is using sets of units designed to be arranged in different ways (Belle & Kissinger, 1999). • Physical adjacency, energy transfer, information transfer and material exchange can be used to group elements together so they are treated as modules (Pimmler & Eppinger, 1994).
X	X	X	*Abstract characterization* • Modules are cooperative subsystems which (i) can be combined and configured with similar units to achieve different outcomes; (ii) have one or more well-defined functions that can be tested in isolation from the system and that (iii) have their main functional interactions within rather than between modules (Marshall et al., 1998).

For those designing physical products it might be most natural to think of interfaces in terms of physical structure or geometric fit. However, interfaces can also be realized in nonphysical ways and the interactions need not be determined by geometry. Examples of nonphysical interfaces include standards, protocols, agreements, languages, signals, and processes.

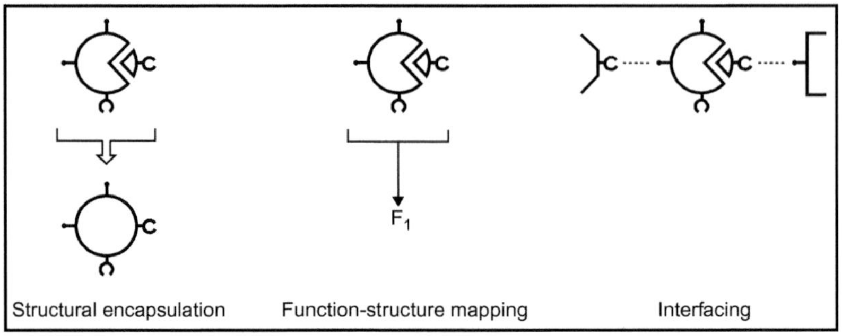

FIGURE 4.23

Three aspects of modularity: structural encapsulation (the module is defined by its composition and structure relating its elements to each other, as indicated by the block arrow), function–structure mapping (the module is defined by the collective mapping of a structured set of elements to a function—in this case, F_1, as indicated by the arrow), and interfacing (the module is defined by how a set of elements interacts with other systems, as indicated by the dotted lines).

Which aspect(s) of an element is treated as its interface depends on the characterization adopted, which defines the set of elements with which interaction occurs. This might also mean that multiple interfaces are identified for an element. Indeed, in some cases the interfaces might even be determined by function–structure mapping. For example, what makes the geometry of a given system element its interface might be the requirement of physical fit for the formation of a composite structure to realize a mechanical or chemical function. In such cases, there is an inextricable link between structure and function.

4.6.14 Two Abstractions From Modularity

From the three core aspects outlined above, we can derive two further abstractions that also pervade the modularity literature: function-driven encapsulation and interface compatibility.

4.6.14.1 Function-Driven Encapsulation

We use the term "function-driven encapsulation" to describe cases where the criterion for encapsulation is the fulfillment of a function (see Fig. 4.24). What connects elements within a group to each other is that they collectively map to a function, and what makes this set of elements disconnected from other elements is the fact that these other elements do not participate in the fulfillment of that function (being "connected" or "disconnected" might also be a matter of degree, and the mapping to a function is specific to a particular level of abstraction and scope).

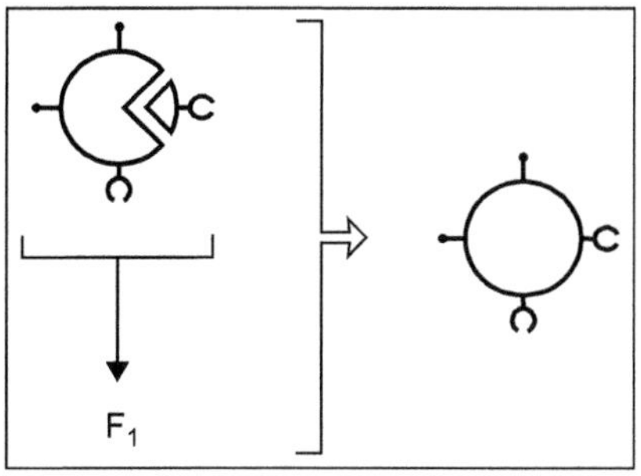

FIGURE 4.24

An example of function-driven encapsulation: the structural encapsulation of the module is determined by function–structure mapping.

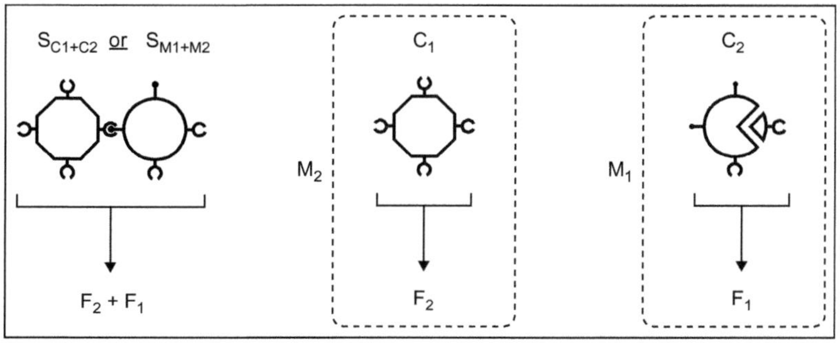

FIGURE 4.25

An example of interface compatibility, leading to a modular architecture. The system type S_{C1+C2} has a completely modular architecture since all its elements (both C1 and C2) belong to or constitute modules (M_2 and M_1, respectively). In this case, the modules are defined by function–structure mapping.

Function-driven encapsulation can be seen as one of a set of many different forms of encapsulation, each of which is distinguished by the kind of criteria that determines encapsulation. For example, we might also have property-driven or behavior-driven encapsulation where elements are seen as "connected" when they collectively realize a particular property. However, since our definition of encapsulation to be function-driven, modularity is concerned with the relationship between elements and functions, we require encapsulation to be function-driven.

We say that a system architecture is "completely modular" if every element in the system belongs to a functional group and fulfillment of the system's overall function is completely accounted for by these function–structure mappings (Fig. 4.25). In the design and management of systems, encapsulation has been said to provide a means of "managing complexity" by hiding the intricacies of certain regions of the system so that characterizations of them can be separated from the characterization of the relationships that exist between them and other regions of the system.

4.6.14.2 Interface Compatibility

Interface compatibility refers to the compatibility between different components of the system. This compatibility might be a matter of degree and characterized as the strength of interaction. Interface compatibilities between system components determine how different groups of system elements are able to interact with each other, thus providing a characterization of the system's architectural constraints. In a completely modular architecture, since all the elements would be modules or would belong to modules and hence be encapsulated in components, interactions between elements in different components would always be via their interfaces. Well-defined interfaces permit components and subsystems with different structures and functions to occupy the same "position" as each other in the system.

If all modules (components mapped to functions) in a system had the same mutually compatible interfaces with each other, there would be no architectural constraints at the module level since any module would be able to interact with any other module, i.e., architectural degrees of freedom would be maximized, and every component could be "repositioned." This is known as "sectional" modularity, where every component in the system has the same set of interfaces.

At the other extreme, where interfaces minimize architectural degrees of freedom and each component has a specific "position" or "role" in the system, we have "slot" modularity. In "slot" modularity, each component has a unique set of interfaces, which implies that it has a unique set of interactions with other components in the system and hence can only be located in a single specific position with respect to them. Interface compatibilities can provide a means of controlling which parts of the system can vary.

In a given system architecture, different elements of different types (possibly mapping to different functions) can interact with the same set of other elements, so long as they have the same interface compatibilities. In a modular architecture (where the system can be decomposed into components mapped to functions), interface compatibilities determine which components can be swapped or substituted for each other. The terms "component-sharing," "substitution," and "standardization" are used in the literature to refer to cases where, at a particular level of abstraction, different component types have the same interfaces (i.e., they are compatible with the same set of other component types). This "component-sharing," together with overall architectural similarity between products, can be the basis for establishing product "families."

The term "component-swapping" is used to refer to cases where, at a particular level of abstraction, component types are mapped to different functions but have the same interfaces and therefore can be substituted for each other architecturally (see Fig. 4.26). If these differences in component function have implications for a product's overall function, they provide the basis for the different product variants in product "families." Interface compatibility provides us with a formal means of characterizing and analyzing architectural variety in terms of elements' compatibilities with each other and the different architectural configurations they permit.

4.6.15 Aspects of Modernization Complexity

The term "complexity" is used in different ways in the design literature and is often used interchangeably with "complicated." We treat these as two distinct concepts. Characterizing a system as "complicated" is to understand it as having many components, subsystems, and interactions; however, as with a simple system, it is theoretically possible to map functions to components and subsystems in a one-to-one fashion, and to describe the interactions between them.

By contrast, characterizing a system as "complex" is to understand the system in a way that does not allow this kind of one-to-one mapping or full description of the interactions between components and subsystems. The three aspects and two abstractions of modularity discussed above can be used to distinguish between different aspects of complexity.

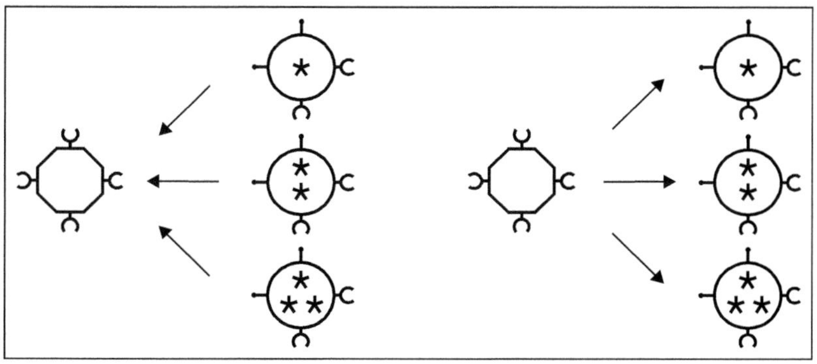

FIGURE 4.26
"Component-swapping" always implies "component-sharing," and vice versa. When we are taking the perspective of a component (here, the octagon) that can interact with a variety of other entities, the architecture is characterized as "component- swapping" (different components can be swapped "in or out" of the octagon). When we are taking the perspective of different entities that can all interact with the same component (the octagon), the architecture is characterized as "component-sharing" (the octagon is a component that can be shared "between" different entities).

4.6.15.1 Complexity as Non-One-to-One Function–Structure Mappings

Function-driven encapsulation ensures one-to-one mapping between function and architecture. Complexity arises when, at some level of abstraction, the mapping is not one-to-one.

4.6.15.2 Multistructural Function Realization and Architectural Robustness

We use the term "multistructural function realization" to describe cases where a function maps to more than one architecture (more than one component type). In design and engineering, the possibility of realizing a function with different architectures offers the opportunity for robustness and reduction in cost. Robustness comes from the fact that if one architecture mapping to a function is not realized, others may be able to realize it instead. Cost reduction would come from the fact that the number of components required for a given level of robustness might be lower than if this robustness were achieved through duplication of components (see Fig. 4.27).

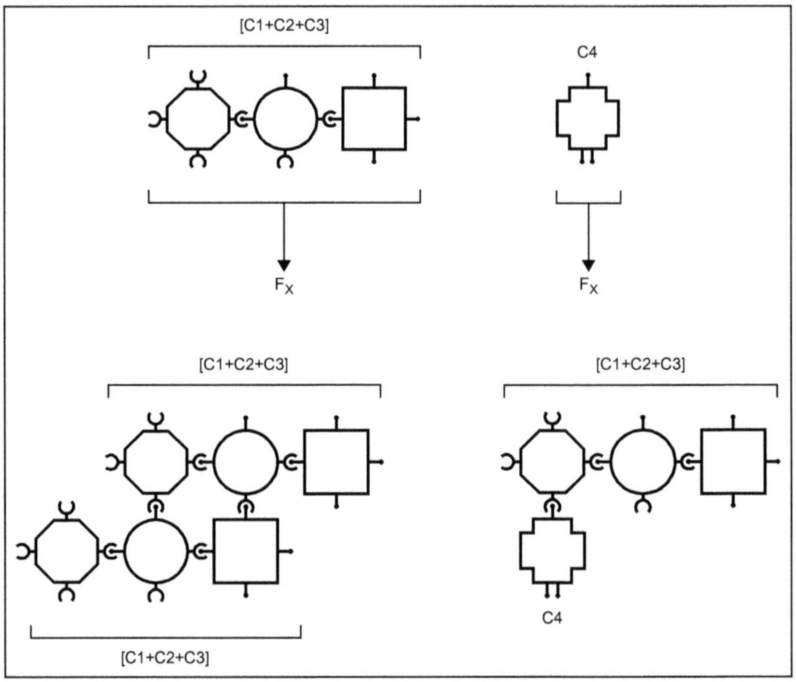

FIGURE 4.27

Redundancy through duplicated architectures and distinct architectures. Top row: Both the [C1 + C2 + C3] architecture and the C4 architecture map to F_X. Bottom left: Redundancy in F_X is provided by an architecture with duplication of [C1 + C2 + C3]. Bottom right: Redundancy in F_X is provided by two distinct architectural realizations, [C1 + C2 + C3] and C4.

In engineering design, the term "principle redundancy" describes cases in which multiple architectures realize the same function. In biology, the term "degeneracy" describes cases where, when a particular element is not able to fulfill its function, other means of fulfilling that function are possible. For example, a function that was previously associated with a single element might also become distributed among multiple elements.

Compared to duplication, multistructural function realization offers a more robust form of redundancy when the different architectures able to realize the function have different points of fragility and strength (see Fig. 4.28). On the other hand, it makes the function–structure mappings harder to analyses, and when there is system failure, it can be difficult to identify the elements involved.

We say that a system is "architecturally robust" if variety in function is low with respect to architectural variety (the ratio of the number of functions to the number of architectures is low). Architectural robustness is positively

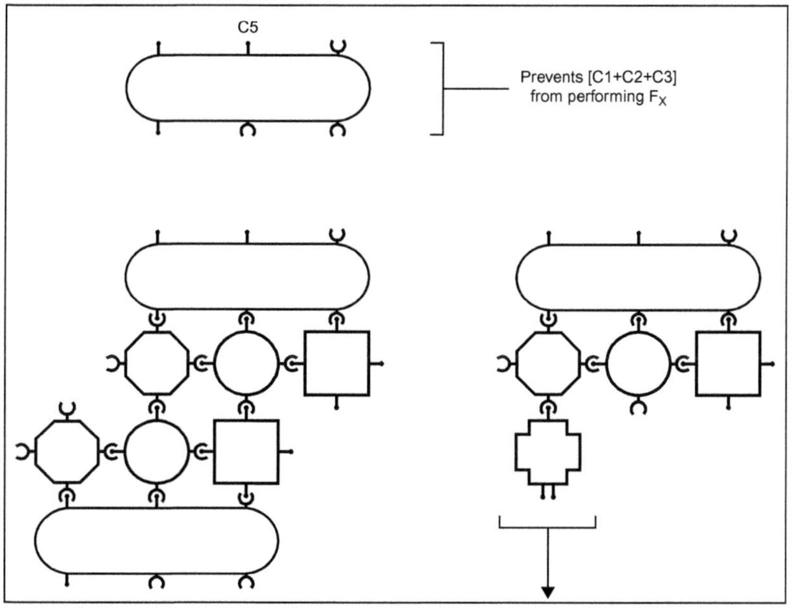

FIGURE 4.28

An example of how multistructural function realization provides robustness. As in Fig. 4.27, [C1 + C2 + C3] and C4 are both mapped to F_X. Top row: C5 prevents the architecture [C1 + C2 + C3] from realizing F_X. Bottom left: The presence of C4 prevents [C1 + C2 + C3] + [C1 + C2 + C3] from realizing F_X. Bottom right: The multistructural function realization architecture of [C1 + C2 + C3] + C4 allows it to be more robust than [C1 + C2 + C3] + [C1 + C2 + C3] with respect to realizing F_X since it can do so in the presence of C5.

associated with evolvability since the greater the architectural variation with respect to a function, the larger the set of possibilities to be selected from, and the greater the evolvability. Selection pressures can also be characterized in terms of function realization. For example, referring back to the architectures in Fig. 4.27, having both [C1 + C2 + C3] and C4 as possibilities would make the system both architecturally robust with respect to F_X (see Fig. 4.28) and more evolvable with respect to F_X compared to the case where only one of the architectures could be realized. If the system found itself in an environment requiring F_X to be realized, there would be a selection pressure in favor of the architecture [C1 + C2 + C3]. We might also say that the evolvability of the system with respect to F_x is in virtue of its adaptability with respect to F_X. To some extent, this is simply a question of the level at which we are considering the system. For example, a production process might permit a change in parts supplier which then allows a manufacturing firm to resist changes in supplier prices; an organism's ability to change its behavior in response to different temperature conditions allows it to operate in different environments.

4.6.15.3 Context-Dependent Multifunctionality and Architectural Flexibility

We use the term "context-dependent multifunctionality" to refer to cases where an architecture maps to different functions based on the wider system architecture it is part of. In systems terms, this means a subsystem realizes different functions based on which other systems it is connected to (its environment), i.e., the supersystem it is part of. Fig. 4.29 shows how C2 can be

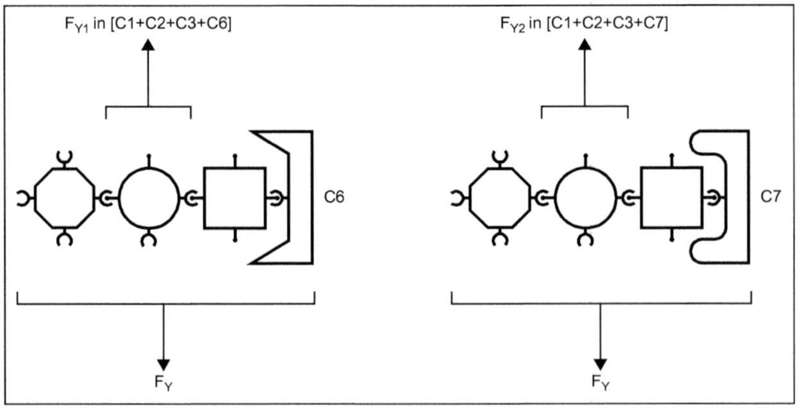

FIGURE 4.29

An example of context-dependent multifunctionality. The same component (in this case, the circle, C2) realizes different functions by participating in different architectures, even if those architectures realize the same function.

characterized as context-dependent multifunctional. When it is connected to C1 and [C3 + C6], it realizes F_{Y1}, and when it is connected to C1 and [C3 + C7], it realizes F_{Y2}.

In design domains, repurposing of products, product parts, and processes are examples of context-dependent multifunctionality. For example, a steel rod realizes different functions depending on the wider physical structure it is part of; in software, the same data can have different functions depending on the sections of the program that they flow into; the biochemical function of a protein can depend on the other molecules present; the economic impact of one consumer's purchase depends on the purchasing activities of other consumers.

When the contexts in which different functions are realized are not well-understood, functions may be realized unexpectedly or "emerge" (sometimes resulting in nonfulfillment of other functions). On the other hand, if the context-dependencies are well-understood, multifunctionality can be exploited to get (desired) functional variety from a given architecture.

We say that a system is "architecturally flexible" if variety in function is high with respect to architectural variety (in the limit, every architectural variation would be functionally relevant and the ratio of functions to architectures would be unity). This has the potential advantage of allowing a system to realize a greater variety of functions with a relatively small number of elements, but also makes it more difficult to analyses and predict with respect to these functions.

4.6.15.4 Complexity as Ill-Defined Interfaces and Shifting System Boundaries

A modular system has subsystems (the modules) with well-defined interfaces, resulting in a perfect compositional hierarchy; each module can be treated as a "closed" system. "Complexity" arises when interfaces are ill-defined or changing, and the boundaries between the subsystems are constantly changing so that subsystems are "open" systems. Of course, as with function–structure mappings, this is really a question of characterization.

In a "closed system" characterization where the system has a well-defined boundary, given knowledge of all the possible characterizations within the boundary, it would be theoretically possible to define all the relationships between all the characterizations. However, when the number of characterizations and/or relationships between them is extremely large or not yet known, an idealized "open system" characterization may be used. For example, in design domains, the realization of a product requires the realization of an intricate set of connections between physical components, processes, people, and organizations; in complex systems science domains, models of entities often consist of a web of interdependencies between a large number of system elements. An "open system" characterization of these scenarios would see the system as interacting with itself (as it would with its environment), and would see the interdependencies between the elements of the system as constantly changing.

4.6.15.5 Complexity as Overlapping Levels

Nonoverlapping hierarchies are those in which a related set of characterizations do not overlap with respect to their supersystem–subsystem or supertype–subtype relationships. In the case of overlapping hierarchies, this no longer holds.

4.6.15.6 Multilevel Characterizations and Heterarchy

Heterarchical characterizations are ones where several hierarchies *overlap* in a single characterization. These should be distinguished from characterizations which integrate multiple *nonoverlapping* hierarchies. For pragmatic purposes, heterarchies are decomposed into such nonoverlapping characterizations, such as in "system of systems" (SoS) characterizations, which integrate different resolutions without overlap in scope.

Heterarchies can represent cases where different domains work together to understand a single entity, since different domains might emphasize different system aspects and consequently "carve up" the entity in ways that overlap. Fig. 4.30 shows an example of a complex systems characterization of the entity "e" based on the heterarchy in Fig. 4.22B. In the real world, these different mappings might represent characterizations from different domains, e.g., programmers, software architects, and business analysts working on the same software; cognitive psychologists, neuroscientists, cell biologists, and molecular biologists studying the brain.

4.6.16 Endogenous and Exogenous Functions

In both design and scientific domains, the functions being considered in function–structure mapping often relate to different aspects of the system or even to different systems (with different boundaries), resulting in modular architectures which differ substantially from one another. For example, in product design, function–structure mappings may be defined with respect to the product's overall function in use (which is typically linked to the satisfaction of user needs and preferences), but they can also be defined with respect to the product's manufacture or contribution to firm strategy. In biology, one set of functions might relate to an organism's survival; another might relate to its development or to its role in evolution.

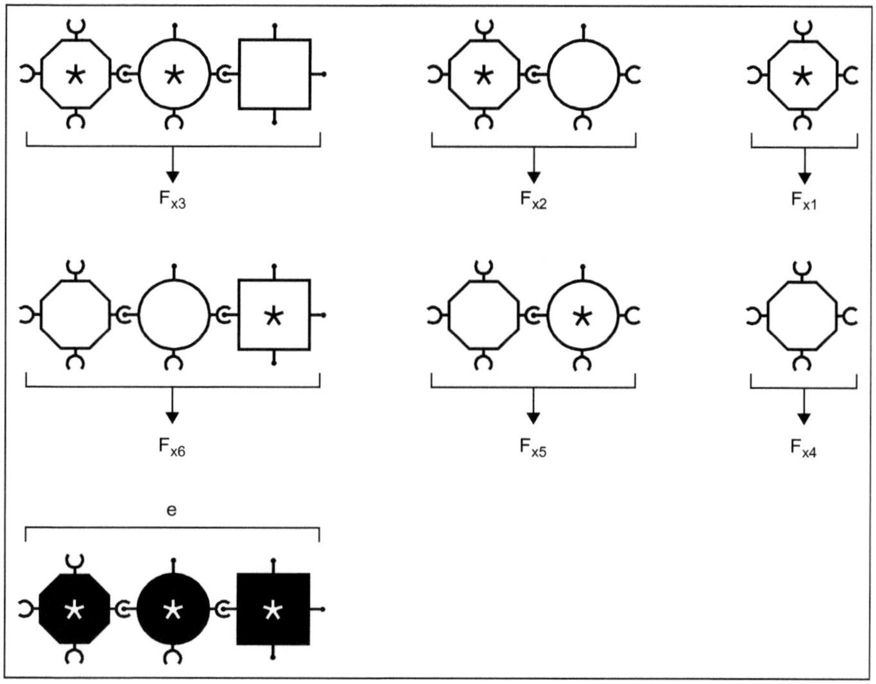

FIGURE 4.30

A complex systems characterization of entity e where functions are mapped to architectures specified at different levels. For example, [C1* + C2* + C3] maps to F_{X3} but [C1 + C2 + C3*] maps to F_{X6}. The complexity comes from the fact that in order for the realizations of all the functions to be characterized, different levels of abstraction and scope overlap, i.e., heterarchy.

To generalize, the functions in a function–structure mapping might originate from the consideration of different systems, and we can dissociate (i) the system for which the architecture is defined (e.g., the product; organism) from (ii) the system determining the functions to which this architecture maps (e.g., user; ecosystem). In the case of (ii), we might draw a distinction between "endogenous" functions, which are defined with respect to the system in question, and "exogenous" functions, which are defined with respect to the supersystem in which it operates.

The distinction between endogenous and exogenous functions is important because they can be associated with different levels of uncertainty. Failure to realize endogenous functions lies in improper realization of the system type (e.g., a system part failing). Failure to realize exogenous functions on the other hand, can be attributed to the system's environment, which can change the function–structure mapping. For example, changes in user preferences might mean that elements of the system that could previously satisfy a particular preference no longer can; a new set of conditions in an organism's

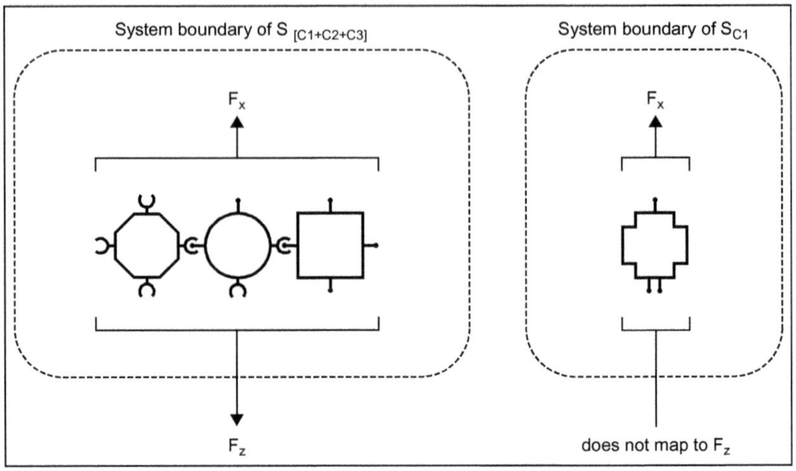

FIGURE 4.31
Endogenous and exogenous functions. The architectures [C1 + C3 + C1] and [C2 + C3] map to the same endogenous function F_X but only [C1 + C3 + C1] maps to the exogenous function F_Y.

environment might mean that certain functions of the organism no longer map to the biological elements they were previously mapped to. If knowledge of the system's environment is inferior to knowledge of the system itself, component types mapping to exogenous functions will have higher levels of uncertainty associated with them in terms of function fulfillment (e.g., user preferences compared to product specifications; organism behavior compared to core metabolic functions).

Fig. 4.31 shows how different architectures might map to the same endogenous function but to different exogenous functions. In many cases, endogenous functions and exogenous functions might also be dependent on each other. For example, the realization of the endogenous function F_X might be dependent on the realization of F_Y, or vice versa.

4.6.17 Behavioral Robustness and Flexibility

Although entity change and entity variety can be seen as two distinct concepts, change can also be seen simply as variety observed through time. For example, with an atemporal view, demands to the system due to alterations in physical conditions or consumer preferences become the same as those made by an environment with a wide range of physical conditions or a market with highly diverse consumer preferences.

While state *transitions* describe the behavior of a system *instance*, state transition *rules* describe the behavior of a system *type*. State transition rules define the set of state transitions that are realizable (or that must be realized) by instances of the type, thus determining the states that the system can instantiate, its "state space," depending on its initial state, which also determines the behavioral trajectories it can take. The rules mean that in a given system instance, transitions between states can be "guided" and "mutually constraining," so that they follow particular "trajectories" depending on previous states. This can result in *behavioral* "robustness" and "flexibility."

In the same way that change can be recast as variety, we can give system behavior (state transitions) an architectural characterization. In the case of "behavioral robustness," it is very difficult to get the system to deviate from a particular behavior. In the case of "behavioral flexibility," there are few constraints on the states that can be realized by the system, and the architecture of the behavior has few regularities. Such a system would be chaotic and difficult to manage, predict or understand.

Terms such as "positive feedback" and "negative feedback" are used to describe the mechanisms which constrain or "guide" behavior. In the case of positive feedback, a particular state or behavior increases the likelihood or extent of states or behaviors of the same type, while in the case of negative feedback, it diminishes their extent or likelihood. These two mechanisms and interactions between them form the basis for the "emergence" of behaviorally robust "self-properties" such as self-replication or self-assembly. They also form the basis of homeostasis or "autopoiesis," the ability of the system to maintain itself in a viable condition.

In some complex systems characterizations, the system's *environment* can put the system into a state in which different rules apply or even directly affect which rules apply, thus making different behavioral trajectories available. In even more complicated cases, the system can itself influence its environment to make it more likely to realize particular states, which then reinforce the above effect. Identifying such scenarios is a key endeavor in the complex systems sciences. The environment might also determine the wider implications of the relationship between functional variety and architectural variety, which in turn can be used to further distinguish between different change-related capabilities. For example, flexibility (on our definition) can mean that a system is fragile in particular types of environment (which might also be characterized in terms of environmental states of a single environment type) because many of the possibilities it has available to it render it nonviable or functionally deficient at some other level of description. On the other hand, in a different set of environments (which might be characterized as different environmental states of a single environment), the system's flexibility might make it resilient because its functional variety allows it to survive.

Table 4.3 Types of Complexity and Means for Their Reduction

Type of Complexity	Description	Means of Reduction
Overall complexity	Degree of manifoldness, interrelatedness, and consequential impact of a decision field	Specialization, risk compensation, levels of complexity, adaptation
Design complexity	Large number of design steps and design specializations	Integrated management system, modularization, scrum agility methods, collaborative software, robust configuration management
Organizational complexity	Numerous groups with substantial interdependence among them	System engineering, integrated management system, scrum agility methods, collaborative software, robust configuration management
Task complexity	Density of activities in a time/space segment	Decentralization, scrum agility methods, collaborative software
Social complexity	Number and diversity of stakeholders	Trust, commitment, scrum agility methods, collaborative software
Cultural complexity	Diversity of the cultural software of the mind	Sense-making, scrum agility methods, collaborative software
Modularity complexity	Interface compatibility, fabrication timing, context-dependent multifunctionality, multiple nonoverlapping hierarchies, endogenous and exogenous functions, behavioral robustness and flexibility	System engineering, one-to-one mapping between function and architecture, scrum agility methods, collaborative software, robust configuration management

Architectural robustness/flexibility and behavioral robustness/flexibility address different aspects of complexity. In the case of architectural robustness and flexibility, it is the relationship between architecture (which might be the architecture of a system, system type, state, or behavior) and function that we are concerned with. By contrast, in the case of behavioral robustness and flexibility, we are concerned only with the architecture itself (characterized as regularities in behavior).

4.6.18 Means of Reduction of the Complexity

It is possible to differentiate between the overall and three further types of complexity. In megaprojects, ICJVs find very specific ways to deal with them. These are strategies of choice and they are pursued regardless of the national background of the ICJV partners with the aim to reduce the project complexity (Table 4.3).

4.7 MEGAPROJECT MANAGEMENT CHALLENGES

In megaprojects, the workforce is composed of numerous work groups that operate in parallel due to the large size and complexity. With a single work entity, the managerial duties are limited to the daily duties such as cost control, time control, progress reporting, communication, correspondence, and workflow direction. The operational level would be consumed with daily workforce activities in order to stick with the project milestones. However, with multiple work entities, the managerial duties would differ substantially. Management has to conduct the daily duties with each of the work groups independently. Furthermore, management has additional duties in order to organize and integrate a complex web of systems over an enormous construction site. It has to manage and integrate work flow, schedule, and resources among all working groups. Consequently, the management team would be overwhelmed by the number of interactions needed to manage all these elements leading to a bottleneck at the management level and an inactivity of work crews at the operation level. Fig. 4.32 illustrates the megaproject management system.

The following is a description of the management difficulties imposed by the characteristics and the corresponding managerial problems.

The increase in size would impose:

- An increase in the number of work groups to be managed leading to a multiplication of the daily efforts to follow up with each of the work groups;
- Work coordination and managerial efforts to integrate the work flow and schedule of the work groups;
- Increased coordination efforts among different work groups;
- Labor workforce management among different work groups;
- Material and equipment management efforts among different work groups.

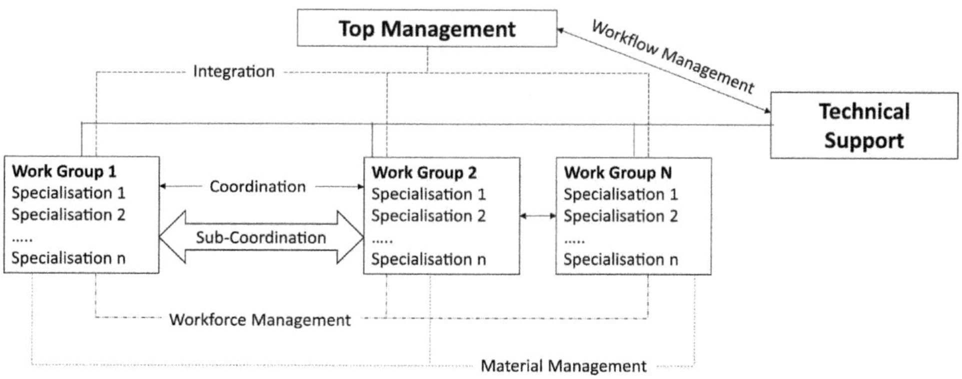

FIGURE 4.32
Megaproject management.

The increase in complexity, namely, design complexity would impose:

- Management efforts between subgroups or engineering specializations within each group. The subgroup management efforts would be rigorous since the works should be coordinated and integrated according to time and space constraints;
- Technical and functional support which would add to the managerial efforts required to handle a complicated design. The support required is summarized by quality control, quality assurance, means and methods, and design support.

The increase in size and complexity is often underestimated by construction practitioners and is usually countered by an increase in resources (equipment, labor) and an increase of the number of working hours. The inability of conventional work practices to handle the increase in managerial load would lead to:

- Unutilized working teams because of the large number to be integrated;
- Slow progress and mistakes in uncoordinated operations because of the complicated sequence of activities and operations that demand coordination;
- Unavailability of resources because of the large load of material to be managed Mismanagement of finite resources such as the labor workforce;

Improper coordination that would lead to the inactivity of some engineering trades and time-consuming error corrections.

Literature

1. A Consistency-Checking Consensus-Building Method to Assess Complexity of Energy Megaprojects
2. Analysis and Improvement of Megaprojects Performance
3. Complexity of Megaprojects
4. Chen, C., & Crilly, N. From Modularity to Emergence - A Primer on the Design and Science of Complex Systems. https://doi.org/10.17863/CAM.1227.
5. Death Of The Megaproject? How to Solve the Problems Involved in the Oil & Gas Industry
6. Spotlight on Oil And Gas Megaprojects
7. Deloitte Oil & Gas Reality Check 2015
8. Deloitte Oil & Gas Reality Check 2014
9. Oil and Gas Industry Megaprojects: Our Recent Track Record

Blog

1. Visualizing Complex Systems Science (CSS)

Handbook

1. INCOSE Handbook-2002
2. INCOSE Handbook-2006
3. INCOSE Handbook-2012
4. INCOSE Handbook-2015
5. Guide to the Systems Engineering Body of Knowledge (SEBoK)
   - SEBoK v. 1.7
   - SEBoK v. 1.6
   - SEBoK v. 1.5.1
   - SEBoK v. 1.5.
   - SEBoK v. 1.4
   - SEBoK v. 1.3.1
   - SEBoK v. 1.3
   - SEBoK v. 1.2
   - SEBoK v. 1.1
   - SEBoK v. 1.0

Guidance

1. INCOSE Guide for the Application of Systems Engineering in Large Infrastructure Projects

CHAPTER 5

Modeling and Simulation: The Essential Tools to Manage the Complexities

5.1 BACKGROUND

Previous chapters introduced the system thinking, system engineering, and complexity management in a glance. It has been shown that a holistic, multidisciplinary, and collaborative approach to designing and maintaining of the modern industrial facilities is the only way to manage the complexity associated with technical, organizational, and people issues. Fig. 5.1 illustrates how traditional physical assets emerge to the smart and connected Industry 4.0 process manufacturing facilities.

Progresses made on the industrial internet of things (IIoT), Big Data, virtual reality (VR), advanced robotics, and artificial intelligence enable to radically change the way we manage and operate the complexities of the industrial facilities. Fig. 5.2 summarises the outcomes of the four industrial revolutions.

The speed, scope, and impact of those technological developments herald a promising era of agile-reliable-safe smart manufacturing, compliance with laws and regulation and standards, production management, less routine jobs with better training, megaprojects realization, and more efficient supply chain management. The technologies which will drive Industry 4.0 are as follows:

- Cheap sensors will connect every object to the digital world, laying the bedrock of the internet of things.
- Artificial intelligence will generate self-learning machines and even robots capable of teaching one another.
- The next generation of mobile internet (5 G) will facilitate the transfer of immense data at the speed of light, supporting up to 100 devices per square meter.
- Supercomputers, Big Data analysis, and cloud computing will allow the collection of vast quantities of data to be exploited.
- Augmented reality (AR), virtual reality, and 3D printing will improve the production process and also our leisure experiences.

A Systems Approach to Managing the Complexities of Process Industries.
DOI: http://dx.doi.org/10.1016/B978-0-12-804213-7.00005-0

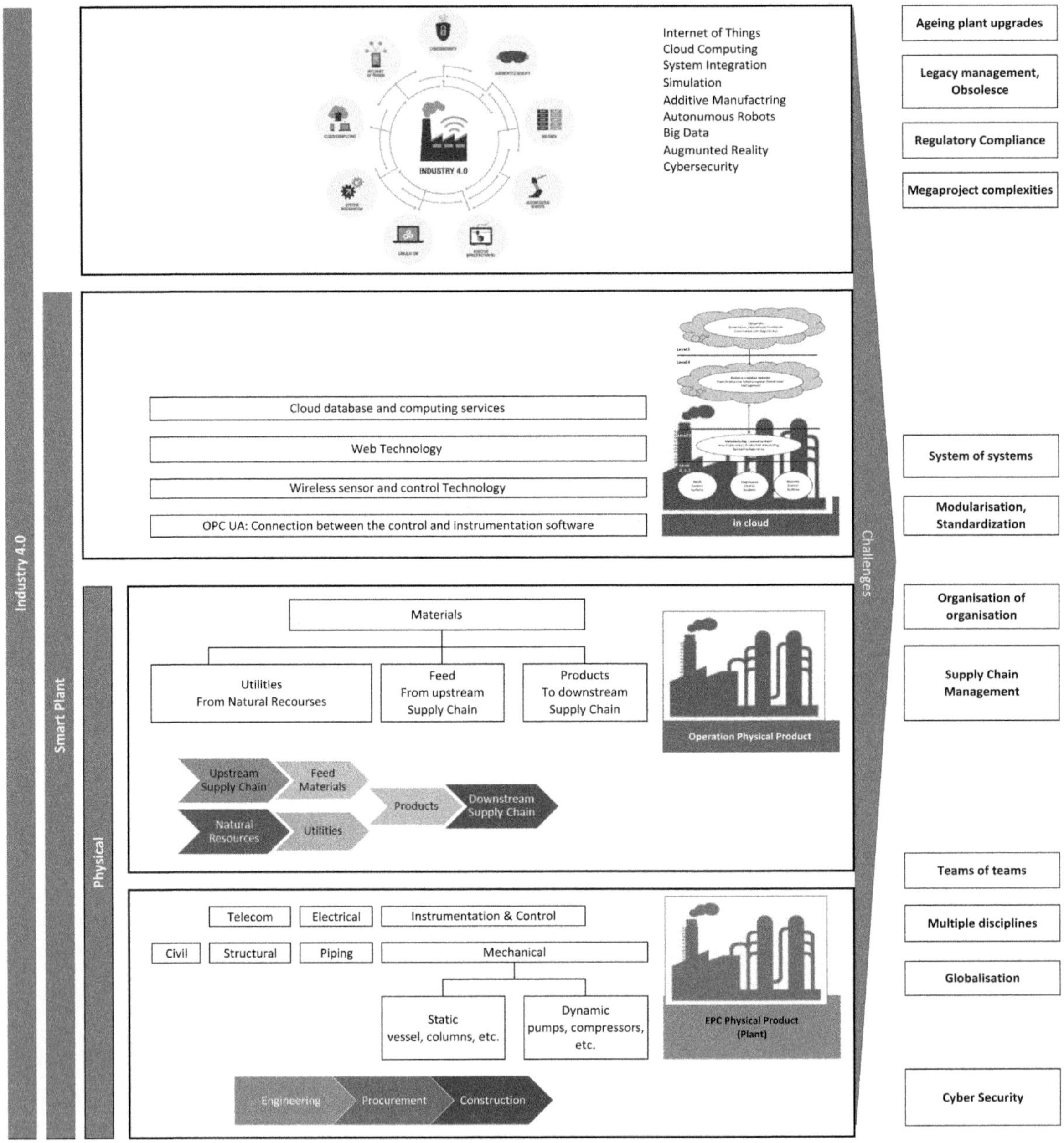

FIGURE 5.1
Smart, connected plant architecture.

1st Industrial Revolution	2nd Industrial Revolution	3rd Industrial Revolution	4th Industrial Revolution
• Mechanisation • Steam engines • Water/steam power • New manufacturing • Mining & Metallurgy • Machine tools • Steam factories	• Technological • Electrification • Production line • Mass Production • Globalisation • Engines/Turbines • Broad adoption of telegraph, gas, water supply	• Computer /Internet • Digital manufacturing • PLC/Robotics • IT and OT • Digitalisation • Automation • Electronic /digital • Networks • Digital machines	• Convergence IT/OT • Autonomous machine • Advance robotics • Big Data /Analytics • Internet of Timings • Digital ubiquity /cloud • Smart factory • Machine learning & AI • Cyber-Physical

FIGURE 5.2
Four industrial revolutions.

Table 5.1 Key Elements of the Industrial Internet

Intelligent Machines	Advanced Analytics	People at Work
Connect the world's machines, facilities, fleets, and networks *with* advanced sensors, controls, and software applications	*Combines* the power of physics-based analytics, predictive algorithms, automation, and deep domain expertise	*Connecting* people at work or on the move, any time to support more intelligent design, operation, maintenance, and higher service quality and safety

Slideshare.net/GE_MC_India/industrial-internet-pushing-the-boundries-of-minds-and-machines.

During the second Industrial revolution, decision-makers had time to study a specific issue and develop the necessary response or appropriate regulatory framework. Today, the ability of the decision-makers to adapt with the agile governance and management processes will determine their survival.

The rigid top-down approach is no longer valid in the hyperconnected world of Industry 4.0, where lower parts of the value chain are empowered (bottom-up), mirroring the agile integration of the manufacturing process.

Industrial internet often used interchangeably with the industrial internet of things enables us to see the more holistic picture of "connected devices," "data and meaning," and "people." The key elements of the industrial internet are summarized in Table 5.1.

When we talk about the internet of things we most often mean the *internet of everything,* which Cisco defines as follows: "The Internet of Everything (IoE) brings together people, *process,* data, and things to make networked connections more relevant and valuable than ever before-turning information into

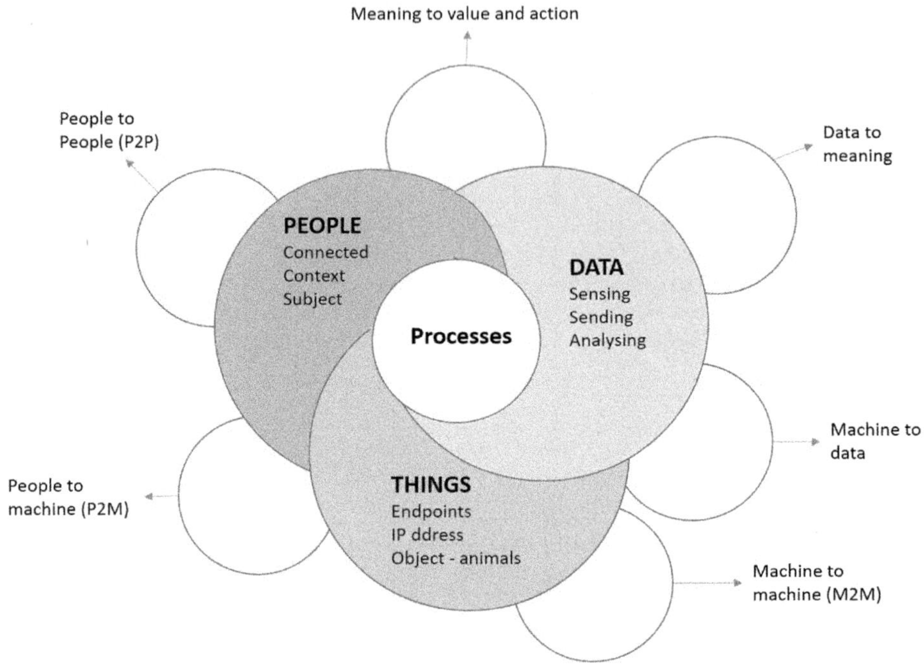

FIGURE 5.3
Internet of everything, processes are the heart of the integration of the technical, organizational, and people management systems.

actions that create new capabilities, richer experiences, and unprecedented economic opportunity for businesses, individuals, and countries."

Fig. 5.3 illustrates how "things," "people," "data," and "*processes*" are woven together in a real system. Processes are the center of integration and assure the consistency of any management system. Without a careful attention to the processes, none of the physical, organizational, and people integration can be effective.

When we design the management system we should ask the right ontological questions about the physical parts and immaterial concepts. For example, are physical objects such as shoes (things) more real than the concept of walking (processes)? In terms of what exists, what is the relationship between shoes and walking? The quality of our ontological questions and the way we answer them indicates our ability of system and complexity thinking.

5.2 EVOLUTION OF WEB TECHNOLOGY

Web is the largest transformable-information construct whose idea was introduced by Tim Burners-Lee in 1989 at first. Much progress has been made

Table 5.2 Comparison Between the Web Generations

Web 1.0	Web 2.0
Reading	Reading/writing
Companies	Communities
Client–server	Peer-to-peer
HTML, portals	XML, RSS
Taxonomy	Tags
Owning	Sharing
IPOs	Trade sales
Netscape	Google
Web forms	Web applications
Screen scraping	APIs
Dialup	Broadband
Hardware costs	Bandwidth costs
Lectures	Conversation
Advertising	Word of mouth
Services sold over the web	Web services
Information portals	Platforms
Web 2.0	**Web 3.0**
Read/write web	Portable personal web
Communities	Individuals
Sharing content	Consolidating dynamic content
Blogs	Lifestream
AJAX	RDF
Wikipedia, google	Dbpedia, igoogle
Tagging	User engagement
Wikipedia, google	Dbpedia, igoogle
AJAX	RDF

about the web and related technologies in the past two decades. Web 1.0 as a web of cognition, web 2.0 as a web of communication, web 3.0 as a web of cooperation, and web 4.0 as a web of integration are introduced such as four generation of the web since the advent of the web. Table 5.2 compares the different web generations.

Web 3.0 is also known as *semantic web*. Semantic web was thought up by Tim Berners-Lee, inventor of the World Wide Web. There is a dedicated team at the World Wide Web consortium (W3C) working to improve, extend, and standardize the system, languages, publications, and tools that have already been developed.

Semantic web is a web that can demonstrate things in the approach which computer can understand. The main important purpose of semantic web is

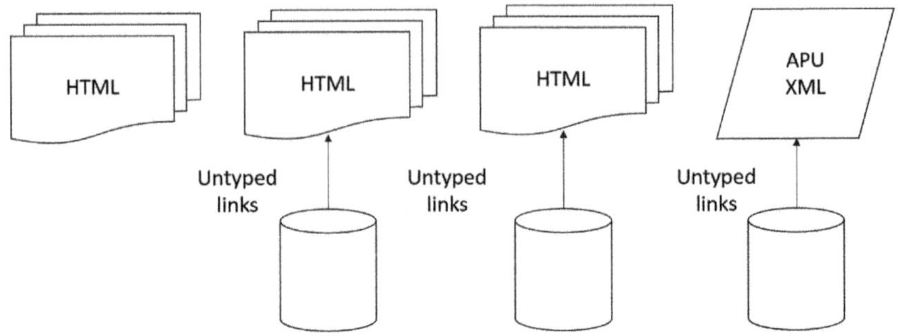

FIGURE 5.4
Web of documents.

to make the web readable by machines and not only by humans. The current web is a web of documents, in some ways like a global file system that the most important problems about it are included: The web of documents was designed for human consumption in which primary objects are documents and links are between documents (or parts of them). Semantics of content and links are implicit and the degree of structure between objects is low. Fig. 5.4 represents the structure of web of documents in simple.

Semantic web is being developed to overcome the problems of current web. Semantic web can be defined as a web of data, in some ways like a global database that most of its features are included: The aim of design web of data is machines first, humans later. Primary objects are things so links are between things. Semantics of content and links are explicit and the degree of structure between objects is high based on resource description framework (RDF) model. In Fig. 5.5, the structure of web of data is shown for simplicity.

The main difference between web 2.0 and web 3.0 is that web 2.0 targets on content creativity of users and producers while web 3.0 targets on linked data sets. Tim Berners-Lee proposed a layered architecture for semantic web that is often represented using a diagram, with many variations since. Fig. 5.6 gives a typical representation of this diagram.

The layers of the semantic web architecture are briefly described as follows:

- *Unicode and URI*: Unicode is used to represent any character uniquely whatever this character was written by any language, and uniform resource identifiers (URIs) are unique identifiers for resources of all types. The functionality of unicode and URI could be described as the provision of a unique identification mechanism within the language stack for the semantic web.

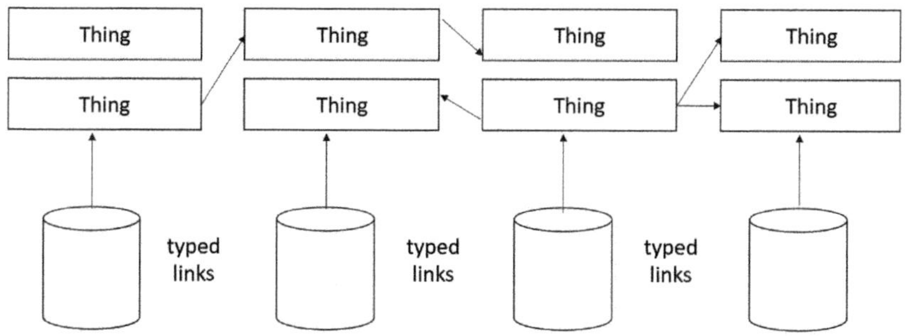

FIGURE 5.5
Web of data.

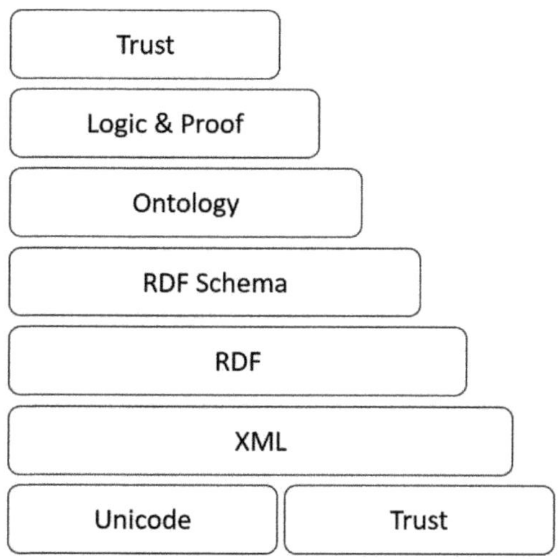

FIGURE 5.6
Semantic web layered architecture.

- *Extensible markup language*: XML and its related standards, such as namespaces (NS), and schemas are used to form a common means to structure data on the web without any communication between the meanings of the data. XML is used as a base syntax for other technologies developed for the upper layers of the semantic web. NS is used to identify and distinguish different XML elements of different vocabularies. It supports mixing of different elements from various vocabularies to do a specific function. XML schema assures that the received information is according to the sent information when two applications at this level exchange information with together.

- *Resource description framework*: RDF is a simple data model that uses URIs to identify web-based resources and describes relationships between the resources in terms of named properties and values. Generally, the RDF family supports interoperability at the semantic level. RDF developments consist of the base web language, so that agents are able to make logical inferences to perform functions based on metadata.
- *RDF schema*: It provides a predefined, basic type system for RDF models. It describes classes and properties of the resources in the basic RDF model. RDF schema provides a simple reasoning framework to infer types of resources.
- *Ontology*: The ontology layer described properties and the relation between different properties. Ontology can be defined as a collection of terms used to describe a specific domain with the ability of inference.
- *Logic and proof*: This layer is on top of the ontology structure to make new inferences by an automatic reasoning system. The agents are able to make deductions as to whether particular resources satisfy their requirements by using such reasoning systems.
- *Trust*: The last layer of the stack addresses trust in order to provide an assurance of quality of the information on the web and a degree of confidence in the resource providing this information.

Insurance	**Assurance**	**Quality**
– Show me the data – Prove the operation	– Conformance – Do it right the first time	Explain the "one-offs" World-class operation
	Requirements	

Web 4.0 is still an underground idea in progress and there is no exact definition of how it would be. Web 4.0 is also known as symbiotic web. The dream behind the symbiotic web is interaction between humans and machines in symbiosis. It will be possible to build more powerful interfaces such as mind-controlled interfaces using web 4.0. In simple words, machines would be clever on reading the contents of the web, and react in the form of executing and deciding what to execute first to load the websites fast with superior quality and performance and build more commanding interfaces.

Web 4.0 will be the *read-write-execution-concurrency* web. It achieves a critical mass of participation in online networks that deliver global transparency, governance, distribution, participation, collaboration into key communities such as industry, political, social, and other communities. Web 4.0 or webOS will be such as a middleware which will start functioning like an operating system. The webOS will be parallel to the human brain and implies a massive web of highly intelligent interactions.

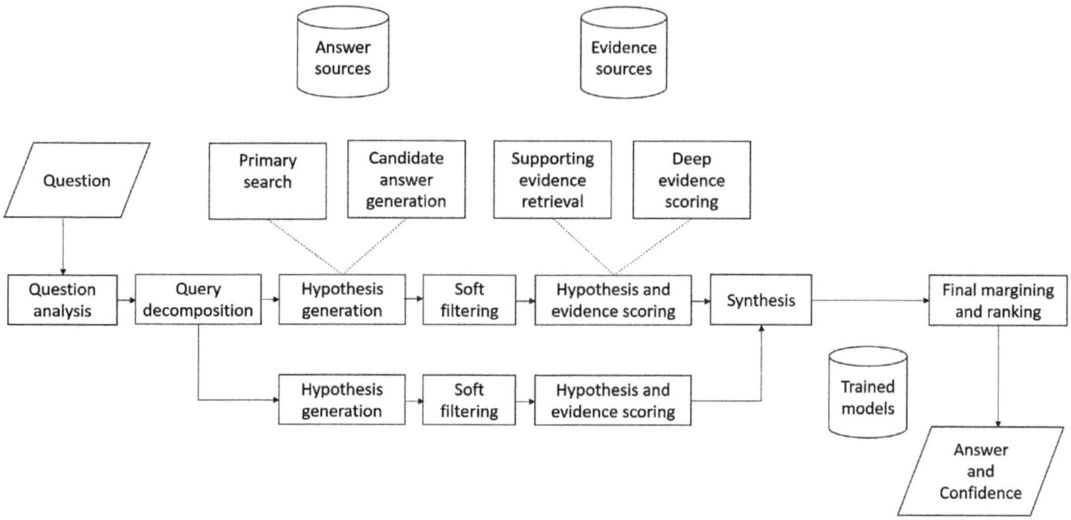

FIGURE 5.7
The high-level architecture of IBM's DeepQA used in Watson.

"Watson" is a question answering (QA) computing system that IBM has built to apply advanced natural language processing, information retrieval, knowledge representation, automated reasoning, and machine learning technologies to the field of open domain question answering.

The key difference between QA technology and document search is that document search takes a keyword query and returns a list of documents, ranked in order of relevance to the query (often based on popularity and page ranking), while QA technology takes a question expressed in natural language, seeks to understand it in much greater detail, and returns a precise answer to the question. Fig. 5.7 summarises the high-level architecture of IBM's DeepQA used in Watson.

According to IBM, "more than 100 different techniques are used to analyze natural language, identify sources, find and generate hypotheses, find and score evidence, and merge and rank hypotheses."

Although there is no exact idea about web 4.0 and its technologies, but it is obvious that the web is moving toward using artificial intelligence to become an intelligent web.

5.3 EVOLUTION OF IIoT

The IIoT is everywhere now and connect millions of devices, machines, sensors, and systems throughout the world. Over the past 50 years, technological milestones big and small, from large-scale system architecture breakthroughs to modest "Eureka" moments, have led to today's IIoT and are still informing predictions for tomorrow's industrial landscape. Table 5.3 summarizes the key dates and developments which lead to the IIoT.

There are distinct differences between the IIoT and the broader Internet of Things (IoT). In some cases, the elements are identical, but the usage varies. Using a smartphone to observe the temperature of your home environment is an example of the IoT, but if you use that same smartphone to monitor the temperature in the controlled environment of a clean room used in production, it is now an IIoT use case. It could be the same systems and the same equipment with only the environment being different. In other situations, there are more distinct differences between the IIoT and the IoT.

For the astute observer, many of the IIoT ecosystem partners are actually leading IoT or cloud providers already. Since the LNS definition of an IIoT platform is more detailed and requires broader functionality many vendors that offer IoT platforms or elements of an IoT solution set fall outside the scope of what we consider a complete IIoT platform. Two examples of this are:

1. Public cloud infrastructure suppliers: Many public cloud services, such as Microsoft's Azure or the Amazon AWS offerings are clearly IoT enabling platforms. When they are leveraged by companies that provide IIoT platform capabilities, additional functionality builds upon these cloud platforms.
2. Communications services suppliers: Since many classic IoT use cases are like *Uber*, where consumers use a smartphone application to request and monitor service delivery, it is natural that cellular infrastructure companies have become part of the IoT platform community. Also, Wi-Fi and wired local area networking are integral to many IoT use cases, so companies like Cisco that are well-established networking infrastructure players have added capabilities in support of IoT. In many cases, companies that are offering fully featured IIoT capabilities have incorporated this communications infrastructure into their IIoT platforms.

In these two cases, the functionality provided in the context of IoT are necessary for an IIoT platform but not sufficient to fully deliver that capability. Fig. 5.8 illustrates the IoT for extended manufacturing enterprise value chains.

In 2016 we have witnessed widespread IIoT solution development. Many organizations are going to make IIoT a strategic priority. These companies are actively exploring new opportunities and identifying the best approach to reap the benefits of their connected product solutions. Whether companies

Table 5.3 IIoT Timeline Inspired From: *The Industrial IoT: A Timeline of Revolutionary Technology*[a]

Year	Event
1968	**PLC is born**
1968-A	On January 1, 1968, Dick Moely hand a hangover. It was not an auspicious start to the day; he would draft a memo that led directly to the invention of the *programmable logic controller (PLC)*
1968-B	*Machine-to-machine (M2M) devices that combined telephoney and computing* were first conceptualized by Theodore G. Paraskevakos while working on those caller line identification system.
1983	Ethernet standardized
1986	PLCs are linked to PCs
1989	*World Wide Web* is invented to share research
1989-A	*Tim Berners-Lee* implemented the first successful communication between a *hypertext transfer protocol (HTPP)* client and server via Internet
1989-B	*Becoming standardized:* – Human–machine interface (HMI) – Supervisory control and data acquisition (SCADA)
1992	Ethernet and transmission control protocol/Internet protocol (TCP/IP)Connectivity for PLCs is introduced.
1994	*OPC Foundation forms* to secure data exchange in industrial automation.
1995	MS Windows become the mainstream OS in the factory.
1996	*OPC data access (DA)* protocol is released.
1997	*Wireless M2M technology* becomes prevalent in industry.
1998	Ethernet turns 25 and become ubiquitous in industrial environments.
1999	*Internet of thing*
1999-A	IOT is coined by Kevin Ashton.
1999-B	*Legacy devices connect to the Internet*, extending connectivity to industrial things.
2002	*Cloud technology*: Cloud computing takes hold with the launch of Amazon Web Services (AWS)
2006	*OPC unified architecture (UA) protocol* is released, enabling secure remote communications between devices, data sources, and operations.
2006-10	Devices get smaller and smaller. Historian adoption rises. Battery and solar power become more powerful and economical.
2010	Sensors drop in price.
2011	Bring your own device (BYOD) becomes mainstream.
2011–2016	IT standards enter industrial automation (Ex: MQTT, REST, HTPP).
2016	*IIOT vision emerges*
Challenges for the coming years	Data scientists move into leadership. *Augmented reality and virtual reality* become standard user Interfaces for IIOT The *security* of industrial data and assets remains critical. IT and operations *collaborate* to drive business-level decisions

[a]*Data from The Industrial IoT: A Timeline of Revolutionary Technology*

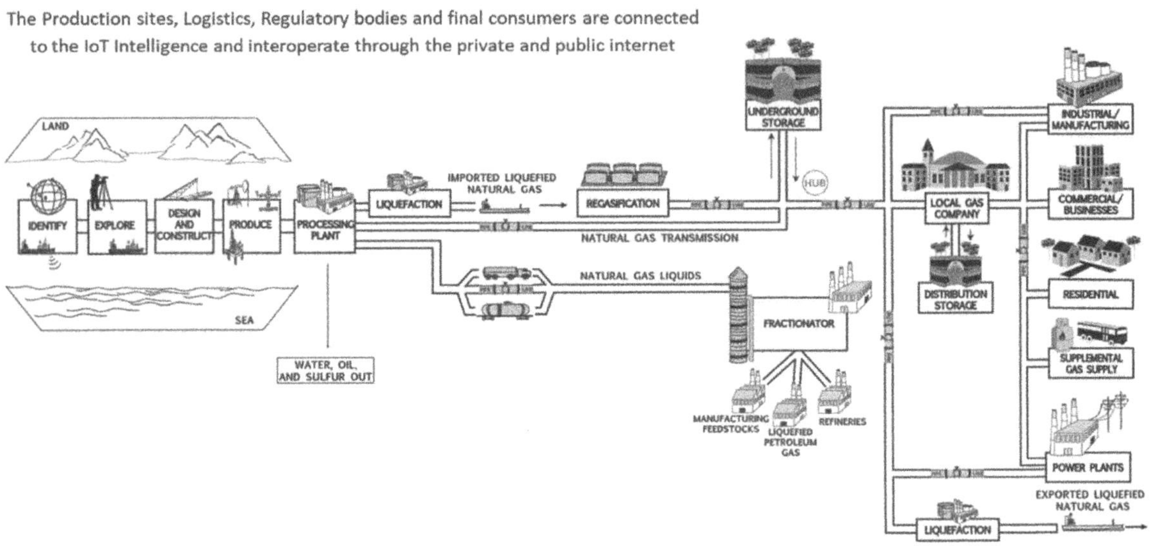

FIGURE 5.8
IoT for extended manufacturing enterprise value chains.

are connecting elevators to the cloud for predictive maintenance or whether they are introducing remote monitoring of industrial machinery—firms are starting to develop compelling business cases for IIoT.

Among the early adopters of IIoT, we're seeing a clear trend of IIoT implementations as limited pilots focused on a specific use case or objective. For example, improving the efficiency of a particular area of production. The goal is usually to demonstrate the business value of IIoT in solving real problems. But these projects are often not scalable to other parts of the business. What's often missing is a broader, unifying principle that places IIoT at the heart of the automation value chain. There are some good reasons for this. In fact, we see three key challenges that are acting as barriers to more widespread, enterprise-wide IIoT implementations:

- Lack of a clear IIoT blueprint
- Legacy mindset and skills
- Uncertain funding

Some of the investors assert that the IIoT is another manifestation of the hype cycle like the so-called Dot Com crash where at the turn of the century we had Internet companies going through the roof producing zero value and it took a decade later before viable products and companies started appearing. Fig. 5.9 illustrates the Gartner hype cycle.

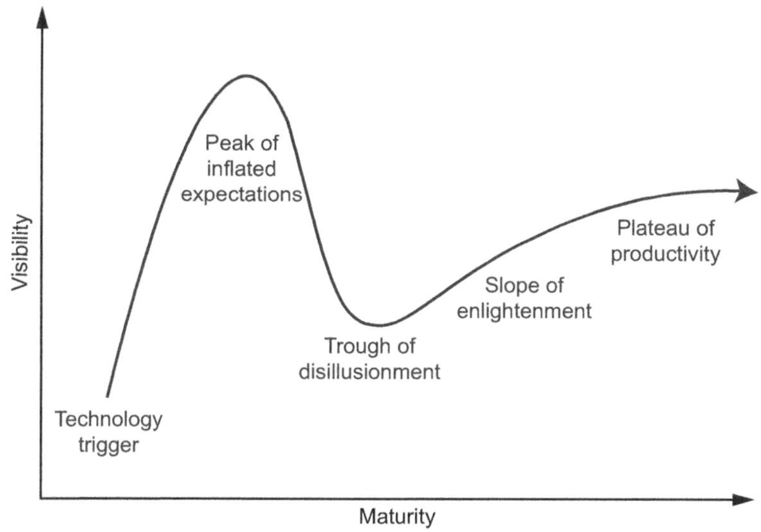

FIGURE 5.9
The Gartner hype cycle.

Associated with the hype of the IIoT is the concept (popularized by governments particularly) that there is a ferocious shortage of STEM (science technology engineering and math) professionals. Quite clearly if there was a shortage there would be sky high salaries for these types of jobs bearing in mind the enormous amount of effort—time, money, and hard work particularly—and experience required—before one can successfully work in this area.

It is not hard to answer these questions. The collaborative web-based learning solutions can speed up the training of both users and developers of the IIoT solutions. There are already many good online contents and tool available. To achieve the effective and quick learning programs a universal platform should be established to gather the experts and available contents.

Network-connected remote terminal units, programmable logic controllers (PLCs), and human–machine interfaces (HMIs) are *nothing new*. Furthermore, the IIoT suppliers and stakeholders create the partnerships to standardize the optimal solutions. For example, since 2011, the FDT Group (1998) and OPC Foundation (1994) have worked together on a technology initiative to provide greater access to critical information throughout the industrial enterprise. OPC provides a uniform interface for many different client applications, whereas FDT provides network/device configuration and access to devices. Fig. 5.10 illustrates the Integrated OPC UA and DTM (device type manager) protocols for industrial internet of things.

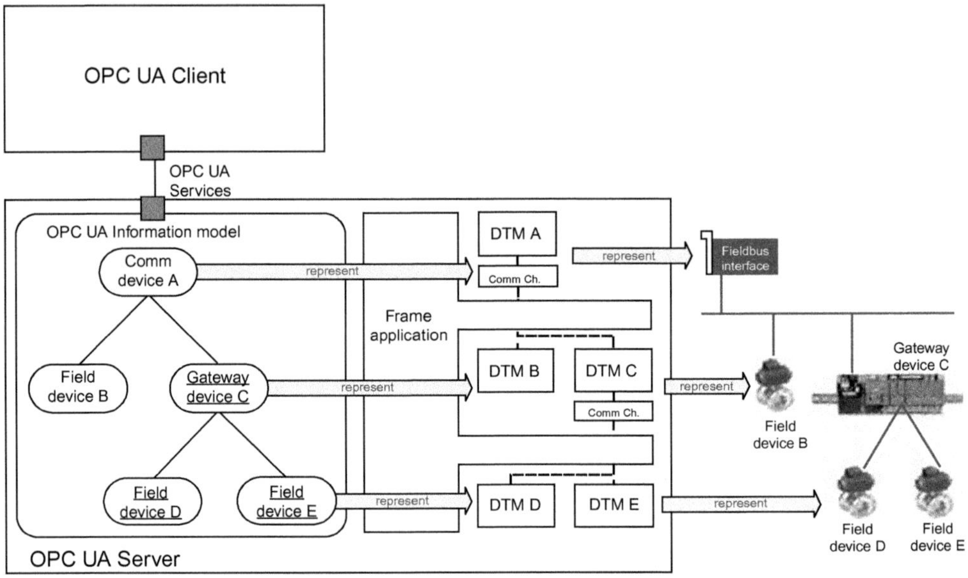

FIGURE 5.10
Integrated OPC UA and DTM (device type manager) protocols for industrial internet of things.

Together, FDT and OPC technologies simplify the transition to the next generation of connectivity for industrial facilities. Defined by an FDT-OPC UA device information model, the recently announced FDT/OPC Annex enhances the configuration of networks and devices, and gives the enterprise access to data without the complexity of protocol-specific handling.

With release of the FDT/OPC Annex, sensor-to-cloud, enterprise-wide connectivity is granted with seamless data communications. The objective is a fully enabled IIoT architecture supporting mobility, on-the-wire security, and comprehensive interoperability. This solution will also be platform-independent and deployable in standalone, client/server, or cloud architectures. Fig. 5.11 illustrates the Data exchange with other IIoT.

Fig. 5.12 illustrates operational architecture of IIoT. Integration of IIoT goes further than integration of the systems on the plant floor. It offered corporate, customer, and partner-wide connectivity on a whole new scale. As a result, it had the potential to unlock tremendous value in the manufacturing chain and transform the way a company does business. Like all new ideas, IIoT is not without its challenges.

IIoT solutions threats include increased security risks, the potential for information overload, a shortage of staff with the needed skill sets and experience, and unexpected effects on corporate culture. If IIoT is deployed incorrectly,

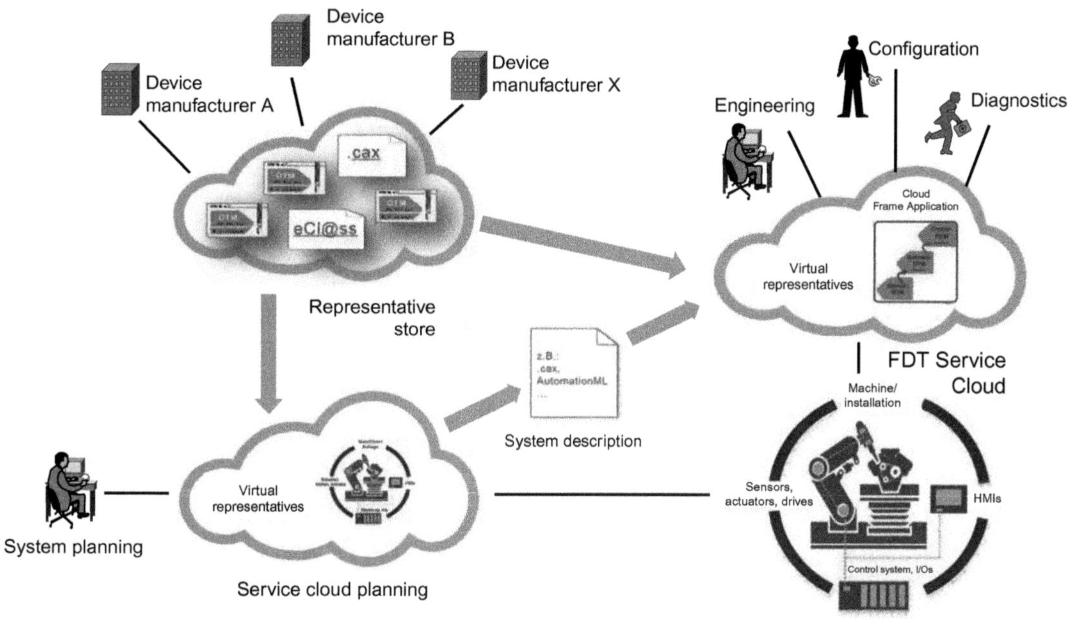

FIGURE 5.11
Data exchange with other IIoT.

we can have a big mess on our hands. The following best practices should be applied to assure the success of an IIoT project:

- Whole company involvement
- Focus on business value
- Design security and robustness in from the start
- Embrace secure connectivity
- Leverage virtualization
- Make it reliable and available
- Prioritize simplicity

2015 was the year of introduction for the IIoT. LNS Research of over 1000 surveyed industrial companies showed that 40% of companies at that time did not know or understand the IIoT. 2016 was the year of education, based on the hard work of analysts at LNS, along with many other voices in the marketplace, the number of companies that didn't know or understand the IIoT dropped to 19%, while the share of companies planning to adopt the IIoT increased by 10%.

Based on this trend, *LNS Research* expects that 2017 will be the year of adoption for the IIoT. They have already predicted that many of the leading

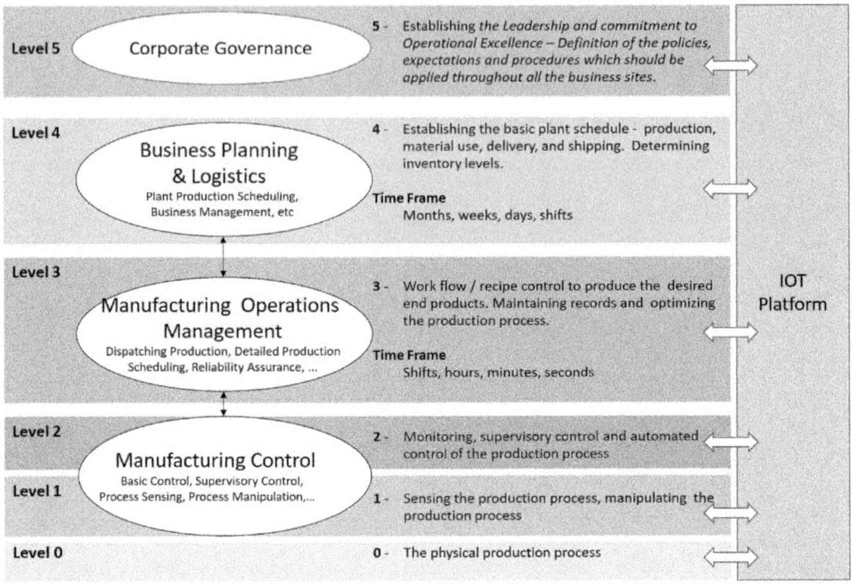

FIGURE 5.12
Operational architecture of IIoT.

industrial companies will be moving out of the planning/pilot stages this year and making enterprise commitments to deploying IIoT platforms.

Developing end-to-end IoT solutions involves multiple layers that fuse together various components. On a high level there are five major layers of an IoT solution including one cross-layer: device, communication, cloud services, applications, and security as shown in Fig. 5.13.

1. *Device layer*: Adding microcontroller units (MCU) and firmware to basic hardware (e.g., sensors and actuators) creates *simple* connected devices. Adding microprocessors (MPU) and operating systems (OS) makes these connected devices *smart*.
2. *Communication layer*: Enabling communication to the outside world through various connectivity networks gives the devices a *voice*.
3. *Cloud services layer*: Ingesting, analyzing, and interpreting the data at scale through cloud technologies generates *insights*.
4. *Application layer*: Connecting and enhancing these insights to the greater ecosystem through a system of engagements enables *action* through a vast range of new applications and connected services.
5. *Security cross-layer*: Securing an IoT solution is an element of such importance that it merits an established *foundation* in each of the other building blocks.

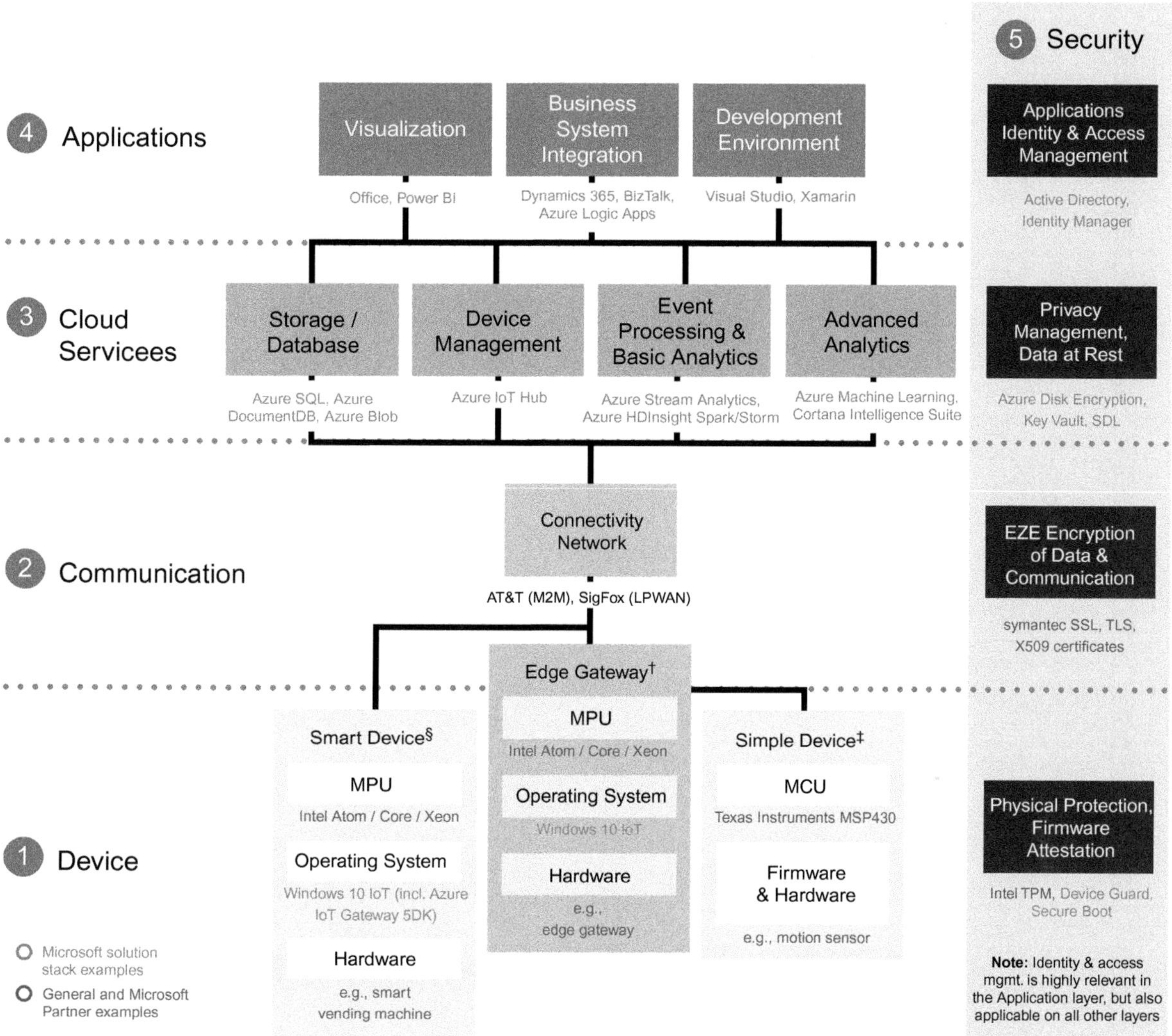

FIGURE 5.13
Generalized five layer model of an end-to-end IoT solution with a focus on software components.

Each component demands a particular array of competences and proficiency to function within its own realm, not to mention the varied skillset required in bringing the end-to-end solution seamlessly together across all of these components.

Correctly assessing the capabilities of each possible vendor against your requirements definition is crucial for your selection. While the use case at

Layer	Components	Definitions	Microsoft	IBM	Google	amazon	ptc	intel	GE	SAP
Application (End-2-end stack)	Visualization	presents device data in rich visuals and/or interactive dashboards	✓	✓	✓	✓	✓		✓	✓
Application (End-2-end stack)	Business System Integration	enables integration with existing business systems	✓	✓	✓	✓	✓		✓	✓
Application (End-2-end stack)	Development Environment	offers an integrated development environment with SDKs for creating apps/services	✓	✓	✓		✓	✓		✓
Cloud Services (End-2-end stack)	Storage / Database	cloud based storage and database capabilities (not including on premise solutions)	✓	✓	✓	✓			✓	✓
Cloud Services (End-2-end stack)	Device Management	enables remote maintanance, interaction and management capabilities of devices at the edge	✓	✓		✓	✓	✓	✓	✓
Cloud Services (End-2-end stack)	Event Processing & Basic Analytics	processes events and handles big data analytics	✓	✓	✓	✓	✓		✓	✓
Cloud Services (End-2-end stack)	Advanced Analytics	preforms advanced stream analytics and machine learning	✓	✓	✓	✓	✓		✓	✓
Communication (End-2-end stack)	Connectivity Network/ Modules	offers connectivity network/ hardware modules enabling air interface connectivity						✓		
Communication (End-2-end stack)	Edge Analytics	enables analytics, time-sensitive decisions and local compute on a smart / edge device	✓	✓			✓	✓	✓	✓
Communication (End-2-end stack)	Edge Gateway (hardware based)	enables manageability security identity interoperability based on a cloud enabled hardware device						✓	✓	
Device (End-2-end stack)	Operating System	offers low-level system software managing hardware, software and runs applications	✓		✓			✓		
Device (End-2-end stack)	Modules and Drivers	offers adaptable modules, drivers,source libraries that reduce development and testing time	✓	✓		✓	✓	✓	✓	✓
Device (End-2-end stack)	MPU / MCU	offers multi-purpose programmable electronic devices at microprocessor or microcontroller level						✓		
Other	Specific Proprietary Enterprise Applications	offers existing in-house enterprise systems (potentially connected to IoT)	• ERP • CRM	• BPM			• CAD, PLM • ALM, SLM		• MES	• ERP • CRM • EAM
Other	Augmented / Virtual Reality	offers an integrated mobile vision platform / related hardware for an IoT solution	✓				✓			✓
Other	Ready-to-go Solutions (i.e.,pre-configured templates)	offers precorfigured IoT solution templates enabling quick setup, testing and data visualization	✓	✓			✓		✓	✓
Other	Dedicated IoT Implementation Teams	offers in-house team with dedicated purpose of integrating IoT solutions	✓	✓			✓		✓	✓

✓ denotes an internal solution offering available i.e, not through the use of a partner. See footnotes below and Appendix for clarification.

FIGURE 5.14

Key providers with in-house offerings across the IoT solution stack—Microsoft, IBM, Google, Amazon, PTC, Intel, GE, and SAP. *Source: IoT Analytics.*

hand determines your solution requirements, the vendor selection process largely depends on the components the vendors offer and how they fit into your solution. Fig. 5.14 presents a high-level comparison of selected vendors across the layers and components.

Table 5.4 IoT Partner Network—a Number of Stakeholders are Required to Develop End-to-End IoT Solutions

Partner Type	Partner Strength
Device hardware vendors	Enable quick start prototyping
Original design manufacturers	Offer reference PCBs for "white label" solutions
Silicon/chip vendors	Customize specific designs for embedded systems
Gateway partners	Deliver security-enhanced data flows
Communication partners	Support connectivity across multiple networks
Cloud platform providers	Offer a cloud computing backend and supporting infrastructure
Independent software vendors	Bring holistic solution development expertise
Analytics professional service providers	Offer unique expertise in Big Data analytics and data science services
System integrators	Play a central role in stitching solutions together

Providers listed cover a major part of the overall IoT solution. Other large IoT vendors include Cisco, Salesforce.com, Vodafone, and Ericsson among others.

When evaluating vendors for a pilot and commercial deployment of a connected IoT solution, you should try to deeply understand how each vendor together with their partner network will support your project, specifically in relation to the learnings outlined in this paper.

A vendor with a strong partner network may be able to minimize solution development problems by building on existing experience and leveraging trusted expertise. A strong partner network can be a key differentiating factor for vendors. You should assess the breadth of the vendors partner network across a number of segments as outlined in Table 5.4.

At its core, IIoT is not a new technology. It takes advantage of some new technologies, but it is actually a new way of examining an old problem. We have always had the data test results, analytics, asset management information, maintenance information but it has often been inaccessible, overlooked, or obscured in our operating procedures. IIoT lets us rethink the way industry integrates the data buried in our manufacturing process.

Effectively implementing IIoT is a continuous process that demands strategic planning, a focus on company goals, the coordination of teams with diverse skill sets, and an investment in quality security measures. The adoption of IIoT brings immediate benefits, such as improved reliability and reduced downtime. Simultaneously, it also enables long-term benefits by establishing a platform for continuous development and offering a greater ROI by integrating information quantity and quality.

By creating a forward-thinking company culture, by maintaining corporate focus, and by designing IIoT systems with appropriate security measures, your business can overcome obstacles and strategically implement IIoT best practices to gain an immense competitive advantage in the digital future.

5.4 OPEN PLATFORM COMMUNICATIONS (OPC)

OLE for process control is a series of standards and specifications for industrial telecommunication. An industrial automation industry task force developed the original standard in 1996 under the name OLE for process control (object linking and embedding for process control). OPC specifies the communication of real-time plant data between control devices from different manufacturers.

After the initial release in 1996, the OPC Foundation was created to maintain the standard. As OPC has been adopted beyond the field of process control, the OPC Foundation changed the name to open platform communications in 2011. The change in name reflects the applications of OPC technology for applications in building automation, discrete manufacturing, process control, and many others. OPC has also grown beyond its original OLE (object linking and embedding) implementation to include other data transportation technologies including Microsoft's .NET Framework, XML, and even the OPC Foundation's binary-encoded TCP format.

The OPC specification was based on the OLE, COM, and DCOM technologies developed by Microsoft for the Microsoft Windows operating system family. The specification defined a standard set of objects, interfaces and methods for use in process control and manufacturing automation applications to facilitate interoperability. The most common OPC specification is OPC data access (OPC DA), which is used to read and write real-time data. When vendors refer to OPC generically, they typically mean OPC DA. OPC DA itself has gone through three major revisions since its inception.

In addition to OPC DA specification, the OPC Foundation also maintains the OPC historical data access (HDA) specification. In contrast to the real-time data that is accessible with OPC DA, OPC HDA allows access and retrieval of archived data. Fig. 5.15 illustrates the OPC UA connections.

The OPC alarms and events specification is also maintained by the OPC Foundation, and defines the exchange of alarm and event type message information, as well as variable states and state managemen. The OPC UA information modeling framework is shown in Fig. 5.16.

OPC unified architecture (*OPC UA*) is a machine-to-machine communication protocol for industrial automation developed by the OPC Foundation

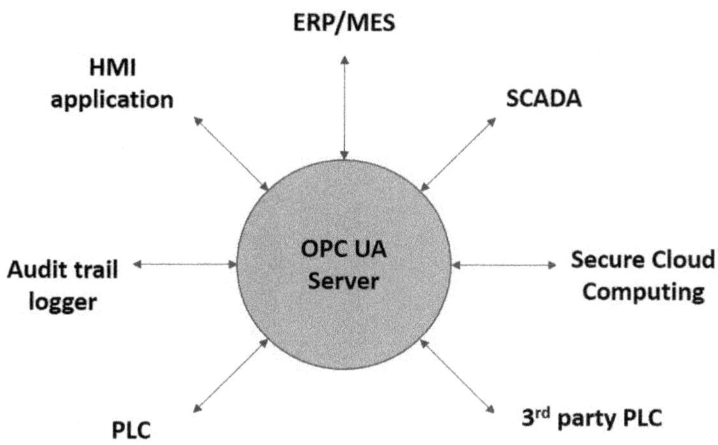

FIGURE 5.15
OPC UA connections.

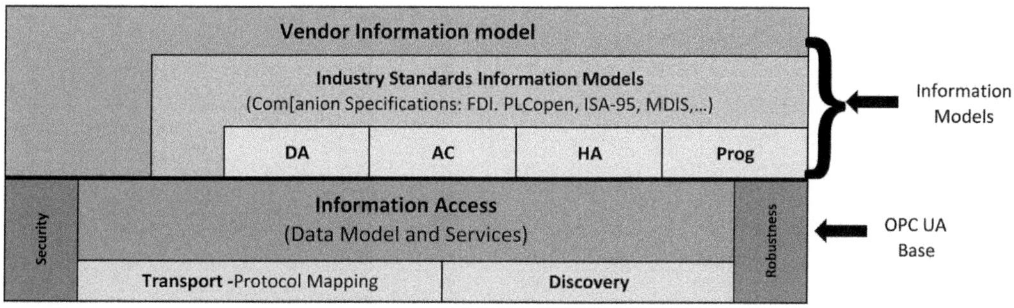

FIGURE 5.16
OPC UA information modeling framework.

released in 2008. It is a platform-independent service-oriented architecture that integrates all the functionality of the individual OPC classic specifications into one extensible framework. This multilayered approach accomplishes the original design specification goals of:

- *Functional equivalence:* all COM OPC classic specifications are mapped to UA
- *Platform independence:* from an embedded microcontroller to cloud-based infrastructure
- *Secure:* encryption, authentication, and auditing
- *Extensible:* ability to add new features without affecting existing applications
- *Comprehensive information modeling:* for defining complex information

The OPC UA information modeling framework turns data into information. With complete object-oriented capabilities, even the most complex multilevel structures can be modeled and extended. Data types and structures are defined in profiles. For example, the existing OPC classic specifications were modeled into UA profiles which can also be extended by other organizations.

5.4.1 ISA-95 Information Model

ISA-95 which is also known as IEC/ISO 62264 is an international standard that has been developed to address the problems encountered during the development of automated interfaces between enterprise and control systems. This standard can be applied in all industries, and in all sorts of processes, such as batch, continuous, and repetitive or discrete processes.

Fig. 5.17 illustrates the different levels of a functional hierarchy model of ISA-95: business planning and logistics, manufacturing operations and control, and batch, continuous, or discrete control. The model defines hierarchical levels at which decisions are made. OPC is interface between the manufacturing execution system (MES) (Level 3) and physical assets and associated sensors and control systems (Levels 0, 1, and 2). MES is connected to the enterprise resources planning via the web XMLs.

The terms MES and manufacturing operations management (MOM) are often used interchangeably and define a different functional space to each and every manufacturing professional:

- Some think they are synonymous and interchangeable.
- Some think MES is a subset of MOM.
- Some believe we should get rid of one and stick to a single term.
- Some are just confused (this is the biggest group).

The term MES is used in so many ways that it is hard to come up with an agreed upon definition. However, we do not want to exclude its use in normal discussion as many manufacturers refer to MES systems used in operations and vendors use MES in naming products. Therefore, we will use MES in all its casual meanings when appropriate, and where there is no confusion. Where there might be confusion we will tend to clear it up by using the term MOM as defined above.

Fig. 5.18 demonstrates how "safety engineering" at levels 0,1,2 is related to "process safety management system" at level 3 and "HSE management system" at level 4,5.

Scope and evaluation methods of the process safety at (L0, L1, L2) are very different from the EH&S at Level L4. During operation process safety at

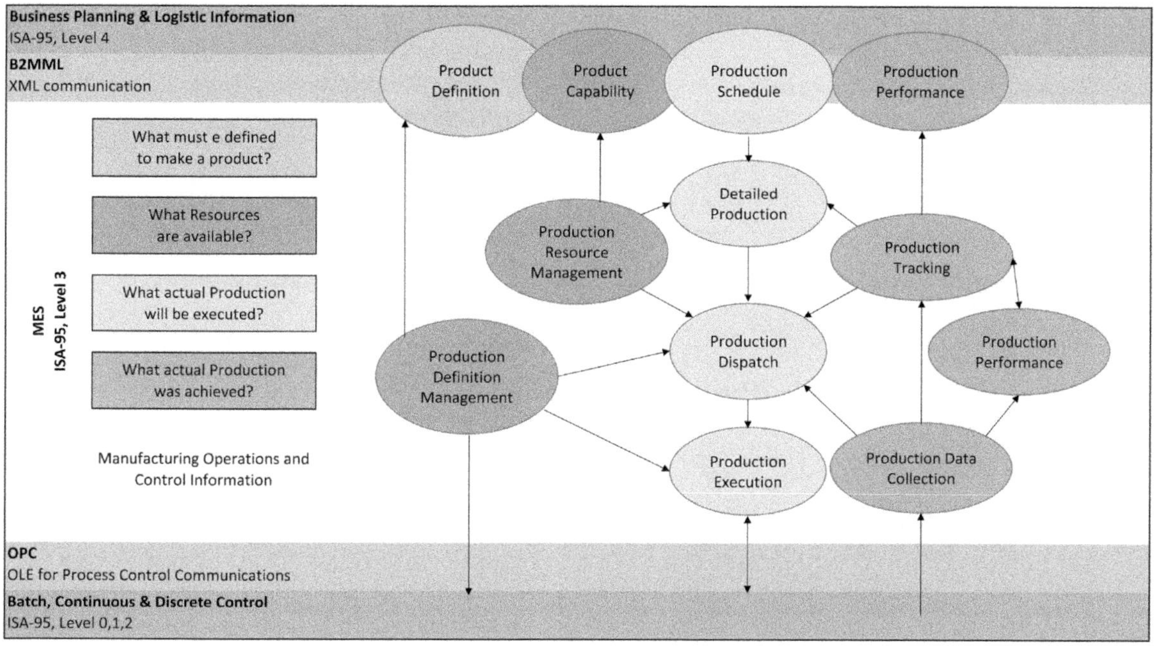

FIGURE 5.17
ISA-95 functional hierarchy model. From: "Alphabet Soup Improves Business Visibility". Dave Harrold. Conrol Magazine, July13, 2007.

(L0, L1, L2) can be improved by effective predictive maintenance (PdM) and optimized human factor engineering (HFE).

5.4.2 Process Safety IIoT Scope

Research conducted by Abnormal Situation Management (ASM) Consortium member companies showed a major cause of incidents was human error, along with process and equipment factors. If it costs $1 to fix a usability problem during design, it will cost $10 to fix once the system is developed, and $100 once it is operational (Fig. 5.19).

Fig. 5.20 illustrates the amount of time that it takes to detect a potential failure interval for each of the four maintenance models commonly used today. PdM enables us to save time and money by detecting the failure based on data sources before damage to the machine occurs.

Start by analyzing available data to define the parameters of normal operation for a machine. This enables the creation of rules through condition monitoring

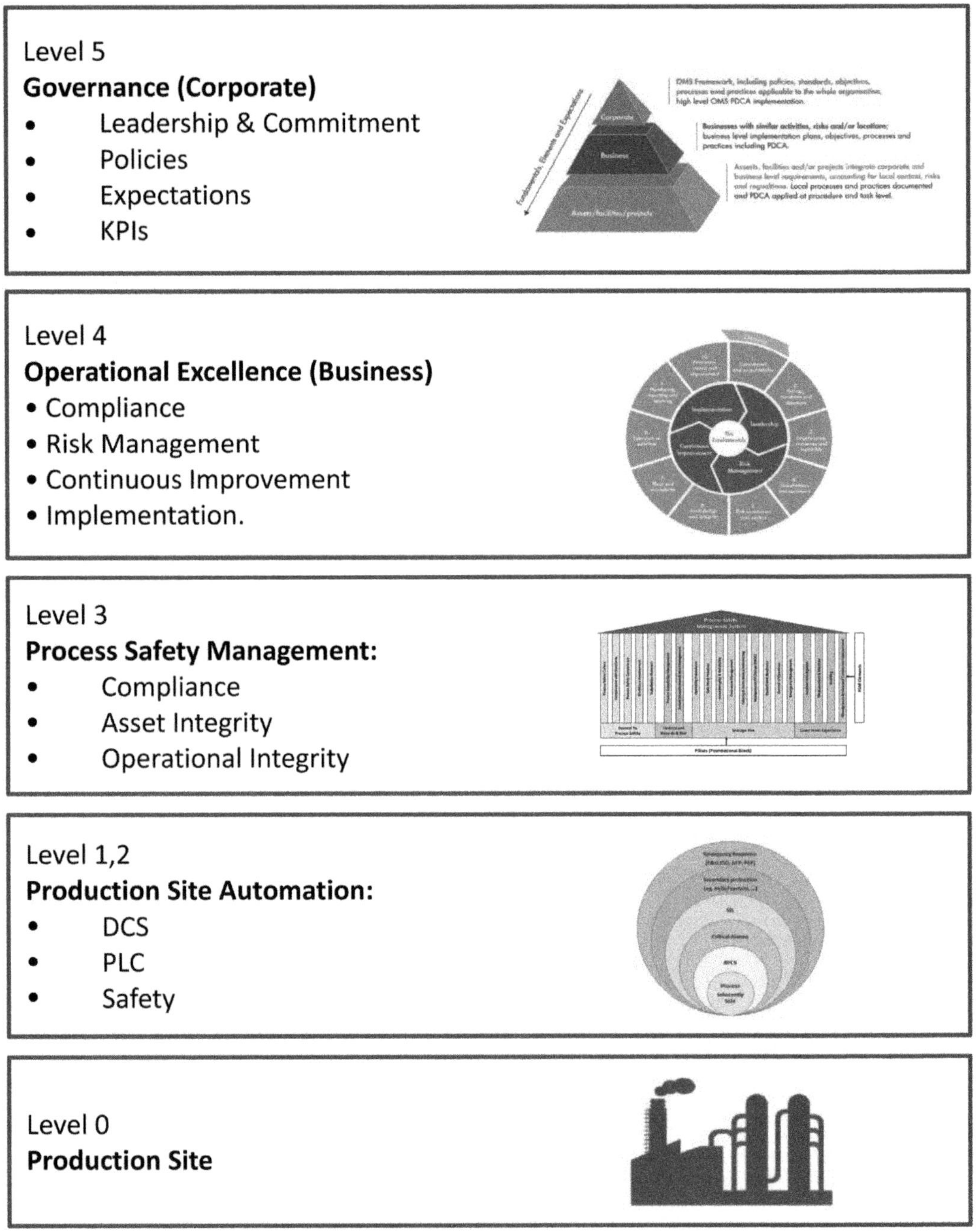

FIGURE 5.18
Safety Engineering & Safety Management System in the ISA-95 functional hierarchy model.

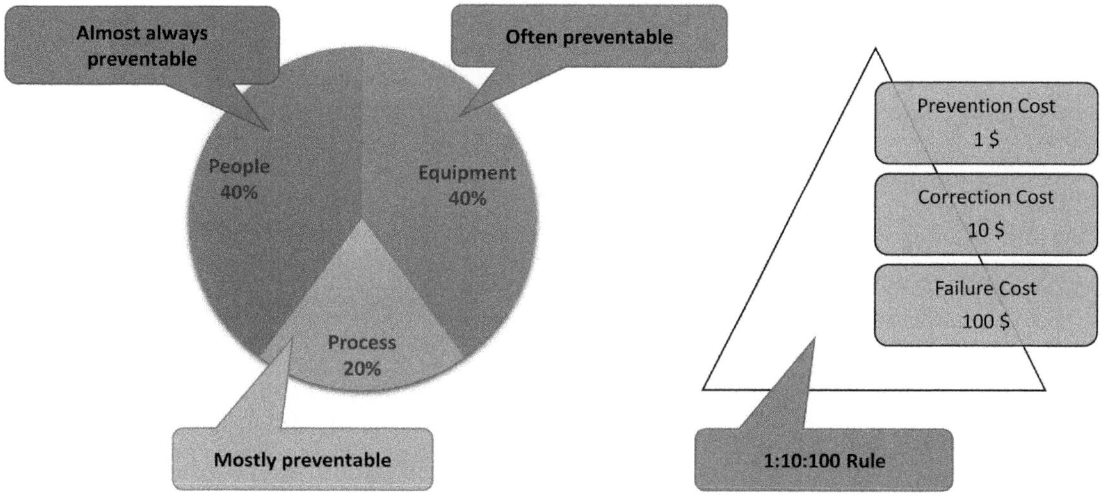

FIGURE 5.19

Causes and cost of the failures.

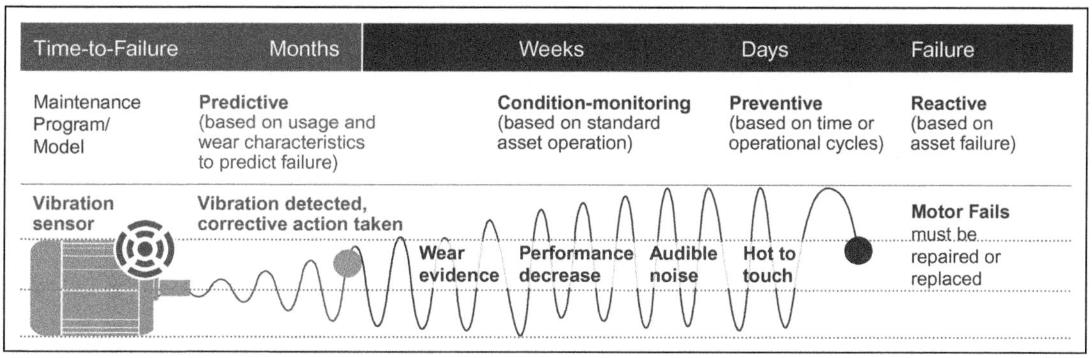

FIGURE 5.20

PdM motor vibration analysis example. *Reproduced with permission from © 2017 Dell Inc. All Rights Reserved.*

for analyzing the real-time data coming directly from machine sensors on the Edge Gateway. With the Edge Gateway, analytics can happen as close to the machine as possible with the native I/O to collect data from industrial equipment and the ability to operate in harsh environments. After analyzing the real-time data, add historical and third party data such as reliability models and logs to uncover meaningful correlations, patterns, and trends with the anomalies generated by the real-time data rules, to signal potential failures. The patterns can be used to further refine your rules and other actionable insights in real time. Fig. 5.21 illustrates how thanks to IIoT the real-time data

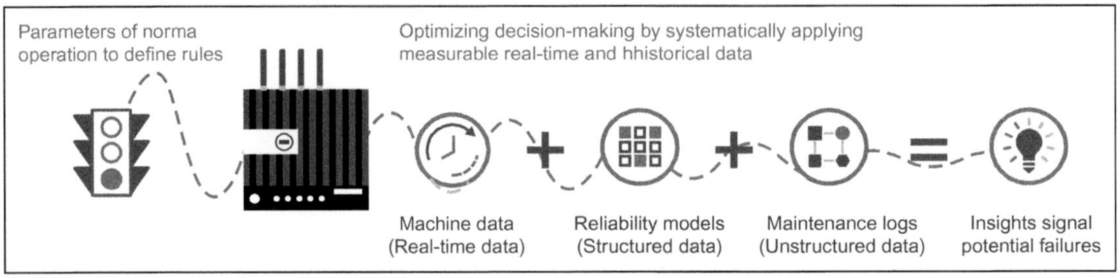

FIGURE 5.21
Safety improvement at Levels 0, 1, and 2. *Reproduced with permission from © 2017 Dell Inc. All Rights Reserved.*

can be combined with structured reliability data and unstructured maintenance log data to improve safety by providing the insight signal for potential failures.

5.4.3 Process Safety Management IIoT Scope

The environment, health, and safety (EHS) business function has been relatively slow to change over the years. EHS management system can benefit the internet of things features to improve its performance by collecting and managing the Big Data using the smart connected devices, mobile application, and collaborative cloud computing capabilities. Fig. 5.22 summarises how the main components of IIoT can contribute in EHS Performance (HSE performance is a better term in this book).

In the smart factory area, our main businesses are MES-oriented businesses, facility control businesses, integrated management, and integrated engineering businesses.

In these industries, various applications and solutions are used depending on the business area and the most important concern in operating them is application integration, modeling, and data exchange in a simple and consistent standardized way while maintaining high performance and stability.

OPC UA enables application integration, modeling, reunion, and data exchange in a simple and consistent standardized way while maintaining high performance and stability. Fig. 5.23 illustrates the Smart factory in compliance with ISA-95 functional hierarchy model.

With OPC UA, businesses can develop products that are ultimately going to be deployed to extend across facility control and the MES domain via a standardized and consistent approach for smart factories. It will be used in the integrated engineering business by providing device modeling and meta information exchange function on the server side, as well as basic data collection, monitoring, analysis, and control functions on the client side.

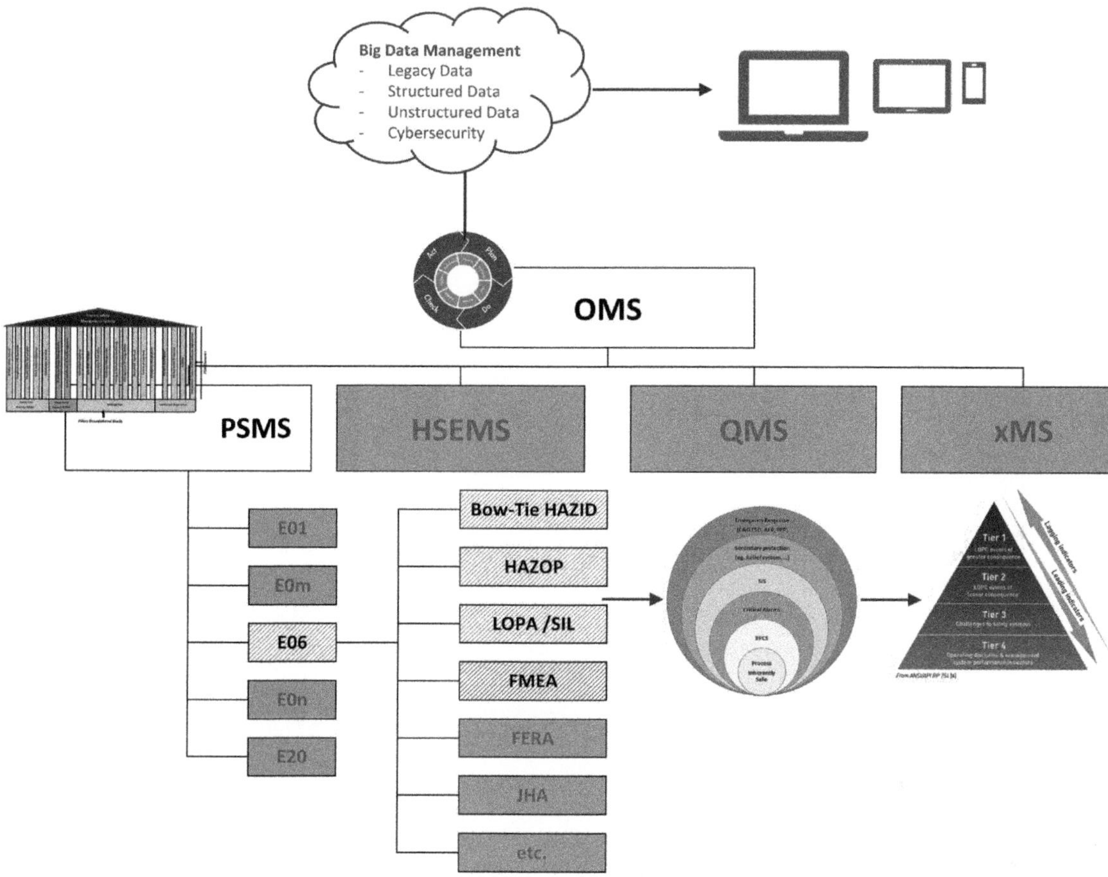

FIGURE 5.22
The road map of the effective implementation of the Operating Management System.

It would also be used for mass data collection to generate models for analysis and forecasting to meet the latest industrial needs of IoT, cloud, and Big Data management.

5.5 BIG DATA MANAGEMENT

Big Data is characterized by the three V's of Volume, Velocity, and Variety as shown on (Fig. 5.24).

In the industrial space we have typically had to deal with large volumes and velocity of data. According to Boeing, the 787 produces approximately half a terabyte of data per flight. What we have not had to deal with is variety.

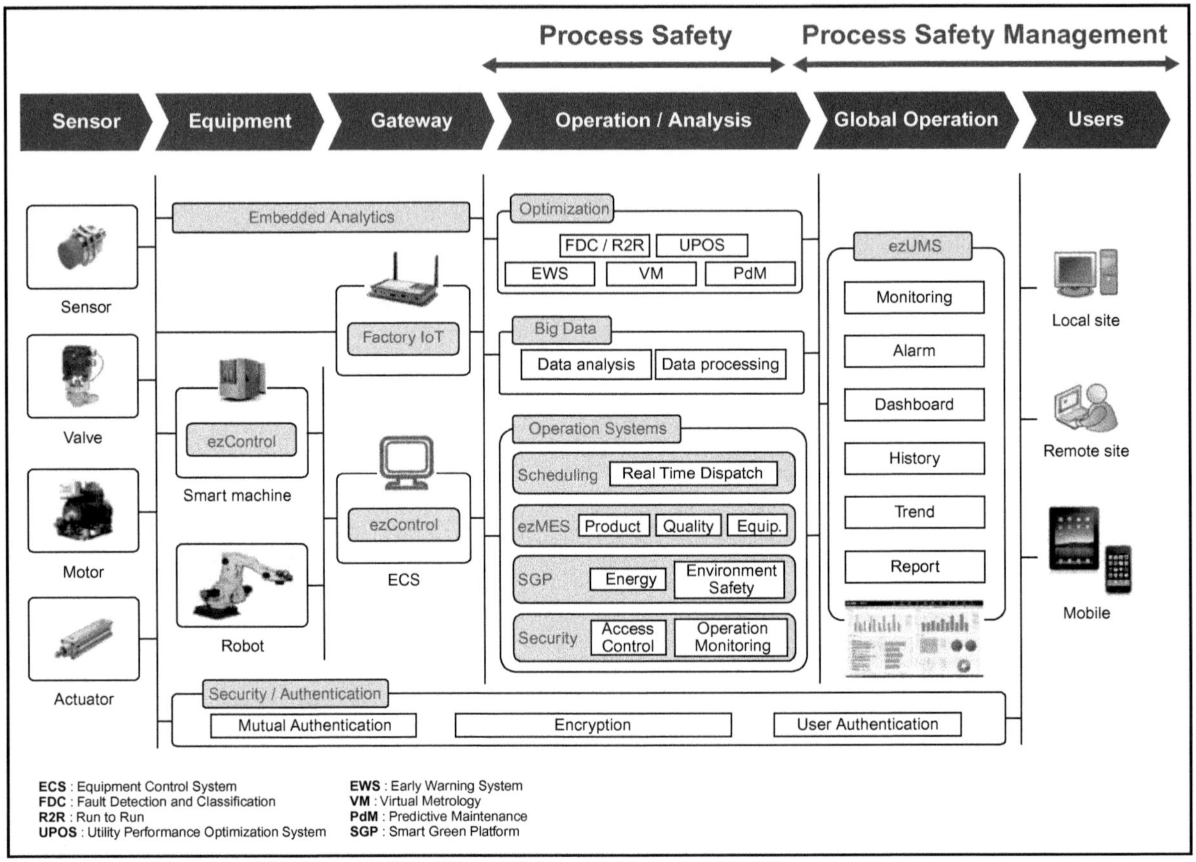

FIGURE 5.23
Smart factory in compliance with ISA-95 functional hierarchy model.

All of these data have been relatively well-structured process data stored as time series or transactional data stored as structured data in enterprise applications. With the advent of the IIoT, data might include images, video, unstructured text, spectral (such as vibration), or other forms, such as thermographic or sound. As all of these data types come together, industrial companies will truly have to deal with Big Data in manufacturing, which will bring together a whole new set of analytics opportunities and challenges.

Big Data is relevant when you see some evidence of a nonlinear or nonstationary generative process that varies with time (or at least, sampling time), on the spectrum of random drift to full blown chaotic behavior. Nonstationary behaviors can arise from complex (often "hidden") interactions within the underlying process generating your observable data. If you observe nonlinear relationships, with underlying stationarity, it reduces to a sampling

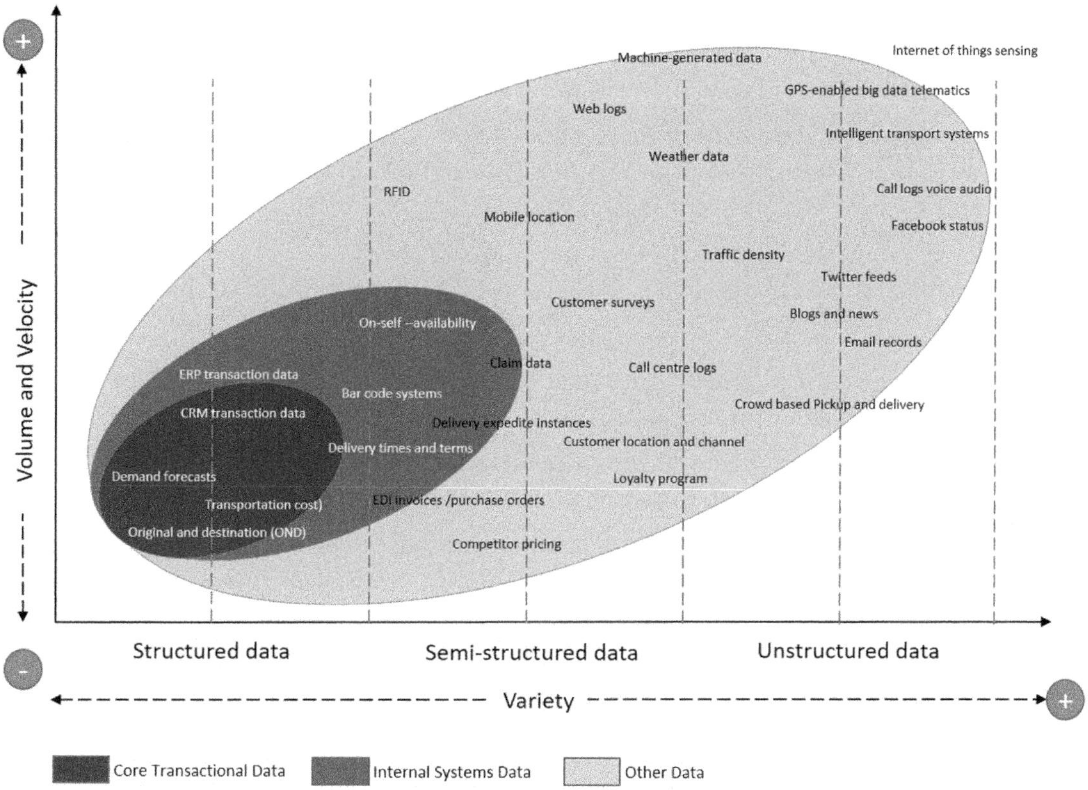

FIGURE 5.24

Supply chain management (SCM) volume and velocity versus variety.

issue. Big Data implicitly becomes relevant when we are dealing with processes embedded in a high-dimensional context (i.e., what's left after dimension reduction). With higher embedding dimensions, we need more and more well-distributed samples to understand the underlying process. For problems where the underlying process is both linear and stationary, we don't necessarily need much data at all. (Fig. 5.25) illustrates how the data complexity level varies by nonlinearity and nonstationarity.

The wrench here is in knowing when you are dealing with a nonlinear or nonstationary process. So, a little thoughtfulness and discovery work can tell you whether you have a Big Data problem, and then you can go about finding ways of actually collecting the required Big Data. Knowing whether you have a Big Data problem (or not) informs your approach for actually learning from the data.

This is an especially important question for data architects/strategists to think about when building their roadmaps against the kinds of challenges they

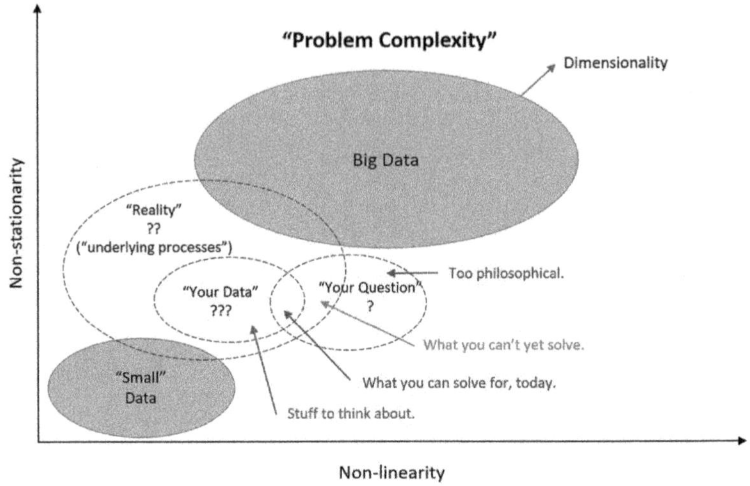

FIGURE 5.25
Data complexity levels.

hope to tackle. Not all paths to good data strategy lead to Hadoop. A good guiding principle of design is to "design with the end user in mind." In this context, the end "user" is actually the algorithm learning from the data. In principle, it is empowering to believe that one can learn anything from an infinite box of all kinds of data (let's call it a "universe box"), in practice, you will want to reduce the data to the essential subset that lets you do something meaningful with it. Just because an algorithm can learn anything, does not mean it should.

Note: The size of the circles do not reflect the frequency of observing any particular type of data. Complex (small, nonlinear, nonstationary) but undersampled data are not rare. However, for complex processes, you need more samples to capture the underlying variability and higher order statistical structure, so the "need" for Big Data is greater. Whether you actually "have" sufficient data is a different issue. Likewise, for a simple linear stationary process, you need very little data.

Any company or industrial site faces the full spectrum of analytics, including descriptive, diagnostic, predictive, and prescriptive data. Traditionally these analytics have been focused on analyzing structured and time series data to address the key drivers in industry: quality, safety, production, assets, delivery, innovation, and more. Examples are given in Table 5.5.

As new solutions have emerged to manage Big Data, new analytics have also emerged that are mainly targeted toward predictive and prescriptive analytics. To add confusion, even though these tools were developed to analyze Big Data, they can be used with any data set: small, large, or big.

Table 5.5 Big Data Analytics Framework

Descriptive	What happened	Metrics and scorecards for overall equipment effectiveness (OEE), on time delivery (OTD), scrap, mean time to failure (MTTF)
Diagnostic	Why it happened	Reliability engineering, quality engineering, root cause analysis
Predictive	What will happen	Modeling and simulation, statistical process control, advanced process control
Prescriptive	What action to take	

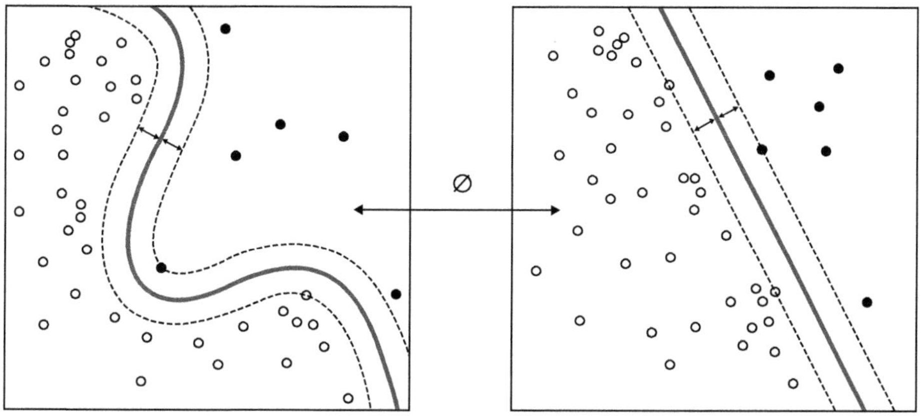

FIGURE 5.26
Kernel machines are used to compute a nonlinearly separable function into a higher dimension linearly separable function.

Big Data management should consider the legacy of these structured data and design the new system in such way that encompass the new structured, unstructured data and provide the access to the legacy data which are ideally processed and organized.

A common example of these new Big Data analytics includes *machine learning,* among many others. These new analytics are all generally data focused, where the traditional tools are model and process specific, which adds to the challenges in bridging the gap between data scientists using Big Data analytics and engineers using traditional model based analytics. As shown in Fig. 5.26 the Kernel machines are used to compute a nonlinearly separable function into a higher dimension linearly separable function.

Next generation prescriptive analytics are really about moving beyond choosing what to do next, and optimizing operations and enabling innovation.

Most companies can do quite well just by collecting MES and PLC data to start. Then as maturity increases, they can add in data from smart devices.

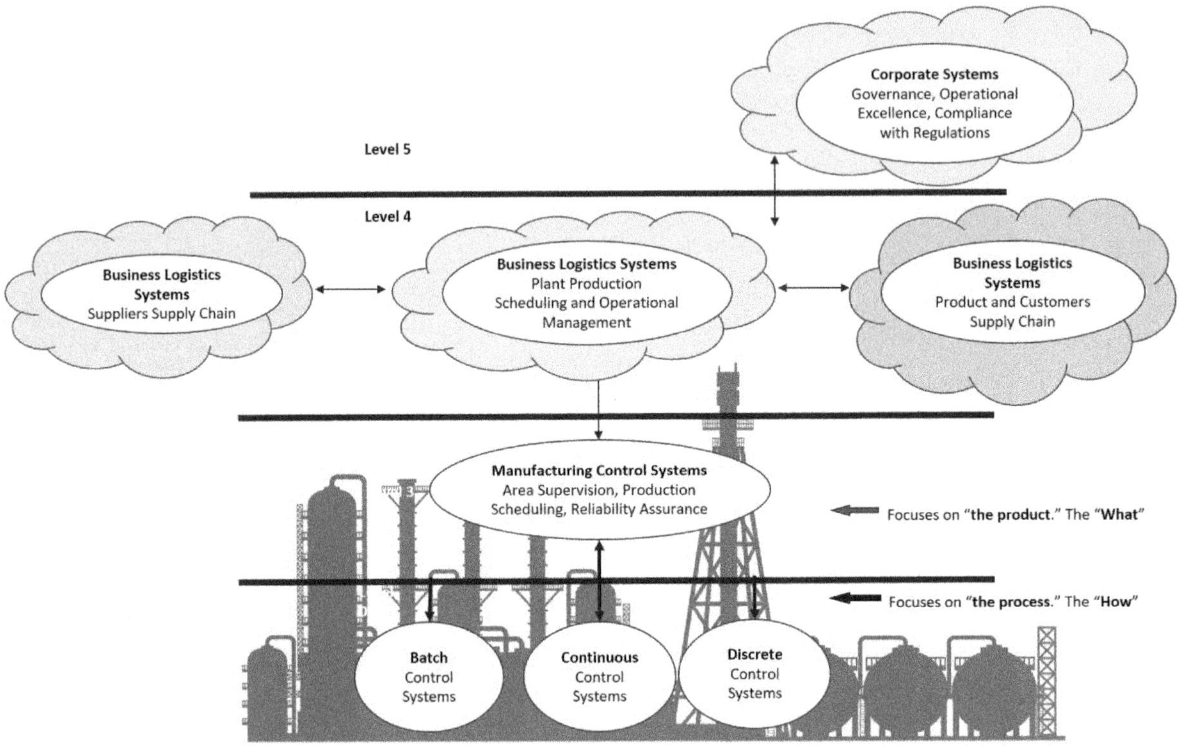

FIGURE 5.27
Smart connected Enterprise according to the ISA-95 model.

Fig. 5.27 illustrates how IIoT assure the connectivity of the different levels of a smart connected enterprise.

Although companies are not ripping and replacing existing control to information system hierarchies, many companies are beginning to deploy gateway to cloud architectures, at least in a limited capacity, to begin delivering IIoT data to higher level enterprise applications and analytics packages.

It remains to be seen if this information flowing through gateways and to the cloud will be enabled on the "thing" side of gateways with direct connections to existing automation equipment, or via a secondary system of sensors and connectivity. There are pros and cons to both and each will likely persist for the immediate future as industrial companies experiment.

As new data sources and systems come online, emerging questions of data ownership and data sharing are becoming critical.

One of the most important questions is "who owns the new machine data?" The maker of the asset/device, or the user? Interestingly, we see the market split on this point today, and there may not necessarily be a verdict or even a single answer any time soon. There are, however, a few points becoming clear.

1. When customers do not own the data, they prefer not to pay for the raw data coming back. Rather, they want to pay for the value services being delivered back to them that may only have been possible through data sharing.
2. Use of the machine matters. In scenarios where the use of the machine has no competitive differentiator, like in compressed air, e.g., data sharing and selling compressed air instead of compressors is not an issue. When the use of the machine does create competitive advantage, like with CNC machines or oil field service equipment, asset users are much more protective of data.

What is more likely, as shown by this companion question, is that most industrial companies are just doing descriptive analytics on structured data sets rather than predictive and prescriptive analytics with Big Data.

To help address this apparent lack of understanding, industrial companies need to invest in both the appropriate technology, but more importantly, process and training as well. Just as Six Sigma and Lean were built into the fabric of continuous improvement initiatives and packaged for subject matter experts to use the appropriate financial modeling, process optimization, and variability reduction analytics without being a statistician, Big Data tools like Hadoop and Machine Learning need to be packaged and made accessible to industrial subject matter experts, not just data scientists.

Capturing the business value of the IoT, we first hypothesized that IIoT platform technologies would precipitate the flattening of the traditional hierarchical model. With this new research, we are seeing the first quantitative evidence of this transformation occurring, and as such there will be several key changes including:

- A transformation of control system hierarchy to move from one of distributed controllers and centralized control to truly distributed control with the enablement of smart connected assets.
- A transformation of MES to become an orchestration and optimization platform for smart connected operations, not simply an integration and analytics middleware layer for execution and compliance.
- A transformation of enterprise applications to more closely map to operations instead of accounting models, and have the ability to work flexibly with operational data and not just structured transactional data.
- The enablement of mash-up applications and analytics that can enable Big Data from anywhere to anywhere and support true end-to-end value chain processes.

5.6 CLOUD COMPUTING

5.6.1 Background

Cloud computing has progressed from a mere data storage solution to a highly complex system which enables more efficient data sorting and resource management. With mobile web usage increasing daily, the Internet has become an extremely important asset for companies wanting to increase profitability and stay ahead of the pack.

With the business models becoming more complex, adopting the cloud to manage their resources more efficiently become more important. Unfortunately cloud computing adoption is slow in the process manufacturing sectors due to regulation, security, and traditionalism.

To compete in today's complex, global environment, manufacturers must be proactive about simplifying complex supply chains, or risk falling behind their competitors. They need solutions that provide the functionality necessary to adapt quickly and run efficiently. According to McKinsey, companies that are quick to implement digital strategies are more likely to have "disproportional gains" over those who do not. But manufacturers need not be daunted by the task of "going digital;" cloud technology can provide them with the always-modern functionality and flexibility they need to react quickly to changing market conditions, cut hardware costs, and use data-rich analytics to strengthen customer relationships and better manage their businesses.

A smart network of vendors, suppliers, and partners supported by cloud deployment's flexible architecture, allows manufacturers increased visibility into every step of the supply chain process. Cloud technology and its flexible scale and scope allow manufacturers to integrate with partners and maintain a constant flow of communication with their extended teams, and keep track of products at every point along the supply chain. Modern ERP solutions deployed in the cloud bring increased flexibility and interoperable functionality to manufacturers contending with disruptive changes in their industry, region, or go-to-market tactics. End-to-end visibility and data science empower the manufacturer with the actionable insights needed to make strategic decisions that benefit customers and the bottom line. Fig. 5.28 illustrates the high-level architecture diagram of a typical IoT-enabled automated demand response.

In an era of uncertainty about foreign policy and international trade, it's important for manufacturers to have increased flexibility in the supply chain. Cloud solutions with vast storage abilities allow manufacturers to leverage internet of things technology and predictive analytics to help manufacturers

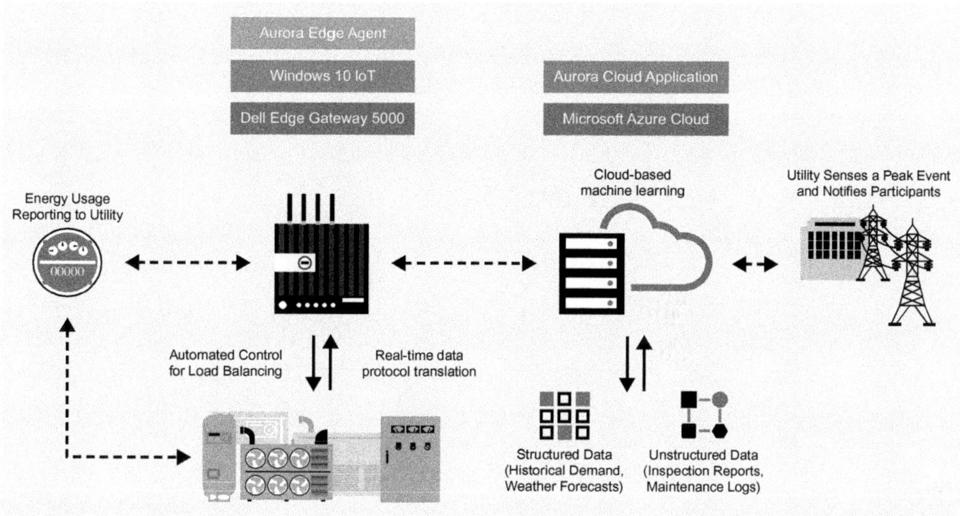

FIGURE 5.28
IoT enabled automated demand response—high-level architecture diagram.

anticipate the market's changes and cope with evolving market pressures. Companies that can adjust to market shifts will avoid decreased speed-to-market and inventory shortages. And when problems arise within the supply chain, cloud-empowered manufacturers are more aware of options for handling the issue.

For example, if a customer goes to purchase a product and finds it is out of stock, manufacturers armed with data collected from sensors about locations of shipping containers and alternative resources available, can help the procurement team decide the best way to stay on schedule, get the product to the customer quickly, and have the best chance of retaining that customer's business.

Long gone are the days of using out dated and heavily modified applications to run your business, and worry about upgrades. Cloud solutions provide the manufacturer with always-modern functionality needed to remain competitive. Manufacturers can engage with innovative technologies with vast storage, visibility, and reporting abilities. The modern capabilities also help manage forecasting, demand planning, tracking, and traceability within one uniformed platform.

Manufacturers can't afford to get left behind. In today's marketplace, a failure to meet customer demands could lead to a decline in business. By making the move to the cloud, manufacturers will reap the benefits of end-to-end modern functionality free of cumbersome modifications and flexible architecture for improved connectivity and visibility into their entire supply chain—cutting costs, increasing efficiency, and gaining the flexibility required to make the best decisions for their global businesses.

5.7 FOG COMPUTING

5.7.1 Fog Computing in an Industrial Context

A fog-computing strategy provides decentralized computing by bringing the intelligence of the cloud closer to the end devices. It is one of many options that machine builders and factories can consider. A central server requires high costs for hardware and software, and cloud-computing services offer new possibilities and connectivity options. Local data processing captures more process data but requires data transport, exchange, storage, analysis, and security; however, it reduces data traffic. The fog stack gets data from sensors to the machine controller, but you can't do analytics locally.

Local condition monitoring has been around for many years, and the model of having processing power on the machine enables the user to bring in and analyze temperature, acoustics, vibration, voltage, or current data. Combining this with IIoT and doing data analysis locally enables threshold monitoring of digital and analog values, timing analysis, lifecycle monitoring, RMS calculation, state analysis, and energy calculation.

Previously, a server–client model needed a VPN connection to tunnel back to the PLC or device in an end user facility. Now, we can use a message broker, and the outside devices don't touch the internal device. It could even be on a cloud server. The device can publish to the broker, and it doesn't care about the end device.

Whether you're using a public cloud service, such as Azure or your own private cloud, you can use an IIoT protocol for communication. If you have a legacy system or third-party device, you can pull data from it in its native language and convert to an IIoT protocol, such as MQTT, AMQP, or OPC-UA. For example, Kepware's IoT gateway shown in Fig. 5.29 supports REST and MQTT, allowing our customers to get PLC data into new IoT platforms and open source tools like Node-RED. From a machine builder's perspective, you want to know the varying differences in time between states of the machine. These analytics calculations can run in real time without another platform. If you don't want the data to leave your enterprise, you can use a message broker in your enterprise network or private cloud, which can include logging to a database and analytics platform, or you can use a public cloud. This enables the machine builder to access the

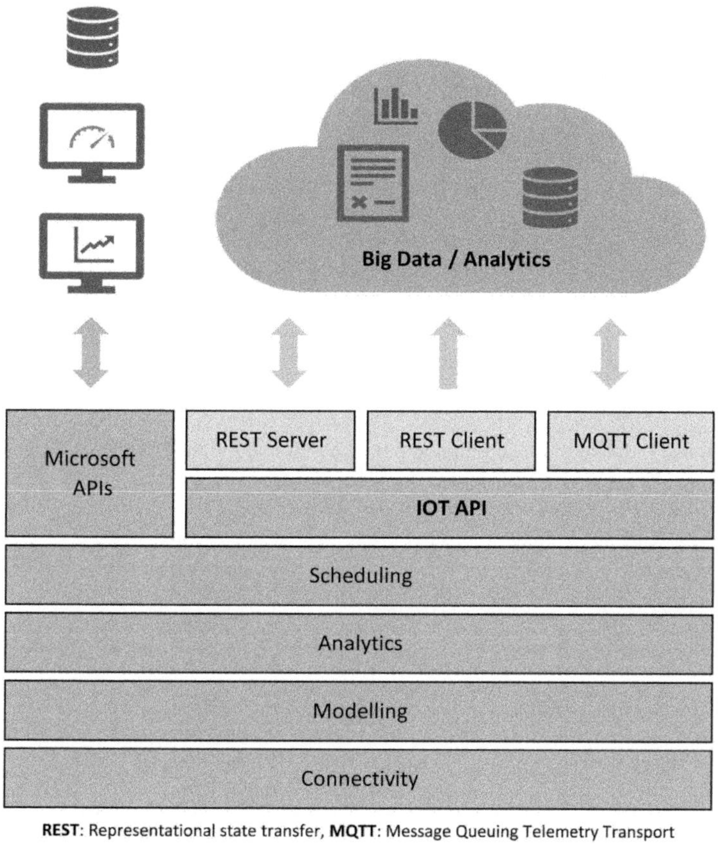

FIGURE 5.29
Kepware's IoT gateway supports REST and MQTT, allowing our customers to get PLC data into new IoT platforms and open source tools like Node-RED. *REST: Representational state transfer, MQTT: Message queuing telemetry transport.*

data and analytics, as well. The infrastructure depends on the application, whose side you're looking from and the business goals.

5.8 CYBER SECURITY RISK MANAGEMENT

With industrial cyber-security events like Stuxnet, Saudi Aramco, ThyssenKrupp, Unnamed German Steel Mill, and many more occurring since 2010, there will eventually be an industrial cyber-security event that changes the consciousness of society; creating a "new normal" where the above survey results become completely unacceptable.

It may be tomorrow, it may be in a year, or it may be longer; but one day we will all realize that without a robust, risk-based, approach to industrial

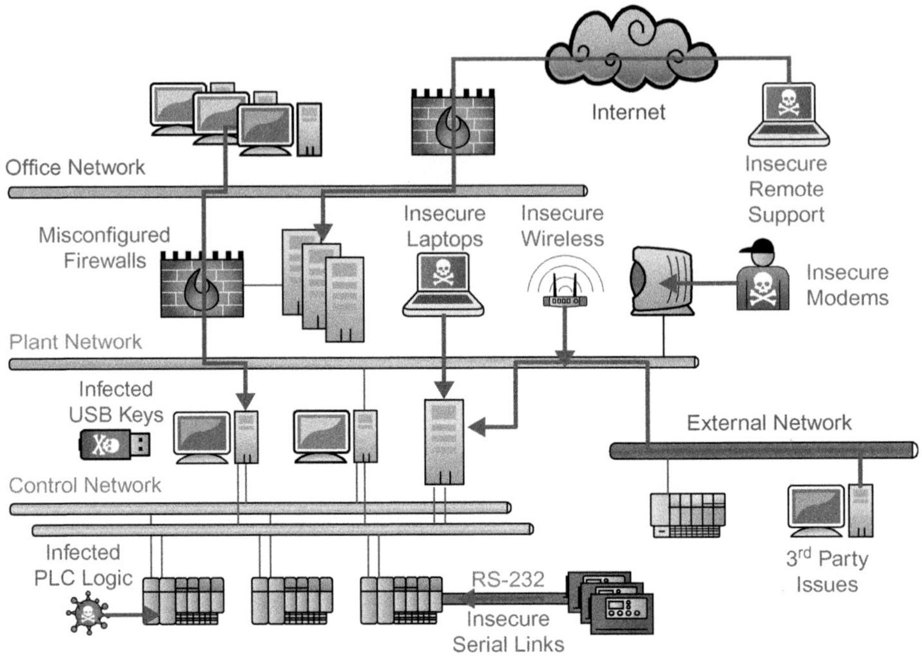

FIGURE 5.30
Possible pathways into the control system. *Reproduced with permission from Eric Byres, ICS Secure.*

control system (ICS) that is bottoms-up and not bolted-on, we will not be able to ensure safety in our industrial facilities and we will lose our social license to operate.

As industrial companies move to aggressively deploy the IIoT over the coming years; if they do not take a platform approach that incorporates ICS capabilities into the very fabric of these solutions, the entire industry is at risk. If new IIoT capabilities are instead just added on to legacy automation solutions without consideration of architecture and ICS, when the brewing storm hits, most companies will not be able to respond in time to withstand without significant damage (Fig. 5.30).

As corporate networks have converged with ICS networks, there have been many integration projects where proprietary networks were replaced with commercial off-the-shelf equipment using ethernet-TCP/IP technology. This shift in technology has greatly increased the complexity and "interconnectedness" of control systems. As a result, they now have many of the same vulnerabilities that have plagued enterprise networks. In addition, the controllers in these networks are now subjected to new threat sources that they were never designed to handle. The result has been a significant increase in the number of plant disruptions and shutdowns due to cyber security issues in the control networks. The Repository for Industrial Security Incidents (RISI1) is the world's largest database of security incidents in control and supervisory control and data acquisition (SCADA) systems. An analysis of the data from 1982 to 2010 found that the type of incidents affecting control systems breaks down as follows:

- 50% of incidents were accidental in nature
- 30% of incidents were due to malware
- 11% of incidents were due to external attackers
- 9% of incidents were due to internal attackers

Studies on the incidents databases indicate that problems arising from three common sources:

1. *Proliferation of "soft" targets*
   SCADA and ICS devices such as PLCs, DCS controllers, IEDs, and RTUs were designed with a focus on reliability and real-time I/O, not robust and secure networking. Many ICS devices will crash if they receive malformed network traffic or even high loads of correctly formed data. Also, Windows PCs in these networks often run for months at a time without security patches or antivirus updates, and are even susceptible to outdated malware.
2. *Multiple points of entry*
   Even without a direct connection to the Internet, modern control systems are accessed by numerous external sources. All of them are a potential source of infection or attack. These include:
   - Remote maintenance and diagnostics connections
   - Historian and MES servers shared with business users
   - Remote access modems
   - Serial connections
   - Wireless systems
   - Mobile laptops
   - USB devices
   - Data files (such as PDF documents or PLC project files)

These pathways are underestimated and poorly documented by the owners and operators of industrial systems. In testimony by Mr. Sean McGurk, the Director of U.S. National Cybersecurity and Communications Integration Center (NCCIC) at the Department of Homeland Security noted:

> *In our experience in conducting hundreds of vulnerability assessments in the private sector, in no case have we ever found the operations network, the SCADA system or energy management system separated from the enterprise network. On average, we see 11 direct connections between those networks. In some extreme cases, we have identified up to 250 connections between the actual producing network and the enterprise network.*

As the Stuxnet worm showed us in 2010, these pathways can be readily exploited by malware and other disruptive elements. Stuxnet used at least eight different propagation mechanisms, including USB drives, PLC project files, and print servers to work its way into the victim's control system.

3. *Poor internal network segmentation*
   Control networks are now more complex than ever before, consisting of hundreds or even thousands of individual devices. Unfortunately the design of many of these networks has remained "flat" with virtually no segmentation. As a result, problems that originate in one part of the network can quickly spread to other areas.

The best IIoT systems are those designed from the very beginning with security and robustness in mind. They include elements such as automated failback features, an increased tolerance for short-term failures, and security monitoring within the system operations plan; "If the Internet connection goes down, your system still needs to function."

Experts recommend reviewing the ISA/IEC 62443 standards (formerly known as the ISA99 standards) as a preliminary road map toward partitioned architectures for the ICS and supervisory control and data acquisition domain. The complete security life-cycle program for industrial automation and control systems, ISA/IEC 62443 consists of 11 standards and technical reports. It introduces the concepts of zones (groupings of logical or physical assets that share common security requirements based on criticality, consequence, and other such factors; equipment in a zone should share a strong security level capability) and conduits (paths for information flow between zones). ISA/IEC 62443 standards provide requirements based on a company's assessment of cyberattack risks and vulnerabilities (Fig. 5.31).

Within the oil industry, a large refinery created a security architecture that effectively protected its operations based on these standards. In its oil refinery process facility, the company had multiple operations (with basic control, safety, and HMI/supervisory systems). It also had considerable wireless and remote communications needs, both for maintenance and for communications to downstream customers.

To secure its operations, the company divided its systems into zones and subzones—depending on operational function, security capabilities and requirements, perceived risk, and process level—to best adjust the security requirements for each particular operation. After analyzing potential threat sources, the company relocated the safety integrated system in each operational unit to its own zone (instead of being part of a basic control zone). Conduits were defined and documented, breaking down the overall system into manageable chunks. The zones and conduits were then implemented with industry security appliances, including firewalls and virtual private

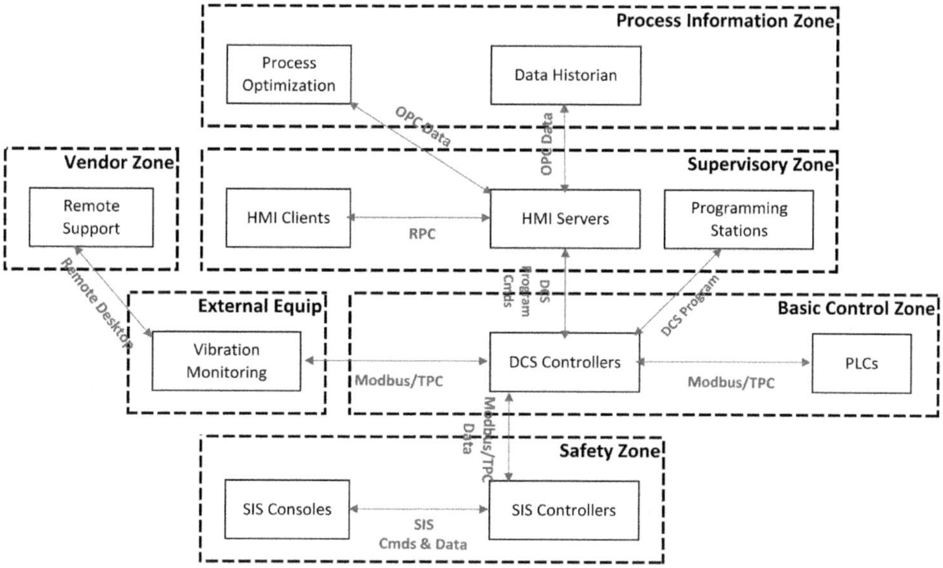

FIGURE 5.31
Example data flow diagram. *Reproduced with permission from Eric Byres, ICS Secure.*

networks, which were tried and tested. The technologies introduced into the control systems also made significant improvements to plant performance and productivity, and the company successfully continued to use and maintain ISA/IEC 62443 standards as a framework for security improvements (Fig. 5.32).

In your IIoT security checklist, strategize to ensure and implement the following proactive and protective measures:

- *Design security in from the start.* Never leave it as an afterthought.
- *Enlist expert help.* Fuse a team of senior management and security specialists who can communicate and work together to design protective strategic measures that work seamlessly with the plant's (and whatever products or services therein) functionality and features.
- *Compartmentalize IIoT solutions* into security zones to prevent the spread of malware throughout the plant. In tandem, integrate security best practices during each phase of the developmental process on the plant floor.
- *Monitor your IIoT system continuously* to understand vulnerabilities and manage emerging threats. It is essential to detect issues as early as possible.

IIoT should not be a raw or experimental practice. It must be designed reliably and with evolving security systems that are punctually followed and

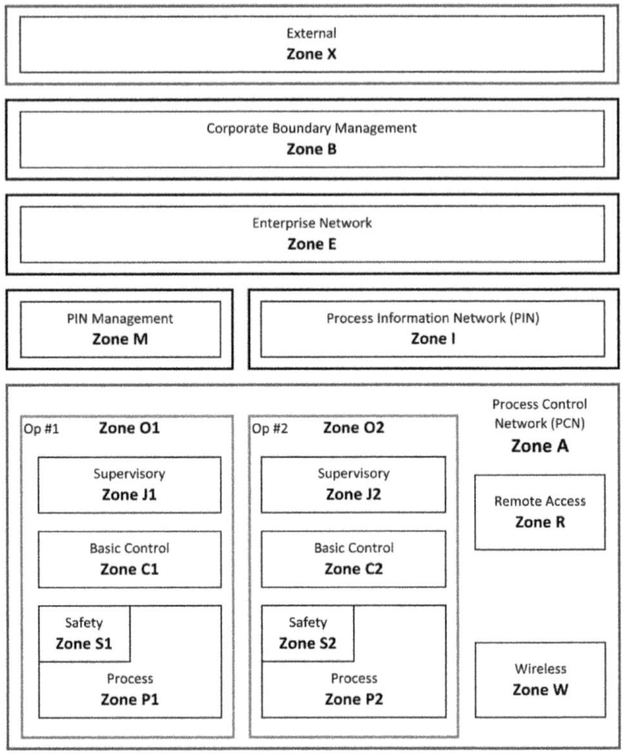

FIGURE 5.32
Refinery zone diagram. *Reproduced with permission from Eric Byres, ICS Secure.*

updated. Otherwise, it is no different than installing a burglar alarm system in your house... and never bothering to turn it on.

Industrial automation and control system (IACS), such as SCADA, are relied upon to monitor and control the operation of industrial machinery and associated devices. Because most IACS are not designed to ensure resilience against cyberwarfare, an IACS cyberattack can impair and disable safe operations of industrial facilities. The consequences—which can include plant shutdowns, widespread power blackouts, explosions, chemical leaks, and more—can place national and economic security as well as lives, personal safety, and the environment at risk. Fig. 5.33 shows a zone definition and assessment worksheet.

ISA/IEC 62443 enables owners and operators of critical infrastructure to achieve and maintain IACS security improvements through a lifecycle that integrates design, implementation, monitoring, and continuous improvement.

While the United States does not legally require implementation of industrial cybersecurity standards and best practices, the government has

Zone Name: Unit 1 Hydrocracker Safety System

Definition of this Zone: This zone includes all safety integrity systems for the Unit 1 Hydrocracker.

Controlling Agency: Process Automation Department, SIS Team.

Zone Function: The systems in this zone provide safety functions to the Unit 1 Hydrocracker.

Zone Boundaries: The Safety Integrated System as defined by the Unit 1 Hydrocracker HAZOP.

Typical Assets: The Safety Integrated System controller, engineering station and communications hardware

Inheritance: This zone inherits attributes from Zone C1 (Unit 1 Hydrocracker Basic Control System)

Zone Risk Assessment: This is a low to moderately secure zone with extreme consequences if breached.

- **Security Capabilities of Zone Assets:** All assets are assumed to be incapable of withstanding low level attacks (i.e. those launched by unsophisticated attackers or malware) on their availability or confidentiality. This is a result of the protocols in use and system design. Assets are assumed to be capable of withstanding medium level attacks (i.e. those launched by moderately sophisticated attackers or malware) on their integrity.
- **Threats and Vulnerabilities:** The vulnerabilities of this zone are assumed to be typical of legacy industrial control devices using MODBUS for communications. The principal threats are:
  - Network-based Denial of Service to SIS communications.
  - Internal or External unauthorized access to the SIS engineering station.
  - Spoofing of MODBUS/TCP control commands.
  - Spoofing of MODBUS/TCP responses to the process system.
  - Reprogramming of safety functions.
- **Consequences of a Security Breach:**
  - Loss of production >6 hrs from false trip of emergency shutdown system.
  - Loss of production <6 hrs due to loss of visibility to safety system.
  - Disabling/manipulation of emergency shutdown resulting in fatality or major community incident.
- **Business Criticality:** Extreme

Security Objective: To protect the integrity and availability of the Unit 1 Hydrocracker Safety System.

Acceptable Use Policy: I/O and Fieldbus communications is allowed to Zone P1 (Unit 1 Hydrocracker Process). Read access to published data is allowed to approved systems in the Zone C1 (Unit 1 Hydrocracker Basic Control System). All Write access to this zone is forbidden. All system management and programming functions shall be internal to this zone.

Inter-zone Connections: Conduits to this zone may be established from Zone C1 (Unit 1 Hydrocracker Basic Control System) and from Zone P1 (Unit 1 Hydrocracker Process).

Security Strategy: All connections to this zone must be controlled using type S conduits. Access to these systems must be approved by the Controlling Agency.

Change Management Process: All changes to this zone or any of its connecting conduits must follow the approved change management process of its corresponding Controlling Agency (see above). This includes, but is not limited to, the installation or replacement of equipment, modification of security policy, and exceptions to security policy or existing practices.

FIGURE 5.33

Example zone definition document for the safety zone. *Reproduced with permission from Eric Byres, ICS Secure.*

developed a voluntary plan to follow. The plan, known as the U.S. Cybersecurity Framework, serves as a how-to guide for American industry and operators and owners of critical infrastructure to strengthen their cyber defenses.

Representatives of both ISA and its affiliate, the Automation Federation, served as expert consultants to the National Institute of Standards and Technology (NIST)—an agency of the U.S. Department of Commerce—as it coordinated the development of the framework. The ISA/IEC 62443 series of IACS security standards are key components of the framework recommendations, which were made public in early 2014.

Furthermore, the Automation Federation is the host organization for the *LOGIIC (Linking Oil and Gas Industry to Improve Cybersecurity) Program*, an ongoing collaboration of major oil and natural gas companies and the U.S. Department of Homeland Security, Science and Technology Directorate. LOGIIC undertakes collaborative research and development projects to improve the level of cybersecurity in critical systems of interest to the oil and natural gas sector. ISA has harnessed the ISA/IEC 62443 standards to develop a comprehensive set of industrial cybersecurity training courses and aligned certificate programs—covering the complete lifecycle of IACS assessment, design, implementation, operations, and maintenance. ISA's suite of industrial cybersecurity courses includes:

- *Introduction to Industrial Automation Security and the ISA/IEC 62443 Standards (IC32C)*
- *Using the ISA/IEC 62443 Standards to Secure Your Control Systems (IC32)*
- *Using the ISA/IEC 62443 Standard to Secure Your Control Systems (IC32E, Online Version)*
- *Assessing the Cybersecurity of New or Existing IACS Systems (IC33)*
- *IACS Cybersecurity Design and Implementation (IC34)*
- *IACS Cybersecurity Operations and Maintenance (IC37)*

ISA cybersecurity certificate programs are awarded to those who successfully complete the requirements of ISA's related cybersecurity courses. Individuals who complete all four ISA certificate programs earn the designation of ISA/IEC 62443 Cybersecurity Expert. For more details on the four certificate programs and their aligned courses, visit www.isa.org/CYBERcertificate

In addition, ISA has developed a certification program—ISASecure—that ensures that control systems conform to relevant ISA/IEC 62443 cybersecurity standards and apply to the security lifecycle concept that forms the basis of the standards.

Asset owners and integrators that include the *ISASecure* designation as a procurement requirement for control systems projects have confidence that the selected products are robust against network attacks and free from known vulnerabilities.

5.9 MODEL-BASED SYSTEM ENGINEERING (MBSE)

5.9.1 Introduction

Fig. 5.34 illustrates the evolution of the way that we work in any business. To preserve the competitivity, the business should be agile. Using the MBSE increases agility because a robust system engineering simulator dramatically improves the design and validation of complex systems by:

- Multidiscipline collaboration
- Standards-based digital system design
- System level simulation
- Modeling for IoT and systems of systems
- System product lines
- Engineering smart connected products

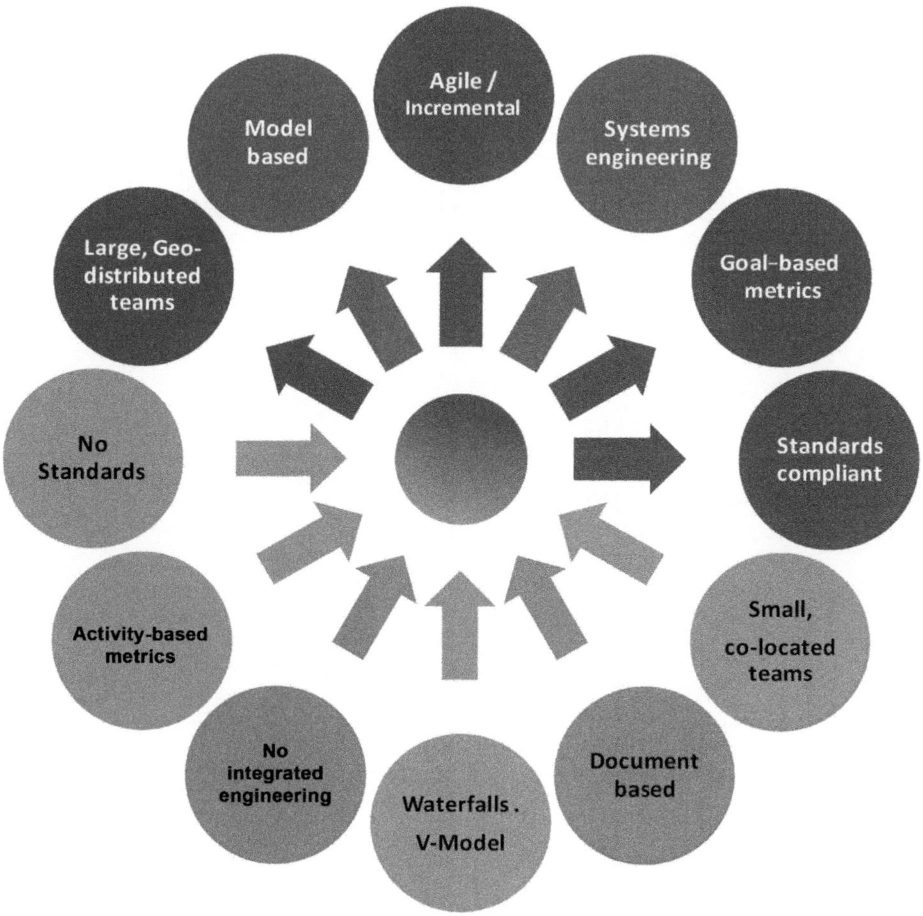

FIGURE 5.34

Technological developments improve the way that we are working.

5.9.2 Systems Engineering Drivers and Inhibitors

The current state of systems engineering practice is evolving under the influence of the respective drivers and inhibitors called out in Table 5.6. Through a series of workshops experts in systems engineering and related fields were brought together to hypothesize about the future of systems engineering, beyond the next decade.

Table 5.6 Drivers and Inhibitors Influencing the Practice of Systems Engineering

Area	Drivers	Inhibitors
Global environment	• Ability to create, produce, and operate on a global basis, supported by rapid and extensive transportation of people and goods, plus high-bandwidth, globally integrated communications capability • Increasing collaboration among professional groups and other stakeholders, worldwide	• Lack of mature standards and their uniform adoption • Increased global tensions; local customs and practices and, decreasingly, language • Heritage of systems engineering impacts its perceived application to social and nontraditional systems, and small-to-medium enterprises
Nature of systems	• Nonlinearly increasing ability to incorporate a greater capability for a given cost; to augment the embedded intelligence in systems and to provide systems with global dispersion • Emerging basic conceptual and technological areas with systems applications, such as chaos theory, nanotechnology, and genetic engineering	• Human limits in comprehension and control of manmade systems • Immature view of the roles of people in (increasingly complex) systems, natural entities, processes and procedures, and facilities • Lag in manufacturing and sustainment capabilities • Slow grasp of the significance of new technologies and concepts
SEprocesses	• Evolution of standards that give a consistent view of process sets that embrace all systems engineering functions on a life cycle basis • Emerging understanding of software engineering and systems engineering synergies	• Lack of tools that enable use of existing standards • Excessively complex process sets and burdensome formalism create a barrier to acceptance of systems engineering practices, especially in small and medium enterprises
MBSE	• Emergence and maturation of modeling languages and information standards • Continuing evolution of information technology as an enabler of modeling techniques	• Inherent difficulty integrating models across organizational, lifecycle, and other boundaries • Limitation of model/data exchange capabilities within the modeling tools • Limited MBSE skills
SEeducation	• Recognition of need to inculcate systems thinking at earlier stages in individual educational experiences and with a broader context • Distance education putting the classroom in the home • Transition to transdisciplinarity in engineering education	• Lack of an accepted set of theory and principles • Lack of an overarching vision or guiding framework for research • Lagging posture of curricula, exacerbated by funding shrinkage • Challenge of educating current employers in new concept

In a best case scenario, the experts anticipate that systems thinking and systems engineering will guide the way people think about solving problems in the next decade and systems engineering will become an established international "interdisciplinary connector" or "metadiscipline."

Increased diversity within the stakeholder population creates the need for shared sensemaking through shared mental models; a need that systems engineering will satisfy. In the future systems engineering will be used to address the significant social, economic, environmental, and planning issues of the day.

Additional systems engineering capabilities will be created to support the adaptation of system engineering methodology to the agile and robust operations of extended enterprises and businesses of all sizes. This will include:

- Advanced systems theory application
- Increased use of analytical methods and tools
- Advances in engineering education with an emphasis on interdisciplinary integration
- Improved use/integration of engineering specialties
- Improved understanding of psychology, languages, and culture
- Improved shared understanding of systems engineering concepts among all stakeholders

Stovepipes between disciplines will begin to blur as the benefits—diversity of ideas, lower organizational costs, and greater development efficiency—of interdisciplinary integration are realized. Stakeholders will take the time to identify the characteristics (capability, culture, communication and decision-making style, appetite for risk, ethics, etc.) of their global business partners.

5.9.3 MBSE Overview

MBSE is an umbrella term that describes an approach to systems engineering that emphasizes a system architecture model as the primary work artifact throughout the system development life cycle as shown in Fig. 5.35. It combines traditional systems engineering best practices with rigorous visual modeling techniques to enhance the ability to capture, analyze, share, and manage the information associated with the complete specification of a product, resulting in the following benefits:

- *Improved communications* among the development stakeholders (e.g., the customer, program management, systems engineers, hardware and software developers, testers, and specialty engineering disciplines);

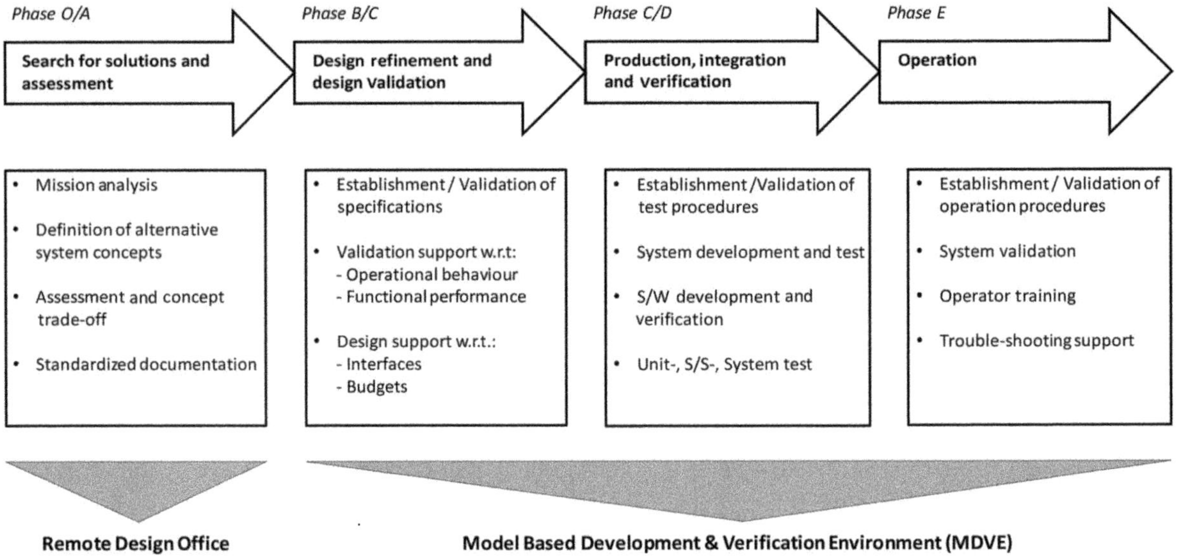

FIGURE 5.35
MBSE across the system life cycle.

- *Increased ability to manage system complexity* by enabling a system model to be viewed from multiple perspectives, and to analyze the impact of changes;
- *Improved product quality* by providing an unambiguous and precise model of the system that can be evaluated for consistency, correctness, and completeness;
- *Enhanced knowledge capture* and reuse of the information by capturing information restandardized ways and leveraging built-in abstraction mechanisms inherent in model-driven approaches. This in turn can result in reduced cycle time and lower maintenance costs to modify the design.
- *Improved ability to teach and learn systems engineering fundamentals* by providing a clear and unambiguous representation of the concepts.

MBSE is expected to replace the document-centric approach that has been practiced by systems engineers in the past and to influence the future practice of systems engineering by being fully integrated into the definition of systems engineering processes.

MBSE is part of a long-term trend toward model-centric approaches adopted by other engineering disciplines, including mechanical, electrical, and software, as depicted in Fig. 5.36.

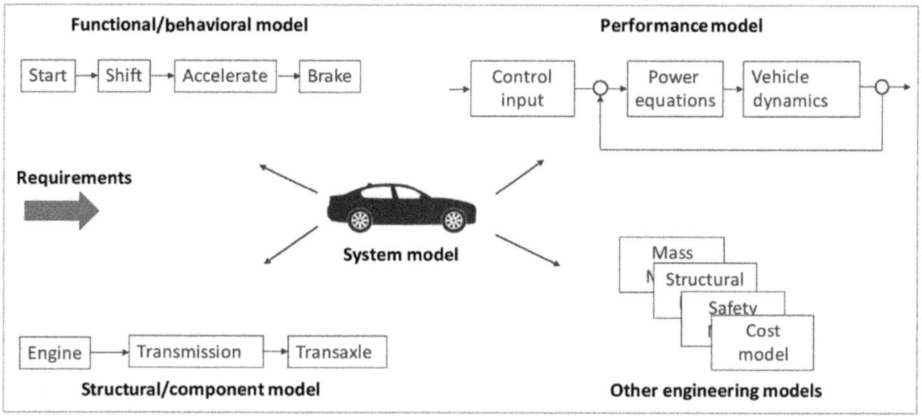

FIGURE 5.36
MBSE model-centric approach.

Table 5.7 Technical and Business Advantages of a MBSE Approach

Technology Drivers	Technology Advantages	Business Advantages
Model requirements	Ensure that the requirements are an integral part of the model and all other parts of the model can be traced back to requirements	Validate that you are "building the right system" (Boeham, 1984)
Model analysis and design	Provide a precise architectural blueprint organization by the views that are meaningful to all systems stakeholders	Verify that you are "building the system the right way" (Boeham, 1984)
Model simulation	Automate system verification and validation	Reduce errors and costs early in the life cycle
Model code	Automate generation of production quality code	Accelerated time to market
Model test	Automate testing	Ensure system implementation is correct and reliable

Advantages of the MBSE over traditional document-centric systems engineering approaches are summarized in Table 5.7.

MBSE typically uses the SysML as a standard visual modeling language. SysML not only supports rigorous modeling techniques among systems engineers, but also serves as a lingua franca among various other kinds of engineers (software engineers, electrical engineers, mechanical engineers, etc.) and stakeholders who need to understand MBSE work products, such as visual requirements analysis and verification, CONOPS, functional analysis and allocations, performance analysis, trade studies, and system architectures.

Per INCOSE Systems Engineering Vision 2020, the projected state of MBSE practice in 2020 will extend MBSE to modeling domains beyond engineering models to support complex predictive and effects-based modeling. This will include the integration of engineering models with scientific and phenomenology models, social, economic, and political models and human behavioral models. The key characteristics of MBSE in 2020 include:

- Domain-specific modeling languages and visualization that enable the systems engineer to focus on modeling of the user domain
- Modeling standards based on a firm mathematical foundation that support high fidelity simulation and real-world representations
- Extensive reuse of model libraries, taxonomies, and design patterns
- Standards that support integration and management across a distributed model repository
- Highly reliable and secure data exchange via published interfaces.

Domain-specific modeling languages built on the general purpose systems modeling language will increase the abstraction level to represent the user domain. Validated and specialized model libraries for specific domains will be established that can be reused across organizations and evolved over time. A system engineer familiar with the domain will be able to rapidly search the distributed model repository and evaluate a broad trade space of solutions based on an understanding of the user requirements and measures of effectiveness. Thus, the systems engineer will readily perform what-if analysis to assess the requirements, design, and technology impacts, and optimize the solution, using multidimensional visualization capabilities.

The conceptual depiction in Fig. 5.37 indicates a desired goal and direction toward advancing capabilities for MBSE. The integrated capabilities shown will dramatically increase the application of MBSE to support marketing research, decision analysis, integration with biological system models, environmental impact analysis, and the design of social systems in support of urban planning and government social programs, to name a few.

Collaborative environments will be set up quickly for new project teams and extended across global government, industry, and academic teams. Management of product data throughout the lifecycle will provide improved logistics and operations and maintenance phase support, since design data are retained in standards-based repositories. Virtual development environments will minimize the need for physical prototypes and will accelerate new product development while providing realistic verification against customer-driven requirements.

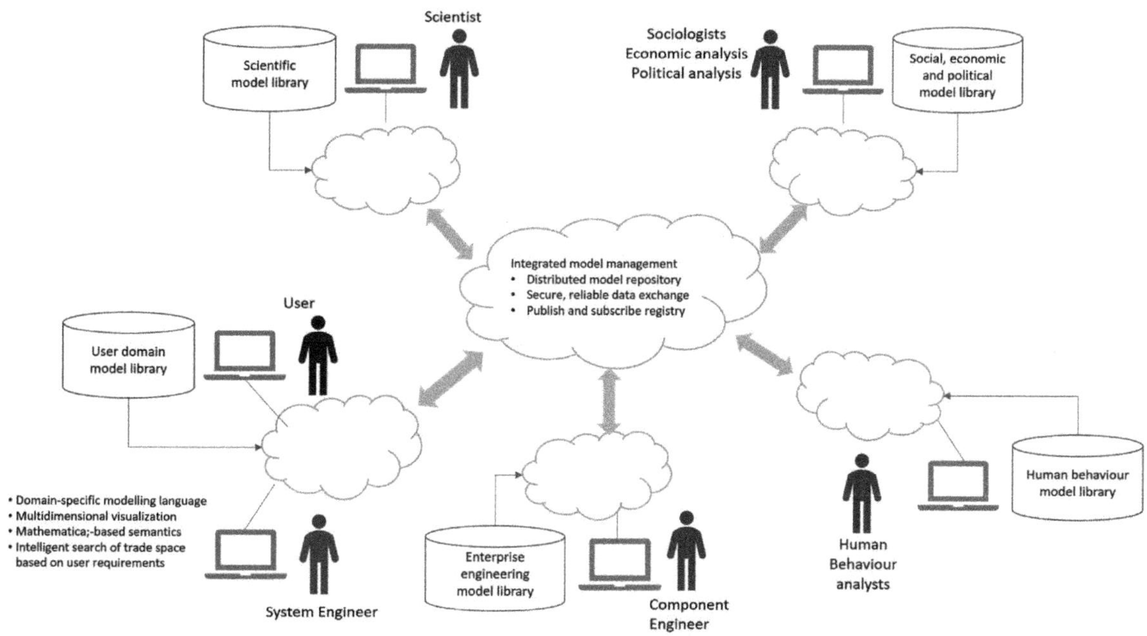

FIGURE 5.37
Cross-domain model integration.

System development times will be substantially reduced relative to current practice, while improving overall system quality and availability. This will be accomplished by a combination of failure-mode avoidance and knowledge-based engineering. The application of increased computer power will enable rapid system design with models in virtual development environments, greatly reducing the need for physical prototypes.

The breadth of applied models will not be confined to the traditional extent of dealing with the immediate system interfaces. Relationships will emerge between individual system models that are built to analyze systems across life cycles spanning decades and the social and economic models that estimate the impact of cumulative decisions on the economy and our environment. "Intelligent systems" will add significant effort to up-front systems engineering with the benefit of enabling planned change at an industry, domain, or societal level. System models and simulations will explore the impacts to our environment, society, and the stakeholder organizations in ways that are ignored today and discovered by analysis of impacts after systems become operational.

A complete discussion of the MBSE vision must also be accompanied by realistic concerns about inhibitors to its progress such as those described in

Table 5.6. Difficult technical and cultural challenges remain to be overcome in order to realize the many facets and benefits of the envisioned MBSE evolution. Meaningful progress will require both market forces and motivated visionaries inclined to take some risks in pushing the envelope to demonstrate value, exploiting opportunities and setting an example for others to follow.

5.9.4 OMG System Modeling Language

The *Object Management Group* (OMG) is an international, open membership, not-for-profit technology standards consortium, founded in 1989. OMG standards are driven by vendors, end users, academic institutions, and government agencies. OMG Task Forces develop enterprise integration standards for a wide range of technologies and an even wider range of industries. OMG's modeling standards, including the *Unified Modeling Language* (UML) and *Model Driven Architecture* (MDA), enable powerful visual design, execution, and maintenance of software and other processes. OMG also hosts organizations such as the user-driven information-sharing *Cloud Standards Customer Council* (CSCC) and the IT industry software quality standardization group, the *Consortium for IT Software Quality* (CISQ). OMG also managed the *Industrial Internet Consortium*, the public-private partnership that was formed in 2014 with AT&T, Cisco, GE, IBM, and Intel to forward the development, adoption, and innovation of the Industrial IoT.

The OMG Systems Modeling Language (*OMG SysML*) is a general-purpose graphical modeling language for specifying, analyzing, designing, and verifying complex systems that may include hardware, software, information, personnel, procedures, and facilities. The language provides graphical representations with a semantic foundation for modeling system requirements, behavior, structure, and parametric, which is used to integrate with other engineering analysis models. It represents a subset of *UML 2* with extensions needed to satisfy the requirements of the UML for systems engineering RFP as indicated in Fig. 5.38. SysML leverages the OMG XML Metadata Interchange (XMI) to exchange modeling data between tools, and is also intended to be compatible with the evolving ISO 10303-233 systems engineering data interchange standard.

The UML for systems engineering RFP was developed jointly by the OMG and the International Council on Systems Engineering (INCOSE) and issued by the OMG in March 2003. The RFP specified the requirements for extending UML to support the needs of the systems engineering community. The SysML specification was developed in response to these requirements by the diverse group of tool vendors, end users, academia, and government

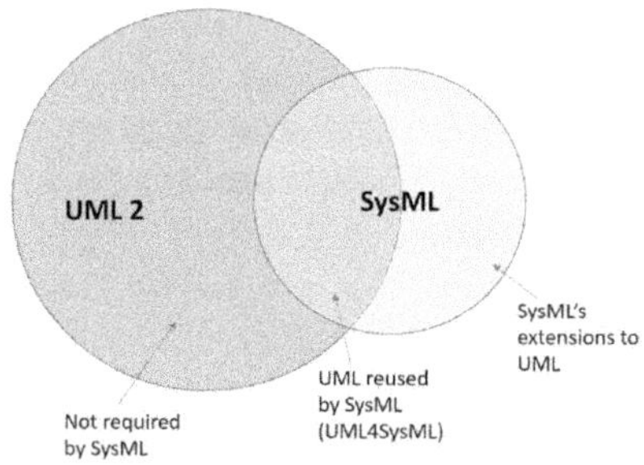

FIGURE 5.38
Relationship between SysML and UML. *SysML is a trademark of the Object Management Group, used with permission.*

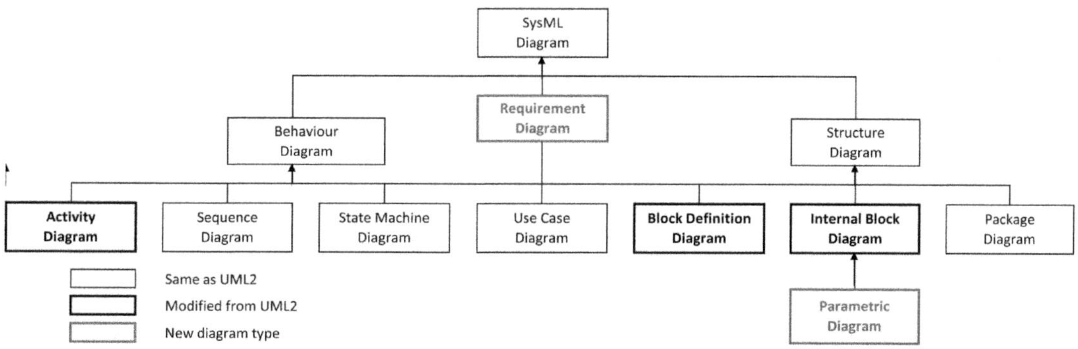

FIGURE 5.39
SysML diagram taxonomy. *SysML is a trademark of the Object Management Group, used with permission.*

representatives. The Object Management Group announced the adoption on July 6, 2006 and the availability of OMG SysML v1.0 in September 2007.

The SysML diagram types are identified in Fig. 5.39 and summarized below. Refer to the OMG SysML Tutorial for an overview of the language or the APL MBSE with SysML course material for a more detailed description. (Note: Because these are large files, it is recommended that you save to your desktop by right clicking and save target.)

The is the basic unit of structure in SysML and can be used to represent hardware, software, facilities, personnel, or any other system element. The system structure is represented by block definition diagrams and internal block diagrams. A block definition diagram describes the system hierarchy and system/component classifications. The internal block diagram describes the internal structure of a system in terms of its parts, ports, and connectors. The package diagram is used to organize the model.

The behavior diagrams include the use case diagram, activity diagram, sequence diagram, and state machine diagram. A use-case diagram provides a high-level description of functionality that is achieved through interaction among systems or system parts. The activity diagram represents the flow of data and control between activities. A sequence diagram represents the interaction between collaborating parts of a system. The state machine diagram describes the state transitions and actions that a system or its parts perform in response to events.

SysML includes a graphical construct to represent text based requirements and relate them to other model elements. The requirements diagram captures requirements hierarchies and requirements derivation, and the satisfy and verify relationships allow a modeler to relate a requirement to a model element that satisfies or verifies the requirements. The requirement diagram provides a bridge between the typical requirements management tools and the system models.

The parametric diagram represents constraints on system property values such as performance, reliability, and mass properties, and serves as a means to integrate the specification and design models with engineering analysis models.

SysML also includes an allocation relationship to represent various types of allocation, including allocation of functions to components, logical to physical components, and software to hardware.

A simple example of some of the key diagram types is highlighted in Fig. 5.40.

INCOSE proposes an object-oriented systems engineering method (OOSEM) as follows:

- Integrates top-down (functional decomposition) approach with model-based approach
- Leverages object-oriented concepts
  - Uses *OMG SysML* (formerly UML) to support specification, analysis, design, and verification of systems
- Intended to ease integration w/object-oriented S/W development, H/W development, and test
- OOSEM includes the following activities:
  - Analyze stakeholder needs
  - Define system requirements
  - Define logical architecture
  - Synthesize candidate allocated architectures
  - Optimize and evaluate alternatives
  - Validate and verify system
- Tool- and vendor-neutral

Figs. 5.41 and 5.42 illustrate the OOSEM Pyramid and activities.

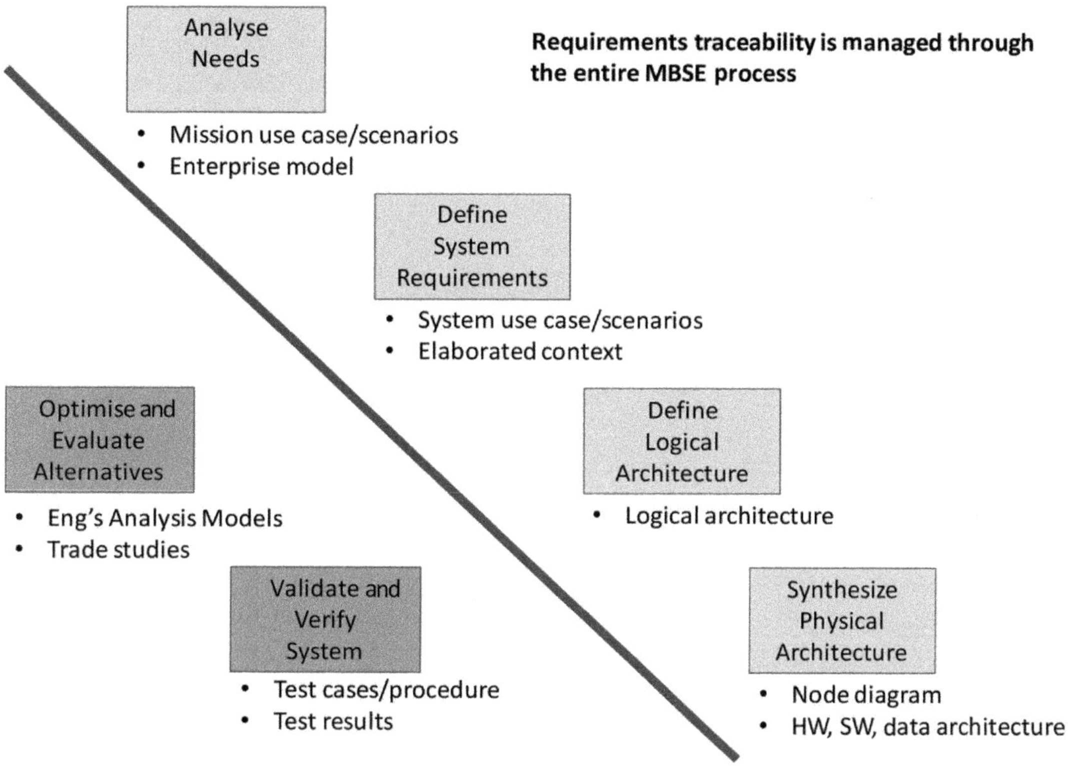

FIGURE 5.40
OOSEM activities—integration w/object-oriented S/W development, H/W.

5.9.5 MBSE in Practice

MBSE uses models to define the system. Among the benefits, MBSE can help in three ways:

- Address complexity
- Encourage reuse
- Manage product lines

Complexity is often the root cause of systems engineering challenges. By using a visual representation of the system, relationships between different parts of the system are easier to see and manage. This visual reference supports the design, analysis, verification, and validation of the entire system. The model provides a common reference across engineering disciplines, so that teams may more easily communicate and collaborate during the development process. Improved communication leads to greater efficiencies,

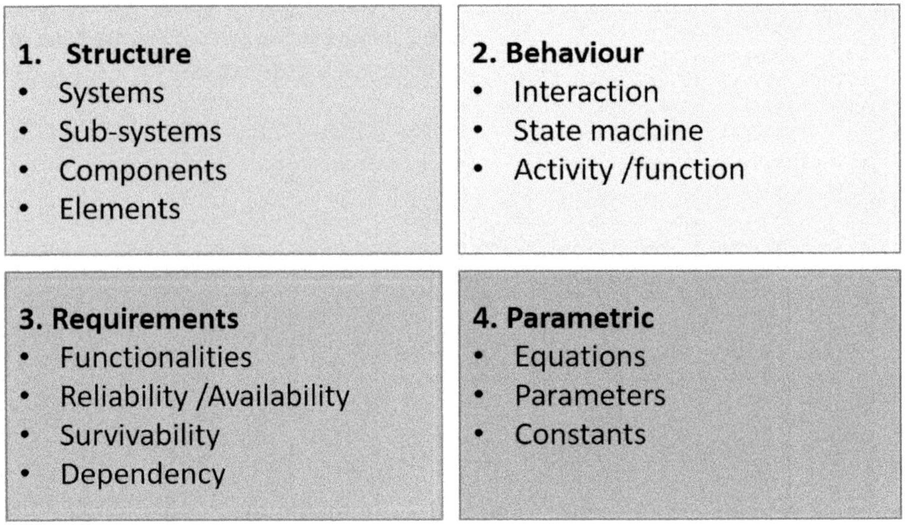

FIGURE 5.41

The four pillars of SysML. *SysML is a trademark of the Object Management Group, used with permission.*

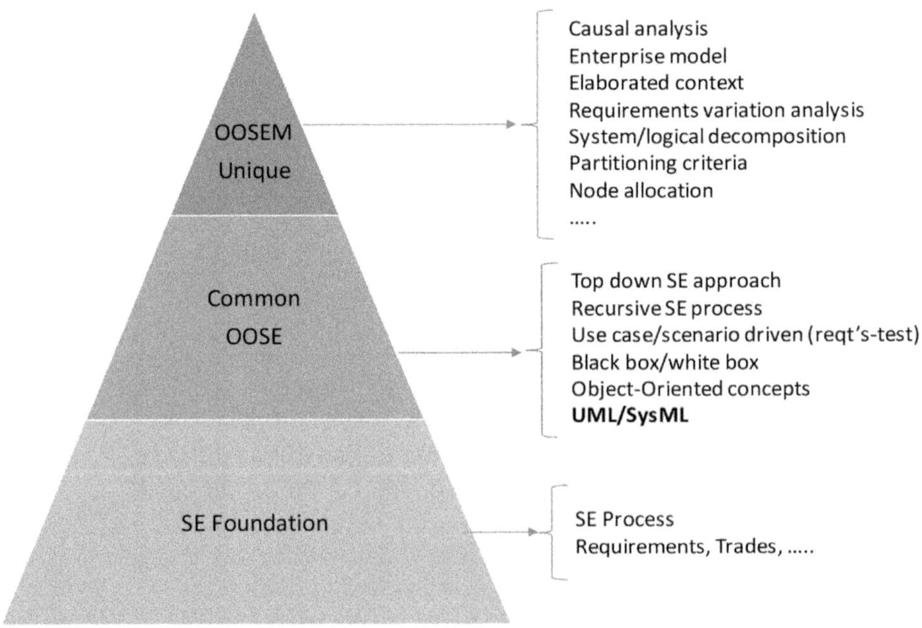

FIGURE 5.42

OOSEM pyramid—top-down (functional decomposition) approach development, and test.

improving time-to-market. Also, because the system model makes it easier to visualize the interdependencies in the system, the risk of errors and quality problems is reduced, especially when making changes.

Reuse is important because it saves design, development, and test time. Also, reusing proven and validated components, subsystems, and digital assets will reduce the risk of errors. The challenge is that system complexity and interdependencies make it very difficult to identify what can be reused. MBSE supports a modular approach for developing systems. With a *modular system*, the system is broken down into smaller "sections." These smaller sections become much easier to reuse in other systems. As the design evolves, it is also easier to isolate the small sections so that they can be individually validated to catch problems sooner.

The embedded software provides a design platform. However, designing platform variants is a top challenge. With MBSE, engineers can more easily assemble different reusable components to create unique models for each variant. Fig. 5.43 illustrates the PCT MBSE, ALM, and PLM interoperability.

For example, PCT MBSE software has created the communication ports between MBSE and embedded application lifecycle management (ALM) and product lifecycle management (PLM) softwares.

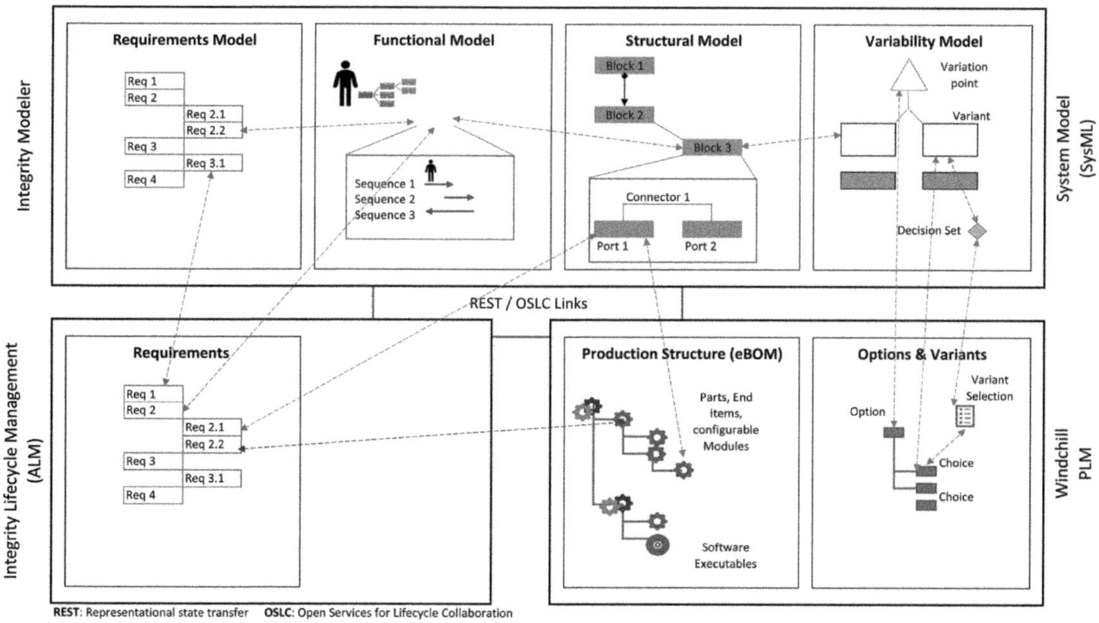

FIGURE 5.43

PCT MBSE, ALM, and PLM interoperability.

The PCT integrity modeler can create the codes for the objects and be linked to the mashups and data at PCT IoT platform called ThingWorx.

ThingWorx machine learning is a learning technology that automates advanced and predictive analytics. The technology uses proprietary artificial intelligence and machine learning technology to automatically learn from data, discover patterns, build validated predictive models, and send information to virtually any type of application or technology. ThingWorx machine learning is built to create intelligent systems by tightly integrating into applications, processes, and technologies already in place.

ThingWorx machine learning delivers automated predictions, pattern, and signal detection for manufacturers operating complex production processes. The platform is designed to constantly feed existing production planning and Software configuration management (SCM) systems with easy to interpret proactive information. It enables quick decisions to avoid production downtime, defective finished goods, and poor production yield.

5.10 APPLICATION LIFECYCLE MANAGEMENT

ALM is the product lifecycle management (governance, development, and maintenance) of computer programs. It encompasses requirements management, software architecture, computer programming, software testing, software maintenance, change management, continuous integration, project management, and release management.

ALM is a set of predefined processes that start somewhere in the business as an idea, a need, a challenge, or a risk and then pass through different development phases such as requirements definition, design, development, testing, deployment, release, and maintenance spanning across an entire life cycle of a product. Throughout the ALM process, each of these steps is closely monitored and controlled, followed by proper tracking and documentation of any changes to the application.

ALM is a broader perspective than the software development life cycle (SDLC), which is limited to the phases of software development such as requirements, design, coding, testing, configuration, project management, and change management. ALM continues after development until the application is no longer used, and may span many SDLCs.

Modern software development processes are not restricted to the discrete ALM/SDLC steps managed by different teams using multiple tools from different locations. Real-time collaboration, access to centralized data repository, cross-tool and cross-project visibility, better project monitoring and reporting are the key to develop quality software in less time.

This has given rise to the practice of integrated application lifecycle management, or integrated ALM, where all the tools and tools' users are synchronized with each other throughout the application development stages. This integration ensures that every team member knows who, what, when, and why of any changes made during the development process and there is no last-minute surprise causing delivery delays or project failure.

Today's application management vendors focus more on API management capabilities for third party best-of-breed tool integration which ensures that organizations are well-equipped with an internal software development system that can easily integrate with any IT or ALM tools needed in a project.

There has been talk to change the term ALM to ADLM (application development lifecycle management) to include DevOps.

5.11 PRODUCT LIFECYCLE MANAGEMENT

PLM has been introduced by the automaker American Motors Corporation (AMC) in 1985, when it was looking for a way to speed up its product development process to compete better against its larger competitors. Today, any discrete manufacturing business use PLM as an indispensable tool to power innovation and productivity by connecting people and processes with a single source of product and process knowledge. Fig. 5.44 illustrates a typical PLM portfolio.

PLM helps to manage complex, cross-functional processes, coordinating the efforts of distributed teams to consistently and efficiently develop the best possible products. All aspects of the product development lifecycle, from

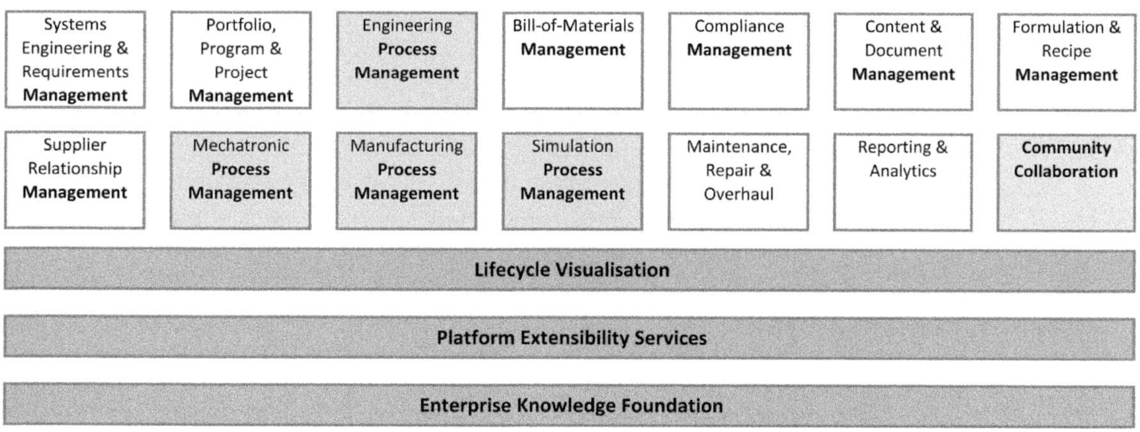

FIGURE 5.44

A typical PLM Portfolio.

concept through service and retirement can be managed by an appropriate PLM software. By optimizing product development processes and using a single, secure source of information, the competitive, cost-effective, and high-quality products can be developed. The typical PLM software features are as follows:

- *Systems engineering and requirements management*: PLM software allows companies to create and communicate the correct set of requirements to downstream decision-makers. By empowering all individuals and functional teams with visibility to each requirement and the knowledge behind it throughout the lifecycle, the PLM software ensures the delivery of purpose-driven products and achieve their strategic intent.
- *Portfolio, program, and project management*: PLM software allows companies to connect their strategic portfolio plans to program and project management and to detailed operational execution. This maximizes investment returns by continuously governing the selection of the right mix of investments, then managing resources and driving activities.
- *Engineering process management*: PLM software provides a single, organized, and secure source of product engineering and process knowledge. By allowing different development groups to work together, as a single entity irrespective of location and development time, quality and productivity are significantly improved.
- *Visual collaboration*: The visual collaboration capability in PLM allows user to easily view and comment on parts and documents, providing security, control, and context to your day-to-day product development discussions. It includes threaded discussions with comments and replies, configurable discussion forums, and the ability to easily view and graphically comment on all your file content, including 3D models, drawings, and Office documents. Visual collaboration provides everyone from engineering, operations, quality, and the supply chain with a way to communicate about products more effectively, make decisions quickly, and solve problems faster.
- *Technical publications*: Technical publications is a content authoring tool that provides component content management capabilities for topics-based, modular documentation. Technical documentation enables users of various disciplines to author, visualize, share, and publish information within a central, controlled, and collaborative environment, where change management capabilities ensure document quality.
- *Document management*: Document management includes version and change control for files in multiple formats in a secure, searchable repository. Users across multiple teams and locations can access the critical information required for the projects they are working on.

The capability allows users of various disciplines and departments to author, visualize, share, and publish information all within a central, controlled, and collaborative environment.

- *Quality management*: Quality management application provides organizations with tools to manage risk, improve quality, and attain environmental, regulatory, safety, medical, and other forms of compliance. The application uses advanced product quality planning (APQP) to help you to control and mitigate critical product risks. Using the capability, you can deploy failure mode effects analysis (FMEA) processes to achieve compliance with regulatory requirements and quality standards. Other features allow user to manage libraries of critical characteristics, failure modes, effects, and control mechanisms, utilize ISO-compliant process flows to drive cross-functional collaboration, and create dashboards, scorecards, KPI metrics, and reports to track your project's status.
- *Requirement management*: Requirement management enables the creation of a single, multidiscipline requirements hierarchy for managing complex engineering systems including planning, development, and realization. Requirements stay in sync as a system or product evolves, providing accountability over the product lifecycle for all disciplines including mechanical, electrical, software, and firmware.
- *Configuration management*: Configuration management provides a comprehensive methodology for managing the configuration (hierarchical set of information) of a product or system throughout its life. The capability provides identification, control, status accounting, and verification of configuration changes throughout the product lifecycle using graphical online business process workflows with e-mail notifications and reminders. These modules should follow the ISO 10007 standard and CMII/CMPIC best practices.
- *Change management*: Cross-discipline change management capabilities provide mechanical, electrical, software, and documentation teams with complete visibility into the change status of a product or system. Change histories are automatically captured, providing upstream and downstream traceability and ensuring users throughout the extended supply chain have easy visibility into the engineering change status throughout the product lifecycle.
- *Component engineering*: Component engineering simplifies component selection, approval, sourcing, and compliance processes. Engineers and procurement specialists have instant access to complete, current technical data, as well as obsolescence and compliance information from leading manufacturers around the world.

- *Parts management*: Parts management provides the complete control over the life cycle of parts, including automated workflows and permissions to help manage the approval processes. Links to computer-aided design (CAD) models, drawings, and documents are automatically maintained, and classification schemes may be implemented to support search and reuse.
- *Bill of materials (BOM) management*: PLM software allows companies to effectively manage essential product definition beyond design and engineering, right across the product lifecycle. By providing tailored visibility to clear, current, and accurate product BOM definition, specific to the needs of teams and users, companies are more able to meet quality, time to market, and cost requirements.
- *Compliance management*: PLM software provides a complete environment to document, enforce, and track regulatory compliance for the entire product lifecycle. With clear, documented regulatory requirements, managed and consistent processes, and an ability to track accountability, regulatory compliance can be embedded into development processes with minimal impact.
- *Supplier relationship management*: PLM software offers a portfolio of configurable solutions that allow companies to more effectively engage with their supply chain through a collaborative environment that supports better cost management and more effective product development and manufacture.
- *Mechatronics process management*: PLM software provides a rich environment for the associative development of mechanical, electrical, electronic, and embedded software technologies in a single source of product and process knowledge that spans all four domains. With a common data model across engineering domains and a common framework to work to, individual development teams can retain their mechanical, electrical, electronic, or software focus, while working in-context and together to meet overall development goals.
- *Manufacturing process management*: PLM software offers a single, scalable, and secure source of manufacturing data; product, process, resource, and plant, fully managed and associated to support lifecycle processes from engineering through production. With a single source of knowledge, manufacturing's influence on product innovation drastically improves profitability, time to market, and quality.
- *Simulation process management*: PLM software provides a single organized and secure source of simulation data and processes that can be embedded into the product lifecycle to better assess product performance and quality while improving development efficiency. By ensuring simulation teams can more effectively coordinate their

activities, simulation processes can be integrated and certified with the rest of the product development process.

- *Maintenance, repair, and overhaul*: PLM software offers configuration-driven service data management and MRO capabilities. Configuration-driven MRO bridges the gap between the logistics, maintenance, and engineering communities, to help drive more profitable growth. leaner operations, and improved asset performance.
- *Lifecycle visualization*: PLM software provides a view of your products throughout their lifecycle with a complete, scalable family of solutions for allowing lifecycle participants to visualize product data in 2D and 3D formats even when these data are created using diverse authoring software.
- *Reporting and analytics*: PLM software's capabilities for reporting and analytics provide a basis on which to establish, measure, and analyze key performance metrics to drive processes across the product lifecycle. By swiftly and accurately extracting, aggregating, analyzing, and disseminating information from multiple sources, more highly effective decisions making can occur.
- *Community collaboration*: PLM software's capabilities for community collaboration offer a platform for sharing information and working together across the product lifecycle. By breaking down the time, distance, and cultural barriers to communication, teams are able to work together more effectively and drive product development success.
- *Enterprise knowledge management*: PLM software brings together a company's product, process, manufacturing, and service knowledge, as well as participants, into a single PLM environment. By bringing relevant product information together into automated processes, efforts are better synchronized, productivity is increased, and greater innovation can be achieved.

5.12 MULTIPHYSICS

To simulate real-world conditions, analysts must consider the impact of a number of different physics that occur concurrently, such as structural dynamics, heat transfer, fluid flow, nonlinear material behavior, and motion. Typically, effects from one physics domain also impact how a product behaves in another physics domain. For example, combined thermal-structural effects are crucial to jet engine engineering. Understanding multiphysics behavior is therefore a major challenge to accurately predicting product performance. Fig. 5.45 illustrates the multiphysics simulation concept.

The 3D multiphysics softwares streamline the process of conducting multiphysics simulation. The integrated modeling environment eliminates the need for error-prone external data transfers to link coupled physics analyses together. Easier multiphysics simulation workflows mean you can spend

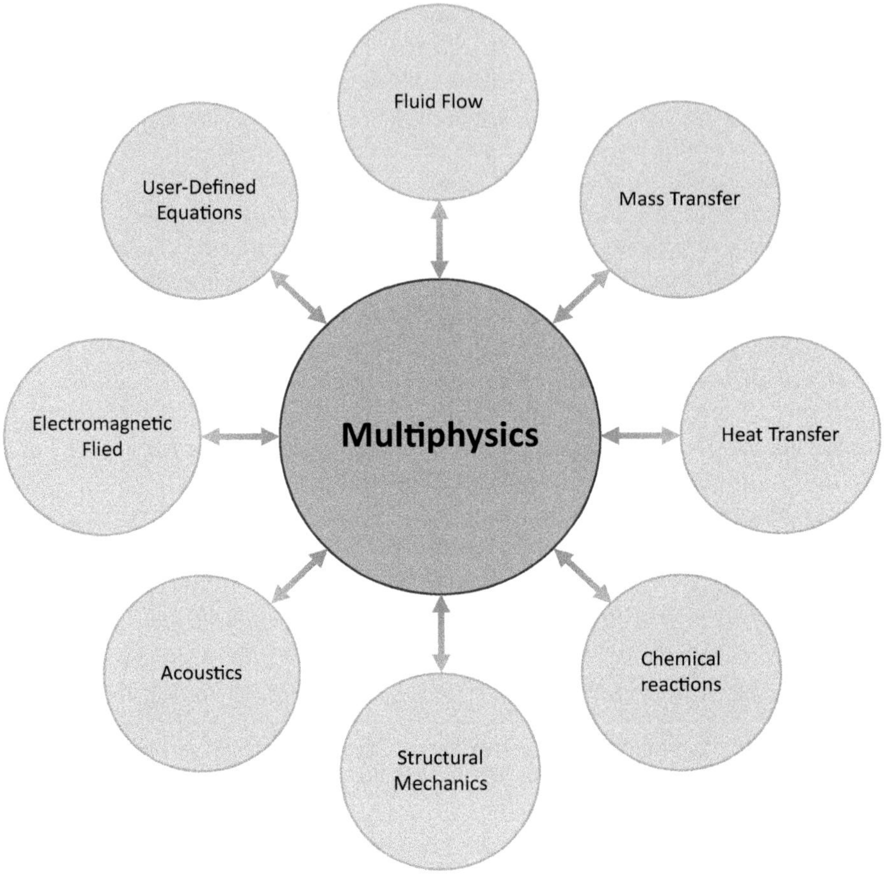

FIGURE 5.45
Multiphysics enable to integrate the simulation of different aspects of a system on the common 3D model.

more time simulating your product under conditions that more accurately represent the real world.

Multiphysics treats simulations that involve multiple physical models or multiple simultaneous physical phenomena. For example, combining chemical kinetics and fluid mechanics or combining finite elements with molecular dynamics. Multiphysics typically involves solving coupled systems of partial differential equations.

Many physical simulations involve coupled systems, such as electric and magnetic fields for electromagnetism, pressure, and velocity for sound, or the real and the imaginary part of the quantum mechanical wave function. Another case is the mean field approximation for the electronic structure of atoms, where the electric field and the electron wave functions are coupled.

Advanced Simulation Library, Elmer, and MOOSE are examples of open source software packages for simulating multiphysics models. Abaqus, ADINA, Altair, ANSYS Multiphysics, Code_Aster, CFD-ACE+, CFD-FASTRAN, COMSOL Multiphysics, FEATool Multiphysics, FlexPDE, LS-DYNA, NEi Nastran, and OOFELIE are some examples of commercially available software packages.

The multiphysics software packages mainly rely on the finite element method or similar commonplace numerical methods for simulating coupled physics: thermal stress, electromechanical interaction, fluid structure interaction (FSI), fluid flow with heat transport and chemical reactions, electromagnetic fluids (magnetohydrodynamics or plasma), and electromagnetically induced heating. In many cases, to get accurate results, it is important to include mutual dependencies where the material properties significant for one field (such as the electric field) vary with the value of another field (such as temperature) and vice versa.

Embedding the multiphysics softwares in PLM structure enables to study the behavior of the parts individually or as an integrated part of system.

Embedded software is changing the nature of, not only products, but also the business of manufacturing. Besides making "smarter products" possible, embedded software enables manufacturers to expand their goals as they can offer value-added services that raise the level of business with the customer. But adding embedded software to the product lifecycle poses numerous challenges. It increases the complexity of the product and the process for creating, manufacturing, and delivering it. There is now a new engineering discipline for software design that must coexist with mechanical and electrical design. Introduction of software also impacts the rate of change and with it, the burdens of change management and quality assurance. Moreover, the growing burden of regulatory compliance—where the impact of each change must be traced and validated—mandates a process discipline to the application lifecycle.

- Software changes the economics of products because it accounts for much of the innovation, uniqueness, and value of products.
- Embedded software development amplifies the challenge of change management and quality management in the product lifecycle because it allows product teams to innovate more rapidly.
- ALM disciplines become more critical than ever for dealing with the new complexities that embedded software brings to the product lifecycle.
- ALM facilitates the traceability that is essential for developing engineered products where embedded software plays a major role.
- Traditional ALM spans requirements management, architecture management, change and software configuration management, and quality management.

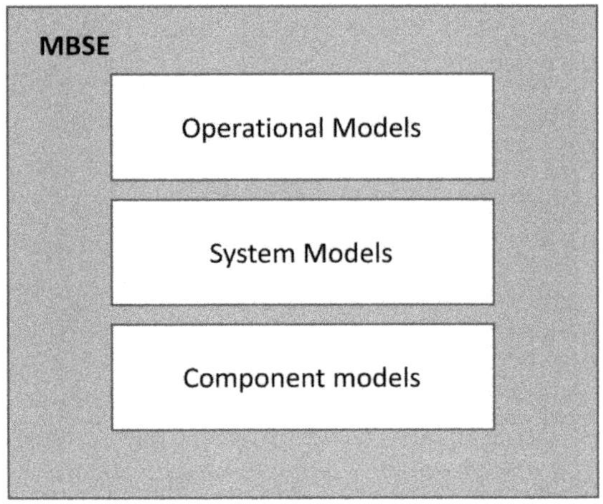

FIGURE 5.46
MBSE modeling at multilevels of the system.

A key ALM discipline is MBSE, which allows distributed teams to collaboratively build digital models to drive better architectural decisions, accelerate design and development cycle times, and facilitate reuse (Fig. 5.46).

As companies increasingly move toward multiphysics and multidisciplinary design optimization to address the performance requirements for tomorrow's complex cyber-physical systems, an open and collaborative approach will be essential.

5.13 "FRONTLOADING" SIMULATION RESULTS IN OPTIMIZED PRODUCTS

Nowadays, computational fluid dynamics (CFD) is intensively used during development of high-quality products in a wide range of industries such as automotive, aerospace, power generation, electronics, process, and many others. However, heavy competition dictates new requirements to be imposed to computer-aided engineering (CAE) in general and to CFD in particular aimed to significantly reduce development cost and the time-to-market. To satisfy these new challenges new methods must be developed.

Many engineering organizations are frontloading simulation, including CFD, by moving it to the beginning of the design cycle. By making early fluid flow and heat transfer simulation the responsibility of designers rather than specialists and manufacturers can slash the design cycle, lower their engineering costs, and achieve a superior return on investment. Engineers design and analyze early in the design process to discover potential design flaws, then redesign, meaning they can iterate quickly to get to the optimal product design.

Specialized analysts can still step in later in the project, but they will be working with parts that need fewer and smaller design changes than they would without the frontloading process. The designs also meet specifications without being overdesigned. The key to *frontloading* is the use of fully integrated CAD and analysis software. These integrated systems allow engineers to move quickly and seamlessly between product design in 3D CAD, performance analysis, and redesign.

For example, the integration of FloEFD from Mentor Graphics with the Solid Edge CAD system from Siemens PLM brings the ideal solution for frontloading CFD into a modern, easy-to-use CAD system. The FloEFD tool helps design engineers conduct CFD analysis early in the design process. Now embedded in Solid Edge, FloEFD offers intelligent technology that helps engineers set up and analyze problems quickly, it helps reduce overall time to a solution by as much as 65%–75% compared to traditional methods and products.

5.14 VIRTUAL REALITY

The Oculus Rift, Microsoft HoloLens, and even Google Cardboard are a far stretch from the first VR headset, created in 1968 by computer scientist, Ivan Sutherland. The concoction was called the sword of Damocles and, because of its formidable size and weight, had to be anchored to the ceiling so it didn't crush the user. Almost 50 years later, we are only now seeing VR and AR being used in manufacturing environments.

Like any disruptive technology with roots in the consumer market, industry viewed VR with a certain level of scepticism to begin with. Many companies questioned the practical applications of the concept, labeling it another gimmick that would not stand up to the rugged manufacturing environment. However, like ethernet, touch screen, and mobile devices proved before it, VR has real-world manufacturing potential.

New concepts and technologies are prone to hype. In 2015, Gartner's Hype Cycle saw VR emerging from the trough of disillusionment into the slope of enlightenment—this means VR started being used for real-world useful applications. In 2016, we are betting the technology will continue its ascent as more manufacturers start to take advantage of its benefits.

Indeed, a PWC survey at the beginning of 2016 found that more than a third of the U.S. manufacturers surveyed, already used VR technology or planned to do so in the next three years.

The oil and gas virtual reality training and simulation market is set for strong growth over the next 10 years as a combination of increasing demand for oil and gas, the retirement of the baby boomer generation, improving

FIGURE 5.47
Virtual reality applications for design and ergonomics optimization. Virtual reality for design applications and virtual reality for ergonomic applications at Ford.

technologies, and increasing safety concerns help virtual reality technologies become more widespread.

5.14.1 Virtual Design

The automotive market tends to be an early adopter of disruptive technologies—including automation, robotics, and now VR.

The U.S. automotive manufacturer Ford built its own immersion lab where designers, engineers, and other employees can don an Oculus Rift headset and walk around exploring the exterior and interior of its cars (Fig. 5.47).

Ford uses VR to test its designs and assess how individual elements of a vehicle look, without having to build a physical car. The VR links directly with the company's CAD software, so engineers can make changes and visualize results quickly and easily.

5.14.2 Virtual Training

Another area in which manufacturers are seeing VR shine, is training. The British engineering firm, BAE Systems, recently revealed that it creates virtual representations of projects, such as ships, for engineers to practice on. Figs. 5.48–5.50 provide the examples of EON Reality VR solutions for oil & gas and emergency response training programs.

BAE's virtualization suites allow engineers to examine the virtual elements of a system, so they can analyze, design, and plot where they need to make changes in the physical world. VR provides a level of test redundancy by giving engineers the chance to try out changes before they make any final alterations.

FIGURE 5.48
Virtual reality for training of hazardous task *With permission of EON reality.*

FIGURE 5.49
Virtual reality for training of the emergency response and investigation labs.

FIGURE 5.50
Virtual reality for training of emergency response situation *With permission of EON reality.*

VR training programmes can simulate realistic and hazardous situations on the manufacturing floor, including chemical spills, dangerous machinery, and loud environments, without putting operators at risk. Should the unavoidable happen, employees have relatable experience and are more likely to react appropriately in an emergency.

Furthermore, VR is an effective way of teaching machine operators or maintenance technicians about a new piece of equipment on the factory floor. Visualizing the inner components of devices allows companies to make detailed maintenance plans. This process is incredibly useful for identifying obsolete components or predicting which parts the original equipment manufacturer will cease to support in the near future (Figs. 5.51 and 5.52).

FIGURE 5.51
Virtual reality for training of complex systems.

FIGURE 5.52
Virtual reality for validation of assembly and disassembly procedures at an earliest stage.

This allows plant managers to create an effective obsolescence management plan, which could involve stocking up on spare parts or opting for a redesign.

5.14.3 Virtual Construction

The construction industry has a disproportionately high level of fatalities and injuries, and considered as a high-risk industry. Virtual reality platforms can be adapted into a suite of modules for safety training in the construction industry.

The virtual experiences enable users to begin to visualize and recognize complex workplace situations, build up knowledge of procedures and skills, and undergo training in a safe and forgiving environment.

The system is similar to a sophisticated, immersive computer game, with each interactive software module recreating environments for various training scenarios that can be displayed in multiple theatre modes. These include the 180-degree iDome for individual training sessions, and the 360-degree platform for group training within the immersive advanced visualization and interaction environment known as the AVIE.

5.14.4 Virtual Factory

Perhaps one of the biggest indicators of the potential of AR and VR for industry has come from a shift in recruitment at major engineering companies. Recently, firms have been very open about actively recruiting graduates with game design degrees. Astute with VR, Android, and mobile technology, this next generation of engineering recruits are helping make Industry 4.0 and IoT applications a reality.

Although VR and AR are years away from mainstream adoption in manufacturing, the technology is being put to good use by a minority of progressive companies, looking for a competitive edge. It seems VR and AR both have a prosperous future in the manufacturing industry; those willing to invest in the virtual world will be rewarded in the physical. Fig. 5.53 shows

FIGURE 5.53
Blending virtual and real for *Virtual commissioning and maintenance applications (from Siemens).*

how the digital twin can be used to manipulate the physical equipment during commissioning.

5.15 MBSE FOR PROCESS MANUFACTURING

The overall model based system engineering for process manufacturing should cover consistently the following cases:

1. *MBSE for EPC phase*: Plant is the final product of the EPC business. Plant is a system of the civil and structural, mechanical (static and dynamic), piping, electrical, instrumentation, telecommunication, and control systems. "Teams of teams" through "organization of the organizations" engineer, procure, and construct the plant.
2. Megaprojects, modularization, and fabrication of them are relatively new subjects for the process manufacturing industries. The exactitude of the MBSE for them is the key to delivering the quality results with predicted cost and schedule.
3. *MBSE for operational phase*: Produced materials are the final product of the operational phase. Maintenance is an enabling system to assure that the "enabling plant system" is fit for service. "Teams of teams" through "organization of the organizations" assure the continuity of the upstream supply chain, production facilities, and downstream supply chain.
4. *MBSE for brownfield changes and modifications*: Changed/modified facility, organization, or people are the final products of such projects. Configuration management of whole system is the most important in these cases.

5.15.1 Four Pillars of MBSE for Process Manufacturing Lifecycle

"Technical" and "organizational" systems are engineered to fulfill the requirements of the duty holder which is defined in the contract. Relevant regulations, codes, standards, and best practices add the constraints to the designed system.

"People" could have managerial or operational responsibilities. The reliability, quality, and safety of the manned operations depend on the behavior of people. Public, contractors, third parties, etc., add to the complexity of people roles and responsibilities. Fig. 5.54 brings these elements together as a first draft of MBSE for the lifecycle of a process plant.

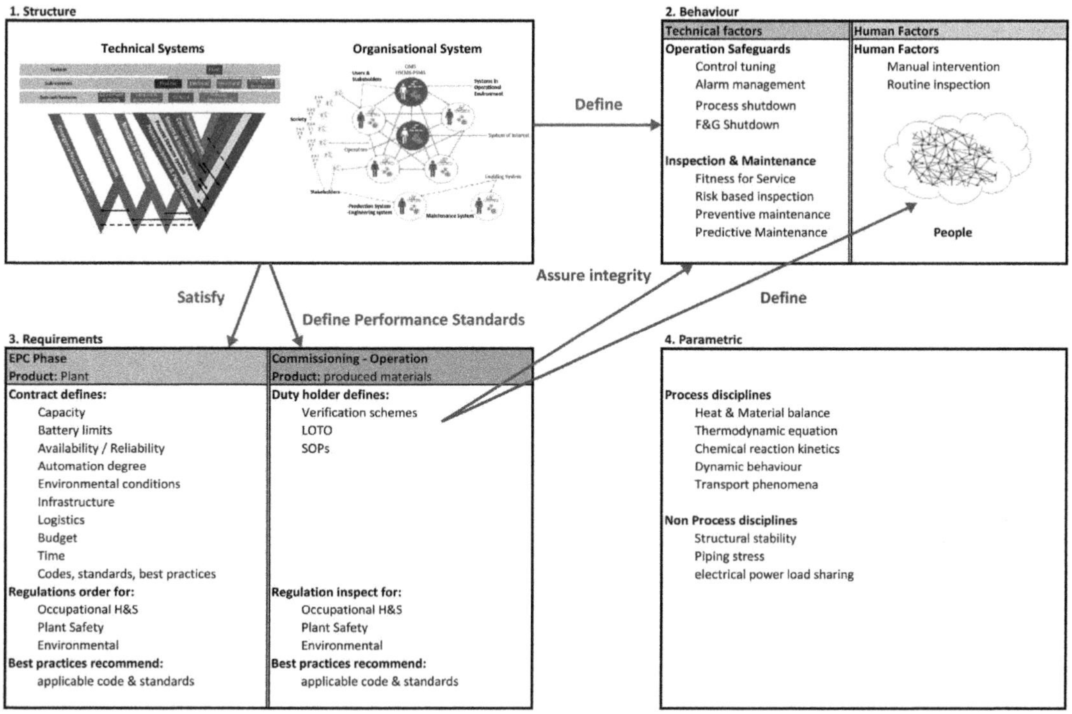

FIGURE 5.54
MBSE for process plant (draft).

The typical integrated tools shown in Fig. 5.55 are required to simulate the MBSE pillars of Fig 5.55. We will discuss the practical use of these tools in next section.

5.15.2 Digital Asset in Process Manufacturing

The block model is used for evaluating different layout options and to establish the location of main equipment and the main pipe routings in the unit, taking into account the required safety distances and accessibility for construction, operation, and maintenance. After approval of the layout by the principal, it can be used as a basis for preparing a more detailed (initial) plot plan drawing (Fig. 5.56).

The block model is used as the basis on which a 3D-CAD model or a physical model will be developed.

The 3D-CAD model is a collection of graphical data which describe the facilities of a plant for visualization purposes. The 3D-CAD model can be accessed

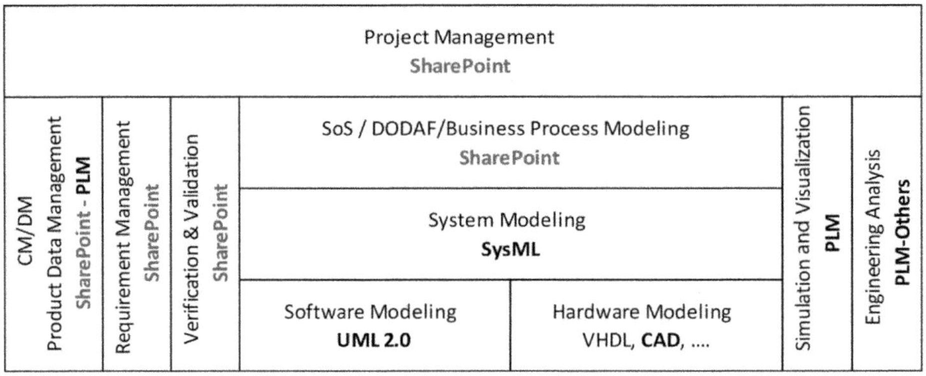

FIGURE 5.55
Typical integrated tool environment for MBSE.

via a 3D-CAD system. The graphical data are often an integral part of an engineering database supporting the design process.

An important advantage of a 3D-CAD engineering database is the possibility to maintain consistency between the produced documents and to assist the coordination of work between different disciplines. It enables the contractor to integrate the different design stages and the different subjects, such as piping, foundations, table tops, steel structures, equipment, instrumentation, electrical equipment, and HVAC equipment into one 3Dmodel. Typical deliverables of a 3D-CAD system include:

- Detailed engineering drawings such as plan and elevation drawings, bird's-eye views, etc.
- Piping isometrics
- Material take-off (bills of material)
- Shaded perspective views separated in logical areas, e.g., units (model review)
- Reports such as line lists, instrument data, interference checks (clashes), and mock-ups.

The CAD system shall be structured in such a way that all disciplines involved can check whether sound design and engineering criteria have been met with respect to safety, ergonomics, operability, and maintenance. This should include accessibility for maintenance and firefighting, location of valves, constructability, process instrumentation, routing of the piping adequate for the process, location of emergency shutdown stations, proper and safe location of vents and drains, location of fire protection and fire-fighting equipment, and escape routes.

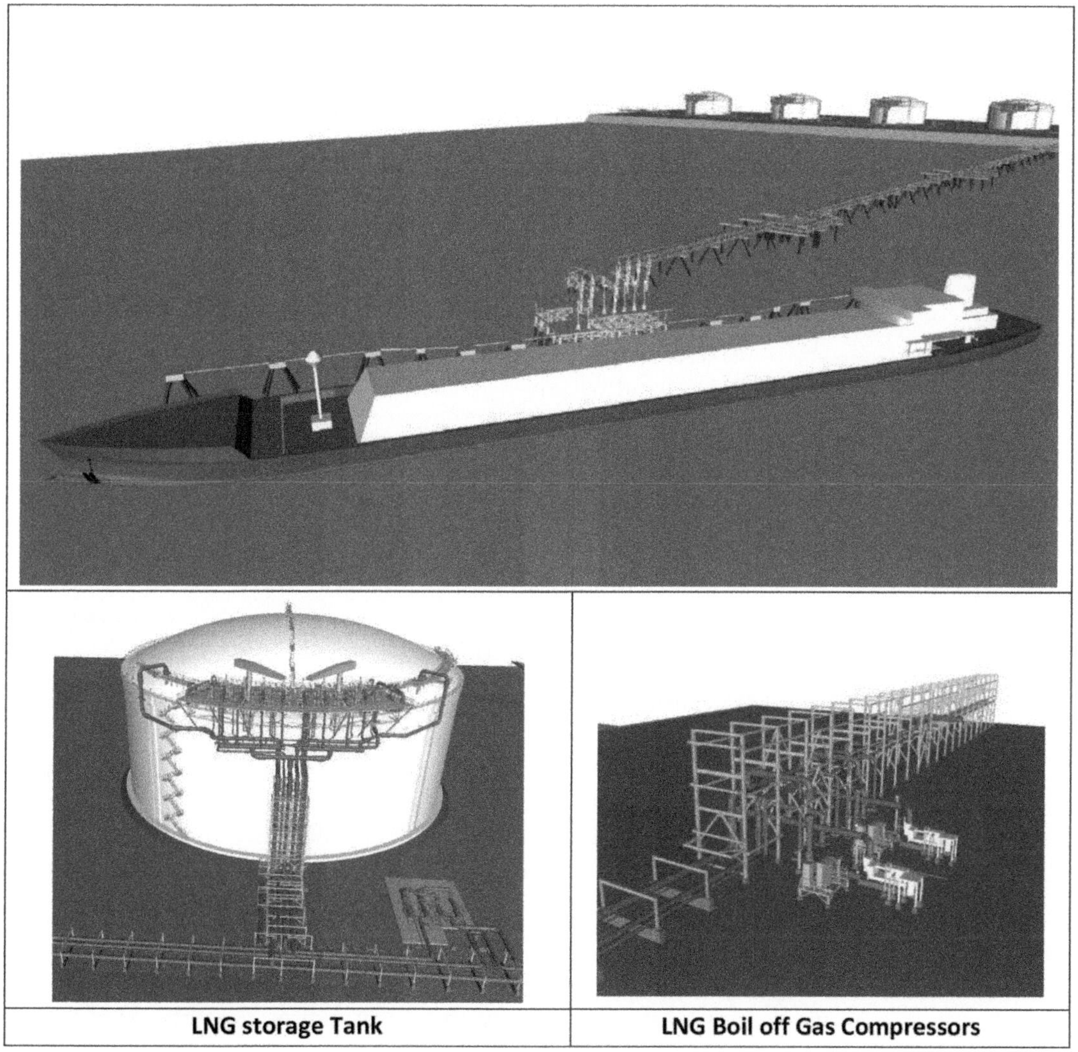

FIGURE 5.56
3D model of a LNG exportation plant: LNG storage tank and LNG boil off gas compressors.

A 3D-CAD system should be made available at site, to provide information during construction for aspects such as transport and erection of equipment, the routing of piping, positioning of structural steel, platforms, cable trenches, trunking, and channels.

The 3D-CAD model may be used for training and instruction of plant operators, construction staff, maintenance staff, and emergency response organization.

The required features are agreed between the duty holder and the EPC contractor before the setup of the 3D-CAD system and engineering database is started. The 3D-CAD system is programmed to provide reports regarding situations where certain criteria are not met. This approach can substantially improve the efficiency of the model review. The application of 3D-CAD technology allows for the following:

- Definition of classified 3D shapes to indicate envelopes for:
  - Safety
  - Area classification zones
  - Head room/tripping hazards
  - Access/escape corridors
  - Valve hand wheels
  - Reach/clearances for activities
- Access envelopes for:
  - Maintenance equipment
  - Operability
- Identification of deviations from normal engineering practices
- Modeling of foundations and underground systems
- Consistency between the 3D-CAD model and the process engineering flow scheme.

The CAD suppliers to the process manufacturing like AVEVA, Intergraph, Autodesk, etc., strive to go beyond 3D-CAD modeling and touch the EPC and O&M issues by offering digital asset solutions. They aim to use the information management technology to bring together all types of data from any source system, validate data automatically against predefined standards, and make this digital asset readily available to users across the entire enterprise.

The digital asset is not a software application or solution. The term refers to the aggregated information that describes a physical asset, and to the technology framework that brings this information together. The digital asset approach to engineering information management is new in that it uses advanced technology which can function with data from any source (it is "data-agnostic"). In fact, every physical asset already has at least some digital information that describes it, and digital information is being created from the very start of a project. The digital asset approach unlocks the potential value of this information and enables further information to be continually added, managed, and made use of.

Common practice is to use point solutions for different engineering and design purposes. Some solutions offer a level of application integration, but most asset information is locked away in separate data repositories ("silos"). This is a result of differences between these individual applications, file formats, data libraries, and so on, and causes difficulties in accessing and connecting (contextualizing) the different types of information.

While many organizations use document management systems, which have made significant improvements to the ability to aggregate information, they do not go far enough to overcome the problem. A single document may contain many items of data that cannot be cross-referenced with other data items of different types, held in other systems or documents. While documents are easy to control, the information held in documents cannot be controlled, unlike information held in databases.

5.15.3 Digital Asset Maturity Level

file:///C:/Users/Fabienne/Documents/AAAA/PLM/ILM/aveva-white_paper-reaching_level_five-2016.pdf

Asset information deteriorates rapidly if it is not rigorously managed and controlled, increasing operating costs and leading to safety hazards.

A mature asset information management enables access to, and robust management of, all types of engineering information throughout the entire operational life cycle and will provide the organization with a digital asset replica of its physical assets. The asset information in this digital asset replica will be highly functional, centralized, accessible, and therefore can be controlled and managed in a collaborative environment.

By integrating the relevant aspects of these industry standards and best practices into its digital asset strategy, the organization is able to ensure that asset information management processes are consistently applied and are as efficient as possible, leading to increased productivity, safety, and regulatory compliance. The following standards and best practices focused on management of asset information:

- *ISO 55000*: Asset management
- *ISO 15926*: Industrial automation systems and integration—integration of lifecycle data for process plants including oil and gas production facilities
- *ISO 14224*: Industrial automation systems and integration—integration of lifecycle data for process plants including oil and gas production facilities
- *CFIHOS*: Capital facilities information handover specification (for process)
- *ODATA APIs*: The open data protocol (OData)

These standards are used in the new products of the 3D-CAD model suppliers to process manufacturing industry. They try to minimize the engineering works by providing all the capabilities needed to design the typical plant, marine, and materials handling facilities, and then maintain their 3D "as-built" representations. Take advantage of data-centric technology, a strong rule- and relationship based architecture, customized automation capabilities, and an integrated reuse approach to execute even the largest and most complex projects with centralized visibility and control.

Large engineering capital projects require many partners, subcontractors, suppliers, EPCs, and other stakeholders who create project data. These companies have invested in software tools and training to support their own processes, workflows, and business objectives. Software solutions, and the designs and data they produce, become the pivotal point of collaboration, communication, and design integrity, and ultimately become deliverables for the project.

Designs and data must move freely during the project, during handover, and for the lifecycle of the facility. However, project data are often locked into silos. It is governed by software format, readability, compatibility limitations, and restrictions. Converting very large project databases is a complex and expensive undertaking for owners and EPCs.

Today, CAD suppliers provide the interoperability tools which walk the user through the process of converting and referencing data from/to one format to another. Interoperability enables users to:

- Reuse 3D data without having to convert and validate an entire model
- Incrementally deal with only the model sections needed
- Reference designs and data from previous formats
- Attach reference files, filter objects, and extract drawings from the original design
- Speed transfer of designs and data using out-of-the-box (OOTB) utilities and automation routines
- Minimize the effort and cost of converting between different CAD products.

The softwares such as Palladio provide the interoperability between static 3D model and dynamic risk based inspection (RBI) and asset integrity management performance standards and verification schemes during operation and maintenance phases of the plant. This provides the information exchange on a user-friendly pictorial platform which accelerate the routine maintenance tasks and trainings.

The "PALLADIO" suite has, among other features, the purpose of documenting operational activities carried out in any type of industrial system (inspections, scheduled maintenance, stops); this documentation, because of safety, planning, and management requirements and the need to

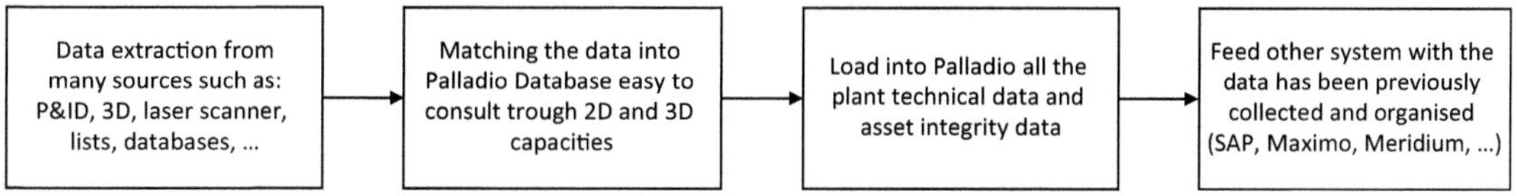

FIGURE 5.57
Palladio workflow. *PALLADIO is a tredemark of Antea srl based in Padova - Italy (anteash.com).*

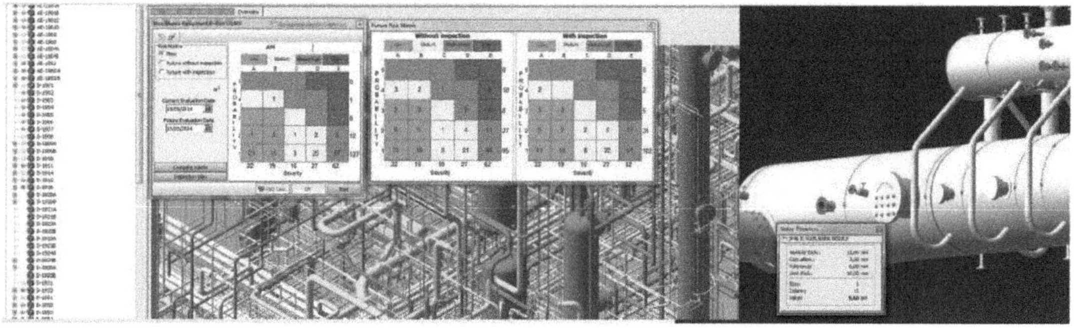

FIGURE 5.58
Palladio user interfaces. *PALLADIO is a tredemark of Antea srl based in Padova - Italy (anteash.com).*

interface with external Institutions, is easily found within "PALLADIO," thanks to its graphic potentialities as well. The Palladio Workflow is found in Fig. 5.57.

The as-is status of the system and the various technical documents acquired over time (construction drawings, test reports, etc.) or products (inspection, engineering maintenance reports) is presented by means of indexes created in accordance with business documents, such as P&ID, or survey operations with laser scanning (3D as-built modeling).

When possible, uploaded data are usable for testing and can also be acquired by other systems being used, such as SAP, Maximo, or other business software after they are synchronized. (Fig. 5.58) illustrates the palladio user interfaces.

The present 3D-CAD models are too heavy. They should be treated before being loaded into Palladio which is based on the interconnection of several modules, always ensuring a broad vision of the whole plant. Palladio 3D asset integrity software performs the following functions

- Manage and monitor inspections
- Visualize inspection points and results on the 3D model
- Visualize corrosion trends

- Monitor cathodic protection data
- Manage inspection plans
- Visualize and calculate RBI compliant API581
- Monitor and manage tanks, pipelines, and platforms
- Manage data of sewage systems
- Manage data of the chemical inhibitors and corrosion probes

Further interoperability between Palladio and the master 3D-CAD models will cover the design and maintenance integrities on the integrated platform.

5.15.4 Digital Asset Maturity Level

Another example of the efforts of the 3D-CAD model suppliers is the AVEVA digital asset maturity model and recommendations based on the information asset management standards and best practices. Tables 5.8 and 5.9 summarize the characteristics of the maturity levels, quality and validation requirements and AVEVA recommendation to take the existing and new systems to the higher maturity level. This model covers three assessment domains: process, technology, and people.

Table 5.8 Characteristics of Each Maturity Level as a Basis for Self-Assessment

Level 1	• Risk of regulatory noncompliances and HSE incidents
	• Inefficient O&M decision-making, unplanned outages, and productivity losses
	• Limited or no asset information supporting infrastructure
	• Poor information quality, scattered storage, and duplication
Level 2	• Operational inefficiencies and risks due to inability to keep information up to date
	• Recognized need for a digital asset strategy and supporting infrastructure
	• Inconsistently enforced and monitored asset information standards
	• Some manual interventions to validate information and reconcile its format
Level 3	• Defined, documented, and implemented processes to manage asset information (class library definition and implementation)
	• Relevant information from systems of record is captured and made available via a central, collaborative, web-based environment
	• Missing, inconsistent, and invalid information can be readily identified
Level 4	• Assurance that asset information fully complies with information standards
	• Improved management and efficiency of O&M processes via access to asset information in context
	• Assurance that information remains indicative of the as-operated state throughout
	• The duration of the facility's operational lifetime
	• Centralized environment for identifying and highlighting information issues as they arise
Level 5	• Ability to make full use of the value of 3D models and other information comprising the digital asset to support O&M processes
	• Rapid identification and determination of change impact, duration, and financial implications
	• Controlled information update processes to support streamlining of MoC processes

Table 5.9 AVEVA Digital Asset Maturity Model

Maturity Level	Level 1: Initial
Process	• Information not kept up to date, causing it to become inaccurate or incomplete • No formal processes in place for analyzing data quality • Limited or no formalized, defined, engineering information standards covering information types including structured and unstructured data, and documents • Limited or no sharing of engineering data between disciplines, and overall lack of awareness of what changes to information have been made. Hence, limited ability to ensure the latest version of content is being referred to.
Technology	• Multiple data repositories exist that are not synchronized, or that communicate via bespoke interfaces • Information exists in a variety of incompatible formats • O&M personnel waste considerable effort finding and verifying information before being able to carry out essential tasks.
People	• No formalized data governance model in place • Limited or no awareness of the need to appoint a member of top management to be responsible for oversight of the asset information strategy • Users have a limited understanding of the importance of data quality and its impact on decision making.
Recommendations	**Description**
Compile an initial content inventory of all electronic and hardcopy information necessary for O&M.	Identify and document information sources. These will later serve as asset information management solution integration points, once the organization has reached level 3.Examples of content that should be accounted for in the inventory include: • CAD drawings, P&IDs, electrical schematics • Operational specifications • Vender documentation • HSE plans and environmental impact reports • Equipment certifications and material safety data sheets • Permits • Operating, inspection, and maintenance procedures • Production planning information
Define an asset information management strategy.	The asset information management strategy should be formally documented and be part of the enterprise-level information management strategy.
Secure an executive-level champion.	Executive-level support is essential to ensuring an asset information. Management strategy is fully adopted and communicated and that adequate funding is allocated for supporting technology and resources. Without executive commitment, no asset information management program is likely to succeed.
Consult all O&M stakeholders.	Individuals representing every O&M stakeholder group spanning business, IT, engineering, and maintenance should be methodically surveyed and interviewed to establish the following: • Information access requirements • Current challenges in accessing information • Future wants and needs regarding access, management, and control of engineering information.

Continued

Table 5.9 AVEVA Digital Asset Maturity Model *Continued*

Maturity Level	Level 2: Defined
Process	• Process are defined and enforced for review and approval of engineering and design changes, but the communication of design changes is largely manual. • The organization is aware of the need for engineering information standards and has started the process for developing such standards. Information standards are defined on an ad hoc basis but are not enforced in a formalized or consistent way. • Some data quality measures are in place but not consistently applied across disciplines or projects, and they do not take into consideration the full asset life cycle. • Due to the modifications that have taken place throughout the asset's life cycle, the quality and availability of information is not indicative of the actual as-operated state of the facility.
Technology	• The organization understands the need to ensure that the information acquired across source systems and repositories is maintained, adequately assured and, where appropriate, controlled. It is in the process of determining the appropriate technical capabilities to ensure this is the case. • The organization is investigating the potential use of a commercially available software solution for efficiently and effectively capturing, organizing, and visualizing information corresponding to its assets. • No standardized tools are in place for controlling changes to technical information and synchronizing the changes to business systems. Hence there is a risk that, when asset modifications are made, required information updates may be overlooked.
People	• Top management understands the need to appoint one of its members to be responsible for overseeing the asset information strategy. • There is some awareness of roles in the context of overall asset information management processes to encourage ownership of data quality. • A data governance model is defined, but is loosely or inconsistently applied across disciplines, or only to a select group of projects or assets.
Recommendations	**Description**
Formally define and implement processes to manage information throughout the asset life cycle.	Once processes are in place, technology solutions can be deployed to make asset information available to improve O&M.
Make full use of existing business systems and technology.	To have clear picture of all information relevant to operations and to fully utilize existing information sources without increased software licensing costs, the asset information solution must be open, flexible, and able to integrate with third-party applications. These will typically include: • Engineering and design systems (2D/3D) • Document management systems • EAM and maintenance management systems • Enterprise resource planning (ERP) systems • Project management systems.
Provide a web-based tool that enables all O&M stakeholders to search, navigate, access, visualize, and collaborate on asset information.	Such a tool should enable stakeholders to access a single, data-rich digital asset, in accordance with defined access requirements. It should provide a clear picture of the current state of information, indicative of the as-operating state of the physical facility.Key benefits of a web browser-based tool include: • It eliminates the safety, security, and legal risks arising from users saving data in an uncontrolled manner on individual PCs or storage, or being able to transfer proprietary information outside of the company • All stakeholders see the same version of information • All data are centralized but is accessible from any computer at any time • Secure, password-protected access maintains security.
Collect and present asset data and information in a single place through defined associations.	Stakeholders have a clear understanding of the information dependencies and can rapidly retrieve and navigate between related information items.
Define a formal information classification scheme.	This should encompass numbering specifications for engineering items and technical documents, as well as asset information standards.
Ensure there is a common understanding of requirements across the information supply chain, including O&M stakeholders, contractors, and vendors.	This is best accomplished via a "top-down," operator-driven approach whereby the operations team communicates standards to project teams in the form of operational acceptance requirements for data and document deliverables and the project team communicates the requirements to the contractor via a formal information handover specification. A valuable resource to support definition of such a specification is the Capital Facilities Information Handover Specification (CFIHOS) initiative. CFIHOS provides a sound basis for establishing the operators.

Continued

Table 5.9 AVEVA Digital Asset Maturity Model *Continued*

Maturity Level	Level 3: Proactive
Process	• Baseline standards are defined but may be document-centric (e.g., limited to document naming and numbering standards) or limited to addressing information requirements for regulatory compliance. • Asset life cycle information management processes are established, documented, consistently enforced and maintained across information systems and critical assets to support the asset management strategy and objectives across all life cycle phases. • Information standards consistently enforced, including specification of acceptance criteria for engineering data, as well as information archiving and retention requirements.
Technology	• Capabilities are in place to facilitate automatic analysis of information submitted to the asset information repository to identify any issues affecting its completeness, correctness, or consistency and to report on these issues. • An integrated, web-based information repository ensures that asset information is accessible when and where required, such that the necessary information is readily available to end users to effectively support their asset management activities. • Integration capabilities are in place to establish and maintain logical relationships between the various information types including both data and document-centric content. • As information requirements change and new applications are introduced, the asset information repository can be easily scaled to facilitate integration of the additional information content types.
People	• Top management has appointed resources to oversee asset information initiatives and has granted these resources sufficient authority to fully execute their responsibilities. • Stakeholders have a clear picture of the current state of information, including an understanding of the level of data quality required to execute a given process.
Report example	**Data quality assurance checks provided**
Tag master data integrity.	• Check integrity of tag master data (e.g., naming and numbering and mandatory attributes) versus information standards defined in the class library.
Tag attributes integrity.	• Check integrity of tag attributes (e.g., mandatory attributes) versus information standards defined in the class library.
Document attributes integrity.	• Check integrity of document attributes (e.g., mandatory attributes) versus information standards defined in the class library.
Data source completions.	• Check integrity of data available by and across sources.
Document source, and then document completions	• Check document completeness status by and across sources, based on defined completeness criteria (e.g., number of documents, expected document attributes, etc.).
Contractor or supplier documents summary report.	• Report indicating contractor or supplier documents received based on what is expected, and verification of document naming and numbering attributes versus information standards specified by the class library.
Provide customizable reporting facilities.	Highly configurable reporting tools enable the creation and running of a wide variety of reports. These should include, not only operational key performance indicators (KPIs), but also indicators of data quality, such as completeness, correctness, and consistency.
Recommendations	**Description**
Enable automatic validation of information/data objects against the asset information class library, prior to making information available.	Information must be accepted in a controlled way and comply with the defined information standards. This will ensure that information created or updated remains trustworthy to support safe and efficient operation.
Provide full reporting of compliant and noncompliant information.	Configurable report and KPIs should be available for both scheduled and on-demand monitoring of information quality.
Provide a means for validating data from vendors and subcontractors	Automated validation should be implemented to ensure that information received is valid and complete, and that it complies with applicable standards prior to acceptance.

Continued

Table 5.9 AVEVA Digital Asset Maturity Model *Continued*

Maturity Level	Level 4: Managed
Process	• A class library is formally defined and is aligned with industry best-practice standards (e.g., ISO 15926, ISO 14224, CHIFOS) and includes criteria for compliance and definition of a taxonomy to organize data with respect to classes, attributes, units of measure (UoM), and associations based on the operator's specific asset information requirements. • Data conforms to all information standards expected. The accuracy of data is known at all times, and there are capabilities to ensure the precision of the data is appropriate for the degree of accuracy required. • Processes are in place to automatically interlink as-built asset information to provide a single source of organized and up-to-date asset information. Procedures are implemented for quality assurance to ensure data completeness and consistency across various sources and to report on and address the issues impacting information integrity.
Technology	• Capabilities are in place to facilitate automatic, proactive validation, and bulk data loading of tag and document metadata into a centralized asset information repository. Validation of tag and document metadata is completely class library driven. • Any data that fails validation checks is rejected and reported on for further rectification such that only the valid data are loaded to the asset information repository, thus establishing a single source of trusted data. • A single, integrated asset information repository is optimized via standardized data structures and is configured against a standard set of engineering information standard. • Functional capabilities are in place to facilitate configuration of metadata templates support population of operational systems, including filtering of maintenance significant tags from a master tag list and assignment of functional location codes (FLOCs)from the enterprise asset management system or computerized maintenance management system (CMMS). • Capabilities are provided for automated checking of data consistency across the asset information base as well as highlighting of inconsistencies and selective electronic updating. • The ability to automatically identify and visualize all information dependencies helps ensure that the necessary information updates are made, in response to minor and major asset modifications.
People	• Information quality dashboard reporting capabilities are in place to allow the current information status and information reliability metrics to be monitored by the data owners and other senior managers. • A formal data governance strategy is defined and implemented. • A designated asset information management team regularly reviews this data governance strategy, such that it is continuously updated and improved as needed. • The individuals charged with overseeing the asset information strategy have full responsibility for ensuring that the organization's assets deliver the required information to fulfill the asset information strategy, objectives, and plans. The appointed individuals have been given the necessary authority to achieve this.

Recommendations	Description
Enable automatic validation of information/data objects against the asset information class library, prior to making information available.	Information must be accepted in a controlled way and comply with the defined information standards. This will ensure that information created or updated remains trustworthy to support safe and efficient operation.
Provide full reporting of compliant and noncompliant information.	Configurable report and KPIs should be available for both scheduled and on-demand monitoring of information quality.
Provide a means for validating data from vendors and subcontractors.	Automated validation should be implemented to ensure that information received is valid and complete, and that it complies with applicable standards, prior to acceptance.

Continued

Table 5.9 AVEVA Digital Asset Maturity Model *Continued*

Maturity Level	Level 5: Optimized
Process	• Class-library-based validation rules are defined and implemented to automatically verify that each tag or document is classified, all mandatory attributes are present, UoMs are properly specified and naming/numbering rules are properly adhered to. • Processes are in place to facilitate controlled information update in response to in-plant modifications and not-in-kind changes. • Escalation routes are in place to ensure that data quality issues are addressed appropriately when detected by automated validation checks. • A reliable digital asset is available and maintained throughout the life cycle of the operating asset such that there is a sound basis for asset modifications reduction in the need for site visits, and high-fidelity asset information is readily available to support asset management decision-making.
Technology	• Automated capabilities are provided to create information change working sets comprised of the tag-level data and documents subject to change. • Whenever asset data are changed, the change data are subject to validation against the asset information standards and a formal approval process, before being synchronized back to the master systems of record. • A dynamic 3D visualization environment is provided where every tagged objected is selectable and supported by the full contextualized set of information in the asset information repository. • Information stakeholders are provided with access to whatever asset information they require including schematics, equipment drawings, single-line diagrams, instrument loops, and layout plans, within the context of the 3D models to support activities such as incident investigation, HAZOP studies and risk-based inspection and analysis.
People	• A culture of data quality improvement is promoted which ensures that staff are aware of the importance of good data quality. • The data governance model has been extended such that the majority of the organization's assets and projects are governed in a consistent manner and adhere to a common set of asset information standards. • Effective governance processes are in place to support new business requirements which may require new or improved data processing activities.
Validation type	**Description**
Functional class validation	• Checks that a given functional class is defined in the class library.
Attribute class validation	• Checks that a given attribute is defined in the class library.
Functional class attributes dataset validation	• Checks that a given attribute is defined as part of the corresponding functional class in the class library.
Enumeration list validation	• Checks that a given attribute is a valid member of an enumeration list (which can be defined in the class library or in an external system).
Unit of measure validation	• Checks that a given unit of measure is defined in the class library. • Checks that given unit of measure is defined as part of the corresponding attribute in the class library.
Attribute data type validation	• Checks that a given attribute has the correct data type defined in the class library.
Attribute existence validation	• Checks that a given attribute value exists, according to whether it is mandatory or not.
Recommendations	**Description**
Enable automatic validation of Information/data objects against the asset information class library, prior to making information available.	Information must be accepted in a controlled way and comply with the defined information standards. This will ensure that information crated or updated remains trustworthy to support safe and efficient operation.
Provide full reporting of compliant and noncompliant information.	Configurable reports and KPIs should be available for both scheduled and on-demand monitoring of information quality.
Provide a means for validating data from vendors and subcontractors.	Automated validation should be implemented to ensure that information received is valid and complete, and that it complies with applicable standards, prior to acceptance.

5.15.5 Installation Lifecycle Management

Most of process plant projects and operations call upon different disciplines within the enterprise and extended supply chain to communicate and collaborate on product development. Digital data created by the various teams involved in a product's lifecycle is as diverse as those involved from engineering and manufacturing to marketing, IT, service, and sales. This information is not only rich and varied, but also evolves over time—and quickly. To ensure that each product satisfies all requirements and is of the highest quality, each stakeholder must be able to access the most up-to-date product information. Fig. 5.59 shows that the BOM is the foundation of a digital product definition.

In discrete manufacturing, PLM is used to manage the product's BOM. BOM is used and adapted by many different stakeholders throughout the product's lifecycle. When these stakeholders are forced to work outside the PLM system to access upstream deliverables, enterprise processes and data management become extremely fragmented and ineffective.

Moreover, these multiple variations or views of the same BOM information are frequently managed in different systems. Sharing these BOMs across teams is inefficient, and introduces the risk of errors if information is not properly distributed. Consider that if the product design changes, downstream teams will be using information that is no longer current.

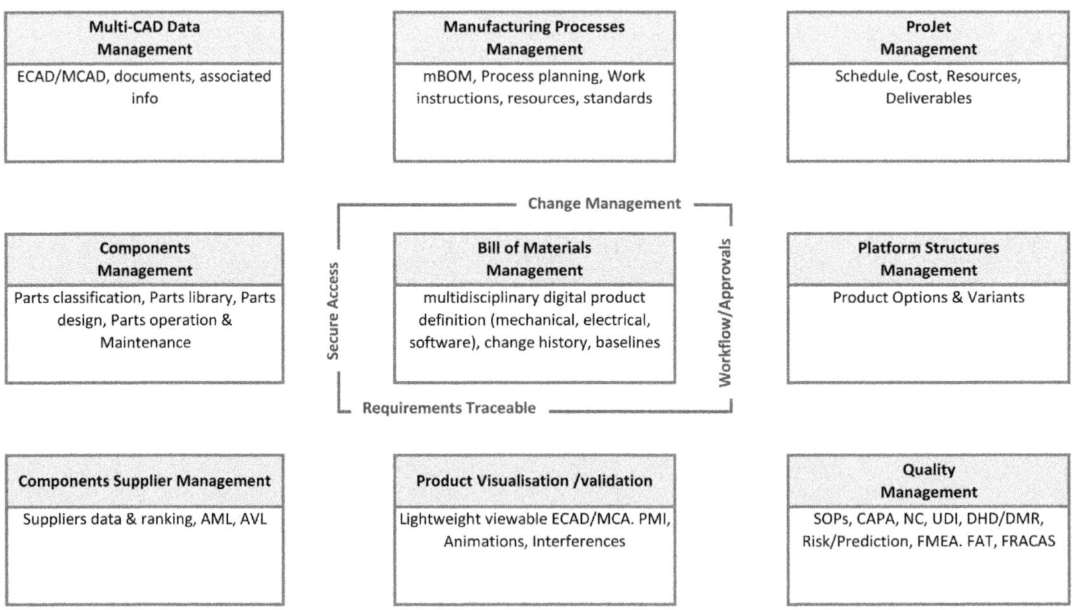

FIGURE 5.59
The BOM is the foundation of a digital product definition.

One way to overcome these issues is to optimize the use of the BOM in a way that enables organizations to realize a complete digital product definition. A digital product definition essentially configures, manages, and stores all product-related content—from final assembly structures to individual components—in a single, central repository. As a result, everyone across the enterprise can easily and instantly access the same product information.

Software suppliers improve multiphysics coupling and compatibility of the software with other tools continuously to enhance their software's ability to tackle multidisciplinary problems. They strive to push the technology curve with multiphysics to accurately predict how products will behave in the real world, while continuing to increase the openness and *interoperability* of their platform with standard engineering tools.

Offloading simulations to designers and nonsimulation experts reduces design cycles and uncovers flaws early. CAE engineering apps give nonsimulation experts the ability to perform these tasks. This frees up experts to work on other, more pressing matters. Fig. 5.60 illustrates how virtual seat solution

FIGURE 5.60

Virtual Seat Solution offers job-specific CAE to assesses the manufacturing, performance, and safety of a seat. *Image courtesy of ESI Group.*

Table 5.10 Simulation democratization Tools

Democratization Tool	Use Case	Target Audience
Simulation templates	Can access all functions of the host CAE platform. Needs direct access to CAE platform to operate.	Trainee, students, designers, and engineers performing repetitive simulation tasks.
Stand-alone	Can access all functions of the host CAE platform.	Internal users: managers, marketers
Simulation apps	App operates with CAE platform under the hood via web access.	External users: customers, clients, partners
Job-specific CAE	Targeted to industry in question. Example: CAE tool that only optimizes molds for casting.	Plant operators and designers

Table 5.11 Vendors Simulation Democratization Tool Options

Vendors	Simulation Templates	Simulation Applications	Job-Specific CAE
Altair/ solidThinking	X		X
ANSYS	X		
Autodesk	X		
COMSOL		X	
ESI Group			X
Siemens PLM Software	X		
SIMULIA	X		

can be applied as the CAE tool assesses the manufacturing, performance, and safety of a seat.

Engineering apps include templates, stand-alone simulation apps, and job-specific CAE tools. They feature industry-specific functions, guided workflows, simple user interfaces (UIs), and protection from human error. Table 5.10 compares the different simulation democratization Tools.

Some vendors offer several tools to build these engineering apps while others supply prebuilt tools. Table 5.11 compares the different vendors simulation democratization tool options.

For process industry, *installation lifecycle management (ILM)* can be developed if the "pervasive engineering simulation" is coupled with the digital asset management features of the modern 3D-CAD products.

Recently, ANSYS has initiated a multiyear trend to ensure that simulation is used throughout the life cycle of a product from design, manufacturing, use, and end of life. With the simulation engineering app builder using the CAE software will be easier for the nonsimulation experts. Fig. 5.61 illustrates the application of the ANSYS Multiphysics to assess the performance of process for both fluids and solids.

Adding the MBSE to an integrated software platform for SCADA systems complete the operational simulation needs such as the real-time status control, alarming, reporting, data acquisition, scripting, scheduling, MES, and predictive maintenance. Fig. 5.62 summarises the Inductive automation ignition SCADA platform architecture as an example of this approach.

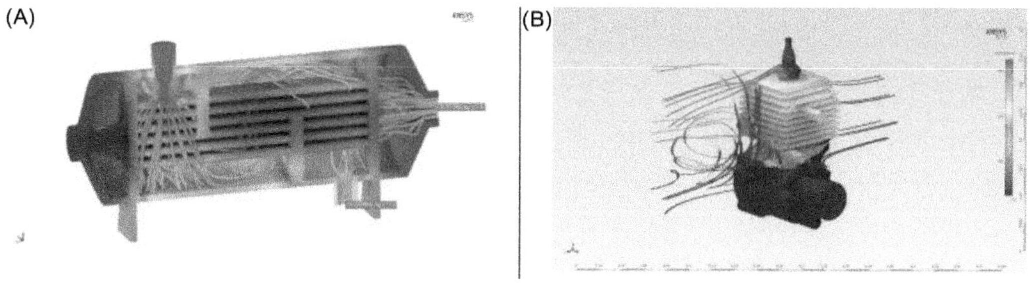

FIGURE 5.61
Application of the ANSYS Multiphysics. (A) Shell and tube heat exchanger simulation shows temperature contours, streamlines, and key flow values. (B) The conjugate heat transfer of an air-cooled engine. Results are included for both fluids and solids.

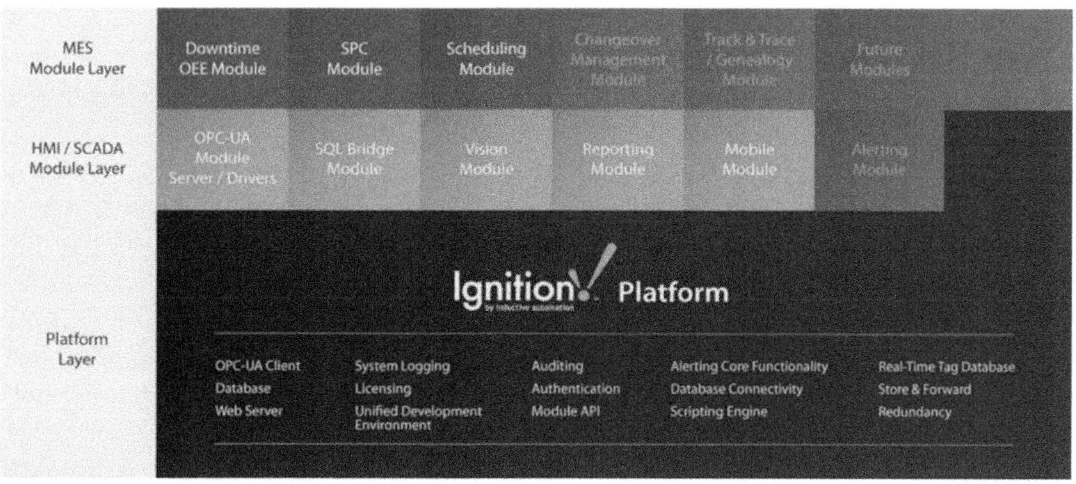

FIGURE 5.62
Inductive automation ignition SCADA platform architecture.

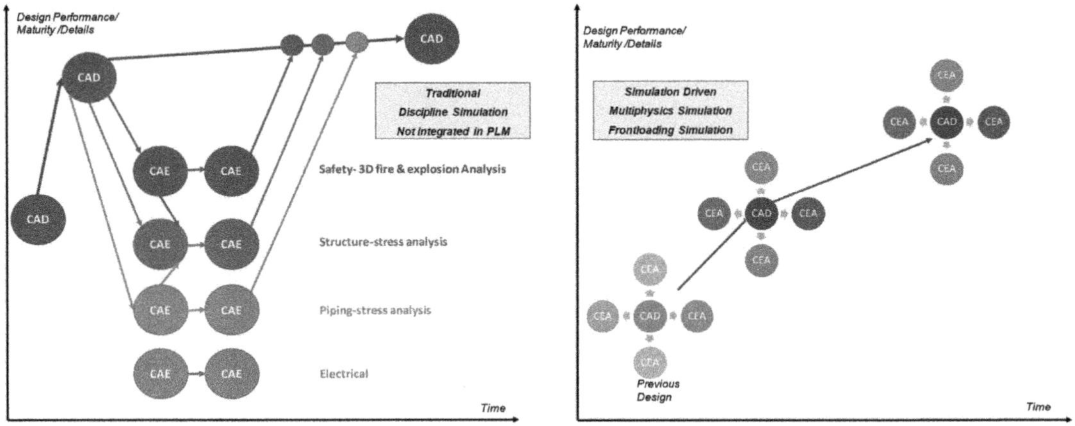

FIGURE 5.63
Comparison between traditional and simulation-driven CAD-CAE approaches.

Today, all the technical tools for integration of the MBSA, 2D and 3D Multiphysics, Frontloading simulations, asset integrity, and SCADA are available. To make ILM a reality, the discipline engineers should acquire the system engineering, CEA and IT skills to collaborate with the software developers and integrators.

Fig. 5.63 Compares the traditional and simulation-driven CAD-CAE approaches. Traditionally, when CAD model is mature enough, the CAE for different studies are performed in parallel, the integration of the results of the CAEs are integrated manually to update and optimise the CAD model. Today, PLM and Multiphysics simulation enable engineers to perform the CAD and CAEs simultaneously and optimise the whole model at the same time. The past simulations can be used effectively in the new project. This will reduce the cost of simulation.

"ILM" enables to perform and manage the consistent and precise design within the expected budget and schedule. Virtual commissioning exercises on this digital twin of plant will assure the safer and more reliable physical commissioning.

Asset integrity, risk-based inspection, and fitness for service models can be configured on the digital when design is matured and reduce the cost and effort associated with collecting data and make them consistent. This approach can also help to optimize design for asset integrity and required maintenance.

During operation, this model can be used for the interactive SOPs, predictive maintenance, optimization of the operation, management of changes, and operator training simulators.

With this model and HTML5 technology many high-quality collaborative e-learning and distant training programs can be produced. These trainings will be available anywhere at any time.

5.15.6 Process Plant Digital Twin

A virtual, digital equivalent to a physical product, or a digital twin, was introduced in 2003 at University of Michigan Executive Course on PLM. At that time, digital representations of actual physical products were relatively new and immature. The information collected about the physical product as it was being produced was limited, manually collected, and mostly paper-based.

Virtual products are rich representations of products that are virtually indistinguishable from their physical counterparts. The rise of MES has resulted in a wealth of data that are collected and maintained on the production form of physical products. In addition, this collection has progressed from being manually collected and paper-based, to being digital and collected by a wide variety of physical nondestructive sensing technologies.

The digital twin concept model, as illustrated above, contains three main parts:

a. Physical products in real space
b. Virtual products in virtual space
c. The connections of data and information that ties the virtual and real products together.

In the decade since this model was introduced, there have been tremendous increases in the amount, richness, and fidelity of information of both the physical and virtual products.

On the virtual side, we have much more information now available. Numerous behavioral characteristics can not only visualize a product, but also can be tested for performance capabilities. On the physical side, we now collect much more data about the physical product. Actual measurements from automated quality control stations and data from the machines that produced the physical part, are now readily available to understand exactly what operations, at what speeds and forces, were applied. Fig. 5.64 illustrates the concept of the Information mirroring model.

Individual bits or "bytes" of "raw" data do not by themselves inform the human mind. However, drawing various data together within an appropriate context yields information that may be useful. Information can be defined as:

Data + *Context* = *Information*

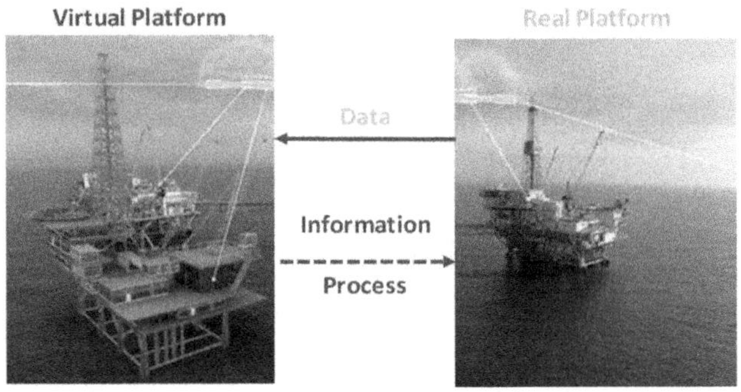

FIGURE 5.64
Information mirroring model.

For example, A value of, say, 10 m^3/h is just a single piece of data. But associate it with a pump, and link that pump with its location in a P&ID, and the data have gained context, which gives it meaning. It becomes information.

In turn, this information helps foster the quality of knowing. Knowledge and experience blend to become wisdom—the power of applying these attributes critically or practically to make decisions. In other words:

Information + Accessibility = Knowledge

By making information easy to access, we turn it into useful knowledge. A maintenance engineer, e.g., can take informed action if he can see that a 20 m^3/h pump output fell to only 10 m^3/h after a filter replacement was completed. Table 5.12 summarizes the content categories.

According to Russell Ackoff, a systems theorist and professor of organizational change, the content of the human mind can be classified into five categories:

Ackoff indicates that the first four categories relate to the past; they deal with what has been or what is known. Only the fifth category, wisdom, deals with the future because it incorporates vision and design. With wisdom, people can create the future rather than just grasp the present and past. But achieving wisdom isn't easy; people must move successively through the other categories.

Fig. 5.65 represents the transitions from data, to information, to knowledge, and finally to wisdom, and it is understanding that support the transition from each stage to the next. Understanding is not a separate level of its own.

The amount and quality of information about the virtual and physical product have progressed rapidly in the last decade. The issue is that the two-way connection between real and virtual space has been lagging behind. Global

Table 5.12 Content Categories

Data	Symbols Data represent a fact or statement of event without relation to other things. *E.g.,* It is raining.
Information	Data that are processed to be useful; provides answers to "who," "what," "where," and "when" questions Information embodies the understanding of a relationship of some sort, possibly cause and effect. E.g., The temperature dropped 15 degrees and then it started raining.
Knowledge	Application of data and information; answers "how" questions Knowledge represents a pattern that connects and generally provides a high level of predictability as to what is described or what will happen next. *E.g.,* If the humidity is very high and the temperature drops substantially the atmosphere is often unlikely to be able to hold the moisture so it rains.
Understanding	Appreciation of "why" Understanding leads to considering options and resilience *E.g.,* We may consider the fixed or hybrid greenhouses because the variation of humidity and temperature are not predictable.
Wisdom	Evaluated understanding Wisdom embodies more of an understanding of fundamental principles embodied within the knowledge that are essentially the basis for the knowledge being what it is. Wisdom is essentially systemic. *E.g.,* It rains because it rains. And this encompasses an understanding of all the interactions that happen between raining, evaporation, air currents, temperature gradients, changes, and raining.

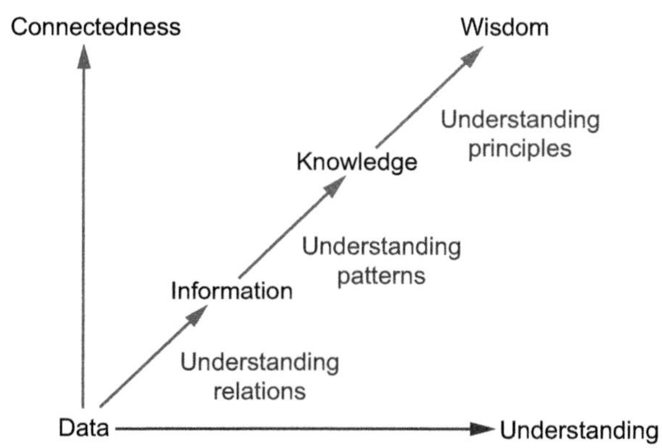

FIGURE 5.65

Transition from data to wisdom.

manufacturers today either work with the physical product or with the virtual product. Historically, we have not developed the connection between the two products so that we can work with both of them simultaneously. This shortcoming, however, may soon go away.

In order to deliver the substantial benefits to be gained from this linkage between virtual and physical products, one solution is to have a unified repository (UR) that links the two products together. Both virtual development tools and physical collection tools could populate the unified repository, creating a two-way connection between the virtual and physical product.

On the virtual tool side, design and engineering would identify characteristics, such as dimensions, tolerances, torque requirements, hardness measurements, etc., and place a unique tag in the virtual model that would serve as a data placeholder for the actual physical product. Included in the tag would be the as-designed characteristic parameter.

On the physical side, these tags would be incorporated into the MES in the bill of process creation at the process step where they will be captured. As the processes are completed on the factory floor, the MES would output the captured characteristic to the UR.

The final step would be to incorporate this information back into the factory simulation. This would turn the factory simulation into a factory replication application. Instead of simulating what should be happening in the factory, the application would be replicating what actually was happening at each step in the factory on each product. Many interesting use cases could then be possible by leveraging this digital twin, which could then contribute to improving overall manufacturing excellence.

Without PLM or ILM, digital twin will be just a virtual image of an asset with some functionalities which are maintained throughout the lifecycle and easily accessible at any time. With PLM or ILM, digital twin will be the central part of the digital asset ecosystem and will enable a new generation of advanced design, operation, and maintenance methods and predictive analytics.

5.15.7 Open Industrial Interoperability Ecosystem

The Open Industrial Interoperability Ecosystem (OIIE) project aims to introduce and standardize the concept of ILM.

The OIIE was born from the collective desire of owner/operators to reduce their reliance on expensive, fragile, custom application integration, and move toward an interoperability paradigm that can be employed across a range of targeted industries. Traditionally, organizations in these sectors have been forced to use large-scale systems integration methods, if they needed their complex applications and systems to work with each other in an integrated manner. This resulted in solutions which were relatively fragile and inflexible, while also being expensive and difficult to sustain.

Related efforts within industry began taking place as early as 2004, when MIMOSA, OPC Foundation, and the National Institute of Building Sciences organized a major standards-based interoperability demonstration at the 2004 International Maintenance Conference in December of that year. This initial collaboration and resulting conversations resulted in the formation of the "Open O&M" Initiative in 2007, in which ISA, MIMOSA, OAGi, OPC Foundation, and the WBF B2MML groups all agreed to cooperate to help enable the development of better solutions properly leveraging true standards based interoperability.

The OIIE is characterized by a solutions architecture framework for developing an enterprise architecture that employs system-of-systems interoperability. The foundation of an interoperability architecture is standards, and the OIIE uses a portfolio approach in leveraging both international and industry standards. The selection of standards is based on their capability to meet the industry-specified requirements, as well as levels of industry adoption and community engagement. Fig. 5.66 illustrates the MIMOSA integration-interoperability architecture framework.

OIIE distinguish between "integration" and "interoperability" as follows:

- Project-specific solutions process: *Integration*
  - Use cases, custom code, and testing are all project specific
  - Integration risks are borne by owner/operator and integrator
  - Expensive and high risk to implement and sustain (20%–25% annual recurring cost)
  - Not repeatable, lower quality due to project-specific code base

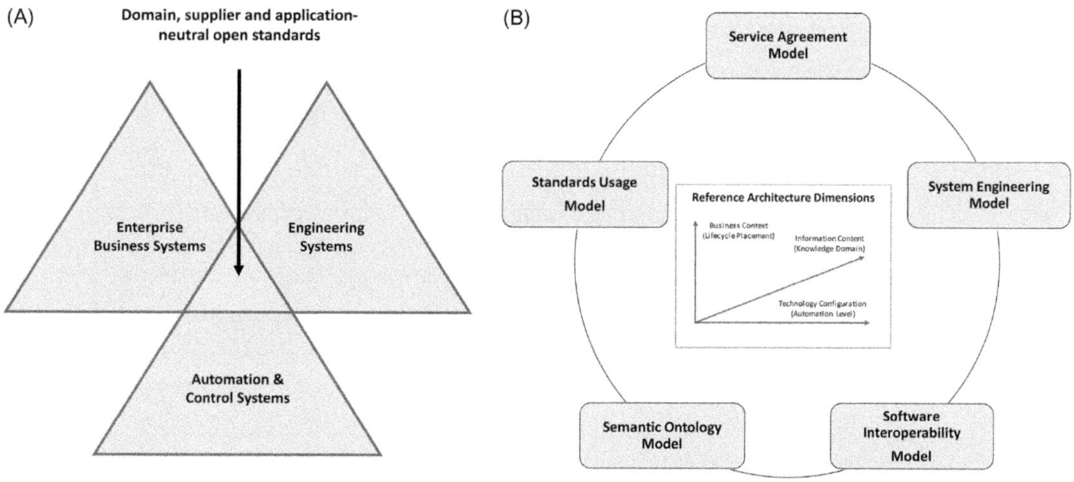

FIGURE 5.66
Integration-interoperability architecture framework. (A) OIIE solutions architecture (B) The PCA-MIMOSA reference architecture framework.

- Industry-driven solutions process: *Interoperability*
  - Requirements driven: The use case approach by Open O&M is a good example of this
  - Industry foundation architecture: Open standards *incorporated by reference*
  - Industry compliance data sets are developed and managed by industry
  - ISO TC 184 OGI TS is an example of an activity that sets forward a "target industry foundation architecture" and codifies piloted industry solution
  - Suppliers develop, maintain, and license compliant adaptors as commercial off the shelf (COTS) products
  - Enables repeatable and scalable industry-driven solutions for *oil and gas* shared with other elements of critical infrastructure
  - Guided by owner/operators with assistance from service companies and standards bodies

The OIIE enables a major paradigm shift from traditional systems integration methods, to standards-based interoperability, in asset intensive industries, including process industries, integrated energy, aerospace and defense, and other key critical infrastructure sectors. Fig. 5.67 illustrates the Simplified OIIE systems architecture.

The OIIE is a supplier-neutral, industrial interoperability ecosystem, which provides a pragmatic alternative to the status quo, enabling the use of COTS applications from multiple suppliers, as well as bespoke applications. It is defined by a standardized industry solutions architecture, which enables implementations of OIIE instances for both owner/operators and their major supply chain partners that are adaptable, repeatable, scalable, and sustainable at substantially lower cost than traditional methods.

The OIIE is an outgrowth of several related industry standardization activities, each of which is addressing a part of the industries requirements for standards-based interoperability. The OIIE brings these individual efforts together, with the direct participation and support of multiple participating industry standards organizations. Major parts of the OIIE include standards associated with the OpenO&M Initiative and with ISO 15926. The OIIE uses these existing standards in combination with each other, to meet the identified systems and information interoperability requirements for use cases which are defined and prioritized by the industries which are served.

The digital twin integrates data from many different software products. Data smart asset solutions are a key to reducing operational costs. The digital twin is a concept that will enhance information management and collaboration,

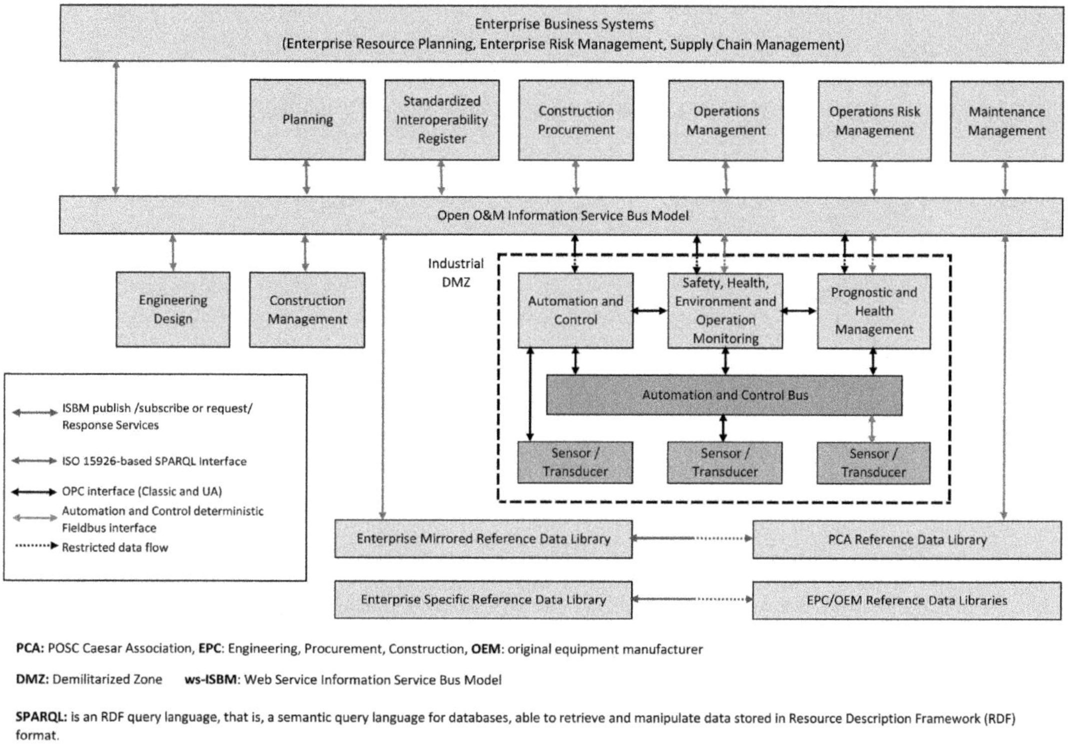

FIGURE 5.67

Simplified OIIE systems architecture. *PCA*: POSC Caesar Association; *EPC*: Engineering, procurement, construction; *OEM*: original equipment manufacturer.

where the experts can work together, preventing costly mistakes and rework. The digital twin will bring experts together by the:

- Cloud-based platform where all data and analyses are available at any times
- Updated throughout the lifecycle with real-time information
- Collaboration platform that prevents working in silos
- Early insight into potential conflicts enhances efficiency

Industrial assets are designed, built, and operated by engineers who rely on many different data sources and various models. They work with enormous amounts of data. Specialized teams create models separately, conducting analyses for their specific tasks. This way of working means that the most current information and calculations may not be readily available for crucial decisions, because the teams operate in silos. It drives cost and inefficiencies, creates uncertainties, and represents a vast amount of time and resources. Getting beyond these challenges means rethinking the way that projects are run.

Using the digital twins, the engineers will work cost-effectively on a collaboration platform that gives early insight into potential conflicts, resulting in fewer errors and less rework. Sensor data, remote monitoring, and analytics made possible by the digital twin enable more profitable, safe, and sustainable operations.

A single unscheduled downtime can cost in the range of $2–5 million per day. Approximately 50% of such incidents are due to various types of mechanical failures, which our products are ideally suited to help customers prevent. Poor information management is often a hidden cost, accounting for up to 20% of operational budgets. At a time when operational margins are low, this can make or break the operational viability of an asset.

An example is lost data, which is costly and sometimes impossible to track down or reproduce. Add to that the cost of replicated data, which over time becomes something of a parallel reality used in many business processes, and often fails to reflect the current state of the asset. Information management is important during changes in major lifecycle stages, e.g., from design and manufacture into the construction phase, and even more critical when the asset is moved into the operational phase. On top of this are the ever-increasing trends in mergers or assets changing ownership, particularly toward the end of design life when asset history information handover is seldom a priority. At this point the need for accurate historical detail is at its most critical in the assessment of fitness for purpose and extension beyond design life.

5.16 APPLICATION OF ILM TO CREATE THE PROCESS SAFETY MANAGEMENT FRAMEWORK

5.16.1 Considering Complexity in Risk Assessment

Highly hazardous chemicals are found in numerous process production facilities. These include toxic chemicals, reactive chemicals, and flammable liquids and gases. Personnel and public safety and environment are protected against the major accident hazards from these facilities by laws and regulations such as SEVESO III regulation in EEC and OSHA PSM standards in the United States.

The governmental inspectors regularly inspect the facilities to make sure if the duty holders comply with the regulatory obligations for controlling of the major accident hazards.

Major hazard management systems should consider all the foreseeable hazardous scenarios during lifecycle of plant. Multiple "technical," "organizational," and "people" failures or domino effects cause the major accidents as shown in Fig. 5.68.

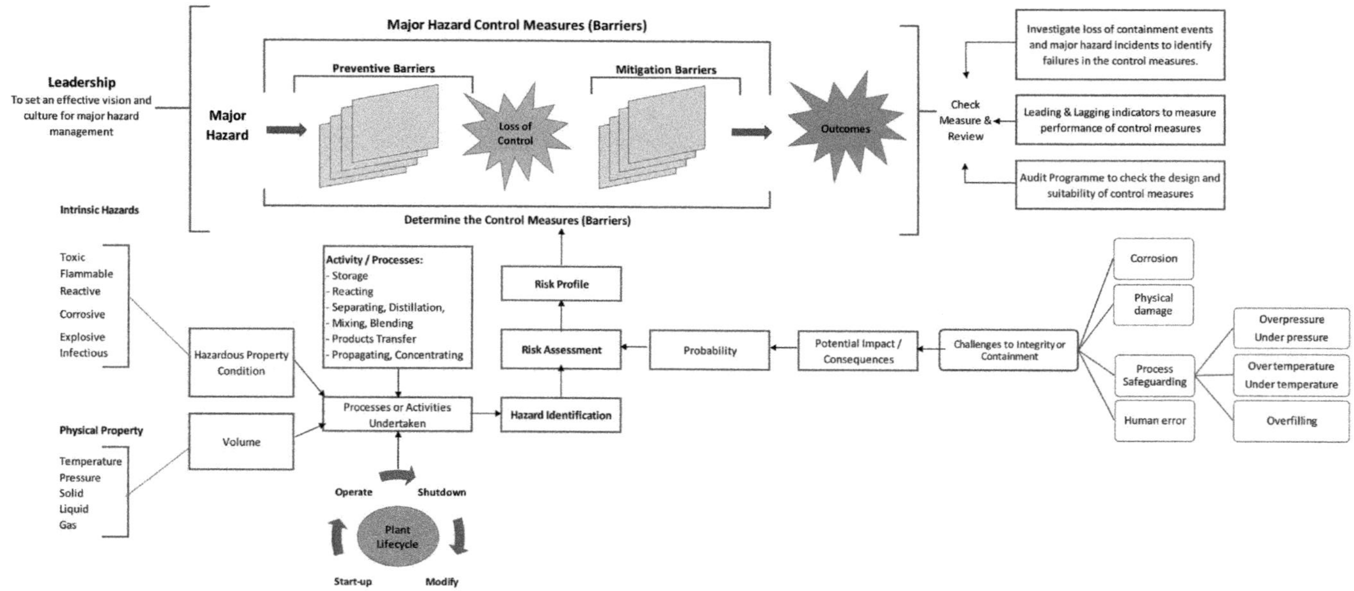

FIGURE 5.68
Major hazard management system: Overview.

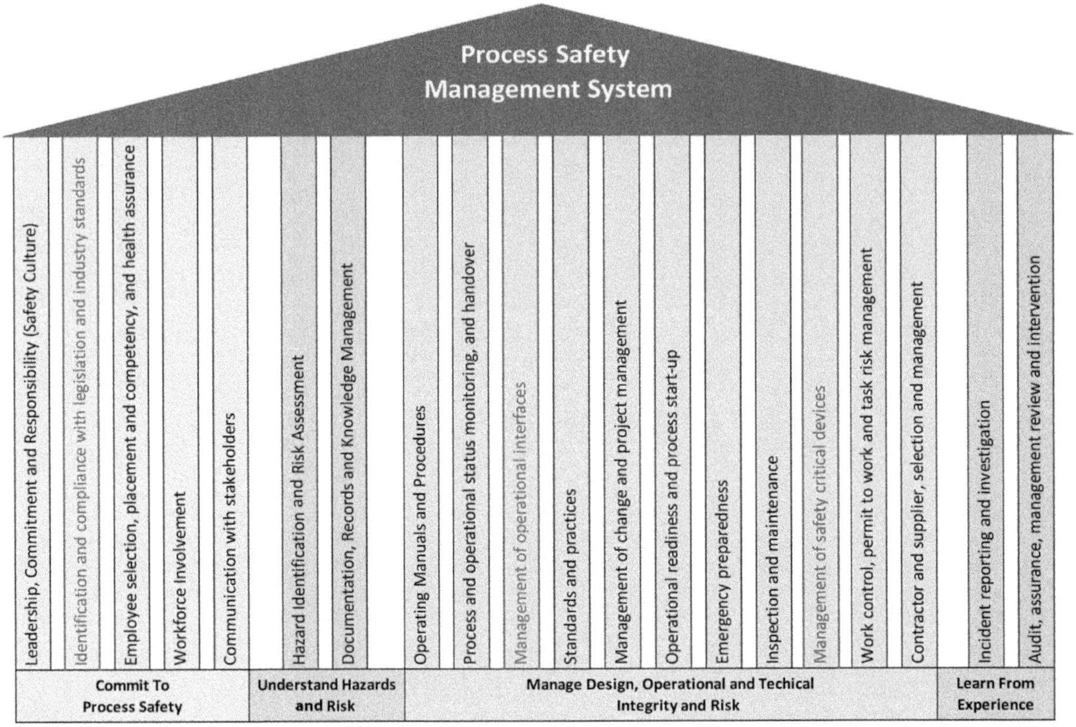

FIGURE 5.69
Energy Institute process safety management elements.

Major hazard risk control has to be systematic. The preventive and mitigation safety measures should be in place to minimize the risk of major accident to as low as reasonably practicable.

CCPS defines the 20 elements for the process safety management (PSM) framework to minimize the risk of major accident hazard. Later, Energy Institute adopted this PSM framework with slight modification in the order and categories of the 20 elements. Fig. 5.69 summarizes the Energy Institute process safety management elements. With a more system engineering oriented vision, the Energy Institute assign the dedicated elements to the following:

- Regulatory compliance
- Management of the operational interfaces
- Management of the safety critical devices

Fig. 5.70 shows that goal setting and risk-based thinking are foundation of the PSM framework. However, there is no guidance about the effect of complexity on the risk assessment approach. Attempt for solving the complex issues with simplified methods could create confusion and lead to disorder.

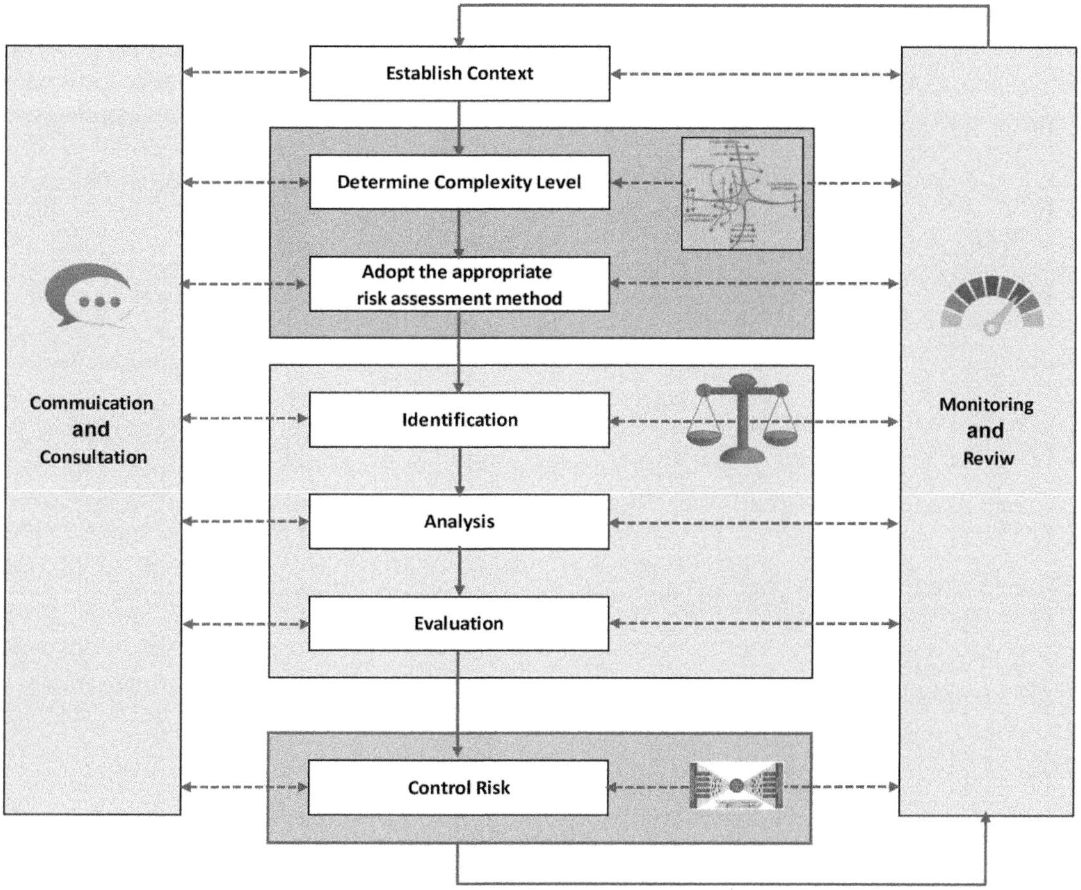

FIGURE 5.70
Modified risk management according to ISO 31000 (p. 8) complexity level should be determined systematically in any risk assessment.

For example, many managers and supervisors to simplify the human errors to operator negligence or disobedience root causes. These complacencies may lead to blindness to the real root causes by the day that much more server accident occurs.

A living example of this blindness is found in the Baker report on the root causes of the Texas BP Refinery in 2005 which suggested that cost-cutting and production pressure from BP executives may have resulted in a lack of needed safety measures across the board.

The authors of this book suggest that "complexity assessment" step be added right after "establishing context" step of the ISO-31000 risk management procedure as shown in Fig. 5.70.

Fig. 5.71 illustrates the Cynefin framework. Cynefin complexity model provides the simple qualitative criteria for determination of complexity levels. Table 5.13 summarizes the Cynefin complexity level criteria and strategies. Risk evaluation method and control measures should be appropriate for the complexity level of the context. For example, the risk associated to an *unordered chaotic* context such as unforeseen emergency scenario cannot be assessed and managed by the methods designed for the *ordered complicated* situations.

The chaotic decision-making during the Piper Alpha disaster is an example of this case. Most of the personnel who had the authority to order evacuation had been killed when the first explosion destroyed the control room. This was a consequence of the platform design, which did not include blast walls. Another contributing factor was that the nearby connected platforms Tartan and Claymore continued to pump gas and oil to Piper Alpha until its pipeline ruptured in the heat in the second explosion. Their operations crews *did not believe they had authority* to shut off production, even though they could see that Piper Alpha was burning. There is still controversy about whether there was sufficient time for more effective emergency evacuation.

Very often the managers believe that they can improve their organization and operational practices with a robust gap analysis performed by the competent external or internal auditors. This approach cannot be effective for the

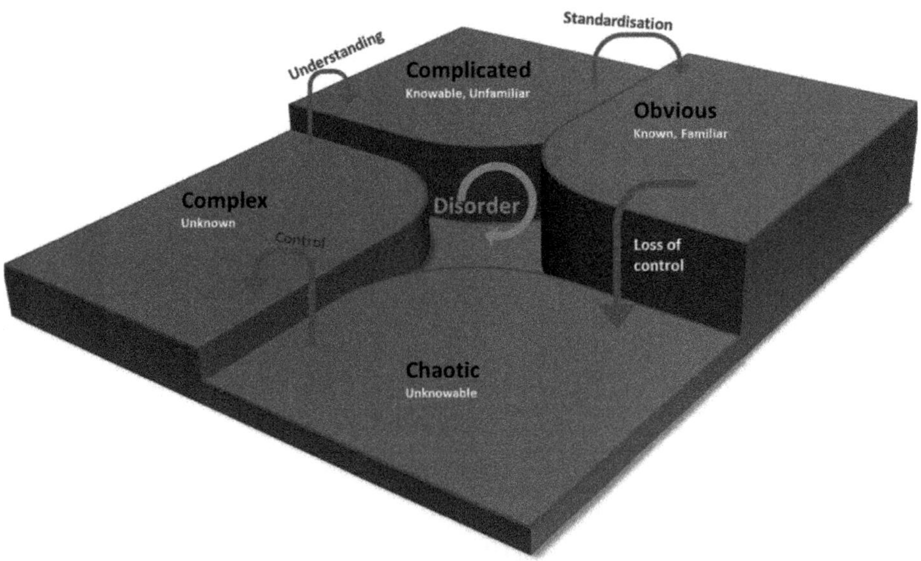

FIGURE 5.71
Cynefin framework.

Table 5.13 Cynefin Complexity Levels

Complexity Level	Characteristics
Known	• Cause-and-effect relationship • Repeatable, perceivable, and predictable • Legitimate best practice • Standard procedures • Process reengineering Sense-categorize-respond
Knowable	• Cause and effect separated over time and space • Analytical, reductionist • Scenario planning • System thinking Sense-analyze-respond
Complex	• Cause and effect are only coherent in retrospect and do not repeat • Pattern management • Perspective filters • Complex adaptive systems Probe-sense-respond
Chaotic	• No cause-and-effect relationship perceivable • Stability-focused intervention • Enhancement tools • Crisis management Act-sense- respond

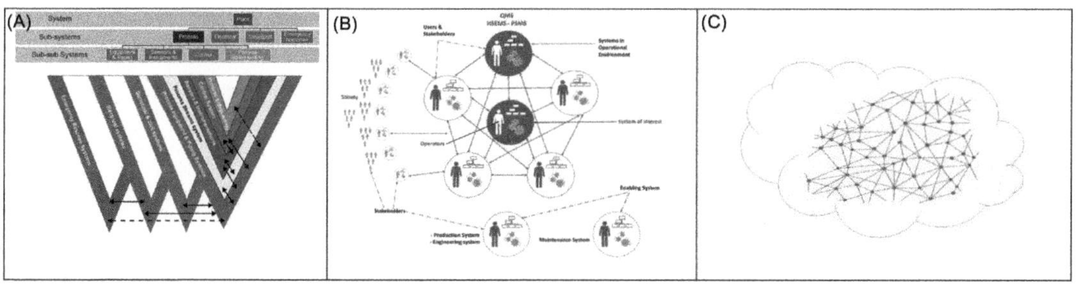

FIGURE 5.72

Complexity change in terms of technical, organizational, and people. (A) Technical system reengineering complexity. (B) Organizational system reengineering complexity (C) People interaction and task management complexity.

"complicated elements" with many technical, organizational, and cognitive human layers and sublayers.

For the existing systems with the heavy legacy issues, the analysts may find themselves in the disorder area of the Cynefin framework. In this case the first thing to do is to bring the context into the appropriate complexity level in terms of "technical," "organizational," and "people" as shown in (Fig. 5.72).

Risk Parameter		Classification
Consequence (C)	C1	Minor injury
	C2	Permanent injury or death to one person
	C3	Death to several people
	C4	Very many people killed
Frequency of, and exposure time in, the hazardous zone	F1	Rare to more often exposure in the hazardous zone
	F2	Frequent to permanent exposure in the hazardous zone
Possibility of avoiding the hazardous event	P1	Possible under certain conditions
	P2	Almost impossible
Probability of the unwanted occurrence	W1	A very slight probability that the unwanted occurrences will come to pass
	W2	A slight probability that the unwanted occurrences will come to pass
	W3	A relatively high probability that the unwanted occurrences will come to pass

Calibration Table

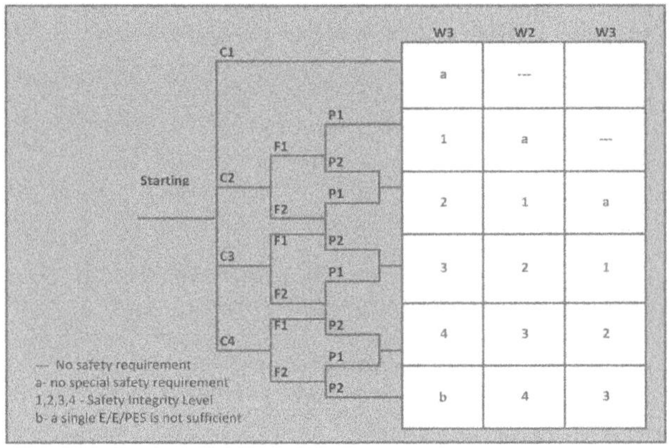

Risk Graph

Complexity Parameter		Classification
Number of Parts	P1	Limited
	P2	Significant
	P3	Unlimited
Diversity	D1	No
	D2	Yes
Connectivity	C1	No
	C2	Yes
Linearity	L1	No
	L2	Yes
Autonomy	A1	No
	A2	Yes
Order	O1	Ordered
	O2	Unordered
	O3	Disorder
Complexity Level	1	Simple
	2	Complicated
	3	Complex
	4	Chaotic
	5	Disorder

Complexity Calibration Table

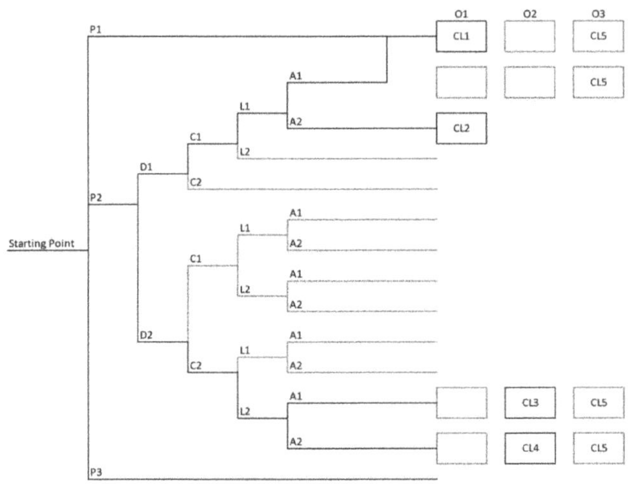

Complexity Graph

FIGURE 5.73

Analogy between SIL assessment and possible complexity level assessment using the graph and calibration table (to be assessed by process industry experts and developed in future).

Then, cautiously reengineer the system for the required technical and organizational changes. A robust communication strategy should be implemented to involve people in these changes. The web technology and cloud computing enable us to design the communication interfaces in such way that the self-organized contribution of the people can be assured. In other words, the complexities in terms of technical, organizational, and people should be assessed and managed simultaneously to implement the required changes effectively and assure the new configuration is sustainable. Fig. 5.73

illustrates the possible analogy between the qualitative SIL assessment and complexity level assessment using the graph and calibration table.

Further works are needed to calibrate with the quantitative measures and the "technical," "organizational," and "people" complexity graphs with the same style as the SIL assessment risk graphs and calibration tables.

5.16.2 Simultaneous Technical, Organizational, and People Complexity Management

A process manufacturing facility consists of many units, modules, and parts. Failure of each part at any phase of lifecycle may cause loss of containment (LOC). Safety systems are designed based on "one failure at the time." Multiple failures or domino effect of a LOC may lead to uncontrolled major accident. Fig. 5.74 illustrates the complexity of the safety barrier system for a typical process facility.

Designers use the codes, standards, and best practices to simplify the technical and organizational complexities of the plant system as much as possible. PSM calls for having a holistic view on the risk assessment but it does not recommend any practical method to implement it. Fig. 5.75 illustrates the horizontal and vertical interrelationship of a process manufacturing value chain.

The authors of this book strived to provide a practical solution to technical complexities of the risk-based approach by developing an innovative decision making tool called ADEPP (analysis and dynamic evaluation of project processes) in 1994. Fig. 5.76 illustrates the ADEPP HSE Toolkit (V1) graphic user interface for the dynamic consequence analysis such as evaluation of jet fire.

The first version of ADEPP has been developed on the WINSIM (today INDISS) software which is the operator training simulator product of a French company called RSI. Institute Français du Pétrole (IFP) welcomed this integration initiative and supported the development of this new concept. Version 1 of ADEPP was an object-oriented software package. The documents and data generated by design and risk studies were used to determine the HSE criticality of safety systems by bow-tie method (Figs. 5.77–5.79).

The GUI were designed and configured by the expert integrators who had the safety engineering and simulation skills. Users could navigate in the user-friendly GUIs and change authorized inputs to study the outcomes of them (Fig. 5.79).

A combination of generic event tree and flow logic was used to assess the contribution of any specific system to the annual fatality rate (AFR), damage

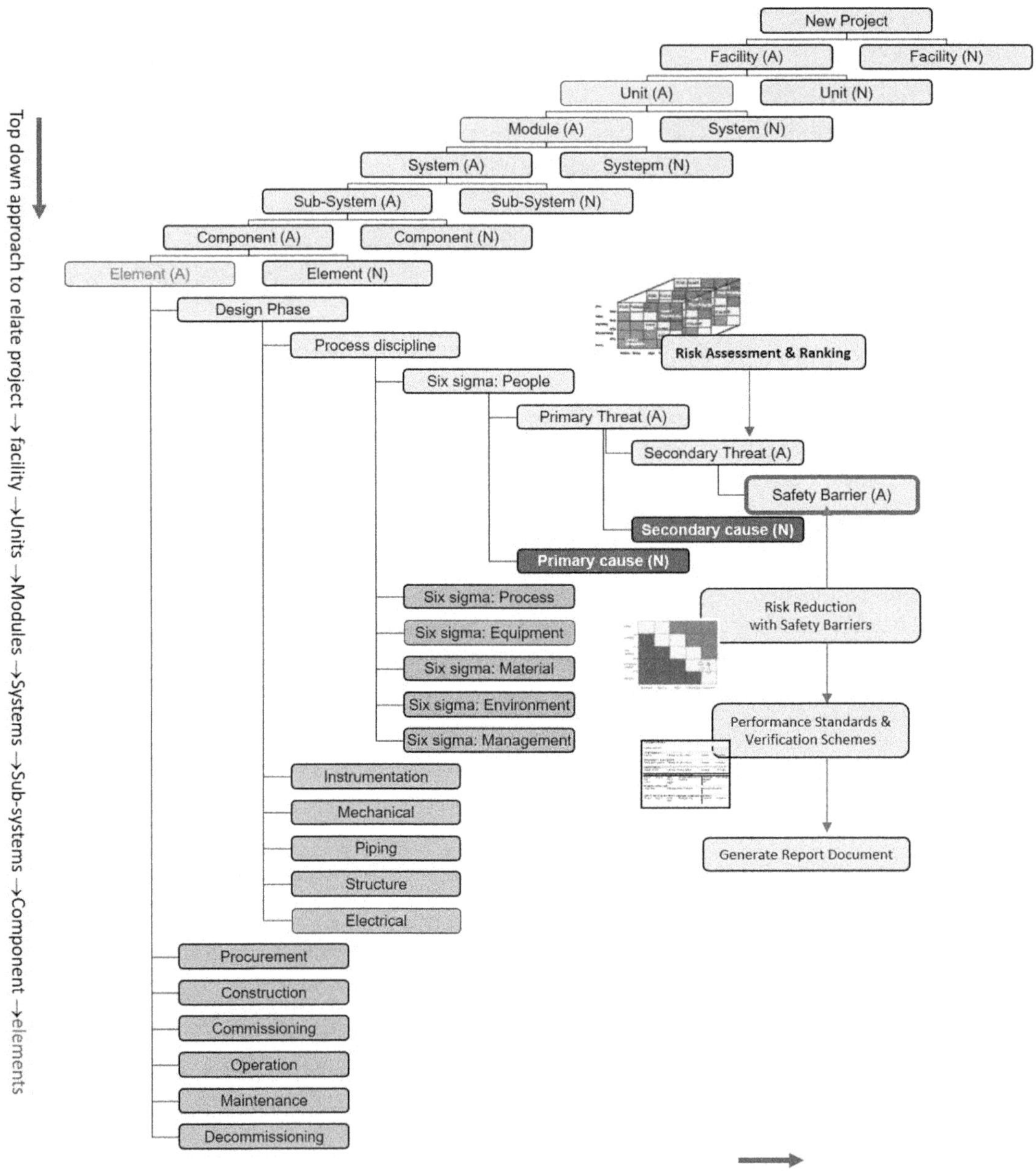

→project Phase → Discipline → Six sigma elements → Primary Causes → Secondary Causes → Failure identification →Consequence & Frequency Analysis & Ranking → Safety Barriers → Performance Standards → Verification Schemes

FIGURE 5.74
System thinking for a process plant.

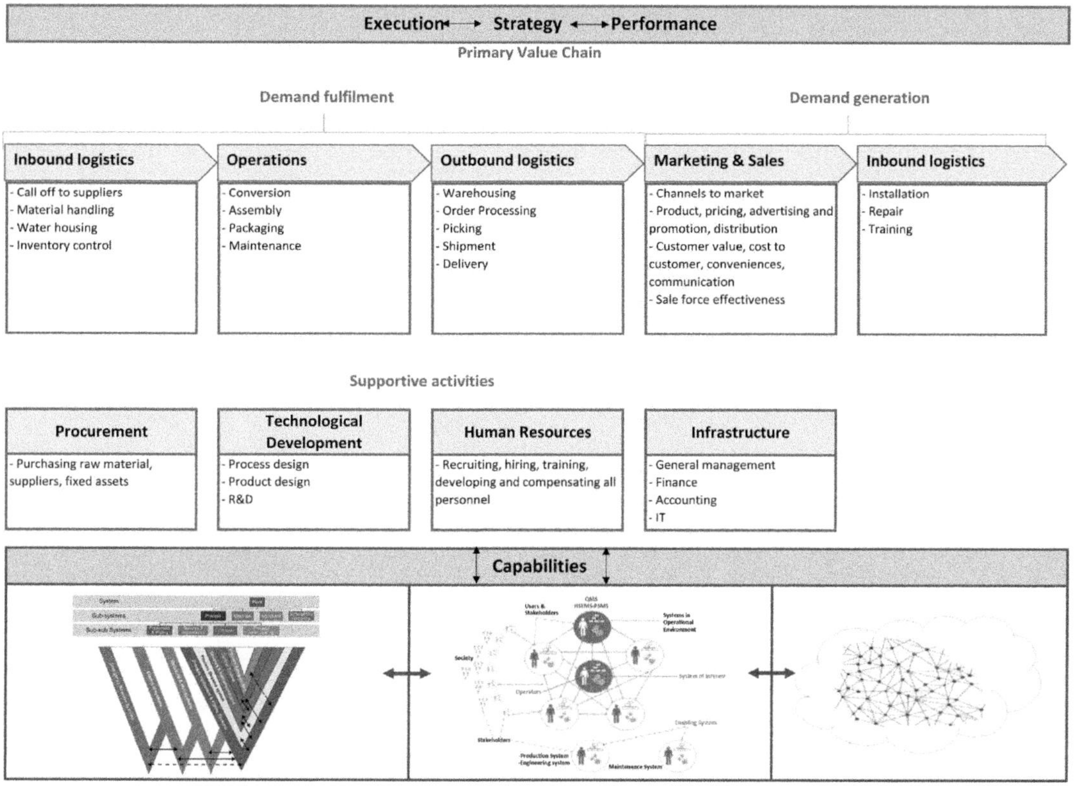

FIGURE 5.75

Horizontal and vertical interrelationship of a process manufacturing value chain.

cost, environmental impact for a range of hazardous events, e.g., gas release, jet fire, pool fire, explosion, etc. The event logic could be customized for a specific installation in closed collaboration between the integrators and client representative. The performance standards were produced for the safety critical elements.

The key features of ADEPP were as follows:

- Combined process dynamic simulation with consequence simulation of the leaks and jet and pool fires
- Combined event tree and fault tree analysis for determination of critical system and safety integrity level (SIL) assessment
- Dynamic information management and electronic HSE case

Although ADEPP (V1) was very well received by the technical safety professionals, it has never been commercialized for an industrial project because at

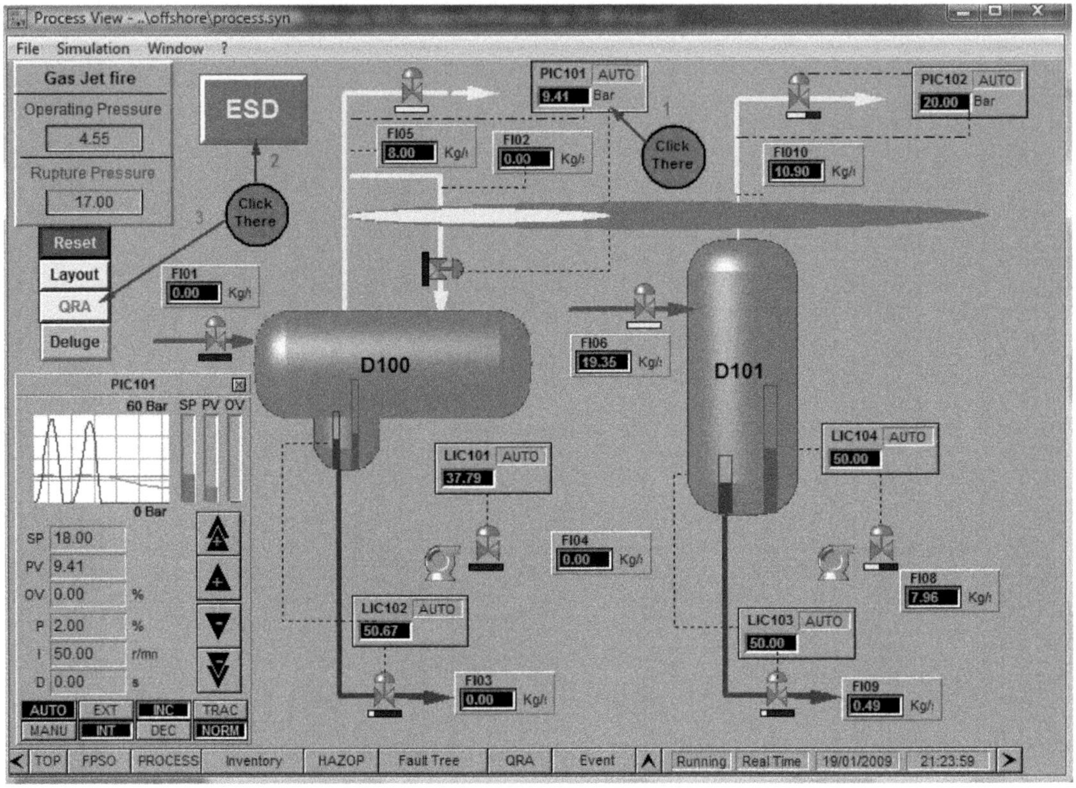

FIGURE 5.76
ADEPP HSE Toolkit (V1) graphic user interface for the dynamic consequence analysis such as evaluation of jet fire.

that time Internet functionalities were limited and considerable manual work was required to create and configure the GUIs. The integrations of the GUIs from the OTS, Open Source database, and 3D model platforms should be performed by the expert integrators. Users had the read-only access and had no power for changing the configuration of the models or GUIs. It was also a stand-alone desktop application which could not be accessible anywhere at any time.

The prototype of Version 2 of ADEPP has been successful and this can be applied for an industrial project. It is a web-based platform which connects the independent process dynamic simulations, PSM framework developed on SharePoint and RBI software thanks to OPC and Internet connections as shown in Fig. 5.80.

Applications developed on the SCADA and SharePoint platforms can become a template and be reused easily. Therefore, customization and development of the new projects can be done cost effectively.

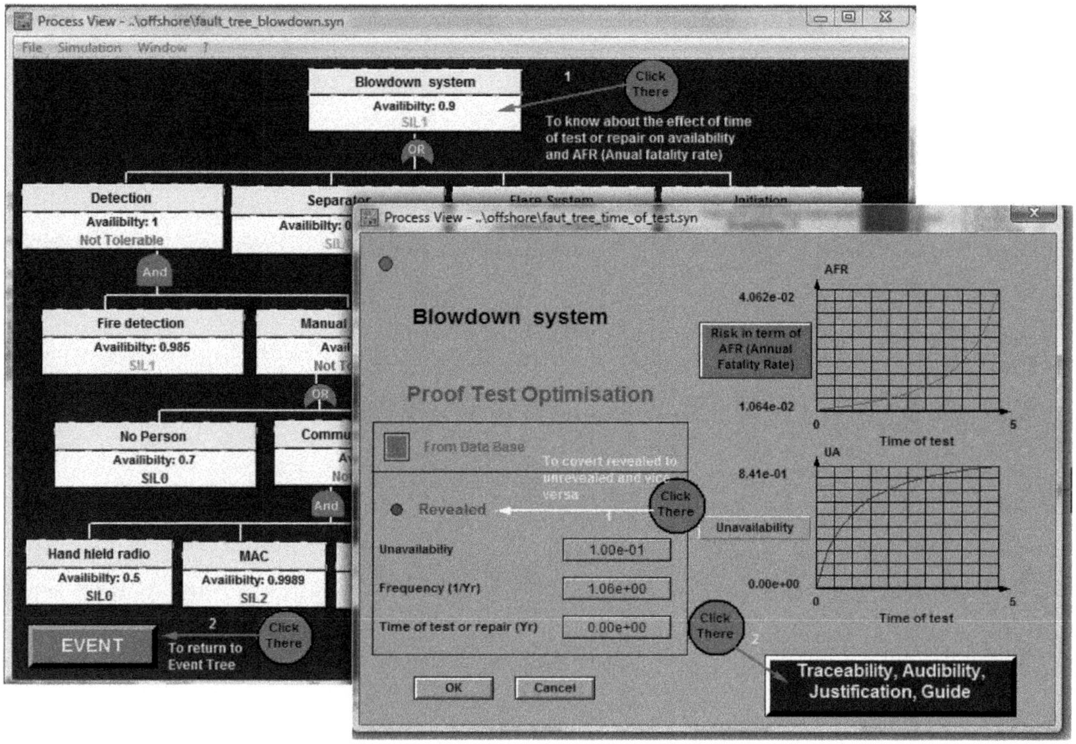

FIGURE 5.77

ADEPP HSE Toolkit (V1) graphic user interface for the SIL assessment.

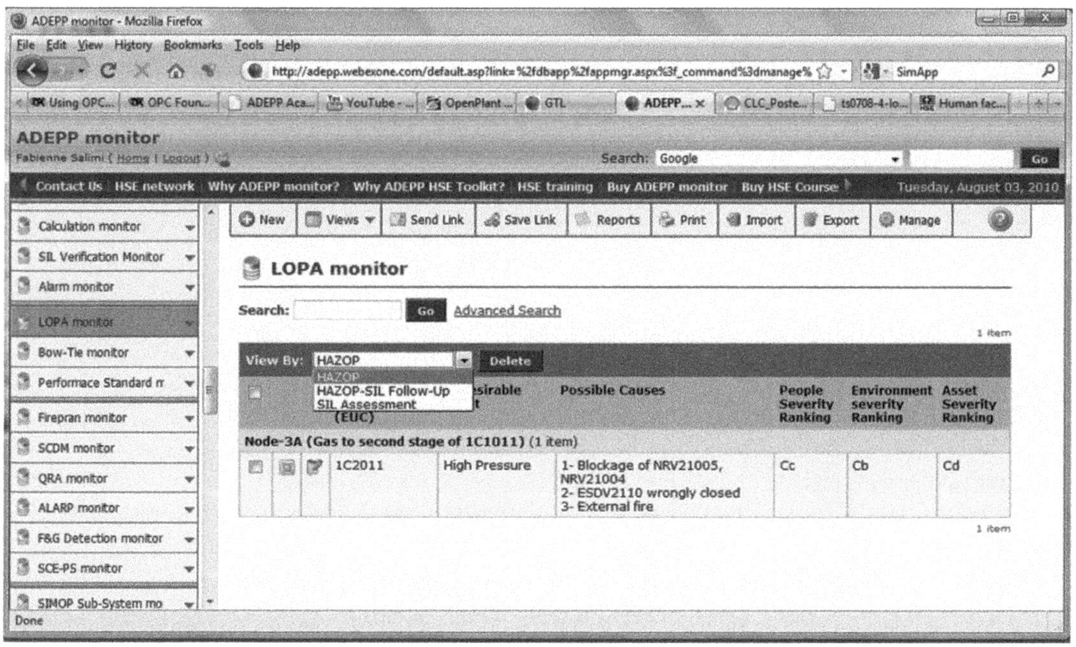

FIGURE 5.78

ADEPP monitors (V1) on Webexone database management.

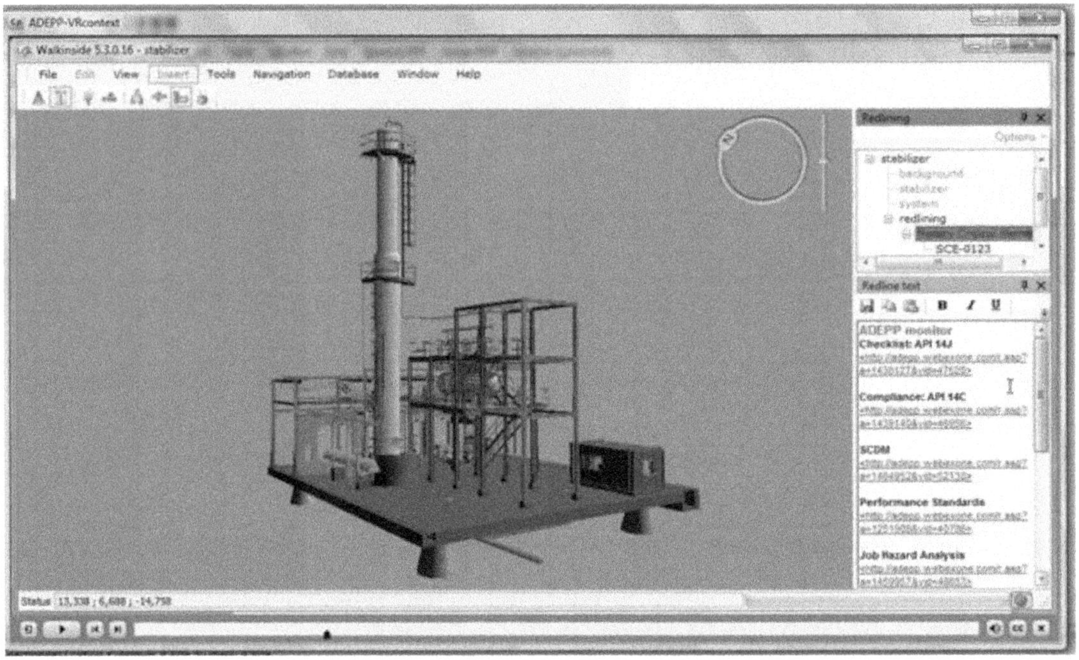

FIGURE 5.79
ADEPP 3D visualization on VRContext. The safety critical elements were marked up and linked with ADEPP monitor and ADEPP HSE Toolkit by "urls".

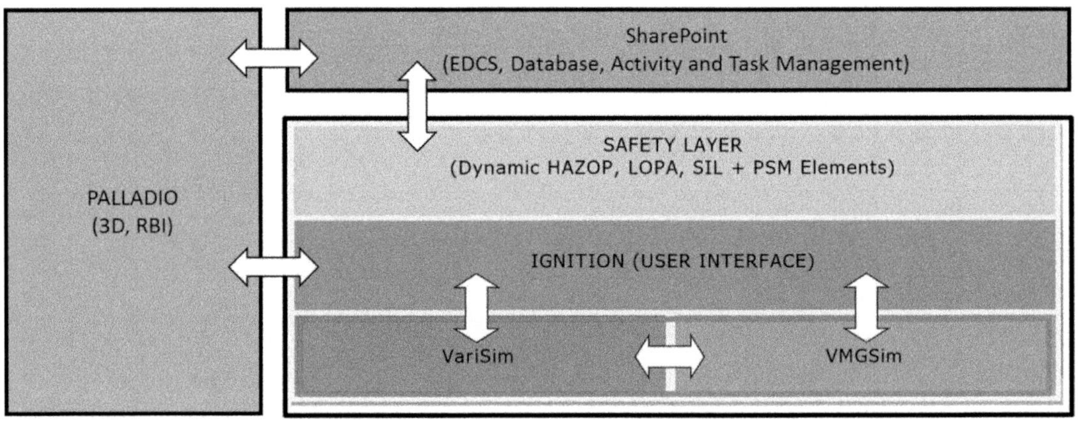

FIGURE 5.80
ADEPP integrated PSM-OTS-RBI architecture. The interoperability between the software is managed by OPC and Internet connections.

In ADEPP (V2) prototype the independent dynamic models of VMGSim and Varisim are successfully connected to simulate a LNG exportation plant. The common GUIs are developed on web-based SCADA platform of IGNITION, product of induction automation (see Fig. 5.81).

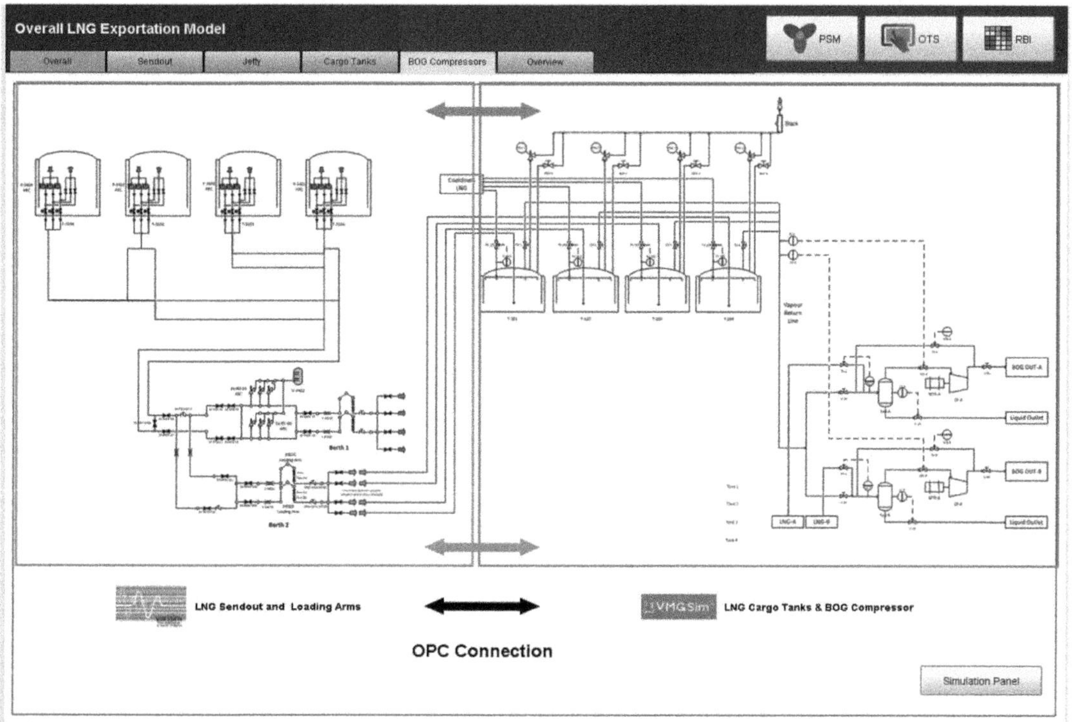

FIGURE 5.81

Varisim pipelines and loading arms and VMGSim BOG compressor are connected by OPC connection to simulate the ADEPP LNG integrated PSM-OTS-RBI platform.

The process safety models for dynamic HAZOP, LOPA alarm management, SIL assessment, leak, etc., are created on the SCADA platform. These models can interoperate with the combined process simulation by OPC connection. The user-friendly and institutive GUIs for these models are designed in compliance with ISA101, human–machine interfaces standard (Fig. 5.82).

ADEPP PSM framework is developed on the SharePoint cloud platform. The workflows, databases, and graphical interfaces of the 20 elements of PSM are designed with a system engineering vision covering all phases of the lifecycle of plant. The complexity is managed as follows:

1. *Technical complexity* by the structured, traceable, and auditable execution plans, disciplines activities, and deliverable breakdown, determination of the safety critical elements and management of them by the lifecycle performance standards and verification schemes.

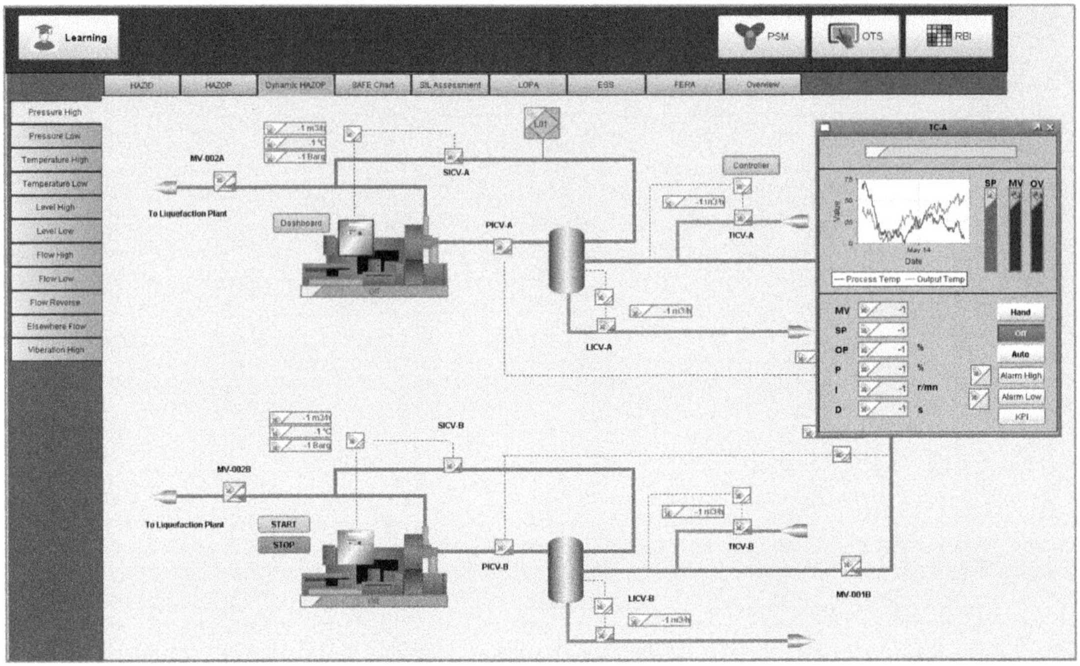

FIGURE 5.82
Dynamic HAZOP on the ADEPP LNG integrated PSM-OTS-RBI platform HAZOP attendees can simulate any failure scenario quickly to have an objective view.

In other words, we manage the technical complexities by *controlled simplification*.

2. *Organizational complexity* by applying the Energy Institute logic flow diagrams, workflows, KPI, and task management.

   In other words, we manage the organizational complexities by combination of the *controlled simplification* and the *self-organization technique*.

3. *People complexity* by engaging the workforce in the collaborative training programs, discussions, problem solving, and incident/nonconformance reporting.

In other words, we manage the people complexities by *self-organization technique* (Fig. 5.83).

Technical and organizational systems are designed by the designers. With a good system engineering practices we should be able to keep the complexity at "complicated or simple" levels. Interactive GUIs and simulations enable to democratize the system and complexity thinking throughout the organization (Fig. 5.84).

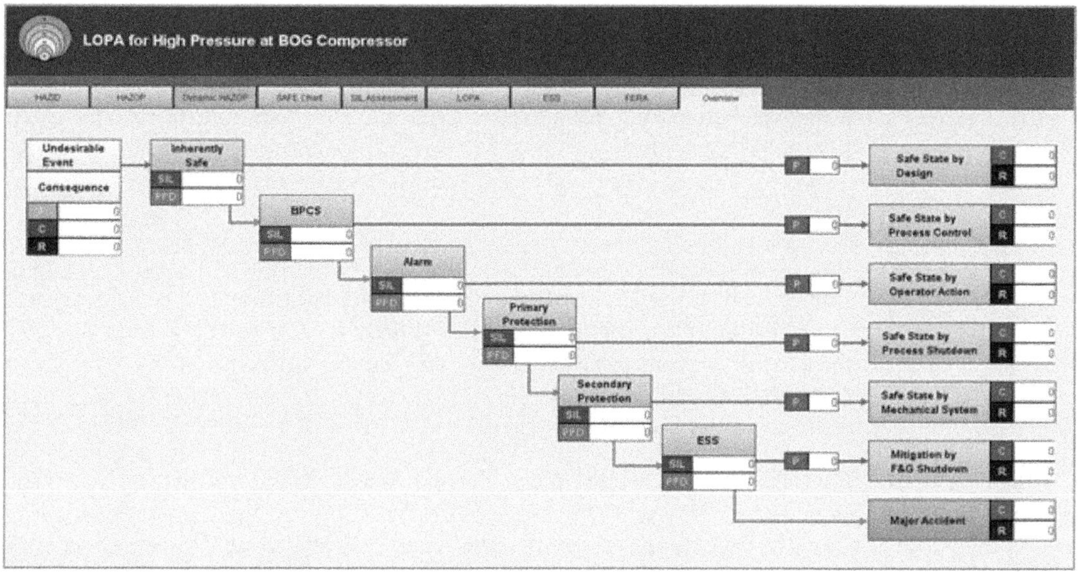

FIGURE 5.83
LOPA analysis for the BOG compressor high-pressure scenario identified in dynamic HAZOP.

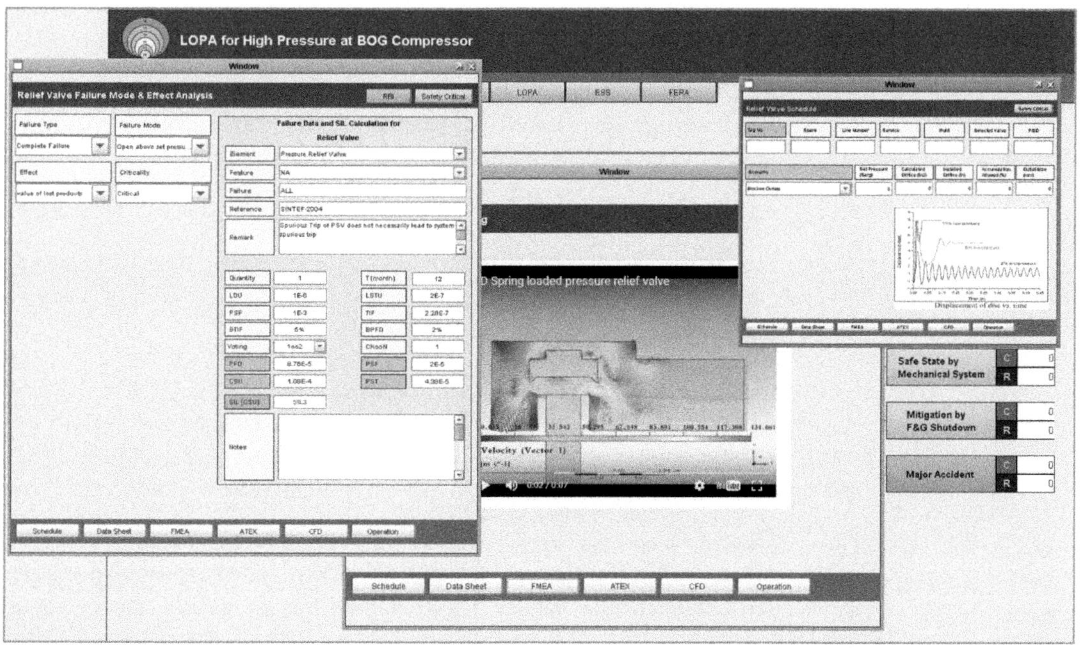

FIGURE 5.84
Studies for the safety barriers are configured in LOPA upon the company policy permission, can be given to the user to interact with the GUIs.

Human is a complex system by its very nature. Analogy between a human and a robot will be an unforgivable misjudgment. People will be engaged in their job and interact with their colleagues with a culture of respect and care if they feel that they are genuinely valued and their contribution is seen and matter. Deployment of a self-organized collaborative platform encourage them to learn every day and continuously improve their performance (Fig. 5.85).

The hierarchy and modular characteristics of SharePoint sites and subsites enable to link the PSMs of two distinct organizations. For example, the PSMs of shore facilities and ship can be linked to assure the robust and reliable ship-shore connection and disconnection procedures as is shown in Fig. 5.86. The reports from these operations can instantly be shared by shore, ship, and regulatory authorities.

The relevant SharePoint pages can be easily embedded in the GUI. User can work on PSM or OTS space just by moving their mouse position. In Fig. 5.87, when users are in PSM zone, everything will be managed through SharePoint and when they move their curser to OTS zone they can perform the dynamic simulations and navigate in the OTS GUIs.

The integrated OTS-PSM dynamic model and simulation is not only used for the operator training purposes but can also be applied to enhance the precision and exactitude of the safety related studies and activities listed in Table 5.14.

At present, the RBI calculation and reports of Palladio are accessible on the stand-alone computer. Therefore, we installed it on a virtual machine and integrated it in the GUI using the Internet connection (Fig. 5.88).

5.16.3 Self-Organization to Manage the People Complexity

A picture is worth a thousand words and a product visual is critical when product information is being shared across the enterprise.

Personnel learn informally and without pressure by engaging with the interactive GUIs with embedded multimedia, animations, and simulations relevant to their day-to-day tasks. The system and complexity thinking can be promoted by online exercises relevant to the day-to-day tasks and decision-making.

With this powerful informal learning platform, learner will be capable to share their experience, discuss a subject, and even define the collaborative projects. Informal learning will stimulate the sense of observation of the individuals, sharing the findings and knowledge with others and desire for continuous improvement. Management can also configure the incentive

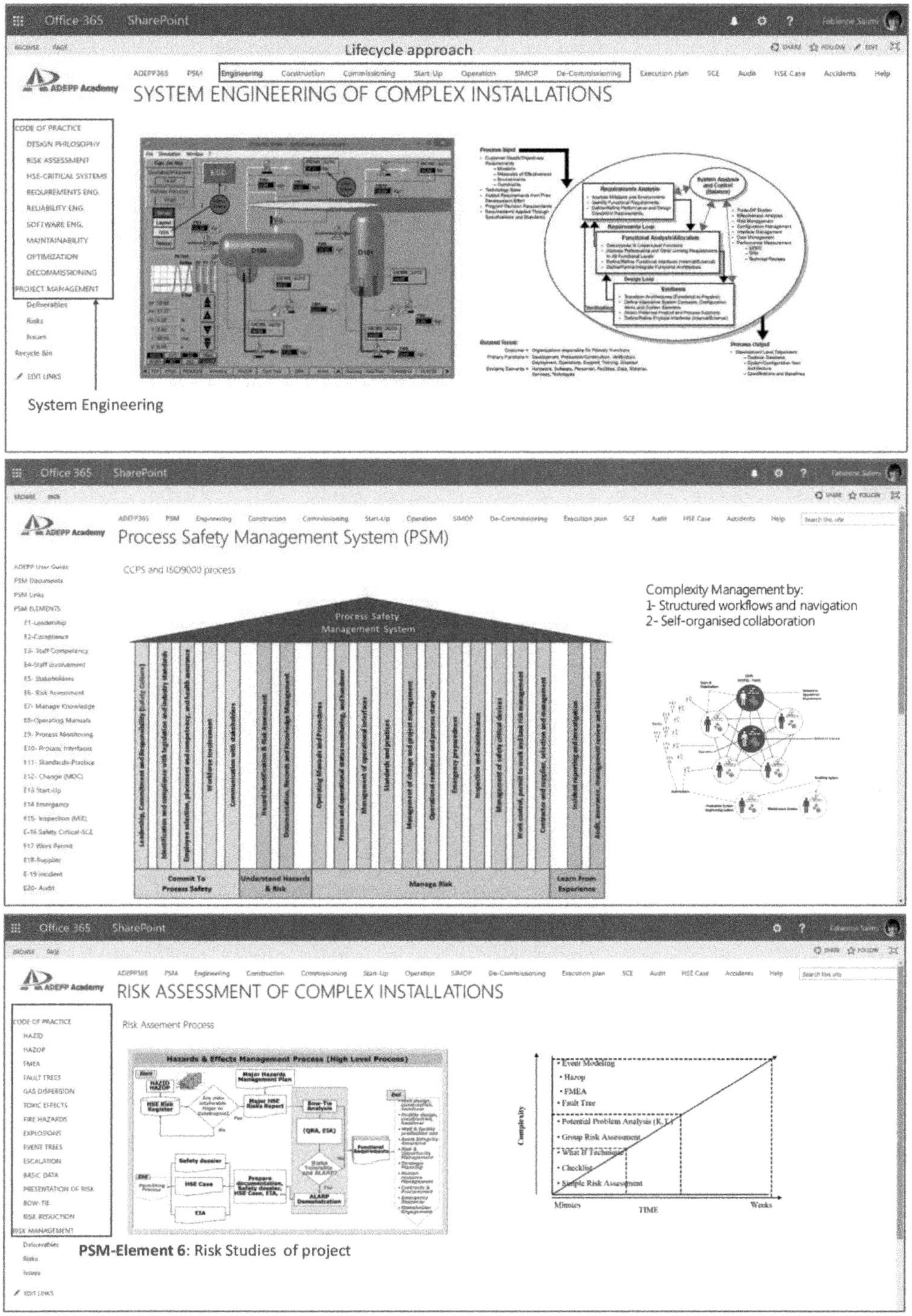

FIGURE 5.85

ADEPP PSM framework on SharePoint encompass the system engineering for lifecycle of plant.

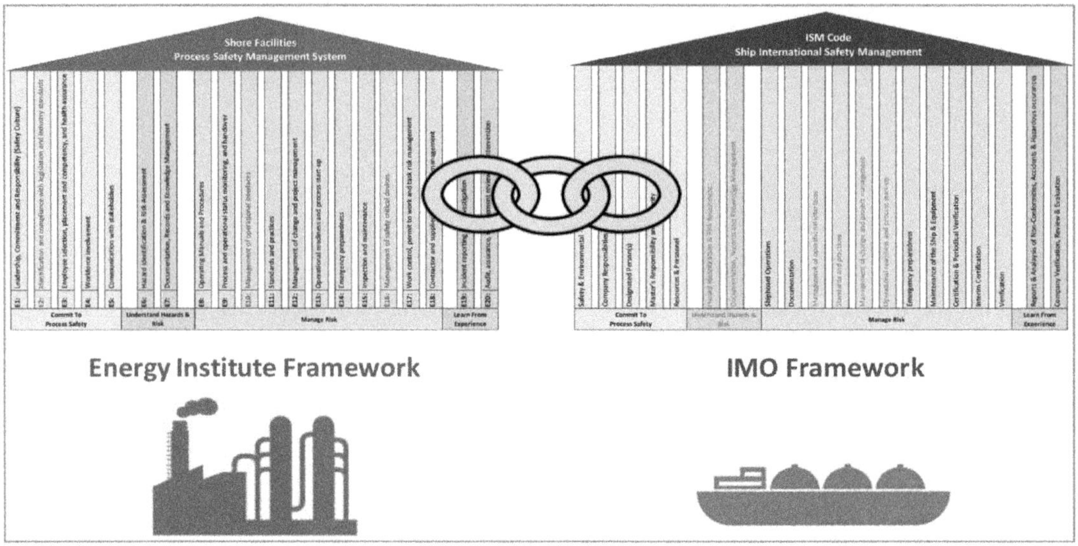

FIGURE 5.86
The shore and ship PSMs frameworks can be linked on SharePoint.

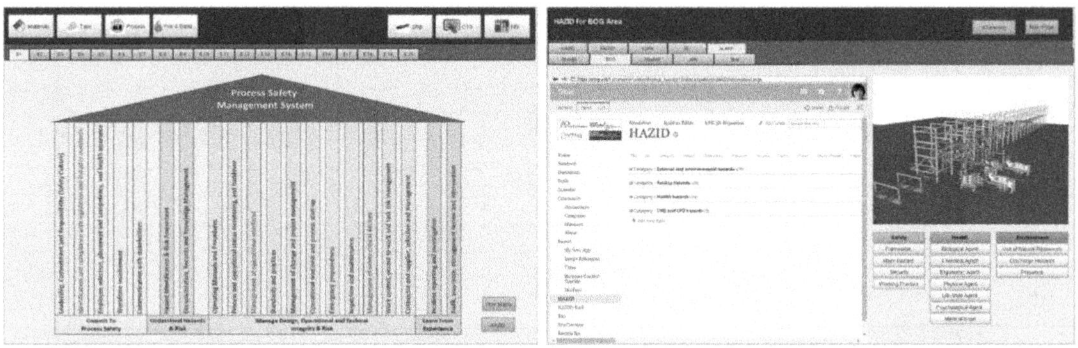

FIGURE 5.87
The PSM workflows, databases, and GUI developed on SharePoint are embedded in the SCADA platform.

initiatives for the quality and quantity of the informal learnings to boost the knowledge management system of the organization (Fig. 5.89).

Informal learning is a pervasive ongoing phenomenon of learning via participation or learning via knowledge creation, in contrast with the traditional view of teacher-centered learning via knowledge acquisition.

Personnel can organize their learning plan by their own free will and without pressure alongside the mandatory learning plan of company. This sense of freedom will boost the creativity and enthusiasm for continuous improvement of their competency and innovation capabilities.

Table 5.14 OTS Added Values to the Safety Activities

Activity	OTS Added Value
Dynamic HAZOP process start-up and shutdown review	Consequences of the: • Single failure such as control loop failure • Multiple failures • Domino effect • The quantified deviation levels can also be applied
Dynamic SIL assessment cause-and-effect matrix review	Consequences of the: • Failure on demand of safety functions • Spurious function of safety systems
Process control strategy	• Effect of upstream and downstream process • Tuning review
Alarm management	Process and safety alarm rationalization
Relief and blowdown strategy review	
Fire and explosion study	Consequence of release in terms of: • Extent of gas dispersion and flame length • Duration of release with ESD and blowdown

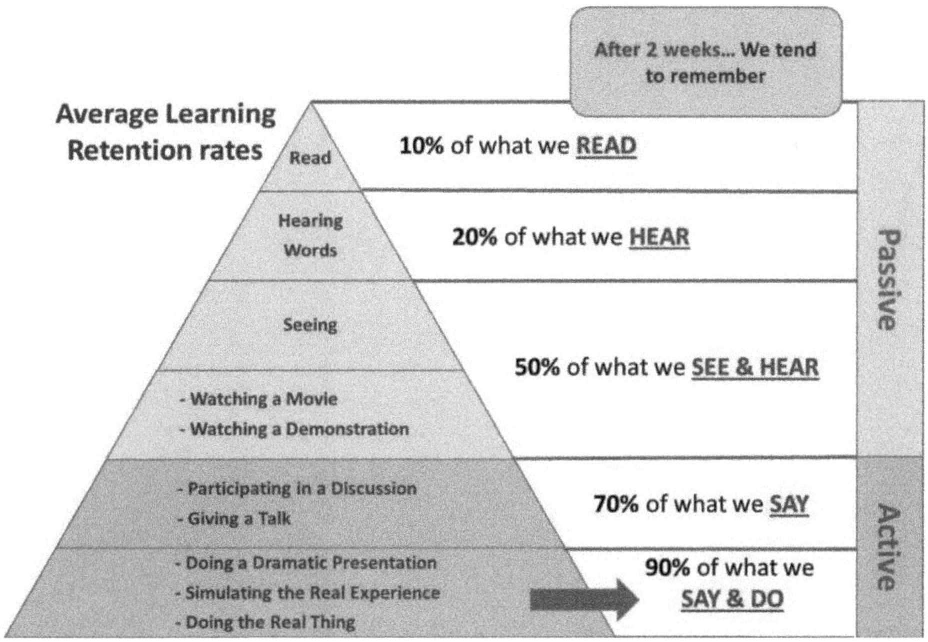

FIGURE 5.88
Average trainee retention rates *Source: Edgar, Dale, National Training Laboratories Institute.*

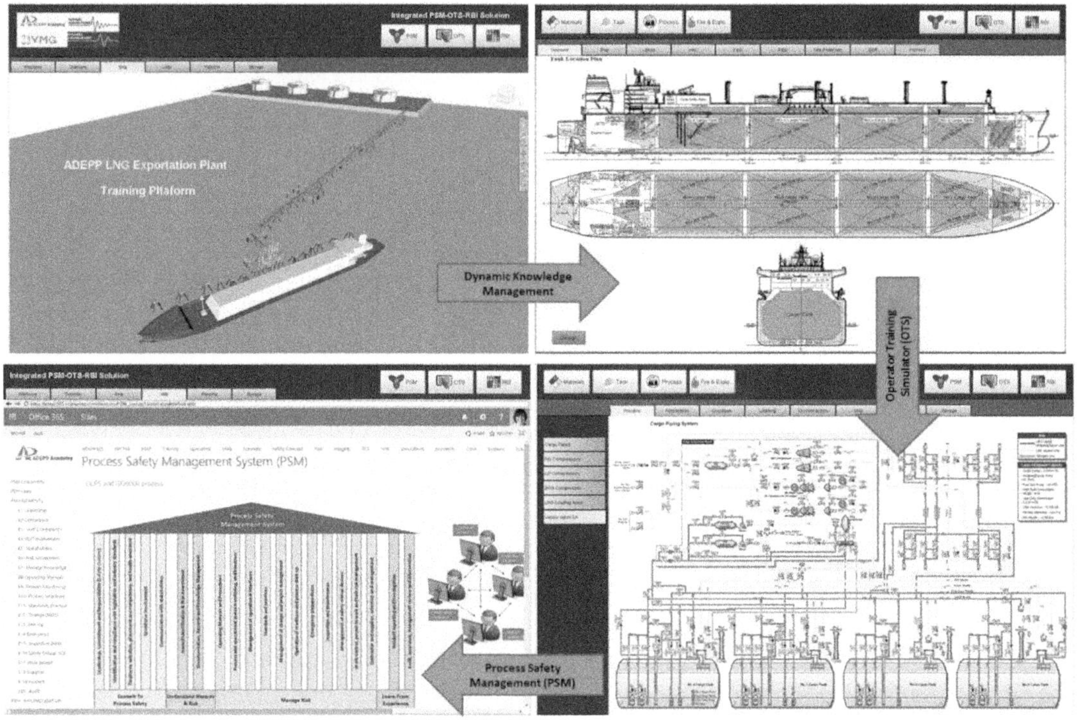

FIGURE 5.89

Example of the web-based interfaces which link the "dynamic knowledge management" with "operator training simulator (OTS)" and "process safety management (PSM)".

Their efforts can be recognized and rewarded by issuing the online certificates and incentive measures of the human resource managers.

Multiple and independent learning platforms can be configured in ADEPP PSM-OTS-RBI platform. Fig. 5.90 illustrates how the ADEPP Academy process safety e-learning platforms can be used alongside the human focus occupational health and safety e-learning site.

For example, for a pig launcher isolation job, operators can learn about the outcome of the wrong isolation sequence or failure of the safety critical system using the ADEPP simulations and e-courses. For the same job, operators use the human focus e-learning site to learn how to work at explosive and toxic atmosphere of sour gas and get the required IOSH certificates to comply with the PTW requirements.

Process manufacturing industries such as oil and gas, and refining and power companies must institutionalize their workforce knowledge in efficient and effective ways. Leveraging virtual reality models to improve time-to-competency

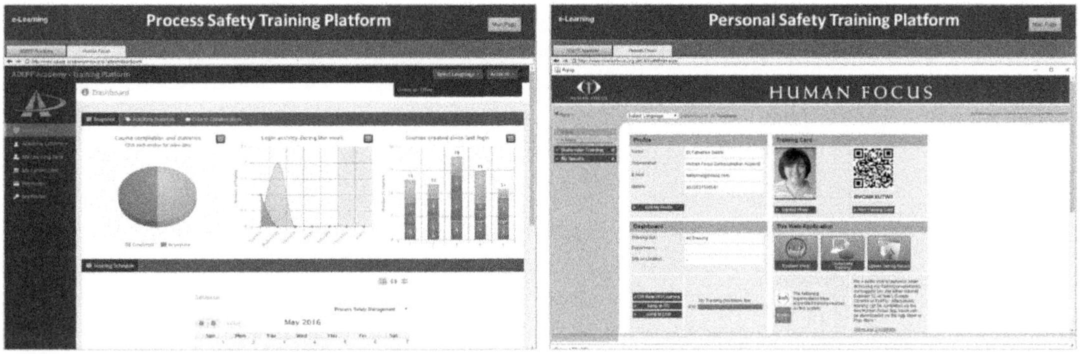

FIGURE 5.90
Integrated PSM-OTS and e-learning platforms.

in critical areas such as safety and environment protection systems, knowledge and performance training, and reliability provide a vehicle to rapidly train this new workforce in ways that align with their interests and skills.

Today, with continuing advances in hardware and software techniques, VR is viewed as the best aid to improving multimedia training, process design, maintenance, and safety, which are currently based around conventional 2D equipment views.

The real-time rendering of equipment views places demands on processor time, with the use of high-fidelity simulators becoming the standard in process understanding and training. For many past VR commercial projects, the results were unrealistically slow or oversimplified to the detriment of the solution effectiveness. Today, as technology evolves, these will no longer be an issue with the new process simulation era.

Virtual reality is a rapidly growing technology that utilizes the increased power of computers to simulate real or imaginary environments and situations with a high degree of realism and interactivity. It is an emerging technology with potential applications in areas such as product design and modeling; process simulation; planning, testing and verification; real-time shop-floor controls; and training and maintenance.

Currently, VR technology is used for training applications in a variety of process industries, and offers the potential to expose personnel to simulated hazardous situations in a safe, highly visual and interactive way. Customized simulations of chemical plants layouts, dynamic process operations, and comprehensive virtual environments can be set up and allow users to move within the virtual plants, making operational decisions and investigating processes at a glance. The consequences of correct and incorrect decisions are sent immediately back to the trainees, giving them the opportunity to directly learn from their mistakes.

Users can interact with the virtual worlds using a variety of hardware devices such as joysticks and data gloves. Special optical and audio devices (such as head-mounted displays, three-dimensional graphics, and surround sound) allow users an enhanced impression of being in the virtual world.

The overall aim in developing an immersive virtual reality plant (IVRP) is the development of novel training techniques that improve the operation efficiency and skills of chemical plant personnel. This includes the development of a large range of training scenarios for application in the chemical process industry. Using IVRP will increase process understanding, readiness, safety awareness, and knowledge of safety procedures, improving production and reducing the plant accident rate.

Augmented reality

Augmented reality is a combination of virtual and graphic three-dimensional images that are transmitted to a user. Augmented reality differs from virtual reality in overlaying images onto the real (virtual) life, engaging the user in an immersive, interactive, and three-dimensional "augmented" environment.

The external environment represents the real (virtual) world and can be linked to additional information systems. For example, in a medical application, it is feasible to overlay a sketch of human organs onto a real (virtual) patient via an optical system. In order to achieve accurate alignment of the real/virtual environments and multimedia data source, it is important to provide the most appropriate link among the display technology, the simulation environment, and the information processing systems. All VR scenarios and applications must relate to the external data source either as simulated or as field data. In fact, the value of VR tools and related benefits for training purposes are strictly related to the "action/reaction" feeling of the trainees, as well as the availability of additional information and data that provide a "supernatural" sense of power and understanding of the process.

By superimposing virtual images, videos, or text onto real (virtual) life, an experience can be heightened or even modified.

Fig. 5.91 shows an example of a trend diagram, which can be activated or deactivated by the user with a simple touch on the hand device at any time during the plant walkthrough or a task procedure. Variables and trends can be selected and customized by type (such as temperature, pressure, and flow) and equipment exactly displayed in the control rooms. The data source will be identical, simulation engine, or real-time database. Alternatively, a different type of augmented reality is possible, where equipment units become transparent and the information related is shown as a video animation of the process behavior.

FIGURE 5.91
Augmented reality—example of a trend popup in the virtual field.

Emergency, risk, and safety

There have been a number of well-publicized accidents in which human error has played a prominent role.

In order to understand the contribution of human behavior to the risk of accidents, it is essential to examine the errors people make and what leads to such errors. The reduction of human error probability can lead to a reduction in the probability of accidents in process industries. A useful classification framework identifies the human errors as slips/lapses, mistakes, or violations. Slips or lapses typically occur through lack of attention or from stressful situations, with the result that individuals fail to achieve what they intended. Slip or lapse human errors include forgetting to turn off power, becoming inattentive to important safety checks, or forgetting important steps in safety procedures, which may cause equipment damage or loss of life.

Training has always been considered an important factor in staying competitive in a global economy. Operators must remain up to date with the latest methods and technology. Training should also involve an introduction to basic hazards and plant safety procedures incorporating fire alarm systems and detailed work safety processes. Therefore, there is a need to create hazardous simulated conditions to provide a real-world (virtual) situation for training without causing harm to the trainee and the environment.

IVRP systems allow users to navigate in any direction within a computer-generated environment, creating crisis conditions, deciding what actions to take and to immediately see the impact of those actions. IVRP also allows trainees to walk throughout the plant, observe all the equipment that constitutes the process, and to have the possibility of starting, running, and shutting down equipment and responding to error conditions.

In theory, every abnormal situation an operator can imagine but never test in reality can now be tested, helping the operator understand atypical plant behaviors. Expected and predictable malfunctions can be tested and forced until the accidental sequence results in a disaster. Learning from a virtual disaster can help avoid a real disaster. The main goal is that safety can now be tested and experimented with for training, and to help risk assessors better identify hazardous scenarios. Most importantly, IVRP can aid operators in making the right decisions in time (Fig. 5.92).

Maintenance and reliability

Plant maintenance is a best practice that requires a team-based approach, where operators perform various equipment maintenance activities and maintenance crews work closely in the daily operation of equipment.

To understand requirements for maintenance training, the first logical step is to review the task the trainee is expected to perform and the outcome of that training. For example, in process operations, the nature of the maintenance task is heavily dependent on the organization's industry sector, range of equipment maintained, and specific culture.

Irrespective of the subject matter, maintenance tasks generally can be broken down into the following subtasks:

- Replication: reproducing the reported fault;
- Identification: accurately diagnosing the source of the fault;
- Rectification: correcting the fault by taking the appropriate action based on the policies of the maintenance establishment;
- Confirmation: verifying that the identified fault has been cleared.

Each of the four stages described above requires a mixture of generic and specific physical and mental skills. Using VR facilities, maintenance operators

FIGURE 5.92
Augmented reality highlighting a dangerous area around a virtual fire.

can be trained to have a full understanding of the maintenance task and the science behind the equipment they are dealing with.

Inherent in VR models is the ability to understand the geometry, layout, and limits of process units and their supporting utility infrastructure. Using a physical representation of an operator or maintenance personnel (avatar), IVRP can be used to optimize gauging and inspection rounds by imitating the operational and maintenance staff behavior.

This is primarily a spatial system analysis where working spaces, escape routes, risky areas, and transportation routes in the plant could be investigated from a logistical point of view. The resulting analysis can be used to optimize maintenance procedures, or make requests to maintenance management for improvements or modifications.

With the aim of preventing production losses, maintenance during operation has high priority in such facilities due to the commercial impact. Therefore, to verify that maintenance is properly conducted and performed on time, maintenance personnel are asked to participate in maintenance training using avatars or in "first person" through VR models of the process machinery and layout. Maintenance issues on diagnostics, timing, and procedures are highlighted and optimized by interactive links to the virtual equipment, as well as to a computer-integrated maintenance and documentation management system. Fault-diagnostic training and rectification training can therefore be based on a comprehensive multimedia tool that transfers knowledge and skills.

Furthermore, model areas can be color-coded to represent areas of the plant that require inspection, as well as establish safety and integrity risk boundaries.

Benefits and value proposition

IVRP can improve design procedures, saves staff time and money in maintenance, and is a superior way to train operations personnel by:

- the provision of more realistic training environments for trainees
- the provision of opportunities for practice in training sessions
- training staff to react quickly and correctly in high-stress situations
- improving skills for rarely performed, but safety-critical tasks (such as emergency shutdown)
- optimizing the transfer of skills from the training environment to the work environment
- reliable and valid evaluation of operational procedures and performance
- team training for the control room, field, shift, operational and safety managers.

Typical return on investment on an IVRP is expected to be more than 50%. The technology's main economic benefits fall within the following main categories:

- *On-the-job training*: Saving 30%–40% on time and costs for new large personnel requirements, familiarization in case of mobility from plant to plant and scheduled replacement of the ageing workforce.
- *Start-up efficiency*: Saving 15%–25% of the time to be "back-on-run" and optimal production for planned or unplanned shutdown, while frequently having all plant crew refreshed on all critical procedures.
- *Maintenance*: Saving 1%–3% on maintenance budgets by improving maintenance operator training and by using IVRP equipment preanalysis for predictive maintenance.

Trainers and trainees can be located throughout the world, and using web facilities, remote workers can work together in the same virtual area on a chemical plant or a building site, resulting in a reduction in travel costs. Interactive programs with multiple sites networked for group learning and communication could also be developed.

5.17 CONCLUSION (CONCLUSION OF THE BOOK)

Process safety management system cannot be designed and implemented without having a deep knowledge and skills for system engineering, system thinking, and complexity thinking.

In the last 25 years, discrete manufacturing industries were focused on development of the interoperability rather than integration methods and tools. The process industry has followed the same trend but these top-level methods have not been communicated and democratized systematically. Therefore, the project managers and discipline leaders during EPC and O&M feel disconnected from the technical safety practitioners and new code of practice about safety critical systems.

The advanced IT and connectivity technologies provide the tools, the standards like ISA-95, ISO-15288, ISO-15926, and ISO-14224 provide the methods and, the nonprofit open source professional organizations such as "INCOSE," "OMG," "Standards Leadership Council (SLC)," "MIMOS," and "Open O&M" provide the required collaboration platform to develop the "ILM" for process manufacturing facilities.

Fig. 5.93 illustrates *OIIE/OGI standard scenarios* which summarize the "interoperability" of "systems" and the "management systems" during lifecycle of process manufacturing plant. Operational excellence including PSM can be achieved if configuration is properly designed and implemented.

FIGURE 5.93

Functions and associated hazards of an equipment can be experienced safely in the virtual field environment.

FIGURE 5.94
Helicopter landing hazards can be studied to optimise the helideck design.

Thanks to joint venture of ADEPP with Varism, Varisim, and Palladio, the prototype of the "ADEPP integrated PSM-OTS-RBI" demonstrates that with a reasonable effort, time, and cost:

1. System engineering, system thinking, and complexity thinking can be implemented effectively for the blocks highlighted in blue with SharePoint.
2. System engineering, system thinking, and complexity thinking can be implemented effectively for the blocks highlighted in red with combination of web-based SCADA platform, SharePoint system engineering, and 3D RBI software.

   At present we use the simulation mode of the SCADA platform. This system can also be connected to the sensors and simulated with real-time data.
3. Connection of all blocks can be achieved with a suitable "MBSE + ALM (IIoT) + ILM (PLM)" platform similar to PTC MBSE shown in Fig. 5.94.

At this stage, a considerable communication and training efforts are required to populate these findings and accompany the businesses in their journey toward the Industry 4.0 revolution.

Literature

1. Evolution of the World Wide Web from Web 1.0 to Web 4.0
2. Established Standards Pave the Way for Industrie 4.0
3. Lifecycle Management Forecasted for the FDT IIoT Cloud
4. Guide to IOT Solution Development
5. Human Factors Engineering Delivers ROI

6. Using ISA/IEC 62443 Standards to Improve Control System Security
7. Cybersecurity Capability Maturity Model
8. The Data Governance Maturity Model
9. AVEVA White Paper, Reaching Level Five, Defining a Digital Asset Strategy by Means of a Maturity Model—A Roadmap Tool for Owner Operators, September 2016
10. Utilizing MBSE Patterns to Accelerate System Verification
11. Survey of Model-Based Systems Engineering (MBSE) Methodologies
12. Murray, J., 2 May 2012. Model-Based Systems Engineering (MBSE) Media Study
13. Model-Based Approach to Automated Hazard identification of Chemical Plants
14. PCA-MIMOSA Reference Architecture Framework for Integrated Engineering and Operations
15. Extending and Automating a Systems Theoretic Hazard Analysis for Requirements Generation and Analysis
16. Industry 4.0: Intelligent and Flexible Production
17. The Democratization of CFD
18. Cybersecurity in the SIS World
19. D.W. Humphrey, Future of Manufacturing: The Value of Integrated Industry Software, ARC View, 16 February 2016.
20. The Evolution of Integrated Drive Systems ARC White Paper, The Evolution of Integrated Drive Systems, June 2016
21. Understanding Process Safety Management

Blog

1. Industry 4.0: The Fourth Industrial Revolution – Guide to Industrie 4.0
2. Internet of Things: The Complete Iot Guide – Benefits, Risks, Examples, Trends
3. The Difference Between IIoT Platforms and IIoT Ecosystem Partners
4. Industrial Cyber-Security (ICS): A Perfect Storm is Brewing
5. The Immersive Virtual Reality Plant
6. Use a "Digital Twin" to Ensure Products are Built to Design
7. Transition from Data to Wisdom

Handbook

1. INCOSE Handbook-2002
2. INCOSE Handbook-2006
3. INCOSE Handbook-2012
4. INCOSE Handbook-2015
5. Guide to the Systems Engineering Body of Knowledge (SEBoK)
    - SEBoK v. 1.7
    - SEBoK v. 1.6
    - SEBoK v. 1.5.1
    - SEBoK v. 1.5.
    - SEBoK v. 1.4
    - SEBoK v. 1.3.1
    - SEBoK v. 1.3

- SEBoK v. 1.2
- SEBoK v. 1.1
- SEBoK v. 1.0

Standards

1. ISA 99/IEC 62443: Industrial Automation and Control Systems Security
2. ISO 15926: Industrial Automation Systems and Integration - Integration of Life-Cycle Data for Process Plants Including Oil and Gas Production Facilities
3. ISO 14224: Petroleum, Petrochemical and Natural Gas Industries- Collection and Exchange of Reliability and Maintenance Data for Equipment
4. CFIHOS: Capital Facilities Information HandOver Specification (For Process)
5. ODATA APIs: The Open Data Protocol (OData)
6. ISO/NP TS 18101: Oil and Gas Asset Management and Operations and Maintenance Interoperability (OGI)

Guidance

1. Guidance on Meeting Expectations of EI Process Safety Management Framework Including:

Element no.	Title	Status
1	Leadership, commitment and responsibility	Available here
2	Identification and compliance with legislation and industry standards	Available here
3	Employee selection, placement and competency, and health assurance	Available here
4	Workforce involvement	Available here
5	Communication with stakeholders	Available here
6	Hazard identification and risk assessment	Available here
7	Documentation, records and knowledge management	Available here
8	Operating manuals and procedures	Available here
9	Process and operational status monitoring, and handover	Available here
10	Management of operational interfaces	Available here
11	Standards and practices	Available here
12	Management of change and project management	Available here
13	Operational readiness and process start-up	Available here
14	Emergency preparedness	Available here
15	Inspection and maintenance	Available here
16	Management of safety critical devices	Available here
17	Work control, permit to work and task risk management	Available here
18	Contractor and supplier, selection and management	Available here
19	Incident reporting and investigation	Available here
20	Audit, assurance, management review and intervention	Available here

Software

1. Predix (GE): The Industrial Internet Platform
2. Mindsphere (Siemens)
3. ThingWorx (PCT)
4. PTC MBSE

Tutorials & Learning Materials (All Chapters)

1. Complexity Labs
2. CIMdata: PLM Leadership Tutorials
3. Engineering.com
4. Kepware Video Centre
5. Automation.com
6. Control Global Knowledge Centers
7. LNS Research - Blogs
8. i-SCOOP
9. IOT Analytics
10. ISA Training
11. INCOSE
12. Future Learning: A Review Of Common Characteristics of Complex Systems

Associations

1. IISE: Institute of Industrial System Engineers (Process Industry Devision)
2. INCOSE: International Council on Systems Engineering
3. CCPS: Center of Chemical Process Safety
4. IChemE: Institute of Chemical Engineers
5. ISA: International Society of Automation
6. ISACA: Information Systems Audit and Control Association
7. MIMOSA: An Operations and Maintenance Information Open System Alliance
8. OIIE: Open Industrial Interoperability Ecosystem™
9. OGI: Oil and Gas Interoperability
10. SLC: Standards Leadership Council
11. REA: Resilience Engineering Association
12. IOGP: International Oil & Gas Producers

Industrial Internet of Things (IIoT) Glossary

Asynchronous Message Queuing Protocol (AMQP) A protocol primarily used to exchange data between gateway technologies in the cloud. It is a connection-oriented protocol over TCP/IP. It supports store and forward and embraces pub/sub data flows from the source to the destination.

Constrained Application Protocol (CoAP) A request/response (or master/slave) protocol intended for use in simple devices for communication over the Internet. It is lightweight, and uses hypertext transfer protocol (HTTP) over user datagram protocol (UDP).

Edge The part of the network that bridges the gap between information technology and operation technology, where the rich resources available in the cloud are not directly available.

Industrial Internet of Things The concept of connecting all hardware and software components within an industrial environment for complete visibility into operations at any time of day and from any location.

Information Technology (IT) A business network (usually public and leveraging the Internet) consisting of hardware and software that allows for the storing, retrieving, and sending information for human consumption.

IPv4 Standard The most widely used version of the internet protocol (IP) that is used to connect devices to the Internet. It is able to uniquely identify 4 billion addressable items.

IPv6 Standard The most recent version of the internet protocol (IP) that was designed to support the growth of the Internet in terms of devices and data traffic. It will be able to identify 340 undecillionth (340×1036) addressable items.

Machine-to-Machine (M2M) The ability for one machine to be able to connect to another and request its information.

Message Queuing Telemetry Transport (MQTT) A protocol geared towards the embedded space that is used over TCP/IP and embraces pub/sub data flows from the source to the destination. Unlike AMQP, it is only capable of storing the last message for future transmission.

Operational Technology (OT) An automation network (usually private) consisting of hardware and software that is responsible for monitoring and/or controlling industrial devices, processes, and events in the enterprise.

Predictive Maintenance This technique includes monitoring operational and maintenance data to predict failures before they occur—providing the ability to schedule maintenance during planned downtime.

Process Optimization This iterative process continually improves operational efficiency—reducing waste and saving costs.

The Cloud There is no clear or narrow definition of what is meant by "the Cloud," and as such a broad "All of the above understanding" of "The Cloud" should be adopted. The Cloud can reference *IaaS, PaaS, SaaS*, public, private, hybrid, hosted single-tenant software, multitenant software, and many more. All of these technologies are credibly referred to as Cloud. With so many differing interpretations, "The Cloud" is almost a meaningless term without additional clarification. Although it may make us all feel old, Wikipedia has had a very useable page on these definitions for almost 10 years, much of which has not changed for at least the last 5.

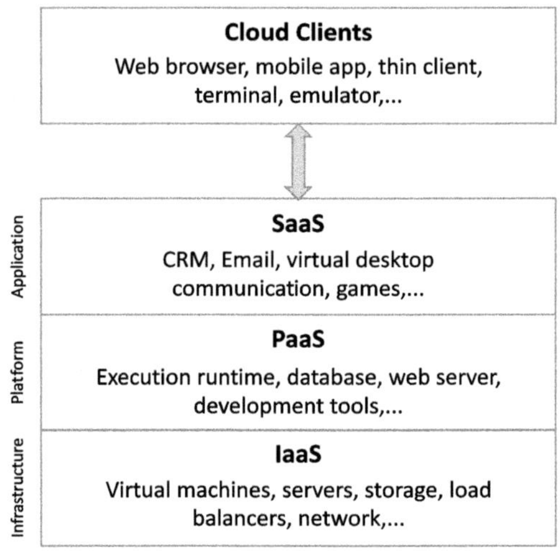

Source: Wikipedia

Cloud Platform Cloud Platforms are consolidating and rationalizing around public IaaS and PaaS for traditional structured databases/data warehouses; like those offered by Microsoft Azure and Amazon AWS, two of the leading offerings in the space.

Big Data Platform Big Data Platforms are still emerging and generally built on top of Cloud Platforms with the ability to handle the volume, velocity, and variety (structured, semistructured, and unstructured) requirements of Big Data. Often Cloud Platform vendors have added Big Data Platforms as an additional offering. In some cases, Big Data Platform providers have developed homegrown Cloud Platforms to offer Big Data Platforms. In other cases, these vendors are offering a Big Data Platform on top of another vendor's Cloud Platform. Generally, Big Data Platforms also have data streaming, an important piece of connectivity, and some analytics capabilities (edge analytics, visualization, collaboration, and predictive/prescriptive analytics), but are often enabling other analytics solutions.

IIoT Platform The IIoT Platform goes beyond just the Cloud and Big Data Platforms to incorporate robust connectivity from the edge (sensors, devices, gateways, assets, etc.); with hardware and software along with application development capabilities to enable *mobility*, collaboration, and mash-ups. For the most part, IIoT Platforms are delivered with an ecosystem of Cloud Platform, Big Data Platform, and other IIoT Connectivity/Application Development providers.

Cloud Software and SaaS SaaS and Cloud Software are two terms that are almost as fraught with misunderstanding and confusion as the general Cloud term itself. If you go back 5–7 years, Cloud Software or SaaS almost always referenced multitenant software that was purchased on a subscription basis. In the consumer space this definition still rings true, but the same cannot be said for enterprise applications.

In the Enterprise Application space there are really two types of Cloud Software or SaaS vendors today:

1. ***Multitenant Cloud and Single-Tenant Hosted Cloud***. Multitenant Cloud vendors (sometimes referred to as true-cloud) take the more traditional consumer approach, where all users share a single multitenant codebase that is generally updated on a continuous basis, free from traditional version and upgrade issues.
2. **Single-Tenant Hosted Cloud** vendors that offer hosted versions of traditional single-tenant software with a subscription model, either on top of a homegrown Cloud Platform or increasingly on top of a third party Cloud Platform like Azure or AWS. To add even more confusion, in some cases vendors have traditionally been Single-Tenant Hosted in nature are beginning to also offer separate Multitenant offerings.

Index

Note: Page numbers followed by "*f*" and "*t*" refer to figures and tables, respectively.

J

K

L

P

CPI Antony Rowe
Chippenham, UK
2017-12-04 18:41

Contents

List of Figures

List of Acronyms

APC	Armoured Personnel Carrier
BBTG	Broad-Based Transitional Government (Rwanda)
BSA	Bosnian Serb Army
CDR	*Coalition pour la Défense de la République* (Rwanda)
CICR	*Comité International de la Croix Rouge*
CFSP	Common Foreign and Security Policy (within EC)
DAMI	*Détachement d'Assistance Militaire et d'Instruction* (France)
DGSE	*Direction Générale de la Sécurité Extérieure* (France)
DRC	Democratic Republic of Congo
EC	European Community
EU	European Union
FAR	*Forces Armées Rwandaises*
FCO	Foreign and Commonwealth Office
ICTR	International Criminal Tribunal for Rwanda
ICTY	International Criminal Tribunal for the Former Yugoslavia
ICRC	International Committee of the Red Cross
JNA	Yugoslav Army
MONUC	*Mission de l'Organisation des Nations Unies au Congo*
MOU	Memorandum of Understanding
MRND	*Mouvement Révolutionnaire National pour le Développement* (Rwanda)
NAC	North Atlantic Council
NATO	North Atlantic Treaty Organization
NGO	non-governmental organization
NRA	National Resistance Army (Uganda)
OAU	Organization of African Unity
RPF	Rwandan Patriotic Front
RRF	Rapid Reaction Force (Bosnia)
RTLM	*Radio Télévison Libre des Mille Collines* (Rwanda)
SC	Security Council
SHA	secure humanitarian area
SG	Secretary-General of the United Nations
UK	United Kingdom
UN	United Nations
UNAMIR	United Nations Assistance Mission for Rwanda
UNGCI	United Nations Guards Contingent in Iraq

UNHCR	Office of the United Nations High Commissioner for Refugees
UNPROFOR	United Nations Protection Force (for Former Yugoslavia)
US	United States
USSR	Union of Soviet Socialist Republics
VOPP	Vance Owen Peace Plan
ZHS	*zone humanitaire sûre*

Acknowledgements

This book would never have seen the light of day without the support and assistance of several people.

I cannot thank enough Professor Adam Roberts and Dr. Richard Caplan, who, throughout the drafting process, offered constructive feedback and challenged me to do better. I have been privileged to work with two such esteemed academics.

Many other people took the time to comment upon or talk to me about my work. These include all the people I interviewed for the book, the two anonymous reviewers of the manuscript, as well as Professor James Mayall, Dr. Guy Goodwin-Gill, Dr. Michael Byers, Dr. Frank Chalk, Dr. Andrew Hurrell, Dr. Catherine Lu, Dr. T.V. Paul, Dr. Mark Brawley, Dr. Michel Fortmann and Dr. Marie-Joëlle Zahar. Many thanks go to Amy J. Harvey and Josephine McQueen, who were painstaking and thorough proof-readers. Paul McQueen helped with the technical side of producing the final manuscript.

This book would not have been possible without the financial assistance of the Rhodes Trust, which funded my studies at Oxford University, as well as trips for research. I am indebted to the Research Group in International Security at McGill University/Université de Montréal for enabling me to work on the book while being paid as a post-doctoral fellow.

My warmest thanks are reserved for my family and friends. My parents, siblings and close friends stood by me during this long and arduous process. I would never have finished the book without their support. My husband Christian Kananura makes it all worthwhile.

The author and publisher gratefully acknowledge permission to reproduce copyright material:

Figure 2.1: Iraq. Map No. 3835, Rev. 4, January 2004, United Nations Cartographic Section. Reproduced by permission of the United Nations Cartographic Section.

Figure 3.1: Safe areas in Bosnia, 1 July 1995. Reprinted from Ivo H. Daalder, *Getting to Dayton: The Making of America's Bosnia Policy* (Washington D.C.: Brookings Institution Press, 2000), p. 65. Reproduced by permission of the Brookings Institution Press.

Figure 3.2: Srebrenica safe area. Reprinted from Jan Willem Honig and Norbert Both, *Srebrenica: Record of a War Crime* (London: Penguin Books, 1996), p. xiv. Copyright © Jan Willem Honig and Norbert Both, 1996. Reproduced by permission of Penguin Books Ltd.

Figure 4.1: Rwanda. Map No. 3717, Rev 9, January 2004, United Nations Cartographic Section. Reproduced by permission of the United Nations Cartographic Section.

Figure 5.1 and Figure 5.2: Approximate location of RPF/FAR front line, September 1993 and May–July 1994. Reprinted from John Borton, Emery Brusset and Alistair Hallam, *The International Response to Conflict and Genocide: Lessons from the Rwandan Experience. Study III: Humanitarian Aid and Effects* (Copenhagen: DANIDA Joint Evaluation of Emergency Assistance to Rwanda, 1996), p. 27. Reproduced by permission of the Danish Ministry of Foreign Affairs.

Figure 5.3: *Opération Turquoise's* Deployment into the ZHS. Reprinted from Jacques Lanxade, 'L'opération Turquoise', *Défense Nationale*, February 1995 (Paris: Éditions CEDN), 10. Reproduced by permission of *Défense Nationale.*

Extracts from all UN documents, including Security Council Resolutions, Security Council Presidential Statements, Secretary-General's Reports, UN special inquiries and reports, and Provisional Verbatim records of the Security Council, were published by permission of United Nations Publications, Department of Public Information. The United Nations is the author of the original material.

Extracts from transcripts from interviews for *The Triumph of Evil* were published with the permission of *Frontline*, a documentary series on *PBS*. This material was originally published on *Frontline's* website (www.pbs.org/wgbh/pages/frontline/shows/evil/) after the program aired on 26 January 1999.

Extracts from transcripts from interviews for *Death of Yugoslavia*, a *BBC* documentary, were published with the permission of *The Trustees of the Liddell Hart Centre for Military Archives*, King's College London.

Extracts from the *Rapport D'information déposé par la Mission D'information Commune sur les Événements de Srebrenica. Tome I (Rapport et Annexes); Tome II (Auditions). Paris: Assemblée Nationale, 11 décembre 2001*; and from the *Rapport D'information par la Mission D'information de la Commission de la Défense Nationale et des Forces Armées et de la Commission des Affaires Étrangères, sur les opérations militaires menées par la France, d'autres pays et l'ONU au Rwanda entre 1990 et 1994. Tome I, Rapport; Tome II, Annexes; Sommaire des Comptes Rendus D'Auditions du 24 mars 1998 au 5 mai 1998; Sommaire des Comptes Rendus D'Auditions du 6 mai 1998 au 3 juin 1998; Sommaire des Comptes Rendus D'Auditions du 9 juin 1998 au 25 juin 1998; Sommaire des Comptes Rendus D'Auditions du 30 juin 1998 au 9 juillet 1998. Paris: Assemblée Nationale, 15 décembre 1998* were published by permission of the webmestre of France's *Assemblée Nationale* website.

The views expressed herein are those of the author and do not necessarily reflect the views of the United Nations.

1
Introduction

This book offers an account and evaluation of the use of safety zones in the 1990s. In April 1991, the United States, Britain and France, with the tacit approval of the Security Council, sent troops into northern Iraq to create a safe haven for the Kurds fleeing from Saddam Hussein's repression. In May 1993, the Security Council designated six 'safe areas...free from any armed attack or any other hostile act' in Bosnia-Herzegovina in an attempt to protect Muslims from ethnic cleansing by Serbs.[1] In May 1994, the Security Council mandated a strengthened United Nations Assistance Mission for Rwanda (UNAMIR) to 'contribute to the security and protection of displaced persons, refugees and civilians at risk in Rwanda through the establishment and maintenance, where feasible, of secure humanitarian areas',[2] though such zones were never created. In June 1994, authorized by the Security Council to launch a temporary operation under national command and control, France chose to set up a '*zone humanitaire sûre*' (ZHS) as a means to contribute to the 'security and protection of displaced persons, refugees and civilians at risk in Rwanda'.[3] In each case, with implicit or explicit backing by the Security Council, some combination of the United States, Britain, France, and their allies intervened, or considered intervening, militarily through the creation of safety zones, variously called safe havens, safe areas or *zones humanitaires sûres*, in order to ensure the security of civilians and displaced persons targeted by extreme violence. These zones encompassed entire towns, as in Bosnia, or large areas of land as in Iraq and Rwanda, and were often implemented without the clear consent of the belligerents responsible for these grave violations of humanitarian law. While they are but one facet of broader instances of international military involvement in humanitarian crises in the post-Cold War era, safety zones merit their own study due to their noteworthy occurrence in three of the cases of humanitarian intervention in the 1990s, and due to their investigative value as precedents for possible future international attempts to protect civilians targeted by belligerents in an armed conflict. In this respect, an inquiry into why, and how effectively, safety zones were employed in the 1990s is of interest to scholars and practitioners alike. Chapters 2, 3, 4 and 5 respectively provide accounts of the Iraqi safe haven, the Bosnian safe areas, the proposed Rwandan United Nations (UN) secure humanitarian areas (SHAs) and France's *zone humanitaire sûre* in Rwanda.

Safety zones in the 1990s: definition and distinctiveness

Historical precedents

Although places of worship have often served as sanctuaries for persons not directly implicated in fighting, a formal safety zone concept[4] only emerged during the Franco-Prussian war. In a letter to the Empress Eugénie of Prussia on 20 August 1870, Henri Dunant, founder of the Red Cross, suggested that certain towns in conflict areas be designated as neutral: therein the wounded would find shelter and receive assistance, and resident civilians would likewise be protected from the ravages of war.[5] The following year, Dunant also worked to institute similar types of zones during the uprisings of the Paris Commune.[6] Although his idea was not adopted in either situation, it elicited interest and several attempts were made to draft new legal principles outlining the basic criteria for the creation of such zones. An initial step was made in the 1907 Hague Convention IV Respecting the Laws and Customs of War on Land, in which Article 27 specified that 'in sieges and bombardments all necessary steps must be taken to spare...hospitals, and places where the sick and wounded are collected, provided they are not being used at the time for military purposes',[7] without making any specific reference to the idea of a demarcated zone or locality that all belligerents would recognize as combat-free. During the 1930s, two additional efforts occurred to incorporate a safety zone concept in humanitarian law. The first was spearheaded by the French *Médecin Général* Georges Saint-Paul who founded an organization called *L'Association des Lieux de Genève:* it lobbied governments to adopt legislation to the effect that they would create areas of refuge for civilians during times of war and grant them protection akin to that provided wounded and sick soldiers *hors combat.*[8] The second initiative occurred within the institutional framework of the International Committee of the Red Cross (ICRC), which convened a Commission of Experts in 1936 and 1938 to develop draft codes for new humanitarian law conventions that were scheduled to be discussed at a diplomatic conference in 1940.[9] The advent of World War II delayed this meeting of states until 1949, but the legal groundwork for both hospital zones and neutralized zones was completed during this period and served as the basis for ICRC lobbying with belligerents to set up such zones during the war. Such pleadings proved fruitless,[10] however, so the first three successful uses of safety zones prior to legal codification of the concept in the 1949 Geneva Conventions would come outside the context of World War II.

In November 1936, General Franco announced his intention to place outside the range of combat the north-east sector of Madrid where the civilian population could seek refuge. Though the ICRC was unable to obtain the official consent of the Republican faction to such a zone, the project nevertheless went ahead and the zone was respected for the duration of the war.[11] In Shanghai in 1937, a private initiative by Père Jacquinot de Besange led to the creation of *la zone Jacquinot* that sheltered 250,000 Chinese civilians during the Sino-Japanese war. Formally agreed to by both parties to the conflict on 6 November, the zone was clearly demarcated, administered by Chinese civilians and provisioned by the French. When the Japanese captured Shanghai in mid-November, the zone fell under the control of Japanese military authorities, but was entirely respected.[12] Finally, in March 1948 prior to the break-out of hostilities between Palestinian and Israeli forces, the ICRC negotiated terms for the creation of three small places of refuge within Jerusalem, each comprising a few buildings and intended for civilians. Although two of these zones were disbanded within a few weeks of the start of the Palestinian/Israeli war in mid-May since their security could not be adequately guaranteed, one of the zones proved successful and endured throughout the fighting.[13] These precedents were vital in convincing states that safety zones designed to shelter wounded, sick and certain categories of civilians would be a valuable contribution to humanitarian law.

Legal codification

Article 23 of the 1949 Geneva Convention I allows for[14] the establishment of 'hospital zones and localities so organized as to protect the wounded and sick [within the armed forces] from the effects of war'.[15] Upon the outbreak of hostilities, the parties concerned may conclude agreements mutually recognizing such zones, and are invited to call upon the good offices of the ICRC or other Protecting

Powers to facilitate their institution. Attached to the 1949 Geneva Convention I is a Draft Agreement relating to hospital zones and localities, which Article 23 of the Convention suggests states consider before setting up such zones.[16] This Draft Agreement further reveals what the Contracting Parties had in mind when they formulated this article, clarifying both that access to these zones should be carefully restricted and that any activity within them of a military character should be prohibited. These zones are not to be defended by military means, and they are open to occupation by the enemy. The Draft Agreement also suggests that hospital zones and localities should not be situated in areas that are important for the conduct of the war.

Article 14 of the 1949 Geneva Convention IV broadens the scope of the hospital zones and localities discussed above to allow for the protection of aged persons, children under fifteen, expectant mothers and mothers of children under seven.[17] As in Article 23 of Geneva Convention I, the belligerents may conclude agreements regarding mutual recognition of the localities they have created, and may call upon the good offices of the ICRC as well as Protecting Powers to facilitate the institution of such zones. Again, a Draft Agreement was annexed to Geneva Convention IV, further outlining what was intended in case such hospital and safety zones and localities actually came into effect.[18] It requires that access to the zones be carefully restricted to only those persons entitled to be there and forbids any activity of a military character within the zone. No defense of the zone is permitted, and it is suggested that the zone not be established in an area likely to be of military significance.

Article 15 of the 1949 Geneva Convention IV lays out specifications for the creation of neutralized zones, envisioned to be 'in the regions where fighting is taking place'. Any party to a conflict may propose to its adversary the implementation of neutralized zones intended to shelter from the effects of war 'wounded and sick combatants or non-combatants'; and 'civilian persons who take no part in hostilities, and who, while they reside in the zones, perform no work of a military character'. Article 15 also requires that all parties to the conflict consent to the geographical position, administration, food supply and supervision of the proposed neutralized zone, and record these latter provisions in a written agreement.

The 1977 Geneva Protocol I contains two articles that are also concerned with establishing areas for the shelter of civilians in armed conflict.[19] Article 59 prohibits parties to the conflict from attacking non-defended localities. A belligerent may declare as a non-defended locality 'any inhabited place near or in a zone where armed forces are in contact which is open for occupation by an adverse Party'. The article specifies that all combatants, as well as all weapons and mobile military equipment must be evacuated; and that activities in support of military operations must not occur within the zone. The declaration of such a non-defended locality must be sent to the adverse party, who is required to respond immediately and to treat the locality as a non-defended zone unless the specified conditions mentioned above are not met.

Article 60 requires belligerents to not 'extend their military operations to zones on which they have conferred by agreement the status of demilitarized zone'. The agreement must be explicit and communicated verbally or in writing, and may be reached in peacetime as well as after the outbreak of hostilities. Demilitarized zones are required to fulfill the following conditions:

(a) all combatants, as well as mobile weapons and mobile military equipment, must have been evacuated;
(b) no hostile use shall be made of fixed military installations or establishments;
(c) no acts of hostility shall be committed by the authorities or by the population; and
(d) any activity linked to the military effort must have ceased.

The parties must clearly delimit the demilitarized zone, and may request neutral supervision by a third party. If fighting draws near to the zone, and the parties to the conflict have so agreed, 'none of them may use the zone for purposes related to the conduct of military operations or unilaterally revoke its status'.

Since this legal codification process was completed, but excluding the episodes in Iraq, Bosnia and Rwanda, safety zones have been established approximately ten times. Space prohibits an extensive review of these historical examples, but they can be listed and certain common characteristics identified.[20] In Dacca during Bangladesh's war of independence in 1971, a hospital, college and hotel were set aside as a safety zone for sick and wounded civilians, as well as foreigners. Three neutralized zones were set up in hotels in Nicosia in 1974 during hostilities in Cyprus. When Saigon fell to the Communists in 1975, Vietnam accorded protection to a small number of persons in a neutralized zone consisting of two buildings. A hospital and safety zone was set up for a very brief period in 1975 in Phnom Penh, Cambodia. A few centers for civilians and disarmed combatants were instituted in Nicaragua during the disorder leading up to the Sandinista take-over in 1979. Argentina and Britain agreed to the creation of a neutralized zone at sea for hospital ships during the 1982 Falklands War. A hospital zone was instituted in 1990 around the Jaffna Teaching Hospital in Sri Lanka during the civil war between the government and Liberation Tigers of Tamil Eelam (LTTE).[21] In 1992, a hospital and a monastery in Dubrovnik, Croatia, were turned into two neutralized zones under ICRC supervision. Although the details of each of these historical instances vary, certain common elements relevant to the current study can be identified. In all cases and corresponding to the provisions laid out in international humanitarian law, the consent of the parties to the conflict was obtained before the safety zones were implemented. In almost every instance, the zone in question was set up and administered by the ICRC.[22] In many of the examples, the zones in question were restricted to one or two buildings, such as a hospital or a hotel, and remained in existence for relatively short times, perhaps only a few days or weeks. In several of the zones, the emphasis was placed on safeguarding the sick and wounded as opposed to protecting civilians in general. In all of these cases and in conformity with the 1949 Geneva Conventions and 1977 Additional Protocol I, demilitarization of the zone was either ensured through inspection or did not represent a problem due to the small size and nature of the zone in question.

Safety zones in the 1990s

A few generalizations about the safety zones envisioned by humanitarian law and implemented in the historical cases listed above can now be made. Hospital, neutralized, non-defended and demilitarized zones share five key attributes. First, they are established usually at the request of one or all of the belligerents, and certainly with the recognition and consent of all parties involved. Second, they are demilitarized and thus preclude the possibility of any kind of military activity taking place within their boundaries. Third, the underlying justificatory premise for creating these zones is the belief that civilians deserve protection because they are, in a sense, outside the realm of conflict. They are unlucky victims of armed confrontation between states. Fourth, as a result of this last assumption that civilians are inconsequential to the war aims of the belligerents, there are no provisions regarding security measures that might be needed to ensure the protection of the civilians or wounded combatants within these zones. Fifth, these zones were envisioned as being relatively small in size and thus achievable within the context of armed confrontation.

With regard to the safety zones in Iraq, Bosnia and Rwanda, they most closely resembled the neutralized zones envisioned by Article 15 of the 1949 Geneva Convention IV or the demilitarized zones of Article 60 of the 1977 Geneva Protocol I: they were intended to protect the civilian population residing there in its entirety, and they aimed to prevent military hostilities from taking place within and around the zones. Since the safe haven in Iraq, the safe areas in Bosnia and the *zone humanitaire sûre* in Rwanda were not intended particularly for the wounded and sick and encompassed the entire civilian population within their ambit of protection, they went beyond the concept of a hospital zone as outlined in Article 14 of the 1949 Geneva Convention IV. Since the international community intervened in each of these cases to safeguard militarily the zones in question, they cannot be considered to constitute non-defended localities, as provided for by Article 59 of the 1977 Geneva Protocol I.

Despite these similarities with neutralized and demilitarized zones, the safety zones created in Iraq, Bosnia and Rwanda represented a significant departure from what was envisioned in existing humanitarian law and from historical precedents.[23] They were distinctive in five critical ways. First, the safety zones in Iraq, Bosnia and Rwanda were, in some measure, imposed upon belligerents by outside powers acting under the implicit or explicit authority of the Security Council. Therefore, consent to the creation of the zones was either not initially obtained, or achieved under duress. Endorsement offered grudgingly, as in Bosnia for example, was withdrawn as new circumstances in the conflict arose, which altered the strategic and political value of the areas in question. Second, the civilians offered protection were, in all three cases, directly targeted by violence. They were not bystanders caught in the crossfire, but victims actively sought out for expulsion or slaughter. Third, these first two characteristics of the safety zones under consideration meant that these areas, and the people within them, had to be protected through force, hence the presence of foreign military troops. Fourth, demilitarization of the safety zones, and continued military activity therein, became issues of contention or confusion. Fifth, the spaces being protected involved fairly large areas of land, either entire villages or specific geographic areas, making the zones much more difficult to monitor than a hotel or hospital, and much more important strategically to the belligerents.[24]

These distinct characteristics of the safety zones in Iraq, Bosnia and Rwanda underline both the significance and approach of this study. It is important to examine these safety zones because they represent a new way of protecting civilians in armed conflict, which may contribute at some stage to the formation of new humanitarian norms. In addition, their imposed character in often difficult circumstances raises the questions of why the international community created them and how it confronted the various problems of implementation when they occurred.

Accounting for safety zones through a broadened conception of interest

This book also serves as an investigation into the role of interests and norms in the implementation of safety zones in the 1990s. The central hypothesis is that the establishment of these safety zones cannot be explained easily in terms of either strategic interests or normative considerations alone. On the one hand, if states were solely concerned with relatively narrow self-interests defined in security terms, then intervening in faraway places and putting their soldiers at risk in defense of beleaguered civilians in Iraq, Bosnia and Rwanda would have been unimaginable. Yet, the United States (US), the United Kingdom (UK), France, and others all did precisely that at various times during the implementation of the safety zones of the 1990s. On the other hand, if states were truly committed to pursuing principled ends such as preventing ethnic cleansing or genocide, as their rhetoric in relation to Iraq, Bosnia and Rwanda often intimated, then inaction in the face of the Rwandan genocide and the fall of Srebrenica would have been unthinkable. Yet, these tragic events occurred in 1994 and 1995 respectively.

Neither willing to engage militarily in a meaningful way to save targeted civilians in Iraq, Bosnia and Rwanda nor to stand entirely aside as massive violations of humanitarian law occurred, states were pulled in different directions, embracing policies that responded in some measure to these competing demands. The adoption and implementation of a safety zone approach in Iraq, Bosnia and Rwanda can thus be adequately explained only by considering both these self-interested and normative influences upon states. This book thus acknowledges, as have other thinkers, that the complexity of international affairs does not permit either an overly idealistic or an exclusively materialistic account of events.[25] Focusing on either of these aspects of International Relations to the exclusion of the other leaves out parts of the puzzle, making an overall assessment of why, and how effectively, states implemented safety zones in the 1990s more difficult. What is therefore needed is a theoretical framework that problematizes the conception of interest and broadens it to encompass both egoistic and normative concerns in order to show how states choose between these different kinds of competing interests in complex, unforeseen ways.

A simple dichotomy between interests and norms is not fruitful since their interrelationship is more intricate than such a division would imply.[26] One way around this problem is to broaden the concept of interest so that it includes concerns that affect others or the community as a whole. Political theorists have for some time now accepted the idea that persons can have interests as social beings, implying a realization that the attainment of individual goals may inevitably be linked to the interests of others. A person may accept, for instance, that some loss of material advantage may be in his or her self-interest for the sake of maintaining the well being of the community as a whole.[27] It would thus be false to argue that self-interest can only mean behaviour that intends uniquely the self as beneficiary.[28] In the field of International Relations, there has been greater reluctance to accept the possibility that state self-interest may involve taking into consideration the welfare of other states or of the international system as a whole: norms and interests are often depicted as being in a dichotomous relationship.

However, a broader interpretation of self-interest does exist in the writing of some prominent International Relation theorists. Hedley Bull, for example, wished to identify a conception of international society that was consistent with self-interest. He argued that, in order to coexist and to maintain order in international affairs, states developed common interests and values and believed themselves to be bound by a common set of rules in their relations with one another.[29] In this way, states form an international society.[30] Andrew Hurrell suggests that the compliance pull of norms and rules is linked to their relationship to the broader pattern of international affairs: 'states have a longer-term interest in the maintenance of a law-impregnated international community'.[31] Legal obligation serves as a means to regulate both present and future state behaviour so that raw applications of power, often expensive and untrustworthy, are avoided, thus promoting stability and leading to common expectations regarding conduct.[32] Keohane too perceives the value of relaxing the assumption of egoism in international affairs, suggesting that states are aware that there are 'solutions to international problems that lead to larger overall value – even at the expense of direct gains to themselves'.[33] He labels this concern for the welfare of others empathetic interdependence[34]

The concept of self-interest can therefore be broadened sufficiently to encompass egoistic factors, which will be labeled state interests, as well as other-regarding factors, which will be called community interests. Community interests embody preferences that are concerned with the well being of international society as a whole. States may not be acting against their self-interest when obeying certain legal obligations or when upholding humanitarian norms: they are simply choosing between different and competing interests.[35] States do not stray away from self-interested behaviour, but their preferences can be broad, long-term and community-oriented, or narrow, short-term and state-centered.

Before turning to a more detailed analysis of how community interests and state interests, as well as their interaction, alter the behaviour of states, let us be clear that the proposed framework draws upon a number of schools of thought in an attempt to devise a useful synthesis of the different approaches to the study of International Relations.[36] Realism, Institutionalism and Constructivism all bring invaluable elements to an effective understanding of state behaviour as seen through the lens of a broadened conception of interest.

Realists shed light on why states, operating in an anarchical international system, prioritize self-interest and power, often to the detriment of collective norms and action.[37] Here, states are mainly concerned with building their power relative to that of others so as to compete favourably in a potentially dangerous, uncertain world; and this power is defined primarily in terms of material capabilities such as military prowess or economic might.

Institutionalists or regime theorists show in particular how the presence of formal institutions and informal regimes push states toward results other than those predicted by power and pure self-interest. Regimes – 'the principles, norms, rules and decision-making procedures around which actors expectations converge in a given issue area' – act as intervening variables between state interests, defined again in material terms, and behavioural outcomes.[38] Since independent, egoistic choice can result in suboptimal outcomes for all states concerned,[39] regime theorists stress the functional benefits of norms and rules, suggesting that only by fostering cooperation can states achieve their goals. Regimes facilitate this interaction between states by easing joint decision-making, lowering transaction costs, reducing incentives for cheating and ensuring that benefits acquired or restraints imposed are reciprocal. Regimes also confer a sense of propriety, enabling states to utilize their power in legitimate ways and thus promoting stability in the international system in general.[40] Regime theory also explains how, over time, the norms and rules of a regime may become decoupled from the state power-structures that created them, and continue to flourish with the help of international institutions. Very close to the institutionalists are the international lawyers, who have in recent years become increasingly interested in how legal rules can sometimes play a distinctive role in influencing state behaviour, mainly due to international law's widespread acceptance by states as the foremost institutional framework in which to explain, justify and persuade others of the legitimacy of their actions.[41] International law presents its own specific internal logic of discursive enmeshment with which states must be familiar if they are to participate in international society: justifications about the appropriateness or inappropriateness of action must be framed in terms of state practice, precedents, judicial reasoning and legal vocabulary.[42]

Constructivists highlight the fact that regimes and norms not only have the capacity to alter outcomes, but also to affect how states determine their interests in the first place. They show how the international normative context shapes the interests and identity of international actors, thus determining their basic character, expectations and behaviour; [43] and how states are embedded in a dense network of social relations, and hence are socialized by the context in which they exist. As a result, their interests are defined in an intersubjective way.[44] In this view, state preferences are not a given, but are malleable since states are sensitive to the international environment in which they operate: states develop perceptions of interest and understandings of desirable behaviour from social interactions with others in the world they inhabit.[45] Agency and environment are thus mutually constitutive.[46] Constructivists also suggest that norms may even shape an actor's interests in ways that contradict the strategic imperatives of the international system.[47]

Community interests and the case studies

Although state and community interests operate in tandem, it will be easier to assess their interactive effect upon state conduct if we begin by analyzing each kind of interest in isolation. Community interests contributed to the way in which the US, Britain, France and the

Security Council as a whole conducted themselves with regard to the adoption and implementation of the safety zones. Embodied in norms, rules and decision-making procedures, community interests were relevant as 1) focal points, as 2) enablers, as 3) constraints, and as 4) aspects of identity. Norms are generalized formulations of shared expectations about appropriate behaviour held by a community of actors.[48] They specify criteria for distinguishing right from wrong; just from unjust.[49] Rules are specific prescriptions or proscriptions for action,[50] and they usually entail some form of legal responsibility, whether as an obligation to respect or to help enforce. This legal obligation arises due to the fact that rules embody the common interests of states as codified in international law. The legitimating premise of the entire international legal system is that states have consented to be voluntarily bound by specific rules so as to preserve patterns of shared understandings and joint expectations.[51] The 'prevailing practices for making and implementing collective choice' constitute decision-making procedures.[52] The aim here is to discuss theoretically how community interests, in the guise of shared norms and rules, matter, and allude but briefly to the case studies.

Community interests as focal points

Community interests can serve as focal points, and thus as means to counteract problems of collective action.[53] In an international system of sovereign states, it is unlikely that the players see eye to eye on all issues. Even if they agree to join together to participate in a regime, they may not be able to resolve their disagreements in all areas. Such disharmony can eventually lead to the disbanding of the co-operative arrangement. Especially in an informal regime that is not yet buttressed by institutional compliance and therefore relies heavily on the basic accord of its key actors, dissension can have crippling effects on the performance of the regime. International rules and norms help to keep the actors focused on what they have in common, i.e. on their community interests. As Kratochwil puts it, even at the basic level of discourse, norms are the means which 'allow people to pursue goals, share meanings and communicate'.[54] A focal point is a locus of convergence where state perceptions about the meaning of events coincide, because this meaning is consistent with predefined normative concepts that states have already accepted. If acts of barbarity conform to the definition of crimes against humanity, then these activities can be labeled as such and thus focus attention on a common denominator that cannot be denied or disputed, at least from a rational point of view. In Bosnia, for example, differences regarding both how to resolve the conflict and what the ultimate political and geostrategic *denouement* should be risked splitting apart Britain, France, the US and Russia. What kept them united, at least in some measure, was the focus on preventing ethnic cleansing and on alleviating the humanitarian catastrophe, leading scholars to suggest that humanitarianism represented a substitute for political action to end the war.[55] This is perhaps true, but without the humanitarian emergency serving as a focal point, it is unlikely that these powers would have remained united behind an effort to halt the conflict.

Community interests as enablers

Community interests enable state action in various fashions. They act as catalysts for action; they create permissive conditions for action; they render action legitimate; and they facilitate action instrumentally.

Since norms and rules set guidelines for appropriate behaviour, it follows that they also embody standards that determine unlawful comportment. Infringements attract the attention of the international community and can sometimes function as a tripwire,[56] setting in motion a response in support of community interests. Violations of international law and the results thereof – refugee flows in the case of ethnic cleansing and genocide – can be perceived as threats to international peace and security by the members of the Security Council, thus potentially serving as a catalyst to the use of its Chapter VI and Chapter VII powers.

While rules and norms create permissive conditions for the support of community interests on the part of states, they do not necessarily determine the precise action to be taken.[57] In other words, they create realms of possibilities and 'options' that would not have been self-evident in the absence of a norm or rule.[58] Examples of legal rules relevant to the case studies in this thesis, which may permit and thus enable third-party response to breaches, will illustrate this point. Certain rules of international law are deemed so important that they are given a generality of standing, meaning that they are to be upheld by the international community as a whole even though only one member of that community may be directly affected by the violation in question.[59] Such generality of standing opens the door to more effective enforcement opportunities since responsibility for preventing violations is widened to include all states.[60]

The 1948 Genocide Convention, for example, contains such a rule. It affirms in Article I: 'the Contracting Parties confirm that genocide, whether committed in time of peace or in time of war, is a crime under international law which they undertake to prevent and to punish'.[61] This article allows for the possibility of third party activity to prevent genocide. Also Article VIII of the Genocide Convention permits any contracting party to 'call upon the competent organs of the United Nations [most probably the Security Council] to take such action under the Charter of the United Nations as they consider appropriate for the prevention and suppression of acts of genocide or any of the other acts enumerated in Article III'.

Similarly, the 1949 Geneva Conventions also contain a common article, which declares that 'the High Contracting Parties undertake to respect and to ensure respect for the present Convention in all circumstances'. Article 1 of the 1977 Geneva Protocol I uses a similar phraseology.[62] There is debate as to whether or not 'the ensure respect' clause really intended that third parties should become involved in preventing violations of humanitarian law in wars in which they are not implicated. Frits Kalshoven has conducted a painstaking study of this clause, looking at legal precedents and at references to it in the drafting process. His conclusion is that states, in 1949 and 1977, never interpreted this clause as implying any kind of positive duty or legal obligation to act to ensure respect for the conventions. Rather, the particular wording related to making sure 1) that states obeyed the rules themselves, even in circumstances in which adverse parties confronting them in battle did not or had not ratified the conventions; and 2) that the contracting states worked in times of peace to ensure that their population and military were aware of the conventions, and to implement appropriate legislation in terms of military codes of ethics.[63] However, recent case law suggests that these legal provisions of the Geneva Conventions have come to be widely viewed as providing a legal basis for a third-party response to infringement, as demonstrated by this extract from an International Criminal Tribunal for the former Yugoslavia (ICTY) judgement:

> The absolute nature of most obligations imposed by rules of international humanitarian law reflects the progressive trend towards the so-called 'humanisation' of international legal obligations, which refers to the general erosion of the role of reciprocity in the application of humanitarian law over the last century. After the First World War, the application of the laws of war moved away from a reliance on reciprocity between belligerents, with the consequence that, in general, rules came to be increasingly applied by each belligerent

> despite their possible disregard by the enemy. The underpinning of this shift was that it became clear to States that norms of international humanitarian law were not intended to protect State interests; they were primarily designed to benefit individuals *qua* human beings. Unlike other international norms, such as those of commercial treaties which can legitimately be based on the protection of reciprocal interests of States, compliance with humanitarian rules could not be made dependent on reciprocal or corresponding performance of these obligations by other States. This trend marks the translation into legal norms of the 'categorical imperative' formulated by Kant in the field of morals: one ought to fulfill an obligation regardless of whether others comply with it or disregard it. As a consequence of their absolute character, these norms of international humanitarian law do not pose synallagmatic obligations, i.e. obligations of a State *vis-à-vis* another State. Rather – as was stated by the International Court of Justice in the *Barcelona Traction* case (which specifically referred to obligations concerning fundamental human rights) – they lay down obligations towards the international community as a whole, with the consequence that each and every member of the international community has a 'legal interest' in their observance and consequently a legal entitlement to demand respect for such obligations.[64]

Both the 1949 Geneva Conventions and 1977 Geneva Protocol I include articles pertaining to safety zones. Although the safety zones in Iraq, Bosnia and Rwanda did not conform to what was set out in these legal documents, these provisions still suggested possibilities: they created possible options for intervention in all three of the case studies under consideration that might not have been envisioned otherwise. When the British Foreign Office looked into the possibility of a safe haven in Iraq, it consulted the existing legal articles on safety zones.[65] In Bosnia, the ICRC had been calling for the use of safety zones before the Security Council formulated the safe area policy.[66]

The fact that norms and rules make action possible does not necessarily mean that states move, in all instances, to uphold the norms and rules under attack. Even if international humanitarian law, as discussed above in relation to the 1948 Genocide Convention and the 1949 Geneva Conventions, grants states the right to ensure respect for certain rules of great importance to the international community, it neither imposes a clear obligation to do so in all cases, nor makes evident what types of specific actions would be permissible. Certainly, any course of conduct would have to respect other important legal rules, such as those laid out in the UN Charter. Whether or not states actually seize such opportunities for involvement, and to what degree, will depend on the configuration of their state and community interests at a given moment in time.

Perhaps the reason why states resort to norms and rules when taking action in support of their community interests is because they impart a sense of legitimacy. Conventionalized behaviour is apt to engender widespread feelings of propriety in the international community since states are socialized to share common expectations about proper conduct.[67] Legitimacy promotes acceptance, and thus effectiveness. If others are to be persuaded to join in collective measures and help defray their costs, then reasons must be provided as to why such conduct is necessary. The legitimacy of these objectives helps influence the extent to which states will be willing to participate, and is measured by the degree to which the proposed action conforms with notions of proper conduct as elaborated by international rules and norms and as recognized by states. In Iraq, Bosnia and Rwanda, the legitimacy that rules and norms bestowed upon state behaviour in support of community interests enabled the emergence of relatively widespread international consensus on the need for and appropriateness of safety zones.

Finally, regimes, as institutionalists maintain, serve a utility function, making possible behaviour in support of community interests that would not have been so otherwise. The cost of intervention in Bosnia, for example, would have been prohibitive for a single country to attempt alone, whereas multilateral involvement was manageable as states pooled their resources. Information gathering and sharing, carried out mainly by the UN Secretariat through their agencies and network of personnel in the field, also facilitates action. Many of the smaller states sitting on the Security Council would have had little access to other sources of information regarding events on the ground in Iraq or Bosnia; and the reports and recommendations of the Secretary-General often provided valuable suggestions for future courses of action. Institutional mechanisms that the Security Council has at its disposal, such as peacekeeping and peace-enforcement, embodied significant tools that states could draw upon as means to implement the zones. Finally, the transparency that the Security Council generated regarding aims and means meant that trust and respect could be achieved amongst actors previously suspicious of each other.

Community interests as constraints

Certain international legal rules not only enable action in support of community interests, but may also create a sense that it is required. As discussed in the previous section, the 1948 Genocide Convention and the 1949 Geneva Conventions allow for third party involvement to ensure respect for the principles enshrined therein. If these international conventions enable parties not directly affected by violations to do something to halt atrocities, it is because their provisions are deemed of utmost importance to the preservation of international society. These rules were given the status of being peremptory in nature, meaning that they are not to be violated for any reason whatsoever unless superseded by another peremptory norm.[68] States thus acknowledged that their community interests demanded that certain rules be absolute so as to ensure that vitally important norms for continued communal co-existence are upheld. These rules serve the highest interest of the whole international community.[69] Although rules of this nature grant states the right in principle to ensure their respect, the legal obligation to do so is not generally recognized by states.[70] However, when grave transgressions occur, a moral incentive or obligation of a kind prods states toward obviating further breaches.[71] States do not always respond to these imperatives for action, but the rules nonetheless exert a constraining effect on their behaviour, often forcing at least verbal condemnations. This constraining effect of community interests was observed in Iraq, Bosnia and Rwanda.

As constructivists suggest, once socially constructed norms are established and promulgated, they tend to feed back on the basic causal variables that gave rise to them in the first place.[72] The norm, and in association the regime supporting it, begin to take on a life of their own, which in turn shapes subsequent action, impelling actors to behave differently than they would if the regime did not exist.[73] This dynamic constrained the international response in each case study. Particularly the US, but also others in the Coalition, for instance, felt added pressure to respond to Iraqi repression of the Kurds because they had espoused a new world order in which intervention to protect human rights was lauded.

Community interests can also have a constraining effect on states when they become influential in internal political debates. Images of massacres of innocent civilians appearing on television around the world often mobilize public outcries against such transgressions of moral norms. Sometimes public sympathy for other human beings thousands of miles away translates into demands that governments do something to stop the slaughter. Interest groups and intellectual elite, concerned with respect for human rights, also play an important role in generating public awareness of the blatant defilement of commonly held values.[74] These processes often lead to the alignment of state and community interests, especially when governments are slow to react and uncertain as to which action to take.[75] Politicians begin to realize how commitment to upholding international norms contributes to their domestic popularity at home.[76]

Community interests as aspects of identity

Identity is a construction that enables recognition by others. It can be characterized by distinctiveness or sameness, and states may possess multiple identities. For example, France has an identity that derives in part from pride in its language, culture and long historical association with human rights. Its identity also reflects its membership in the European Union (EU): it shares the same respect for democracy and civil liberties as other members of the EU do, as codified for example in social charters and human rights documents. Constructivists argue that identity shapes and is shaped by state and community interests.[77] In other words, norms help forge a state's identity, and, in turn, a state's identity will determine which community interests it holds to be important. In the broader international picture, 'configurations of state identity affect interstate normative structures, such as regimes or security communities', and in turn those regimes institutionalize and reinforce identity.[78] Community interests as aspects of identity encouraged the use of safety zones in Iraq, Bosnia and Rwanda. Foremost, America's, Britain's and France's post-Cold War perception of themselves as upholders of 'civilized' values and as the primary states responsible for implementing these values meant that they were inclined to respond to humanitarian emergencies involving massive violations of human rights. By interceding on behalf of innocent civilians, these states could validate their identity, which in turn amplified the importance accorded to these community interests.

The degree to which states perceive their identity to be similar to another's may affect their willingness to respond in times of crisis. For example, when Yugoslavia's Republics sought independence in 1991 and 1992, it was to the EU that they turned for recognition. The EU agreed to grant recognition only if certain conditions concerning the respect for the rule of law, democracy and human rights were met. In essence, these new states would be recognized only if they adopted the type of identity that the EU espoused.[79] Existing members of the EU viewed these states as possible future members of their organization, indicating their awareness that these newly independent states already possessed or aspired to a European identity. Thus, when bloodshed broke out in Bosnia, an added impetus was placed on members of the EU to play their part in ending hostilities. Furthermore, ethnic cleansing and military targeting of civilians raised the specter of Nazi atrocities in World War II. Part of the EU's founding ethos lies in the determination to prevent a recurrence of such terrible crimes on European soil. Thus, the horrors being perpetrated in Bosnia threatened EU members' post-Second World War identity as states committed to ensuring that there would never be another Holocaust on European soil. This fact impelled France and Britain, as well as other EU members such as Sweden and the Netherlands, to action. In contrast, this identity factor was not as strong in the Rwandan case study, perhaps explaining the greater reluctance on the part of European states to become involved there.

Identity is also significant from the point of view of reputation. If reputation is important to states, it is because they have constructed an image of themselves that they believe is valuable and which accords them a certain degree of status and respect from others. Loss of reputation implies that others assess a state's action to be deviant from what rules and norms deem to be appropriate conduct. It also suggests that the actor in question has strayed from the identity it sought to create for itself and which accorded it prestige. The fear of losing its reputation can constrain a state into conforming to the provisions of rules and norms.[80] In the case of Rwanda, for example, international outcries that France was buttressing a genocidal regime tarnished its image as a civilized state concerned with the protection of human rights. France responded by launching *Opération Turquoise.*

State interests and the case studies

State interests, which involve strategic, economic and domestic imperatives, tend to be more straightforward because they illuminate the actions of one state at a time,[81] though various states can have egoistic interests that either conflict or coincide. Three of the aspects of the last section are again relevant here, including state interests as 1) enablers; as 2) constraints; and as 3) aspects of identity.

State interests as enablers

Sometimes state interests cause states to act in ways that also fulfill community interests. A desire to repay Turkey for its loyalty to the Coalition during the 1991 Gulf War, for instance, meant that states would not condemn it for its reluctance to take in fleeing Kurdish refugees, opening the door to the necessity of creating secure conditions for the Kurds within Iraq. Likewise, a desire to permanently weaken Saddam Hussein's control over his territory for geostrategic reasons enhanced America's, Britain's and France's desire to create a safe haven free of Iraqi military personnel and equipment. Avoiding a transatlantic rift encouraged the US and the Europeans to develop a safe area policy in Bosnia. As mentioned earlier, the media and domestic interest groups also play an important role in converting community interests into state interests as governments are concerned with their political survival and popularity. This process no doubt influenced state responses to the humanitarian emergencies in Iraq, Bosnia and Rwanda.

State interests as constraints

As realists predict, states also shy away at times from their community commitments when these are perceived as hindering their pursuit of self-interested concerns. The reluctance on the part of all states, but especially the US, to have any of their soldiers killed in UN enforcement operations greatly hampered the implementation of the safety zone approach in each of the case studies considered. France's state interests in French Africa contributed to its initial unwillingness to halt the genocide in Rwanda, even though its authority and prestige in the country might have enabled it to do so with relative ease. America's negative experiences in Somalia generated a sense, amongst members of the public and the government alike, that it was best not to intervene in internal conflicts in places that had no strategic importance.

State interests as aspects of identity

In the same way that identity can emphasize the importance of community interests, it can also stress the value of egoistic factors. Reputation is again important here, but usually from the point of view of prowess and military might. For example, it was important for France to retain its reputation as a credible military ally in French Africa, partially explaining its continued attachment to the Habyarimana regime in Rwanda despite mounting evidence of genocide. France's sudden change of heart in the form of *Opération Turquoise* resulted from its recognition that conditions in Rwanda had now deteriorated so drastically that inaction would likely tarnish its military credibility more than abandoning a loyal ally. Britain's and France's sense of identity as important international actors meant they were inclined to want to prove their worth as permanent members of the Security Council. They thus played key roles in the implementation of the Iraqi safe haven and the Bosnian safe areas. American unwillingness to allow Saddam Hussein to put into question their military prowess through the expulsion of the entire Kurdish population of Iraq contributed to their adoption of the safe haven policy.

Community and state interests in interaction

In the coming chapters, this book will seek to show how state and community interests intermingled in complex and interactive ways, producing a safety zone approach in each of the case studies under observation. Moving away from the simple dichotomy of self-interest versus normative concerns present in much of the International Relations literature, it will endeavour to illustrate how states grapple with competing interests. There is a spectrum of possibilities linking these two types of interests. State and community interests can be in alignment, for example, allowing states to respond to two distinct but parallel urges at the same time.[82] Community interests may be so strong as to dictate action, outweighing other considerations; or state interests may impede the pursuit of community interests at a particular moment in time. More complicated permutations are also possible. Equally strong opposing state and community interests may be reconciled through compromise measures, often leading to confusing and ambivalent approaches. At one and the same time, state interests may promote and impede the pursuit of community interests, again leading to ambiguous state behaviour. Outcomes are dependent upon this complex interaction of community concerns and state-centered concerns since states seek to satisfy, at least in some measure, both types of interests at the same time. Only an analysis of these various configurations of state and community interests will enable us to understand why states adopted safety zones in Iraq, Bosnia and Rwanda, and to assess how well they implemented them throughout their duration.

The end of the cold war

The Security Council's use of safety zones in the 1990s cannot be fully grasped without an appreciation of the end of Cold War and its effect upon international affairs. The cessation of the Cold War in 1989, like the subsiding of several other major conflicts in modern history, was followed by a period of renewed enthusiasm for collaboration among the major powers and of enhanced commitment to furthering common interests as well as maintaining international order. After four decades of discord and near paralysis within the Security Council, its impotence when faced with threats to international peace and security began to recede in the late-1980s when it authorized five new peace-keeping operations in a five-year period.[83] Although cautious at first, this activity blossomed following Iraq's invasion of Kuwait in 1990: former adversaries, the US and the Union of Soviet Socialist Republics (USSR), cooperated, enabling the Security Council's authorization of the use of force against Iraq.

This rebirth of the Security Council as an active body in the management of international security generated great optimism, as well as a sense that the great powers must not miss a second opportunity to render the UN effective in this area. This conviction in the potential of the UN to operate in a manner consistent with what its founders had envisioned in 1945 culminated in a meeting in January 1992 of the heads of states and governments of the members of the Security Council. At their invitation, UN Secretary-General Boutros Boutros-Ghali prepared *An Agenda for Peace* which suggested ways to strengthen the capacity of the United Nations in the areas of preventive diplomacy, peacemaking and peace-keeping. This document reiterated the 'unprecedented recommitment, at the highest political level, to the Purposes and Principles of the Charter', and confirmed that the Security Council had re-emerged 'as a central instrument for the prevention and resolution of conflicts and for the preservation of peace'.[84] *An Agenda for Peace* also acknowledged that 'the time for absolute and exclusive sovereignty [had] passed', and called upon the members of the Security Council to demonstrate a clear pledge to human rights, especially in regard to minorities.[85]

This post-Cold War context was conducive to the emergence of an informal security regime centered around the UN Security Council. A number of factors coalesced for this outcome to occur. The great powers were confronted with a changing international system in which multipolarity replaced bipolarity and in which new types of threats not seen during the Cold War predominated. The United States was faced with the dilemma of being the world's only remaining super-power, but of not wanting to take on unilaterally the burden of preserving peace and security around the globe. Britain and France were concerned about enhancing their political and military profile in international crises, thus demonstrating the usefulness of their permanent status on the Security Council. The Soviet Union, later Russia, in a state of economic collapse and hoping to secure international respect as a trusted participant in the resolution of crises, opted to endorse the general policy trends advocated by the Western permanent members. China acquiesced, choosing to accept tacitly the general direction of events or to abstain on resolutions it did not openly favour. The result of this collegiality and the new conviction that the Security Council could and should maintain international peace and security was a willingness to tackle complex emergencies, such as civil wars, failed states and massive human rights abuses perpetrated by states. This heightened activity often led to innovative approaches, including intervention in the internal affairs of states as well as the use of safety zones. The Security Council also provided the decision-making infrastructure that made collaboration and multilateralism possible, though implementation was left mainly to the United States, Britain and France, which were the only states capable of enforcing Security Council resolutions in the areas under consideration.

Post-Cold War optimism about the possibilities of the Security Council to redress wide-ranging threats to international peace and security eventually waned as states grew unwilling to make the major military contributions and sacrifices necessary for the Security Council to be at the centre of a global security system. The story of the safety zones in Iraq, Bosnia and Rwanda will also reflect this rising disillusionment and lack of sustained will.

The end of the Cold War also reinvigorated the idea of an international community. No longer polarized by a massive ideological divide between Liberalism and Communism, states came together to form an international community, which was characterized by relatively widespread consensus on certain norms, such as respect for state sovereignty and the illegality of genocide, war crimes and crimes against humanity. Although divisions and misunderstandings persisted within this international community of the 1990s, the case studies nonetheless reveal that key state actors believed that a community of states existed, referred to it regularly when making decisions, and thought it represented a force to be considered when conducting foreign policy. Globalization and the widespread dissemination of TV images also facilitated the emergence of an international community at the sub-state level, making the media, international non-governmental organizations (NGOs) and public pressure groups an ever greater force with which governments had to reckon during the complex humanitarian emergencies of the 1990s. The growing power of this global community will also be evident in the stories of how states came to choose a safety zone policy in Iraq, Bosnia and Rwanda, and of how well these zones were implemented.

2
Providing Comfort at Home: Safe Haven in Iraq

Figure 2.1 Iraq[1]

On 16 April 1991, President George Bush announced that the US, along with its two main Gulf War allies Britain and France, would create a safe haven in northern Iraq so that the Kurds who had fled Iraqi repression could return home again in safety. Following the 1991 Gulf War, both the Shi'ites in the South and the Kurds in the North launched rebellions that succeeded for a time but were eventually brutally crushed by the Iraqi regime. Targeted by Iraqi helicopter gunships and fearing possible chemical weapons attacks, a means used by Saddam Hussein in his 1988 punishment of the Kurds for their involvement with the enemy in the Iran-Iraq war, the Kurds fled en mass toward the Turkish and Iranian borders. Concerned that an influx might exacerbate its own Kurdish minority problem and unwilling to be burdened financially with massive numbers of refugees, Turkey closed its border, leaving the Kurds to languish in severe, freezing conditions along the mountains just inside Iraq. The resulting humanitarian crisis sparked greater and greater international involvement to alleviate the suffering in an incremental process that eventually culminated in a safe haven once it became clear that sufficient comfort could not be provided in the harsh landscape along the Iraqi-Turkish border. Although not explicitly authorized by the Security Council, the safe haven came to encompass almost 10,000 square kilometres of Iraqi territory;[2] and its protection was successfully guaranteed through the presence of coalition troops that remained within Iraq until mid-July and through an effective air power deterrent that prevented Iraq from using military aircraft within a no-fly zone above the 36th parallel.[3]

This chapter aims both to understand why the coalition partners opted for a safe haven approach to assist beleaguered Kurds and to assess how effectively they implemented this security zone. Only an analysis that incorporates a broadened conception of interest will allow us to fully grasp why governments adopted this novel approach; why they did so cautiously after some delay despite the gravity of the situation and the relative ease with which a safe haven could be implemented; and why implementation proved short-lived as intervening states pulled their troops out of Iraq three months after sending them there.

The 1991 Gulf War and its aftermath

Iraq invaded Kuwait on 2 August 1990.[4] As a result of the improved relations between the superpowers, the condemnatory response of the Security Council was immediate: it passed resolution 660 on the day of the invasion, demanding that Iraq 'withdraw immediately and unconditionally all its forces' from Kuwait. Although the US opposed this act of aggression mainly as a result of its state interest concerns about excessive Iraqi control over the world's oil production[5] and about growing Iraqi preponderance in the Gulf region, it nonetheless chose to mount its campaign against Iraq via a multilateral approach undertaken within the UN system, a fact that would later prove to be important in relation to the Kurdish crisis. A coordinated international reaction to the invasion proved possible because Iraq had threatened an important community interest that most states held dear, namely the prohibition against the aggressive use of force as laid out in Article 2(4) of the UN Charter. The US and its key allies in Europe preferred such a collective response for several reasons. It further enhanced cooperation between East and West, and confirmed that the Security Council could be effective in responding to threats to international peace and security. It convinced friends and foes alike of the propriety of the planned military offensive, thus gaining their political and financial support; and it allayed their fears of American hegemony following the decline of the Soviet Union. Finally, it enabled the proposed military operation against Iraq to acquire the necessary domestic support within the western democracies due to its perceived legitimacy.

Once negotiations proved fruitless, the coalition launched *Operation Desert Storm* in conformity with Security Council resolution 678 of 29 November 1990, which authorized all necessary measures to evict Iraq from Kuwait. Due to the virtual collapse of the Iraqi army, the ground war that came after several weeks of intensive bombing[6] only lasted for 100 hours between 24 and 28 February 1991. The end of inter-state hostilities spelled the beginning of active internal dissent in Iraq. On 2 March, a revolt occurred in the Southern Shi'ite city of Basra, initiated mainly by disgruntled and embittered Iraqi army deserters and supported by local residents. The spontaneous revolt spread rapidly through the South of Iraq, but was gradually taken over by more fundamentalist factions, some of whose members returned from exile in Iran.[7] However, as was to become obvious in the coming days, *Desert Storm* had not done as much damage to Saddam Hussein's military machine as might have been expected: almost the entirety of the elite Republican Guard had escaped unharmed, as had numerous tanks and helicopter gunships. Although the ceasefire agreement hammered out between General Norman Schwartzkopf and an Iraqi commander grounded fixed-wing aircraft, it failed to restrict Iraq's use of rotary wing aircraft. A thus unfettered Saddam Hussein employed helicopter gunships to repress brutally first the insurrection in the South, and then the one in the North. He regained full control of southern Iraq on 13 March 1991.

Numbering 4 million in Iraq and representing about 25 per cent of its population, the Kurds did not participate in the Gulf War: they saw little prospect that the western build-up against Saddam Hussein would bring them any direct advantage.[8] Still reeling from the *Anfal* campaign, they were also disinclined to oppose the dictatorship openly,[9] and they believed that neutrality might bring them concessions after the end of the war. However, upon news of the revolutionary success in the South and convinced that the US would support their insurrection, the Kurds launched their own uprising on 4 March, capturing almost 95 per cent of Kurdistan by the 13 March 1991. The prized oil centre of Kirkuk fell into their hands on the 19th. Yet, this success proved short-lived. Soon after quelling the rebellion in the South, Saddam Hussein turned his full attention northward: he brought in helicopter gunships in an all-out offensive against the Kurdish uprising, recapturing Kirkuk on 28 March and massacring indiscriminately as many Kurds as possible. Unable to withstand the brutal onslaught and fearing the possibility of gas attacks as had occurred in 1988, the Kurds fled en mass toward the Iranian and Turkish borders in the last days of March and the first days of April. It is estimated that between 1.5 and 2 million Kurds were in transit from Iraq, with well over 500,000 amassed on the Turkish border.[10]

Turkey's refusal to grant entry to the fleeing Kurds meant that they were stuck on the harsh mountain ranges along the Iraqi frontier, where it was cold, wet, and too remote for access by humanitarian relief agencies. As the death toll climbed to a staggering 1,000 persons a day,[11] a process of gradual international humanitarian involvement occurred, culminating in a safe haven approach. Turkey sent a letter to the Security Council on 2 April requesting that a meeting take place to discuss the mounting humanitarian crisis along its border, stating that Iraq's 'actions violate all norms of behaviour towards civilian populations and constitute an excessive use of force and a threat to the region's peace and security'.[12] Meanwhile France spearheaded the effort to build a consensus among the permanent members of the Security Council that permitted the adoption of resolution 688 on 5 April 1991, which specified that the Council:

1. Condemns the repression of the Iraqi civilian population in many parts of Iraq, including more recently in Kurdish populated areas, the

consequences of which threaten international peace and security in the region;

2. Demands that Iraq, as a contribution to removing the threat to international peace and security in the region, immediately end this repression and expresses the hope in the same context that an open dialogue will take place to ensure that the human and political rights of all Iraqi citizens are respected;
3. Insists that Iraq allow immediate access by international humanitarian organizations to all those in need of assistance in all parts of Iraq and to make available all necessary facilities for their operations;...

On 7 April, the United States launched *Operation Provide Comfort*, dropping food and relief supplies from the air on the Iraqi side of the border. To ensure the safety of these aerial missions as well as to deter further helicopter gunship attacks against the Kurds, the US informed Iraq on 10 April that a no-fly zone was now in effect above the 36th parallel.

However these measures did little to resolve the problem of poor access to the displaced Kurdish population, and Turkey remained adamant in its refusal to allow the Kurds entry into Turkey. The idea that a safety zone might be an effective alternative began to emerge. On 7 April 1991, Turkish President Turgut Özal suggested that the UN take over a strip of land in Iraq for the purpose of providing security and sustenance for the Kurds within the boundaries of Iraq, also hinting that Turkey would be willing to commit troops to such an endeavour.[13] British Prime Minister John Major then adopted the proposal, calling for a militarily protected enclave for the Kurds in Iraq at a meeting of the EC Council of Ministers in Luxembourg on 8 April, where it was speedily endorsed. The US response was at first lukewarm, thus impeding any immediate concrete steps toward implementation, but eventually Bush was moved by international pressure: at a 16 April news conference, he announced that American and allied soldiers had entered Iraq to create a series of six small secure camps, each capable of handling 60,000 refugees, where the Kurds could receive food and shelter.[14] When these camps proved insufficient to entice the Kurdish population home due to its fear of Iraqi retribution, the coalition expanded its protection to a zone that included several important towns and came to within one mile of Dahuk, the capital of Kurdistan.[15] Ultimately, the safe haven enveloped a terrain measuring 160 kilometres east to west and 60 kilometres north to south.[16]

The security of the zone was ensured through the presence of troops[17] and through the air power deterrent guaranteeing the no-fly zone. Although Iraq opposed the formation of the safe haven, it decided not to risk further confrontation with the US and its allies.[18] Two uncoordinated but parallel agreements reached with Iraq further facilitated implementation of the zone. On the one hand, American Lieutenant General John Shalikashvili, commander of *Operation Provide Comfort*, secured agreement from Iraqi military leaders on 19 April that Iraqi ground forces would not be permitted inside the haven and would pull back to an acceptable distance. On the other hand, as a result of the efforts of UN Secretary General Javier Pérez de Cuéllar, the UN and the government of Iraq signed a Memorandum of Understanding (MOU) on 18 April 1991, in which Saddam Hussein agreed to 'a humanitarian presence in Iraq' and to 'the voluntary return home of Iraqi displaced persons'.[19] Such unhindered humanitarian access was to be guaranteed through the creation of UN Humanitarian Centers to be staffed by civilian personnel. On 25 May a further agreement between the UN and Iraq allowed for the deployment of small contingents of UN guards (UNGCI) to these centers.[20] The provisions of the MOU were initially supplanted by the arrangements for the safe haven, but its existence would prove important as the safe haven operation wound down on 15 July when coalition troops departed Iraq. Until September 1991, a 5,000 strong Rapid Reaction Force (RAF) remained stationed at Incirlik air force base in Turkey under *Operation Poised Hammer*, in order to dissuade any renewed Iraqi hostility against the Kurds. When it too was withdrawn, only the no-fly zone endured.

The initial phase of the Kurdish crisis

Constraining state interests mainly explain Turkey's closure of the border with Iraq and the coalition's difficulties in mounting a response to the Kurdish calamity. However, states also faced a dilemma between competing community interests: on the one hand, pressure was increasing for them to alleviate a humanitarian emergency; on the other hand, the principle of non-intervention in the internal affairs of states as well as the previously agreed-upon parameters of Security Council mandated action against Iraq encouraged restraint. In the initial phase of the Kurdish crisis, these latter concerns proved the more dominant.

State interests and the initial phase of the Kurdish crisis

Although Turkey's intransigence toward accepting the Kurdish refugees eventually led to the necessity of protecting them within their country of origin, its predominantly state-centric behaviour was an important cause of the calamity in the first instance. The Kurds who crossed into Iran were appropriately fed and sheltered, thus avoiding the perpetuation of a similarly severe humanitarian emergency there. The constraining effect of state interests best illuminate why Turkey insisted that all aid be deposited on the Iraqi side of the border, and why it dropped leaflets to the fleeing Kurds, warning them not to cross the border. Turkey feared that an influx of additional Kurds from Iraq would threaten its assimilation of its own Kurdish population of 20 million and stoke the fires of separatism, as had been the case in 1988.[21] In order to consolidate control over its vast territory, the Turkish government had instituted a policy of assimilation of its Kurdish minority, which constitutes 20 per cent of the overall population. Although not actively hindered from speaking their own language or maintaining their own identity in their homes, the Kurds were expected to remain loyal to the Turkish state and to partake in Turkish political and social life. Any secessionism was brutally repressed, and Turkey had sought to eradicate the Kurdish guerilla movement for several decades. In a pre-emptive gesture designed to avoid the emergence of an independent Kurdistan in Iraq, President Özal had even met with the two principal Iraqi Kurdish leaders, Jalal Talabani and Masud Barzani, in mid-March at the height of the rebellion in order to encourage them to seek only greater autonomy within Iraq.[22] Turkey also feared that the Iraqi Kurds might seek to continue their war against Saddam Hussein from within Turkish territory, creating the potential for high levels of tension, if not outright warfare, between Turkey and Iraq.

In addition, Turkey was reluctant to bear the financial costs of such a massive influx. Having hosted over 150,000 Kurds following the *Anfal* campaign of 1988, Turkey was well aware of the burden such a flood of incoming refugees could impose upon a country. Although Ankara had been criticized for failing to provide adequate facilities for these Kurds, the Turkish government claimed these complaints were unfair: European states had themselves neglected to furnish sufficient funds, to offer resettlement opportunities elsewhere in Europe or to recognize that Turkey was channeling scarce resources to help the refugees. Once these Kurds were granted entry, it also proved difficult to get them to go home again: 60,000 Kurdish refugees were still in Turkey at the time of the 1991 Gulf War. Accepting a new inflow of refugees risked a permanent increase of the Kurdish population, which was unacceptable to Turkey, and likely meant the government would be saddled with long-term responsibility for their care with little international monetary compensation.

The configuration of American state interests also meant that the one state largely responsible for victory against Iraq in the 1991 Gulf War, and thus the one state capable of influencing the outcome of the uprisings as they occurred, shied away from providing any assistance to the Kurds while they were being brutally repressed by Iraqi forces. Although the US would have liked to see Saddam Hussein removed from power, perhaps through a military *coup d'état*,[23] it did not intend the disintegration of Iraq for a number of reasons. It believed that the 'Lebanonization of Iraq' would 'create a geopolitical nightmare'[24] that might significantly destabilize the balance of power in the region. The US feared the rising influence of Iranian fundamentalism in the Middle East, as did its allies Saudi Arabia and Kuwait, which were more concerned about Shi'ite domination of Iraq than about the continuation of a weakened Ba'athist regime.[25] In addition, the US feared that the disintegration of Iraq into its ethnically homogeneous parts might risk a power vacuum that would lead to instability and recurrent warfare in the region. This American commitment to the *status quo* in Iraq was enhanced by the practice of rewarding one's allies for their loyalty during times of war. US action against Iraq was greatly facilitated by the support it received from the Arab world as a whole, and from Turkey and Saudi Arabia in particular. The US would now acknowledge their state interests in return, not to mention the fact that their continued support was needed for the implementation of the ceasefire provisions then being worked out in the Security Council.[26]

Once the ground war was over, President Bush also faced severe pressure from the American political establishment as well as the public to fly soldiers home as soon as possible and to avoid a Vietnam-style quagmire in Iraq.[27] Having won a spectacular victory in record time with almost no casualties and thus having chased away some of the demons of Vietnam, Bush was not prepared to dent his high approval ratings or to mire the celebrations at home by being lulled into a civil war that was likely not winnable, or would have required 'nothing short of direct U.S. military operations' to be successful.[28] Bush reiterated on 13 April for example: '...I do not want one single soldier or airman shoved into a civil war in Iraq that's been going on for ages.'[29]

The US was also responsible for the week's delay in the creation of the safe haven, despite mounting pressure from Britain and France to take this novel step. It was easier for the European allies to propose such a potentially hazardous plan, because they would not bear as extensively the risk involved with its implementation. With by far the largest number of soldiers in the region and primarily in possession of the military capability needed to move forward with the safe haven approach, America preferred to increase its engagement prudently in incremental stages, all the while assessing the response of the Iraqi regime. When it became clear that a safe haven could be instituted without seriously damaging its state interests, the US took the lead from Britain and France and moved to protect Kurdish civilians within their country of origin.

Community interests and the initial phase of the Kurdish crisis

These constraining state interests were further buttressed by complementary community interests that also inhibited a response to the Kurdish crisis following the 1991 Gulf War. First, states sought to confine their behaviour to that considered acceptable by the coalition as a whole. Once military action was undertaken under the rubric of the UN, its parameters were constrained by the framework to which its participants had agreed in advance. Operating under the authority of the Security Council, the US and its allies could neither alter war aims arbitrarily nor broach new endeavours without consultation. Coalition action was deemed lawful and appropriate, because it was responding to a violation of international law and had set clear UN-approved goals for itself, thus enabling collaboration between states that might not have cooperated under other circumstances. Through resolutions 660 and 678, the coalition endorsed the following war aims: the removal of Iraqi forces from Kuwait, the defense of Saudi Arabia in case Iraq ventured further afield, and the restoration of Kuwait as an independent state. Removal of Saddam Hussein from power or involvement in an Iraqi civil war were never sanctioned by the UN, although ensuring that Iraq would not pose a future threat to international peace and security was part of the ceasefire terms contained in Security Council resolution 687 of 3 April 1991. However, this willingness to weaken Iraq did not amount to permission to invade it, occupy it or interfere in its internal unrest. The US government fully recognized that the legitimacy, and ultimately the effectiveness, of its actions and 'new world order' rested on its respect for multilateral arrangements.[30]

A further example of this concern about adherence to a multilateral framework for action was the American decision to end the ground war after 100 hours. This resolve to limit the onslaught stemmed from the fact that Iraq's army appeared to collapse: the coalition was left shooting at retreating forces along what some journalists labeled the Highway of Death.[31] Pursuing a one-sided bloodbath reflected poorly upon the United States and its allies: it would have intimated at a coalition bent on revenge and uncaring about the loss of life, instead of one committed to upholding international law and respecting the self-imposed restrictions this security regime had placed on itself.[32]

The second community interest that delayed an effective response to the Kurdish crisis was concern about violating the UN Charter rule of non-intervention in the internal affairs of states. Article 2(4) guarantees that states, even small and less powerful ones, will not be subject to invasion or to meddling in their internal affairs by other states, whether they be neighbours or powerful countries further a field. In the context of the 1991 Gulf War, this principle was accorded even greater recognition: the coalition's very *raison d'être* was to restore Iraq's respect for Article 2(4). The legitimacy of coalition action in the Gulf rested on the objective of reversing Iraq's illegal and military interference in Kuwait's internal affairs. Even states dubious about UN collective uses of force, such as China,[33] refrained from impeding coalition action against Iraq because they recognized that violations of Article 2(4) constitute major threats to international order and communal co-existence. For the coalition to become involved in Iraq's internal unrest risked violating, in the eyes of many states, the very principle it had sought to uphold during the war. Likewise, UN involvement in the internal affairs of states is also restricted to certain circumstances through Article 2(7) of the Charter, namely when they threaten international peace and security: it was not self-evident that Saddam Hussein's repression of the Kurdish population posed such a menace.

Although first France and Britain, through resolution 688, and later the US, through the creation of the safe haven, succeeded in convincing Security Council members that intervention in the internal affairs of states can be legitimate in the context of a threat to international peace and security, this process took time, especially if it were to earn widespread international support. States were reluctant to entertain the idea that Iraqi misconduct toward its own population, as well as its repercussions, could constitute acceptable reasons to circumvent Article 2(7) of the Charter. This reluctance was clearly evident in the deliberations of the Security Council on 688 and in the wording of the resolution itself. Of the 15 members of the Council, three states voted against 688 and two abstained.[34] The major reason given by each of these five states was that the resolution violated Article 2(7).[35] All states that voted in favour of the resolution felt the need to indicate verbally their commitment to the principle of non-interference in the internal affairs of states, though they accepted that the tragedy posed a threat to international peace and security and thus permitted Security Council involvement. Although 688 demanded that Iraq cease its repression of its civilian population, its preamble recalled 'Article 2, paragraph 7, of the Charter of the United Nations' and reaffirmed 'the commitment of all Member States to the sovereignty, territorial integrity and political independence of Iraq...' as a concession to states concerned about a new precedent being set that might gradually undermine the restriction on UN intervention in the internal affairs of states.[36] Likewise, the threat to international peace and security that 688 addressed 'was clearly restricted to the transboundary effects of the situation': it did not pertain to the Iraqi civilian repression *per se*.[37] UN Secretary-General Javier Pérez de Cuéllar further echoed this nervousness about encroachment of Article 2(7), stating publicly that 688 violated the spirit of the UN Charter and that it did not authorize a military presence in northern Iraq.[38]

The creation of a safe haven for the Kurds

Despite initial state unwillingness to become directly involved in the Kurdish crisis, a consensus gradually began to emerge within international circles that something needed to be done to alleviate the suffering experienced by the Kurds. Since these earlier concerns dictating a policy of non-response did not disappear, only an approach that addressed them adequately stood a chance of success. In this sense, a safe haven was ideal: it enabled the coalition to ensure the safety of the Kurds within their country of origin without either defending them in a civil war with Iraq or breaching indisputably Article 2(7) of the Charter, since its intervention was temporary and humanitarian in nature. This general virtue of safety zone approaches – namely that they allow states to harmonize competing and often contradictory impulses – nonetheless does not fully account for how this earlier state paralysis toward the humanitarian emergency in northern Iraq was overcome. Only a further assessment of how community and state interests were affected by the crisis will shed light on why and how it became possible, and even necessary, for states to adopt such a novel approach.

Community interests and the creation of the safe haven

Community interests served as strong focal points following the Kurdish uprising, helping to keep states focused on the need to do something in the face of massive repression of the Kurdish population despite their not seeing eye to eye on the permissibility of intervention. Even though discord existed within the Security Council regarding whether or not Iraq's repression of its civilian population constituted a threat to international peace and security, all states could agree that Saddam Hussein's treatment of the Kurds was immoral, reprehensible and illegal. As a result, states allowed the adoption of 688,[39] and failed to voice outright opposition to the creation of the safe haven. That the Kurdish problem would remain insoluble until such time as the refugees could return home in safety leant credence to the safe haven idea, despite its novelty and contentiousness. The safe haven concept thus gradually became a point of convergence for Turkey, Britain, France and finally the US, as well as for other less enthusiastic states that nonetheless acquiesced in the policy.

In a variety of ways, the enabling effect of community interests played an important role in the shift toward international acceptance that something had to be done on behalf of the Kurds. First, community interests helped generate a permissive atmosphere within the Security Council that enabled the adoption of innovative, unprecedented measures. This permissive context had in fact emerged following the end of the Cold War as states sought to create an international order free from superpower rivalry and Security Council paralysis, and it endured throughout the 1991 Gulf War. It meant that states were able to respond to the Iraqi invasion of Kuwait in a multilateral manner, authorizing the collective use of force to reverse the conquest, as well as collective enforcement of the ceasefire terms contained in Security Council resolution 687. In short, the Security Council was able to operate with increasing agility: states were willing to tackle problems as they appeared and to propose original solutions that, even a few years earlier, would have been inconceivable. The adoption of resolution 688 can only be understood in this light of state openness to cooperation and innovation: for the first time, a government's repression of its civilian population, as well as the consequences such nefarious behaviour generated, were labeled threats to international peace and security. Once adopted, resolution 688 engendered further permissive conditions that eventually enabled the pursuit of the safe haven policy. Although the US, Britain and France never sought the permission of the Security Council for their military intervention in northern

Iraq because they feared insufficient support within the Council or even a veto by the Russians or Chinese, these three countries considered that 688 gave them some authority to intervene, as did other states, given their silence on the matter and thus tacit approval of the safe haven.

Second, community norms enabled the adoption of a safe haven approach because they opened the door to third party action in response to breaches of humanitarian law. As discussed in Chapter 1, these rules have been accorded a generality of standing in international law, meaning that third parties are permitted, but not necessarily required, to become involved when transgressions are taking place. In other words, the fact that violations of humanitarian law have occurred is likely to enhance state tolerance toward intervention. In relation to Iraq's treatment of the Kurdish population, France alleged that Iraq had committed crimes against humanity; Britain, Belgium and Germany stated that Iraq had failed to fulfill its obligations under the 1949 Geneva Conventions; and several other states mentioned Iraq's general non-compliance with international human rights instruments.[40] Sir David Hannay, the UK Ambassador to the UN, explained most clearly that violations of this kind 'justified action by the Security Council' since they no longer fell under domestic jurisdiction, but were a matter for the international community to deal with by right.[41]

Third, the coalition pursued the safe haven policy in a manner consistent with community interests, thus increasing the legitimacy of the operation, as well as its level of acceptance among states. Since action perceived to be outside the perimeter of appropriate state behaviour is more likely to encounter opposition, states endeavoured to situate their actions within a framework of acceptable conduct. In order to achieve the adoption of resolution 688, states offered compelling and valid reasons for its necessity: they explained that the consequences of Iraq's repression constituted a threat to international peace and security, thus making a demand for the brutality's cessation appropriate. Although the safe haven proposal was never brought to the Security Council,[42] its proponents nonetheless justified it publicly as being consistent with previous resolutions. By linking the safe haven approach with measures previously endorsed by the UN, the US, Britain and France implied that the safe haven was in line with the type of action already approved by states: the creation of the safe haven did not stray away from legitimate conduct, but was concordant with it. Although Bush acknowledged that not all states would be in favour of the safe haven, he still believed its implementation was necessary and legitimate due to the severity of the refugee crisis: 'Consistent with Security Council Resolution 688, and working closely with the United Nations and other international relief organizations and our European partners, I have directed the US military to begin immediately to establish encampments in northern Iraq...'[43] On a number of occasions, the British also expressed the view that the establishment of the safe haven was compatible with 688. For instance, Major told the House of Commons:

> ...if the relief effort is harassed or frustrated, in my judgement, under Security Council resolution 688, it is clearly the responsibility of the United Nations to protect both helpers and helped. If necessary, the United Nations would have to act on that responsibility and seek from its members whatever assistance, including military assistance, it might need.[44]

In addition, Iraq's signing of the MOU[45] on 18 April also imbued the safe haven policy with a certain legitimacy, even though Saddam Hussein could claim that his acceptance of these UN terms obviated the need for a military enforcement of 688. The MOU attested to some consent on the part of the Iraqi government toward a humanitarian operation within Iraq, thus helping to diminish somewhat international concern about a military intervention in the internal affairs of Iraq. In addition, the safe haven operation could be seen as complementary to the MOU, as one of the MOU's weak points was its lack of security provisions for the Kurds within Iraq. Even though, via the MOU, Iraq had agreed to cease its repression of the Kurds and welcome their return to Iraq, the Kurds remained dubious about these Iraqi promises, given their extreme ill-treatment at the hands of the regime. One way to coax the Kurds to return to Iraq, and thus bring into effect the terms of the MOU, was to create secure conditions for them within Iraq, hence the need for a temporary safe haven protected by coalition troops.

Finally, the multilateral pursuit of community interests throughout the Gulf Conflict period later facilitated from an instrumental perspective the adoption of a safety zone approach in April 1991. The channels of communication developed during the hostilities with Iraq remained in place and eased decision-making about how to proceed with the safe haven policy. The military activities of the three key contributors, the US, Britain and France, were easily coordinated, so the safe haven was mounted effectively in record time. As the costs and tasks of the operation were spread among 30 states, no one state bore too great a burden. The transparency permeating the initial *Desert Storm Operation* against Kuwait was transferred to *Operation Provide Comfort*, so there was little uncertainty about the humanitarian intentions of the intervening states, which soothed the concerns of those states that were not entirely happy about the safety zone.

Not only did community interests and norms have an enabling effect on the safe haven policy, they also had a constraining effect on state behaviour: as the plight of the Kurds worsened, the sense of urgency mounted, eventually driving states to take action to alleviate the suffering. States were compelled to respond to the imperative to do something. This sense of 'obligation' on the part of states arose for four reasons. First, the gravity of the Kurdish crisis – the death toll reached as many as 1,000 per day in the early days of April[46] – meant that states could no longer ignore the moral, and possibly legal, need to do something. For example, US Secretary of State James Baker, who visited on 8 April the border area where the refugees were amassed and 'witnessed the suffering and desperation of the Iraqi people, experiences of cruelty and human anguish that defy description', became convinced that something had to be done to render these Kurds 'free from the threats, persecution and harassment'.[47] Responding to a strong moral imperative to 'do something about this tragic crime', Baker moved to convince the US administration to adopt the safe haven policy already endorsed by the European Community (EC) and 'to devote the same effort to humanitarian relief as we had in putting the international coalition together'.[48]

Second, as community norms gained prominence within the rhetoric of the informal security regime opposing Iraq, they acquired a certain power of their own, putting pressure on states to adhere to the values that they themselves had espoused. Once the coalition identified the protection of certain norms as falling within the scope of its community interests, it became difficult once the new crisis erupted to avoid being answerable to what it had said and done earlier. In essence, the earlier logic of involvement in the 1991 Gulf War fed back upon itself, creating strong expectations that the Kurdish problem would elicit an equally assertive, principled response. During the lead up to hostilities with Iraq, western leaders adopted a lofty rhetoric about their reasons for opposing Iraq, which included respect for international law and the creation of a new, safer world order where the Security Council functioned effectively. On 6 March 1991, for instance, President Bush told Congress:

> Twice before in this century, an entire world was convulsed by war. Twice this century, out of horrors of war hope emerged for enduring peace. Twice before, those hopes proved to be a distant dream, beyond the grasp of man.
>
> Until now, the world we've known has been a world divided, a world of barbed wire and concrete block, conflict and Cold War.
>
> Now, we can see a new world coming into view. A world in which there is the very real prospect of a new world order. In the words of Winston Churchill, a 'world order' in which 'the principles of justice and fair play... protect the weak against the strong.' A world where the United Nations, free from Cold War stalemate, is poised to fulfill the historic vision of its founders. A world in which freedom and respect for human rights find a home among all nations.[49]

Such verbal declarations of intention intensified the expectation in April 1991 that the international community should not return to the previous state of paralysis and indifference toward human suffering that had prevailed during the Cold War. The British Ambassador to the UN acknowledged this sentiment: 'I fear that the Council's recently acquired and well-justified reputation for decisiveness has taken a dent in the last day or two. But now we are moving ahead again...'[50] Likewise, the earlier adoption of resolution 688 contributed to the later sense on the part of the coalition that it had a responsibility to follow through on its provisions: 'If that [humanitarian] effort is harassed and frustrated, those Kurds will not return and the Security Council has a responsibility, which we have fully emphasized for some time under resolution 688, to ensure that that effort succeeds'.[51]

Third, a sense of obligation to do something in the face of the Kurdish calamity arose in part because earlier coalition behaviour in support of community interests had contributed to Saddam Hussein's repression of the Kurds. There was, in other words, a direct linkage between coalition handling of the 1991 Gulf War and Iraqi perpetuation of serious crimes against the Kurds. In the first instance, the Americans made it clear they supported the overthrow of Saddam Hussein. President Bush said on 15 February: 'But there's another way for the bloodshed to stop, and that is for the Iraqi military and the Iraqi people to take matters into their own hands to force Saddam Hussein, the dictator, to step aside and to comply with the UN and then rejoin the family of peace-loving nations'.[52] This message was broadcast repeatedly on *Radio Free Iraq*, a clandestine radio station supposedly funded by the CIA. Though a military coup is what Bush had in mind, the Kurdish leadership interpreted his statement otherwise, repeating often that they had revolted with the expectation of foreign assistance. This plausible assumption on the part of the Kurds placed partial blame for their predicament on the shoulders of the US and its allies.[53] In the second instance, the way the coalition terminated the ground war meant that Saddam Hussein retained sufficient power and means to mount a brutal counterattack against the Kurds. By stopping the ground war after 100 hours, the coalition left the Republican Guard almost intact, enabling it to repress the internal uprisings. In addition, the ceasefire conditions imposed upon the Iraqis by General Schwharzkopf on 3 March only grounded fixed-wing aircraft, and not the rotary-wing helicopters that later carried out the slaughter of the Kurds.[54] Although the US had the military capability to shoot down these gunships when they attacked the Kurds, it decided against such a course of action that 'could suck us into a civil war'.[55] However, these facts suggested a tacit complicity on the part of the coalition in what happened to the Kurds, which encouraged a sense that something should be done to make amends for past mistakes.[56]

The fourth way in which community interests constrained state behaviour during the Kurdish crisis was through their influence upon domestic politics. The media was largely responsible for this impact, since it generated tremendous public support for the plight of the Kurds. Once the public, and by extension the political elite within the media and the government, enter into the decision-making equation, the line between community and state interests is blurred: upholding community norms becomes a matter of political survival and government popularity.[57] As a result of the access to the Kurdish refugees Turkey granted to the media, TV and newspaper coverage of their plight was widespread throughout the western world. The French press was amongst the first to criticize the coalition for its indifference toward the calamity. On 2 April 1991, an editorial graced the front page of *Le Monde* damning the coalition for its lack of response and demanding that effective intervention take place.[58] Over the next ten days, *Le Monde* published dozens of articles about the abandoned Kurds, and several more editorials and OP-ED pieces called for humanitarian intervention.[59] In Britain, the media was at first less supportive of coalition involvement, but changed its tune as the gravity of the Kurdish crisis became apparent. *The Times* initially disapproved of the idea of becoming involved in a state's internal affairs,[60] but altered its position in the wake of the avalanche of public opinion supporting the Kurds and some form of assistance to them.[61] *The Economist*, although recommending caution, endorsed the possibility of intervention in its 13 April edition,[62] and praised the American decision to come to the rescue of the Kurds on 20 April.[63]

In the US, the *New York Times* published innumerable editorials and opinion pieces criticizing Bush for his indifference to the Kurds and calling on him to take military action.[64] It also backed Bush's decision to send troops in immediately and unreservedly.[65] Moreover, as the most powerful member of the coalition and its founder, the American government faced criticism and pressure from foreign media as well. *Le Monde* ran several editorial cartoons contrasting drawings of Bush's fishing holiday with those of Saddam Hussein's murderous inclinations.[66] The British TV media juxtaposed on a regular basis images of dying Kurds with pictures of Bush playing golf or relaxing, thus highlighting both men's similar lack of empathy.[67] The media also suggested that, unless an intervention was forthcoming, the justifications given for the war against Saddam Hussein – namely the defense of international law and the creation of a new world order – were but hollow words.[68] The domestic and international pressure on the US government grew so severe that the US Senate passed resolution 99 on 11 April 1991 that called upon President Bush

> immediately to press the United Nations Security Council to adopt effective measures to assist Iraqi refugees as set forth in Resolution 688 and to enforce, pursuant to Chapter VII of the United Nations Charter, the demand in Resolution 688 that Iraq end its repression of the Iraqi civilian population.[69]

In sum, the media focus on the Kurdish plight helped generate public outcries in Britain, France and the US that governments could not ignore, thus constraining their behaviour in relation to the humanitarian catastrophe in such a way as to make a safe haven approach inevitable. It is noteworthy that the failed insurrection in southern Iraq and its resulting refugee flow into Iran never received anywhere near as much media coverage as the Kurds did along the Iraqi-Turkish border. As a result, the predicament of these refugees did not become a domestic political concern within the West.

Finally, community interests as aspects of identity also played an important role in encouraging the adoption of a safe haven in northern Iraq. In terms of the coalition as a whole, its members shaped an identity for themselves throughout the Gulf Conflict as 'civilized' states that promoted respect for the rule of law and advocated the creation of a more just international order.[70] In order to remain true to that identity, and because its reputation rested in large part upon doing just that, the coalition was inclined to do something to counter Saddam Hussein's 'latest example of inhumanity'.[71] Very often, in order to gain the upper hand over the dictator, the coalition sought to distinguish itself from the Iraqi regime by casting Iraq outside the parameters of international society and of states' socialized sense of what that society's identity was: whereas the dictatorship routinely violated the most fundamental rules of international society, the US, the UK, France and other coalition partners sought to preserve them. The means employed to achieve such a continuous divide between 'civilized states' on one side and 'pariah state' on the other were narrative frames of reference, involving the use of myths, metaphors and stories.[72] For example, one analogy Bush employed frequently was that of Saddam Hussein as Hitler: here was a villain of mass proportions that could only be stopped through an effective coalition of states determined to uphold international law. When the Kurdish crisis erupted, such an analogy acquired even more potency, making international indifference seem all the more like appeasement and an abdication of responsibility. In sum, something was done on behalf of the Kurds partially because the coalition in general, and the US, Britain and France in particular, felt pressure to fulfill the obligations that a 'civilized' identity imposed upon them.

The identity aspect of community interests was also relevant to how Britain, France and Turkey comported themselves in relation to the Kurdish crisis. Britain and France were keen to establish reputations for themselves as prominent champions of community interests that took their status on the Security Council, as well as the responsibility that came with it, seriously. Concerned to make amends for its imperial past in the region, Britain adopted a discourse of global responsibility in the hope of restoring a measure of its former glory and influence;[73] and, in a manner consistent with how it opposed slavery in the 19th century, it proved willing to take on a pioneering role in international affairs.

During the 1980s, France had developed a niche for itself in international affairs as the defender of the idea of humanitarian intervention. In 1987, two influential French academics, Bernard Kouchner and Mario Bettati, published an influential book titled *Le Devoir D'ingérence*,[74] which called for a right to humanitarian intervention to be enshrined in the

UN Charter. This idea was then taken up by France's Foreign Minister Roland Dumas, though more as a rallying cry than as a serious legal initiative. By 1991, Kouchner had become France's Minister of Humanitarian Affairs. He played an instrumental role in convincing Mitterrand that drastic action needed to be taken. France was the key state behind Security Council resolution 688, drafting it and trying to get it adopted as early as 2 April 1991. France had also endeavoured to have the issues of human rights protection and of Iraq's treatment of its civilian population included within resolution 687, as further gauges as to when sanctions against Iraq could be lifted.[75] In a further affirmation of its identity as defender of *le devoir d'ingérence*, France played an important role in the implementation of the safe haven.

In relation to Turkey, being perceived as a recalcitrant state unwilling to respond to a humanitarian emergency would have jeopardized its reputation as a civilized state committed to upholding international law and order, an image Turkey was keen to maintain as it prepared for possible EC membership. So although important state interests dictated Turkey's policy of non-admission to Kurdish refugees, it nonetheless sought to demonstrate empathy for their plight and willingness to find an effective solution to the problem. Turkey claimed its behaviour in relation to the Kurds was consistent with the 1951 Convention Relating to the Status of Refugees.[76] The Turkish foreign ministry issued a statement to this effect on 2 April: 'Iraqi refugees are forcibly directed toward the Turkish border and it constitutes a threat to our security'.[77] Turkey further emphasized the fact that it did not want to be made a buffer zone to hold back refugees seeking asylum in Europe, and made the argument that the current Kurdish crisis should be dealt with through international cooperation and burden-sharing. It was President Özal who initially brought the matter to the attention of the Security Council in his letter of 2 April. It was also him who first suggested the idea of a safety zone within Iraq and offered to contribute Turkish troops to such a project. In sum, Turkey's concern about fostering an identity acceptable to the EC meant that it took the first steps along the path leading to a safe haven in northern Iraq.

State interests and the creation of the safe haven

Although state interests prevented military intercession on behalf of the Kurds during the uprising, they nonetheless enabled a more limited response in the form of a safe haven. In fact, the adoption of a safe haven policy allowed members of the coalition to achieve a number of self-interested objectives. First, Turkey could be repaid for its loyalty during the Gulf War. Though one of Iraq's biggest trading partners and the country where the pipeline transporting Iraq's northern oil to Europe was situated, Turkey nonetheless sided with its North Atlantic Treaty Organization (NATO) allies and helped them enforce the embargo against Iraq. Without this Turkish participation, the blockade would not have been as effective. Turkey's continued strategic importance following the end of the Cold War – it was the only NATO country to border Russia directly and an important buffer between Europe and the Middle East – also meant that coalition states found it more important to support Turkey than to condemn it for not accepting the Kurdish refugees. There was little admonition of Turkey's behaviour in international diplomatic circles. The creation of the safe haven also enabled the US and its allies to prevent Turkey from taking matters into its own hands as it might have done if no collective action proved forthcoming. Any Turkish military involvement within Iraq might have precipitated a war between the two countries, an eventuality that the coalition wished to avoid. Turkey's offer to participate in the humanitarian operation in Iraq was likewise refused for this reason.

Second, members of the coalition could contain the refugee problem by finding a way to protect the Kurds within their country of origin. European states were eager to avert a politically taboo quota-for-resettlement scenario, given Turkey's intransigence to admit the fleeing refugees; and the consequences of such a massive Kurdish exodus from Iraq, namely its potential to threaten international peace and security in the region, could be eliminated.

Third, the US and its allies could legitimize the need for a no-fly zone above the 36th parallel in northern Iraq due to its importance as a means to deter Iraqi attacks against *Operation Provide Comfort* and the coalition troops taking part in it, when in fact this air exclusion zone, as well as its twin erected below the 32nd parallel in August 1992, served important strategic purposes. They permitted the US and its allies to infringe Iraqi air space long after their soldiers had left the country in order to monitor Iraqi troop movements, military activity and Saddam Hussein's suspected nuclear, chemical and biological weapons programmes. Not having access to the skies above the entirety of Iraq also weakened the Iraqi regime without destabilizing it, and guaranteed that it would be less likely to threaten its neighbours.

The configuration of British and French state interests further facilitated the adoption of the safe haven policy. In terms of the UK, PM John Major was able to use the safe haven idea as a means to counteract domestic criticism that he was neither as decisive nor as vigorous as former British Prime Minister Margaret Thatcher who had played an instrumental role alongside Bush, until her replacement by Major in November 1990, in forging a coalition against Iraq in the days after the invasion of Kuwait. Although Major carried on where Thatcher left off when he came to office, the first real test of his foreign policy came when the Kurdish crisis erupted. As the drama unfolded, it seemed that Major was at a loss as to what to do, simply copying the American stance of refusing to become involved in the internal affairs of another state. On 3 April, Thatcher criticized Major in an interview, saying that 'legal niceties' should not be allowed to stand in the way of assisting dying and starving people.[78] This attack from a rival and from someone who continued to enjoy great political clout helped spur Major into action: without prior consultation with the United States, he developed the idea of a militarily protected enclave for the Kurds in Iraq and presented it to the EC Council of Ministers on 8 April 1991. Once Britain took up the concept of a safe haven, it made it more difficult for the Americans to ignore it, as they owed a certain degree of allegiance to their most trusted Gulf Conflict ally.

French state interests dictated that France demonstrate its capacity to take the initiative within a multilateral context, leading to its leadership role in terms of securing the adoption of 688, which proved to be a necessary stepping stone on the road to the safe haven. In particular, France wished to forge a foreign policy separate from that of the Americans, as well as invigorate Europe's and the UN's roles in international affairs. As the Arab world's main European ally, France believed it had the necessary clout in the region to be able to offer an alternative approach to the crisis without appearing to be an unreliable coalition ally.[79] It promulgated its own 4-point peace plan in September 1990, which it brought to the General Assembly and which stressed the importance of finding solutions to other Middle East quagmires as a component of ending the war over Kuwait.[80]

However, none of France's efforts at mediation were successful: it was unable to broker a deal with Saddam Hussein, and several of the Arab states in the coalition distrusted French motives due to its past connection with radical groups in Algeria and Yemen.[81] Certainly, they did not feel that France had the military capability to bring about an alternative peace to the one proposed by the Americans. The UK and the US were unmoved by France's endeavours, since they fundamentally disagreed with the concept of linkage between the Iraqi invasion of Kuwait and the Israeli-Palestinian peace process. As a result of these failures to modify in any substantial fashion the coalition's response to the conquest, France's ability to conduct an effective, independent foreign policy was put into question. The Kurdish crisis offered the French the opportunity to regain the initiative and to show themselves capable of playing a leadership role within the Security Council. The fact that humanitarian intervention was something it felt strongly about enhanced France's diplomatic effectiveness, and enabled it to secure the adoption of 688.

Finally, state interests as aspects of identity also helped make possible the adoption of the safe haven in northern Iraq. The ability to implement a safety zone restored the military credibility of the US and its allies, for Iraq's quashing of the rebellions indicated that the coalition military machine was not as well oiled as was originally perceived following the 1991 Gulf War victory. To let a Third World dictator overshadow the coalition's celebrations with its capability to brutally repress an uprising suggested that the US and its allies were impotent in the face of renewed Iraqi aggression and raised speculation about their military competence. The United States bore the brunt of the criticism for the Kurdish crisis, since it masterminded the Gulf War and labeled the quick defeat of Iraq the most dramatically one-sided and effective military victory in history. Bush told Congress on 6 March: 'tonight in Iraq, Saddam walks amidst ruin. His war machine is crushed. His ability to threaten mass destruction is itself destroyed'.[82] Less than two weeks later, these words proved premature as Saddam Hussein quelled the Kurdish uprising. At issue was whether or not the US had made poor military decisions regarding the cessation of hostilities against Iraq.

First was the question of whether or not the war should have continued beyond 100 hours, as General Norman Schwarzkopf preferred, but as President Bush and Chairman of the Joint Chiefs of Staff, General Colin Powell, opposed.[83] Although perceived to be a humane gesture at the time, in retrospect the early ending of the war appeared less benevolent when the Republican Guard units the US had purposely allowed to escape perpetrated the brutal counterattacks against the Kurds. It also seemed likely that the Americans had overestimated how much damage they had done to Iraq's military arsenal. The CIA later estimated that Iraq lost, of the equipment brought into the Kuwaiti theatre of conflict, 75 per cent of its tanks, 50 per cent of its armoured personnel carriers (APCS) and 90 per cent of its artillery; however, 50 per cent of Iraq's arsenal was left in reserve in Iraq, and 50 per cent of the Republican Guard's major combat equipment escaped destruction in the Kuwaiti conflict.[84] The third issue which caused the rest of the world to question America's military competence was its poor handling of the ceasefire talks that took place on 3 March at Safwan: it seemed strange that Schwarzkopf was willing to make any concessions whatsoever to the Iraqis and that no one in the US administration had foreseen more sinister uses for these helicopters. The fact that the US did nothing to rectify this oversight when the repression started further put American military credibility in question.[85] Although it is unclear how much more the Americans could have achieved militarily without drastically altering their war objectives, the fact that they let slip away an opportunity to render ineffective the war machine of a dangerous pariah state tarnished their military reputation and marred an otherwise perfect victory. *Operation Provide Comfort* and the safe haven enabled the US, as well as it allies, to recapture their earlier success and restore their military credibility.

The implementation of the safe haven

The safe haven in northern Iraq was perhaps the most successful of the safety zones examined in this book: it provided adequate security and sustenance for the Kurds within their country of origin; it successfully deterred through troops and air power any reprisals from Saddam Hussein; and it retained sufficient international support to avoid the fracturing of the 1991 Gulf War coalition. However, its one significant drawback was that its implementation was not linked to a broader solution to the Iraqi problem: the coalition that established it was eager to withdraw, so no clear provisions were put in place to reintegrate the safe haven and its surroundings into Iraq or to clarify the area's status generally beyond autonomy in a *de facto* sense. Although the coalition retained the no-fly zone above the 36th parallel, mainly as a means to contain Iraq, they did nothing about more localized incidents of Iraqi police repression[86] or about the mounting problem of the now penetrable border between Iraq and Turkey. In order to quash the Turkish Kurdish terrorist element that sought refuge in northern Iraq, the Iraqi Kurds were subjected to intermittent bombardment by Turkey,[87] as well as to the imposition of a three mile-wide buffer zone along the border that could be patrolled by the Turkish military. Due to Saddam Hussein's establishment of an economic blockade of Kurdistan, the Kurds had little recourse but to accept these measures, as they were entirely dependent upon the goodwill of Turkey to allow foreign aid and market goods into northern Iraq. An examination of state and community interests will enable us to reflect upon why the coalition effectively implemented the safe haven, but then left behind a troublesome situation.

State interests and the implementation of the safe haven

Effective short-term implementation of the safe haven arose mainly because the configuration of state interests was such that states were willing to assume the costs and risks of the operation, which turned out to be rather minimal. On the one hand, the cost of implementation was relatively minor considering that the military hardware and manpower needed to ensure the security of the haven was easily obtainable as a result of the preparations for the 1991 Gulf War. On the other hand, the fact that Saddam Hussein had just sustained a crushing defeat at the hands of the coalition meant that he proved unwilling to mount a renewed attack against allied soldiers in the North, even though he believed *Operation Provide Comfort* to be a violation of Iraqi sovereignty. As a result, adequate protection of the safe haven was easily ensured: state concerns, about military casualties for instance, which might have hindered effectiveness of implementation, were thus never triggered.

However, once the threat to international peace and security receded following the refugees' return home, the state interests that had initially impeded an early response to Iraqi repression reasserted themselves, making a long-term commitment to the Kurds an impossibility. With the Kurds resettled within Iraq, the specter of Kurdish independence reappeared, threatening once again American and allied geostrategic interests in the region. Soon after the creation of the safe haven, Kurdish leaders entered into autonomy negotiations with Saddam Hussein. At first, these talks seemed to progress well, but it soon became apparent that Iraq was stalling on key issues, such as control over the oil-rich city of Kirkuk, until coalition troops departed Iraq. Not keen to see Iraq dismembered or the region destabilized, the US and its allies showed no interest whatsoever in seeing these negotiations through or in pressuring Saddam Hussein to strike a just bargain with the Kurds. The coalition also failed to offer any guarantees of international protection to the Kurds, were a settlement to have been achieved.

In terms of domestic politics, the US government in particular faced pressure to withdraw from Iraq within weeks of Bush's announcement of the creation of the safe haven on 16 April 1991. Americans were anxious to see their soldiers return home and to avoid any possibility that the US would become mired in a Vietnam-style quagmire. A *New*

York Times editorial on 28 April said the 'Kurds cannot count on permanent military protection from abroad', and suggested that Britain, France and the United States hand over the protection of the safe haven to a UN force as soon as possible.[88] The US considered pulling out as early as mid-June, but was deterred from doing so by its European allies, who did not encounter such strong public opposition to continued military presence in northern Iraq. Britain, in particular, endeavoured to slow American withdrawal from Iraq by laying out four conditions that would need to be met for the safe haven to come to an end: replacement of allied forces with UN forces; conclusion of an autonomy agreement between the Kurds and Baghdad; continued sanctions against Iraq until such time as it was clear that the regime would not renew violence against the Kurds; and a clear warning to Iraq that any attack against the Kurds would be met with military retribution by the West.[89] Although Britain did not secure American backing for these proposals, it did manage to extract one concession from the US: *Operation Poised Hammer*, the 5,000 strong rapid reaction force stationed in Turkey until September 1991 as a deterrent to any misconduct toward the Kurds on the part of Saddam Hussein.

Community interests and the implementation of the safe haven

Once the humanitarian tragedy on the Turkish-Iraqi border was resolved, the appropriateness and legitimacy of the safe haven was put into question. In essence, community interest concerns about possible violations of 2(4) and 2(7) of the UN Charter reasserted themselves. Britain, France and America legitimized their safe haven approach as a temporary humanitarian operation, required as a result of the threat to international peace and security Iraqi repression caused and necessary in order to implement 688. However, once the Kurds were safely back in their homes, these rationales for the coalition's stay in Iraq disappeared, and international concerns about non-interference within the international affairs of other states reappeared. The coalition did contemplate the possibility of having a UN authorized peacekeeping force replace their troops upon departure, but never proposed such a scheme to the Security Council for fear of lack of support. The UN Secretariat adopted another tactic: it secured Iraqi agreement to the deployment of a UNGCI operating under the authority of the Secretary General. This plan was achieved on 25 May in an annex to the earlier MOU.[90] It allowed for the deployment of a maximum of 500 UN guards, with no more than 150 operating in any given area. Lightly armed with pistols, these guards were to protect the UN Humanitarian Centers. Once this alternative arrangement to coalition troops was in place, one that was consented to by the government of Iraq, the legitimacy of the safe haven operation was further reduced, thus increasing the sense that it should be withdrawn. Unfortunately, the UN guards proved unable to match the effectiveness of the coalition at providing adequate security for the Kurds;[91] and Saddam Hussein continuously restricted their numbers and placements. In the end, the Kurds relied on their own *Peshmerga* fighters for protection. In sum, the coalition faced strong external pressure to wrap up its military operation in northern Iraq due to its now perceived infringement of Iraqi sovereignty; and it was unable to implement a viable replacement peacekeeping force due to the same reluctance on the part of the international community to circumvent the legal rule of non-intervention in the internal affairs of states.

Conclusion

Competing and complementary community and state interests played important roles in shaping the safe haven policy in Iraq. Strong state interests that discouraged involvement, buttressed by a strong community acceptance of the importance of Articles 2(4) and 2(7) of the UN Charter, meant that states failed to intercede on behalf of the Kurds when their repression began in March 1991. Gradually, as the crisis deepened and the death toll mounted, the growing concern about the extreme humanitarian need of the Kurds enabled the pursuit of a novel approach to the problem, namely a safe haven. The creation of a safety zone was an ideal solution to the problem because it allowed states to assist the Kurds within their country of origin, while at the same time avoiding the disintegration of Iraq and putting an end to a threat to regional and international security. Short-term effective implementation was ensured because states were willing to enforce the zone and because they encountered no Iraqi opposition to that enforcement. However, once the humanitarian emergency was satisfactorily resolved, the reassertion of state and community interests that discouraged continued involvement in northern Iraq meant that the coalition withdrew promptly: no long-term solution to the Kurdish problem in Iraq was put forward.

3
Only So Far but No Further: Safe Areas in Bosnia

Figure 3.1 Safe areas in Bosnia, 1 July 1995[1]

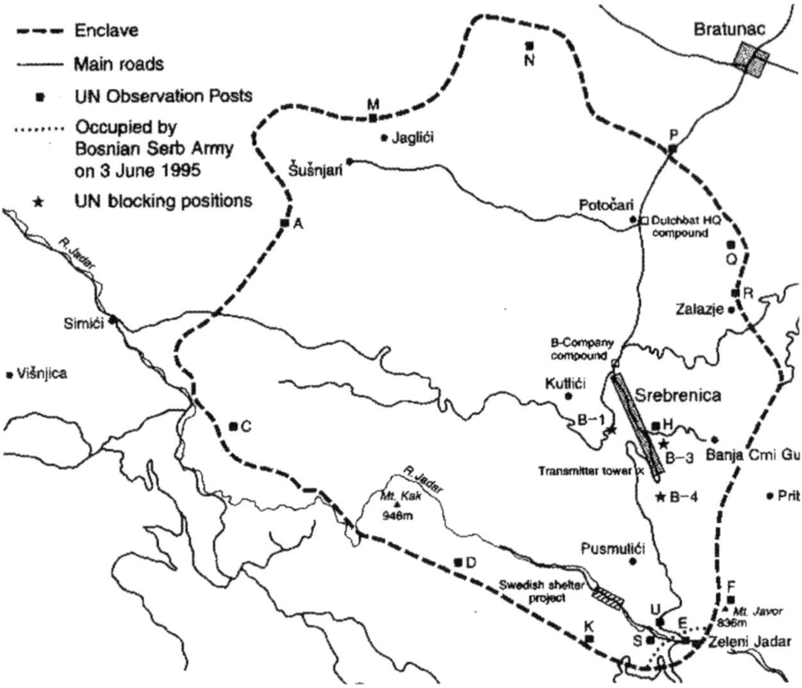

Figure 3.2 Srebrenica safe area[2]

With Srebrenica on the verge of capitulation after weeks of attack and siege by the Bosnian Serbs, the Security Council adopted resolution 819 on 16 April 1993, demanding that 'all parties and others concerned treat Srebrenica and its surroundings as a safe area which should be free from any armed attack or any other hostile act'. Originally introduced by the non-aligned and intended as a temporary measure designed to protect one of the eastern enclaves assigned to the Bosnian government under the Vance Owen Peace Plan (VOPP), then still on the table as a comprehensive settlement to the conflict in Bosnia, the safe area idea was extended at the insistence of France to five other areas through Security Council Resolution 824 on 6 May. It also became a key component of the Joint Action Programme agreed to on 22 May by Britain, France, Spain, Russia and the United States as their common approach to the war in Bosnia following the demise of the VOPP.[3] Clarification as to how the existing peacekeeping force on the ground in Bosnia, United Nations Protection Force (UNPROFOR), would implement the safe area policy occurred during the following weeks, in relation both to the number of troops required for the task and to the changes needed in UNPROFOR's mandate to accommodate the new policy. Although the safe areas were now fully integrated into the UN's mission in Bosnia, it was still hoped that a settlement to the conflict would soon obviate the need for their existence. However, the war dragged on for a further two and half years until the Dayton Accords of November 1995, as did the safe area policy, much to the consternation of the UN, UNPROFOR and the relevant states involved since several of the gravest crises in Bosnia in 1994 and 1995 revolved around the safe areas, their continued violation by the belligerents, the difficulties surrounding their implementation and the reluctance of states to either abandon or fortify the concept. Failure to implement the safe areas adequately eventually culminated in the worst massacres in Europe after World War II following the Bosnian Serb capture of Srebrenica in July 1995, which then resulted in a renewed international commitment to them, especially in Gorazde and Sarajevo, through NATO's *Operation Deliberate Force* two months later.

This chapter aims to understand why the international community opted for a safe area approach in Bosnia and to assess how effectively it implemented the safe areas throughout the remainder of the conflict. Safe areas were created and maintained because states were attempting to harmonize their competing interests, both of a state-centric and a community-oriented nature, and because states sought a way to reconcile differences between them through an ambiguous policy that was able to address a variety of concerns, both strategic and moral. An analysis that incorporates a broadened conception of interest will be helpful in trying to understand why governments adopted a safe area policy in Bosnia, but failed to do more to back it up, and why NATO air power was not used to better effect during implementation when it was the primary means through which attacks against the safe areas could be deterred.

The war in Bosnia prior to the safe areas

Origins of the conflict in Bosnia

The historical origins of the conflict in Bosnia stem from the fact that important fault lines of religious, ethnic and ideological division run through the Balkans, and most particularly through Bosnia, which in many ways was cobbled together from its diverse ethnic mix more as buffer zone between its more dominant neighbours Serbia and Croatia than as a political entity embodying the aspirations of a clearly-defined nation.[4] Bosnia was home to three different ethnic/religious groups: the Catholic Bosnian Croats (17.3%), the Eastern Orthodox Bosnian Serbs (31.4%), and the Bosnian Muslims (43.7%)[5], all living side by side in a veritable patchwork.[6] This last factor explains why a carve-up of Bosnia was exceedingly difficult to achieve as it was almost impossible for any group to achieve control over contiguous territory.

The 1992–95 conflict in Bosnia was integrally linked to the disintegration of the former Yugoslavia after the cessation of the Cold War. The end of communist rule signalled a period of uncertainty and instability in the Balkans, characterized mainly by the rise of nationalism in several of the former Yugoslav republics. Within this context of resurgent national aspirations for statehood on the part of the main ethnic groups within Yugoslavia, a climate of distrust and fear developed between the different republics. Although Serbia's leader Slobodan Milosevic publicly adhered to the idea of the preservation of Yugoslavia with a strong centralized government, his neighbours were apprehensive about Serbia's actual aims and about its dominance within the institutions of the former Yugoslavia, most notably the army. Believing independence would bring them greater prosperity and fearing that Serbia's outward support for Yugoslavia dissembled its hidden pursuit of a re-centralized Yugoslav federation with Serbia at its core, the other constituent republics moved to secede at the earliest opportunity. The 25 June 1991 declarations of independence of Slovenia and Croatia sparked the federal parliament in Belgrade to announce its intention to protect Yugoslav borders though armed intervention, thus precipitating several days of military confrontation between Yugoslav Army (JNA) troops and republican forces. Although the war in Slovenia ended quickly in a Slovene victory, the war in Croatia did not: fighting endured until 2 January 1992, during which time local Serbs with the help of the JNA managed to seize control of nearly one-third of the country. With a ceasefire in place, the Security Council approved the creation of a UNPROFOR on 21 February 1992[7] to help implement the Vance Peace Plan. UNPROFOR's headquarters were situated in Sarajevo.

Although discord existed within the EC as to whether or not to grant recognition to the former constituent republics of Yugoslavia, EC Foreign Ministers eventually recognized Croatia and Slovenia on 15 January 1992. Bosnia had also asked to be considered for statehood at this time, but its request was deferred, as the risk of ethnic conflict was deemed to be too great. The first months of 1992 revealed the extent of ethnic division within Bosnia. On 9 January, an assembly representing uniquely its Serbian population declared the establishment of an autonomous Republic of the Serbian People of Bosnia-Herzegovina. In a session boycotted by Serbian parties, the Bosnian Parliament nonetheless endorsed a decision to conduct a referendum to decide the republic's future. The result of this 29 February–1 March referendum plunged Bosnia-Herzegovina into war: 99.4 per cent of those who voted preferred Bosnian independence from the Yugoslav federation.[8] However, voter turnout was only 62.7 per cent due to the fact that the Serbs boycotted the referendum because they did not want to validate a process they knew would favour independence.[9] Alija Izetbegovic, President of the Republic of Bosnia-Herzegovina, proclaimed independence from Yugoslavia on 3 March, sparking Radovan Karadzic, leader of the main Serbian party, to declare the Serbian Republic of Bosnia-Herzegovina's loyalty to the 'all-Serb state of Yugoslavia'.[10] Although a majority of Croats voted in favour of independence, they did not necessarily endorse the same vision as the Bosnian Muslims who lobbied for a territorially intact Bosnia-Herzegovina with a federal structure of government. Many Croats wished to join the Republic of Croatia. Thus Bosnia was very soon marred by a three-way conflict until the Croats and Muslims agreed to a cease-fire and settlement in March 1994.[11]

Once fighting began in earnest in early April, Serb militias, backed by Serb leaders in Belgrade and armed by the JNA,[12] seized Bosnian towns in the eastern part of the country and began a siege of Sarajevo that would last throughout the war. The Serb campaign proved effective: in a two-week period, the cities commanding the roads into Bosnia from Serbia and eastern Bosnia were taken by Serb forces. With these major towns in their possession, the Serbs went about opening up the roads connecting them, thus creating a supply network that would allow them to persist in the consolidation of territory even though they constituted only a third of the population of Bosnia.[13] By August, Serb forces occupied almost 70 per cent of Bosnian territory,[14] a percentage of land they were to hold until *Operation Deliberate Force* and the Croat-Muslim offensive of September 1995. One reason explaining the battlefield success of the Bosnian Serbs was the help they received from Serbia. Although the Federal Republic of Yugoslavia, now comprising Serbia and Montenegro, denied the involvement of JNA troops in the conflict after their 10 May official withdrawal from Bosnia, the situation on the ground was much more complex. A very significant number of well-trained JNA soldiers were Bosnian Serb, so they remained behind, and the JNA Commander in Bosnia, General Mladic, became the commander of the Bosnian Serb Army (BSA). The retreating JNA also left most of its equipment, including aeroplanes and heavy weapons, in the hands of the Bosnian Serbs, thus swaying the military balance in their favour.[15]

Nature of international involvement during 1992

A few critical observations about initial international involvement in the conflict in Bosnia will suffice here to set the stage for the later analysis of safety zones. First and foremost, America chose to stand aside in Bosnia and let the Europeans take charge. Facing an election and not believing that any American interests were at stake, Bush did not assume a leadership role on Bosnia.[16] This state of affairs was welcomed by the Europeans, most notably France, because it presented the EC with an opportunity to strengthen its foreign-policy making capabilities consistent with ambitions for the development of a Common Foreign and Security Policy (CFSP) as reflected in the Maastricht treaty, and to show the US that it could solve problems in its own backyard. However, as tensions prevented unity within European ranks and as the conflict worsened, the venue for dealing with the Bosnian problem shifted to the UN. Here again the Americans followed the European lead. This characteristic of the initial international response – Europe heavily engaged and the US not taking the initiative – endured throughout much of the safe area policy until *Operation Deliberate Force*.

The second key aspect of the international community's early approach to Bosnia was containment, i.e. preventing the conflict from spreading. This policy was a hold-over from the response to the war in Croatia, namely the imposition of an arms embargo on the former Yugoslavia as a whole through Security Council resolution 713 of 25 September 1991. This embargo was kept in place throughout the entire conflict, thereby cementing into place the imbalance in armament between the Bosnian Muslims and the Bosnian Serbs. International discord over whether or not this embargo should be lifted will factor in important ways in the safe area analysis. A NATO-enforced no-fly zone was also established over Bosnia, and economic sanctions were imposed on Serbia.[17]

The third essential component of the early international response to Bosnia, which was to remain prevalent throughout the entire conflict, was avoidance of military intervention in the war itself. Instead, a two-pronged strategy was adopted. First, an international effort was made to help negotiate a peaceful settlement to the conflict. The International Conference on the former Yugoslavia, sponsored by the EC and the UN, was launched in August 1992 as the main institutional apparatus facilitating negotiations between the belligerents. It was chaired initially by Cyrus Vance and David Owen. Second, peacekeepers were sent into Bosnia in June 1992 in order to open Sarajevo's airport to humanitarian assistance. Although the Secretary-General opposed the deployment of a peacekeeping force in Bosnia at this time, saying that conditions there were not yet suitable,[18] the Council nonetheless authorized first the enlargement and mandate of UNPROFOR to include military observers for the airport[19] and second the deployment of 1,000 UNPROFOR troops stationed in Croatia to Sarajevo airport for the purpose of guaranteeing the safety of humanitarian personnel.[20] From here, 'efforts to alleviate the human suffering caused by the conflict included a progressive expansion of the UNPROFOR mandate to support the delivery of humanitarian assistance to people in need, by land and air'.[21] The same dynamic that engendered this mission creep, namely pressure upon states to do something, would also figure in the later safe area policy.

The creation of the Srebrenica safe area

The start of 1993 looked promising for a conclusion to the conflict in Bosnia. After six months of negotiations, Vance and Owen unveiled a peace proposal in January that they thought would be acceptable to all parties, but nonetheless preserve some semblance of a multi-ethnic unitary state. The VOPP incorporated three key elements.[22] First, it planned for the reorganization of Bosnia-Herzegovina into ten provinces: three mostly Croat; three mostly Serb; three mostly Muslim; and one, Sarajevo and its surroundings, open, mixed and demilitarized. Serbs would control about half of the territory, but figures were not elaborated. A map, dividing Bosnia into ten provinces, was still being worked out. Second, it promulgated constitutional principles for the Republic, allowing for a large measure of autonomy for the provinces within a decentralized state. The central government and the provinces would have separate elected legislatures, and the nine-member presidency would comprise three representatives of each ethnic group. Thirdly, a cease-fire and demilitarized arrangement would be brought into effect under EC and UN supervision. Five major corridors between the provinces were to be established and policed by UN and other forces. The Bosnian Croats signed the entire package immediately.

Both the Bosnian Muslims and the Bosnian Serbs accepted the outlines of the plan, but refused to sign the agreement pending further negotiations regarding the map. In March 1993, the Bosnian Muslims finally relented under great international pressure, but Radovan Karadzic refused to sign it pending a decision of the Bosnian Serb Assembly in late April. In the meantime, the Bosnian Serbs carried on with their ground offensive in eastern Bosnia.

One enclave of particular concern within this area, destined to fall within a Muslim controlled province under the VOPP, was Srebrenica. A few weeks into the war it had initially yielded to the Bosnian Serbs, but the charismatic Muslim military leader Naser Oric wrestled the town back again before the end of 1992. The Bosnian Serbs launched a counteroffensive that erased Muslim gains. By mid-March 1993, the town was entirely surrounded and under siege with the Serbs blocking access to the humanitarian convoys aiming to deliver food to its now swollen population of 60,000.[23] Conditions in the town were horrendous, and the vast majority of the population was exhausted, weak and starving. Fearing the town was on the verge of collapse, General Philippe Morillon, UNPROFOR Commander in Bosnia, decided, unbeknown to his superiors and certainly without their permission, to make a foray into Srebrenica.[24] Once there, the civilians of Srebrenica would not let him leave again. In what the French Inquiry into Srebrenica calls an *'intuition brillante'*,[25] Morillon hoisted the UN flag from the local post-office and told the assembled crowd: 'You are now under the protection of the United Nations. I will never abandon you'.[26] He saw his words as a temporary promise only necessary until such time as the VOPP went into effect. Although Morillon managed to secure an agreement with Milosevic that enabled UN convoys back into Srebrenica and permitted the evacuation of some of the refugees to Tuzla, this small improvement in the circumstances of the town did not endure: the Serbs renewed their attack on the town in early April and once again cut it off from outside aid. The town's plight grew desperate, and it sustained intense, indiscriminate artillery attacks that killed innocent civilians.[27] No longer able to defend their town, Srebrenica's leaders sent a secret message via UN officials responding in the affirmative to Bosnian Serb demands for surrender negotiations, information which was then leaked to the press and made international news. Within days, Morillon had secured a MOU between the parties about Srebrenica's fate and the Security Council had declared the town a safe area.

Community interests and the Srebrenica safe area

In a pattern that became familiar in the international response to the conflict in Bosnia, crisis episodes on the ground involving serious violations of international law would act as catalysts for more concerted action toward the war. Morillon's visit to the town and subsequent media coverage served to highlight the extent of Bosnian Serb barbarity, as well as the fact that another important portion of Bosnian territory was about to fall and be ethnically cleansed due to the aggressive military tactics of one of the belligerents. These threats to important community interests triggered the intensive diplomacy on the part of the non-aligned for a safe area policy, as well as a new openness on the part of the Western powers to consider such a concept. Such blatant transgression of norms considered fundamental to international coexistence was also often a focal point that kept states united in a common effort despite their divergent interests.

Community norms also enabled the development of the safe area policy due to the permissive environment they generated within the Security Council, allowing for novel actions, unlikely in other circumstances, to be undertaken with legitimacy and widespread support. Although all sides in the conflict in Bosnia committed violations of humanitarian law, the Bosnian Serbs were the worst offenders, mainly because they engaged in the practice of ethnic cleansing on a greater scale than the rest, because they undertook mass killings in a manner not seen in Europe since World War II, and because they attempted to destroy militarily the very fabric of a multi-ethnic Bosnia through territorial conquest and siege.[28] From the start of the conflict in the Spring of 1992, there emerged in both the deliberations and the resolutions of the Security Council a growing awareness of and stance against two principal types of violations of international law committed by the Bosnian Serbs: 1) the conquest of territory through the use of force; and 2) significant violations of humanitarian law in the form of ethnic cleansing,[29] obstruction of the delivery of humanitarian assistance, and laying siege to towns and villages such as Sarajevo. The first behaviour violates the spirit of the UN Charter.[30] The latter actions are violations of the Geneva Conventions and their Protocols; and some may, in their worst form, constitute crimes against humanity and possibly genocide. Such violations open the door to third party involvement due to the generality of standing accorded to these rules. When violations occur, not only is the state directly affected permitted to take action to rectify the situation, but other states are as well.

With the precedents of Iraq and Somalia firmly established, and operating within the sphere of influence of the Western permanent members, the Security Council was prepared to go far beyond the cautious nature of resolution 688 condemning Iraq for the repression of its civilian population. In fact, deliberations within the Council and the resolutions it adopted made numerous and clear references to the violations listed above. Almost every Security Council resolution adopted concerning the conflict in Bosnia prior to the safe area policy, as well as numerous presidential statements, either reminded the parties to cease their criminal behaviour or demanded that they do so. For example, resolution 764 of 13 July 1992 reaffirmed 'that all parties are bound to comply with the obligations under international humanitarian law and in particular the Geneva Conventions of 12 August 1949'. Perhaps the most forceful resolution in this regard was 787 of 16 November 1992, in which the Council, acting under Chapter VII of the Charter,

> 2. Reaffirm[ed] that any taking of territory by force or any practice of 'ethnic cleansing' is unlawful and unacceptable, and will not be permitted to affect the outcome of the negotiations...[and]
> 7. Condemn[ed] all violations of international humanitarian law, including in particular the practice of 'ethnic cleansing' and the deliberate impeding of the delivery of food and medical supplies to the civilian population...

787 also specifically singled out the Bosnian Serb paramilitary forces as the key transgressors, and urged them to comply with all resolutions pertaining to the Bosnian conflict.[31]

Not surprisingly, the preamble to Security Council resolution 819, which declared Srebrenica a safe area, also alluded to the serious breaches of international law perpetrated by the Bosnian Serbs: it stated that the Council was 'concerned by the pattern of hostilities by Bosnian Serb paramilitary units against towns and villages in eastern Bosnia, and in this regard reaffirming that any taking or acquisition of territory by the threat or use of force, including through the practice of "ethnic cleansing", is unlawful and unacceptable'. Within this context of clear Security Council acknowledgement of violations of *erga omnes* legal rules, conditions developed in the Council that enabled the pursuit of a policy that would not have been self-evident under other circumstances. Provisional verbatim records of the Security Council also confirm the extent to which states considered it appropriate and possibly imperative for the international community to 'do something' to alleviate further suffering of the civilian population and to find a way to end ethnic cleansing and territorial conquest.[32] For example, Cape Verde confirmed that 'the "ethnic cleansing" and other grave violations of humanitarian law taking place in Bosnia-Herzegovina...are crimes committed against us all, for they violate our very decency and human dignity'.[33]

The fact that such crimes were taking place in Europe also seemed to heighten the importance of community interests as aspects of identity. There was a strong sense that similar crimes perpetrated fifty years earlier by the Nazis had been carefully overcome through the nurturing of a 'civilized' international community with codified and enforceable laws preventing a return to such uncivilized practices.[34] The end of the Cold War was regarded as reinforcing this underlying system of values governing relations between states, which atrocities in Bosnia threatened to undermine, particularly because they implied that if crimes of this magnitude could occur in Europe with impunity, then the very idea of a 'civilized' international community would be at risk. There was also considerable concern that the Security Council's and the UN's reputation and credibility would be irrevocably tarnished.[35] Two groups of states also felt a particular affinity resulting from the identity aspect. On the one hand, the Islamic world felt drawn toward their religious kin, the Bosnian Muslims, and thus was all the more concerned when this group endured massive violations of their rights. On the other, the Europeans, and particularly the French, considered that they had to be present on the ground in Bosnia because they could not not-be.[36] As Bosnia was part of Europe and as Europeans were facing unspeakable brutality, it was unthinkable that other European states would not intervene in some way.

A safety zone concept had been floating around for some time prior to April 1993 as a result of the Security Council being so open to measures that might deter or halt grave breaches of international law. In October 1992, the ICRC distributed a position paper in which it called for the establishment of protected zones for civilians. Concerned that Muslim civilians were being terrorized and that neither neighbouring nor third party states were willing to accommodate refugees, the ICRC suggested that protected zones, grounded in international humanitarian law, but distinctive in important ways, be set up in Bosnia. It recommended that, prior to the creation of any such zones, the parties to the conflict agree to them, and that international responsibility for them, in terms of adequate security and UNPROFOR presence, be established.[37]

The idea was also supported by Tadeuz Mazowiecki, UN Special Rapporteur on the Situation of Human Rights in the Territory of the Former Yugoslavia, in several of his early reports.[38] A few states also championed the idea in the fall of 1992, Austria commenting in the Security Council that 'Austria sees only one possibility to solve this problem, that is to create safe areas under military protection within Bosnia-Herzegovina itself. In these areas the civilian population...could find refuge and live protected by peacekeeping forces'.[39] They effectively incorporated a clause into resolution 787 of 16 November 1992 inviting the Secretary-General to study the possibility of and requirements for a safe area policy.

These early proposals were not pursued in a serious way until April 1993, when state and community interests interacted in such a way as to render safe areas a viable option. The terrible predicament of the town of Srebrenica in March and April 1993 meant that the constraining effect of community norms became more potent. General Morillon's visit to Srebrenica in March 1993 played a key role for a number of reasons. First, it generated sustained media focus on the plight of Srebrenica.[40] Such regular and persistent coverage ensured that public pressure for something to be done would mount and become a domestic political issue. This was true in France where *le devoir d'ingérence* was a strong component of French political culture as discussed earlier,[41] and in Britain where media discussion of the Bosnia crisis was extensive and well-informed. It was also the case in the United States where Clinton's strong rhetoric in the election campaign about taking a tougher stance on Bosnia and preventing ethnic cleansing now encouraged at least symbolic commitment to the enclave,[42] already demonstrated in February 1993 when the US air-dropped humanitarian aid into the Srebrenica pocket. The vitriolic attacks against the VOPP in certain American elite media circles, with the main criticism being that the agreement rewarded ethnic cleansing, indicated frustration at the administration's inability to put an end to abuses and pressured Clinton to do something in aid of the Bosnian Muslims.[43]

Second, the promise Morillon made to the people of Srebrenica put pressure on both himself and the UN to honour that pledge in some way. The Office of the United Nations High Commissioner for Refugees (UNHCR), the lead relief agency in Bosnia, felt acute pressure to do something about the situation, particularly as it was placed in a difficult bind. On the one hand, it was unable to get the convoys of food and aid into Srebrenica due to Serbian intransigence; and on the other, it feared abetting ethnic cleansing if it emptied Srebrenica of its Muslim population, especially as this course of action was strongly opposed by the Bosnian government.[44] This dilemma led Sadako Ogata, UN High Commissioner for Refugees, to send a letter to the Secretary-General on 3 April 1993, which Boutros-Ghali then shared with the Security Council. She stated that either the enclave needed to be protected by the UN, or its civilian population evacuated with the help of UNPROFOR.[45] Likewise two non-aligned members of the Security Council, Venezuela and Pakistan, were greatly affected by the UN commitment to Srebrenica, and succumbed to the constraining effect of community norms.[46] They took it upon themselves to convince the Council that something had to be done, and that the second option proposed by Ogata was inconceivable given that it would reward ethnic cleansing and the conquest of territory by force.

In Bosnia, more so than in Iraq and Rwanda where troops under national command implemented the safety zones, the safety zone approach was facilitated instrumentally by the UN and UNPROFOR. However, having the Security Council in charge of making political decisions in New York and the UN administrative unit, here comprising the secretariat and UNPROFOR, in charge of carrying out activities on the ground led to rather serious coordination problems. Often communication between UNPROFOR and the Security Council was inadequate, the Secretariat not able to provide the Council with sufficiently up-to-date facts for it to make informed decisions, and Council members reaching decisions for state and community interest reasons without making sure with UNPROFOR and the secretariat whether or not such objectives were conceivable in practice. As a result, in this first phase of the safe area policy, there was frequently a sharp deviation between what was envisioned by the Security Council and what was instituted on the ground in Bosnia.[47]

This divergence was already apparent even as Srebrenica was 'saved' from the Bosnian Serbs. With UNPROFOR already present in the field, it was easy for the Security Council to incorporate new tasks into its mandate, but the problem was that demanding that 'all parties and others concerned treat Srebrenica and its surroundings as a safe area which should be free from any armed attack or any other hostile act'[48] was more of an affirmation of what the international community would like to see take place in Srebrenica than an effective means of actually stopping the threat to Srebrenica. Airing mainly the intent behind the policy, 819 was meant to be a strong chastisement of the Bosnian Serbs: it showed the determination of the international community to oppose Serb intransigence, but did not specify how UNPROFOR would be able to implement the resolution on the ground from a position of weakness and reliant upon the consent of the belligerents, including the Serbs. At this early stage of the safe area policy, no mention had yet been made of the possibility of using NATO air power as a dissuasion against attacks on Srebrenica.

While New York moved toward passing 819, Morillon was in Bosnia working to cement a cease-fire deal that would somehow preserve Srebrenica as a Muslim enclave despite overwhelming Bosnian Serb superiority on the ground. He was in a difficult position. On the one hand, it was unthinkable for UNPROFOR to negotiate a surrender at this time, as such an action would be contrary to the diplomatic peace effort under way. On the other, there was as yet no way to deter the takeover of Srebrenica from a military point of view, and little leverage existed to pressure Mladic to abandon his campaign to overrun the town. All Morillon could do was try and recast surrender talks as cease-fire negotiations, as well as turn Srebrenica into something like the equivalent of a demilitarized zone under Article 60 of 1977 Geneva Protocol I. In talks convened at Sarajevo airport on the night of 17 April 1993, Mladic and Bosnian General Sefer Halilovic ironed out the terms of what both men called a surrender, but which UNPROFOR labelled a 'breakthrough toward disarmament'.[49] Depending on how one looked at the MOU that emerged from the meeting, it was either a success, preventing the Muslim enclave from being overrun, allowing for the evacuation of the seriously wounded and sick to Tuzla, guaranteeing access by humanitarian convoys and allowing for the presence of Canadian peacekeepers to oversee the demilitarization of the town; or a failure, placing UNPROFOR in the awkward position of disarming only the Bosnian Muslims and, in a sense, of carrying out demilitarization on Mladic's behalf.[50] This agreement certainly did not embody the spirit of Security Council resolution 819, and even Boutros-Ghali in his report pursuant to 819 was forced to admit that 'the Srebrenica arrangement [could not] serve as a model' for other safe areas.[51]

State interests and the Srebrenica safe area

State interests also played an important role in the passage of the Srebrenica safe area resolution, mainly because they encouraged a stance against Serbian intransigence but curbed more meaningful military action in the war itself. Constraining state interests meant that it was highly unlikely that Western states would intercede militarily in the conflict in Bosnia in order to impose a settlement upon the parties.

Following the end of the Cold War, France, the UK and the US did not have vital interests at stake in Bosnia. Britain, for example, viewed the conflict in former Yugoslavia very much through a historical lens, believing it erupted as a result of ancient animosities that would be difficult to overcome.[52] It was one thing for the UK to take part in a limited peacekeeping intervention as a means to demonstrate international solidarity and British status on the world stage,[53] but quite another to take sides in a civil war. The French concurred. Although France was more open to the use of force for humanitarian purposes than the British were, pushing for UNPROFOR's involvement in the delivery of aid for instance, it never contemplated a military intervention.[54] As French President Mitterrand repeated often enough: France would not 'add war to war'.[55] Finally, the Clinton administration did not believe there existed sufficient domestic consensus, within either Congress or public opinion, to enable the sending of American ground troops to Bosnia. With military intervention unambiguously ruled out, there would be no concerted military action to 'save' Srebrenica.

Although the Europeans and the Americans could not agree upon a formula to end the conflict in Bosnia, in terms of whether to lean toward ethnic partition or preservation of a multi-ethnic state and whether to achieve it through negotiation or coercion,[56] they did see eye to eye on the necessity to contain the conflict and its consequences, namely wider instability and the generation of even larger numbers of refugees. Neither the Europeans nor the Americans wanted to see a Serbian hegemony of the Balkans, and allowing the Serbs to overrun Bosnia risked destabilizing the region. Complete inaction would also set a negative precedent for other ethnically-contested parts of Eastern Europe and the former Soviet Union emerging from communist rule. To find a way not to let Srebrenica fall without having to defend it was thus an ideal solution: the Serbs would not capture yet more territory in Bosnia; the refugees would stay home; and the creation of safe areas would neither commit the West to ethnic partition through negotiation nor to a multi-ethnic Bosnia preserved through force.

The Europeans also considered that declaring Srebrenica a safe area would facilitate the implementation of the VOPP, a peace plan they had officially endorsed. The fall of this eastern enclave would entail one more town to wrest back from the Bosnian Serbs through negotiation and would further weaken the position of the Bosnian government at the table. As the VOPP was already faltering at this point due to Bosnian Serb unwillingness to commit to the plan, it was felt that an impetus or clear indication of the international community's willingness to see a settlement reached might help the process along. Declaring Srebrenica a safe area was seen as a temporary measure 'which had to fill a gap before there was a peace settlement'.[57]

The US voted in favour of 819 even though it did not fully support the Srebrenica safe area concept, mainly because Clinton needed to appear to be engaged on the Bosnian question in the absence of a clearly developed US government position on the conflict. Although Clinton's campaign for office had mostly focused on domestic issues such as improving the economy, the one area of Bush's foreign policy where Clinton felt there was room to offer criticism was its handling of the war in Bosnia. Arguing that the US needed to do more in Bosnia to put an end to the conflict and to stop the slaughter, Clinton was also able to counter charges that, like for past Democrats such as Jimmy Carter, foreign policy would not be his administration's forte.[58] However, once elected, Clinton discovered that the Bosnia question was far more difficult and divisive than he had imagined. Bringing together Vice-President Al Gore, Secretary of State Warren Christopher, Secretary of Defence Les Aspin, National Security Advisor Anthony Lake, Joint-Chief of Staffs Chairman Colin Powell and US representative to the UN Madeleine Albright, the Principals Committee in charge of formulating foreign policy began meeting regularly in January 1993, but failed to reach a consensus. Whereas Albright, Lake and Gore generally favoured a stronger approach to Bosnia, Aspin and Powell[59] were much more cautious: they were not certain that air strikes alone could achieve anything, and favoured containment and a negotiated settlement. Christopher remained undecided and tended to vacillate in whichever direction Clinton seemed to be leaning.[60] Clinton found it impossible to set a course in favour of one path or another and kept postponing a final decision.

So US policy on Bosnia was initially framed mainly in terms of negatives – what the administration would not do as opposed to concrete measures it would adopt. Through a preliminary statement made by Christopher on 10 February 1993, the US made clear that it intended to contribute directly to the humanitarian relief effort, to continue enforcing the no-fly zone and to tighten sanctions against Serbia. America also stated its intention at this time to provide ground troops to help enforce a peace agreement that was acceptable to all parties to the conflict. However, on a number of other points, the administration was less precise. The statement hinted at serious reservations and ambivalence toward the VOPP. First, Clinton appointed an envoy to the peace talks the Europeans considered to be a done-deal,[61] suggesting that he was not pleased with the final outcome. Second, the US rejected the possibility of imposing a settlement that was not voluntarily accepted by all parties, thus weakening the likelihood that the VOPP would prevail since coercion was apt to be necessary to induce the Bosnian Serbs to relinquish some of the territory they had seized. Christopher later acknowledged that although the US 'admired the Vance Owen effort, we thought the plan too complex to be implemented'.[62] Yet, nothing in the February statement indicated what alternative the US envisaged as a replacement, mainly because it did not have one. In the absence of clearly defined state interests, the US government acquiesced in the Srebrenica safe area concept mainly because it allowed the US to engage in a new initiative and fulfill, in some measure, its community interests. Yet, its vote in favour of 819 was at best lukewarm: on the day Srebrenica was designated as a safe area, Clinton said at a press conference, 'At this point, I would not rule out any option except the option I have never ruled in, which was the question of American ground troops'.[63]

State interests as enablers also played an important role for Islamic countries and the non-aligned. To them, Bosnia was seen in equivalent terms as an alliance partner in need of assistance. If they had been strong enough to mount a military operation on their own, or influential enough to convince the international community to do so, they would have. As an alternative, their primary aim was to see the arms embargo imposed upon Bosnia by Security Council resolution 713 lifted so that the government could exercise its right to self-defence as laid out in Article 51 of the Charter and thus reverse territorial losses. But here again they were frustrated by the veto-holders in the Security Council. The next best option was a safety zone that at least reduced the likelihood that Srebrenica would fall to the Serbs. Not surprisingly, Venezuela and Pakistan were the two members of the Security Council that lobbied the hardest to get resolution 819 adopted.

The extension of the safe area concept and strategy for implementation

Designating Srebrenica a safe area was an *ad hoc* response to a crisis: community interests dictated that something be done, and state interests enabled acceptance, albeit hesitant, of the policy, perhaps explaining why no clear enforcement measures were considered at the time 819 was adopted. In contrast, the subsequent safe area policy was thought-out and deliberate. It was spearheaded by France with the support of the other major powers. Although much remained constant in terms of the configuration of community and state interests, new elements entered the equation that permitted the expansion of the concept to other areas, the determination of how it would be implemented in terms of troop numbers and UNPROFOR's mandate, and the adoption of enforcement measures in the form of a NATO air power deterrent. Despite this strengthening and clarification of the concept, states remained divided and ambivalent toward the new policy, laying the foundations for its eventual inadequacies in terms of coherence and enforcement.

The adoption of the safe area policy coincided with the demise of the VOPP. In May 1993, one last attempt was made to convince the Bosnian Serbs, dissatisfied with the territorial arrangements outlined in the settlement, to consent to the plan, but it ultimately proved futile: they overwhelmingly rejected the plan by 96 per cent in a 15–16 May referendum.[64] In the midst of this policy vacuum a broader safe area policy emerged. First, at the behest of the non-aligned but more strongly supported this time by others, the Security Council adopted resolution 824 on 6 May 1993 which declared that 'the capital city of the Republic of Bosnia and Herzegovina, Sarajevo, and other such threatened areas, in particular the towns of Tuzla, Zepa, Gorazde and Bihac, as well as Srebrenica, and their surroundings should be treated as safe areas by all the parties concerned and should be free from armed attacks and from any other hostile act'. As a result of a mission to the enclave by six Security Council members in late April, the Council was now well aware of the circumstances of both the Srebrenica agreement between the parties and of the gap between the spirit and intent of resolution 819 and actual conditions on the ground.[65] So it added a clarification of its intention in relation to the safe areas. 824 specified the following should be observed:

> The immediate cessation of armed attacks or any hostile act against these safe areas, and the withdrawal of all Bosnian Serb military or paramilitary units from these towns to a distance wherefrom they cease to constitute a menace to their security and that of their inhabitants, to be monitored by United Nations military observers.

Second, the Joint Action Programme sponsored by the Americans and endorsed by the French, British, Spanish and Russians on 22 May included safe areas as one of its main policies. It announced that British and French troops would serve in the safe areas, as well as forces from a number of other countries, including Spain and Canada. Although no US troops would be made available for this task, the Programme specified that 'the United States is prepared to meet its commitment to help UNPROFOR forces in the event they are attacked and request such action', a pledge that was to play a key role in the endgame phase of the conflict.[66] Finally, on 4 June, the Security Council enacted UNPROFOR's new mandate in relation to the security zones through the adoption of resolution 836. The Council, acting under Chapter VII of the Charter, stipulated several points, three of which will feature in the discussion below and will be quoted here:

> 5. Also decides to extend to that end the mandate of the United Nations Protection Force in order to enable it, in the safe areas referred to in resolution 824 (1993), to deter attacks against the safe areas, to monitor the cease-fire, to promote the withdrawal of military or paramilitary units other than those of the Government of the Republic of Bosnia and Herzegovina and to occupy some key points on the ground, in addition to participating in the delivery of humanitarian relief to the population as provided for in resolution 776 (1992) of 14 September 1992;
> 9. Authorizes the Force, in addition to the mandate defined in resolutions 770 (1992) of 13 August 1992 and 776 (1992), in carrying out the mandate defined in paragraph 5 above, acting in self-defence, to take the necessary measures, including the use of force, in reply to bombardments against the safe areas by any of the parties or to armed incursion into them or in the event of any deliberate obstruction in or around those areas to the freedom of movement of the Force or of protected humanitarian convoys;
> 10. Decides that, notwithstanding paragraph 1 of resolution 816 (1993), Member States, acting nationally or through regional organizations or arrangements, may take, under the authority of the Security Council and subject to close co-ordination with the Secretary-General and the Force, all necessary measures, through the use of air power, in and around the safe areas in Bosnia and Herzegovina, to support the Force in the performance of its mandate set out in paragraphs 5 and 9 above;...

State interests and the extension of the safe area concept

New developments in terms of state interests mainly encouraged a shift toward a broader safe area policy, so these will be discussed first. Britain, France and the United States remained opposed to military intervention to impose a settlement, so continued to favour an approach in which they could act against the Serbs without taking military action against them, something an extended safe area policy made possible. However, the real impetus for resolutions 824 and 836 was a whole new set of state interests as focal points and as enablers that appeared when the US finally launched a diplomatic initiative to end the war in early May, which risked causing a major rift in the Atlantic Alliance. Preventing that rupture caused the US to return to a policy of indecision and acquiescence, and encouraged France to take the lead on a safe area policy designed to mend, at least superficially, the divide.

Recognizing that the US had to assume a more assertive role in the conflict, Clinton and the Principals Committee finally hammered out a Bosnia policy on 1 May 1993, after which Christopher flew to Europe to sell 'lift and strike' to America's alliance partners. The policy proposed that air strikes be used to roll back Bosnian Serb conquest of territory, and that the arms embargo be lifted so as to level the playing field and ensure the Bosnian government could defend itself in renewed fighting. Christopher's mission was an absolute failure: the Europeans were aghast. PM John Major even said his government would fall if he endorsed the proposal.[67] Although they could recognize the sentiment behind the policy, they considered it totally unacceptable in pragmatic terms. First, it would place their troops on the ground in danger of retaliation by the Bosnian Serbs.[68] Second, it might jeopardize their aim of containing the conflict as Serbia, Russia, the Islamic world and others might be sucked in on opposing sides. And finally, it was not clear that the policy would achieve its stated aim anyway: it would take time for the Muslims to reach the military strength of their Serb counterparts, and the latter might push for an all-out victory in the short-term.[69] Clinton's policy was so disliked by the Europeans, it was often referred to derogatorily as 'lift and pray'.[70]

This rift between the Europeans and the Americans, as well as a commitment to healing it, had a number of consequences that enabled the development of a safe area policy. In the first instance, the Americans backed down. A cornerstone of Clinton's approach to Bosnia was multilateralism. He did not want to adopt a unilateral role in Bosnia, as that would entail America being saddled with the

Bosnian problem all on its own, a responsibility that might hinder Clinton's main domestic policy objectives.[71] Even before going to Europe, Christopher was coached by Clinton to take a conciliatory approach and not present the initiative as a *fait accompli* shift in American foreign policy.[72] So, when the Europeans balked, Clinton did not press. Related to this preference for multilateralism was the American administration's goal of strengthening and expanding the NATO Alliance.[73] Part of Clinton's foreign policy platform, this belief in NATO's importance as the institution most capable of guaranteeing security in Europe and North America post-Cold War further ensured that Clinton would not urge a showdown. In essence, America's interest in NATO proved stronger than its interest in ending the war in Bosnia, which remained true as long as action taken to achieve the latter aim threatened to damage NATO unity. Important to the success of the NATO policy was also the maintenance of good relations with the Russians, which again encouraged the US to not proceed with 'lift and strike' as they also opposed it. Finally, Clinton himself, as embodiment of the government's executive power, was not able to overcome nagging doubts that the US might not have a 'dog in that fight' after all. Reading *Balkan Ghosts* by Robert Kaplan as Christopher was in Europe, Clinton moved toward the position that Bosnia constituted a century's old ethnic conflict where each of the parties was as nationalistic and as blameworthy as the other, and where foreign involvement was likely to backfire.[74] Fearful of being drawn into a quagmire, Clinton had already reneged on 'lift and strike' before Christopher returned to Washington.[75] The US had reverted to a position of indecision and acquiescence, ready to go along with what the Europeans, and particularly France, proposed as the next best option, safe areas.

In relation to the opening of Sarajevo airport, the French took the initiative following the collapse of the VOPP and the failure of 'lift and strike'. Such assertiveness stemmed from a number of state interests. First was the sense that Europe needed to play a key role in solving disputes in its own backyard. With the Germans constrained by their constitution and the British more cautious about too great an involvement in the Balkans, it was up to the French to step forward.[76] The critical factor encouraging their engagement at this juncture was the imperative to suggest an alternative to the American proposal of 'lift and strike'.[77] Rejecting 'lift and strike', especially in the context of the policy vacuum following the demise of the VOPP, meant that another option had to be put forward, one that made concessions to the American position if the rift in the transatlantic community were to be overcome.[78] So France began to work out the mechanisms for how a safe area policy could be implemented. The significant concession to the Americans was the idea that air power could be used to deter attacks against the safe areas, which opened the way toward the possibility of air strikes against the Bosnian Serbs. America's concession to the Europeans was its acquiescence in a safe area policy and its promise to come to the aid of their troops in Bosnia if the occasion arose.[79]

On 13 May, France delivered a non-paper to the US, UK and Russia elaborating their thinking on the safe area concept, with the hope of moving toward a new resolution that would outline the modalities of how the policy could be implemented. As David Owen explains, the French 'underlined that there was no question of reconquering territory occupied by the Serbs, nor of providing complete armed protection around the designated safe areas, since that would require a massive UN presence of 50,000–60,000 troops on the ground. But 10,000–12,000 would suffice to cover – sanctuarize – the areas',[80] especially if air power were added to the equation. A more refined version of the non-paper was then shared with the Security Council in a memorandum of 19 May 1993. It stated that the aim of the safe areas should be 'to stop territorial gains by the Serbian forces and to achieve a negotiated settlement', and it included three options, a very light option, involving only military observers, a light option of about 10,000 troops and a heavy option of about 45,000 troops. Only this heavy option could 'oppose aggression' against the safe areas. The light option, the most likely scenario, could 'deter aggression, monitor the cease-fire, occupy some key points on the ground, and participate in relief operations'. The criteria for triggering air power were defined as 'shelling of the safe areas, armed incursion into the safe areas and impeding the free movement of UNPROFOR'.[81] This memo indicates that France played a key role in formulating Resolution 836's underlying elements, such as the notion of 'deterring attacks against the safe areas' and of air power being triggered under certain conditions. Relating this use of force to self-defence also limited the scope of the resolution in conformity with general state interest preferences for non-military involvement.

A final state interest that enabled a broader safe area policy was preventing the spread of refugees. UNHCR reckoned that, by 27 March 1993, 1.8 million Bosnians (40 per cent of the population) were already refugees or internally displaced persons.[82] Croatia and Slovenia had closed their borders to further influx in July 1992[83] and the EU was worried about taking in any more refugees than it had already.[84] In the previous year, Germany had proposed a quota system for acceptance of refugees, which was firmly rejected by Britain and France among others who argued that refugees should be accommodated and given assistance as near as possible to their place of origin.[85]

Community interests and the extension of the safe area concept

Community interests also encouraged the expansion of the safe area concept. In the wake of the transatlantic rift, as well as the policy vacuum stemming from the VOPP failure, the harmonizing effect of normative concerns helped states, namely the Europeans and the Americans, overcome their differences. More importantly, community interests as enablers and as constraints played a significant role in the drafting process of 824 and 836, as well as in the practicalities of developing a workable mandate for UNPROFOR on the ground.

The group that expressed the strongest commitment to community norms in the context of the extended safe area policy was the non-aligned: for them it was imperative that something be done to protect the Bosnian Muslims from ethnic cleansing and their enclaves from aggression. The Venezuelan and Pakistani delegates in particular, having travelled to Srebrenica in late April with the Security Council mission, were determined to make certain that the safe area policy lived up to the spirit in which the resolutions were adopted – primarily as a stance against barbaric practices of the Bosnian Serbs.[86] Relying on the fact that it would be difficult for states to deny the Bosnian government the right to self-defence as outlined in Article 51 of the Charter, the non-aligned, this time in conjunction with the US, were able to insert particularly telling phraseology into 824 and 836 that singled out the Serbs for enforcement measures. 836 specified for instance that UNPROFOR was to 'promote the withdrawal of military or paramilitary units other than those of the Government of the Republic of Bosnia and Herzegovina'. In theory, UNPROFOR was not to disarm or touch the paramilitary units of the Bosnian government. To do so, and at the same time to only deter attacks against the enclaves as opposed to protecting them, would have been a violation of the Bosnian Muslims' right to self-defence. Similarly, further weakening the military strength of the Bosnian Muslims, in the context of the unfair arms embargo, also jeopardized their right to defend themselves.[87] The Clinton administration also felt strongly that the arms embargo was an infringement of the Bosnian government's right to self defence,[88] so Albright supported the non-aligned on this aspect of 836.[89] This component of resolution 836 favouring the Bosnian Muslims over the

Bosnian Serbs arguably most blurred UNPROFOR's impartiality on the ground, which would have important consequences during the implementation phase. As the main instigators of resolutions 819 and 824, the non-aligned also felt a responsibility to make sure the safe areas did not lull the Bosnian Muslims into a false sense of security. The non-aligned lobbied to see 836 strengthened in another respect: it wanted 'deterring attacks against' the safe areas to be strengthened to 'defending' the safe areas, since the first type of action would not provide the Bosnian Muslims with the protection that they needed. The use of the less forceful phraseology was a clear abdication of international responsibility toward the Bosnian government, as was linking UNPROFOR's right to use force predominantly to its own self-defence. However, Britain and France could not be persuaded along these lines due to their reluctance to be drawn into the civil war.[90] As a result, Venezuela and Pakistan ultimately abstained when voting on resolution 836.[91]

Finally, the organizational structure of the UN again facilitated the expansion of the safe area concept. Both UNPROFOR and the Secretary-General played an important role in this process. Though the UN Secretariat had been sceptical about the safe area concept from its inception, considering it to be almost impossible to implement in practice,[92] the Secretary-General nonetheless prepared a report for the Council following the adoption of 836 outlining two different options for implementation of the safe areas. If UNPROFOR were to carry out effectively the tasks contained in 836 and in the French proposal, namely '(a) deterrence of attacks; (b) monitoring of the cease-fire; (c) promotion of the withdrawal of military or paramilitary units other than those of the Government of Bosnia and Herzegovina; (d) occupation of key points; and (e) protection of humanitarian relief, delivery and distribution',[93] then it estimated that the number of additional troops it needed would be 34,000. Even this size of contingent would not provide protection for the safe areas, but only obtain 'deterrence through strength'.[94] Recognizing that it was unlikely that Security Council members would agree to such a large deployment, the Secretary-General did suggest an alternative 'light option' of 7,600, but made it quite clear that this type of emplacement would not be able to meet entirely the objectives of 836.[95] He specified that the light option could not 'in itself, guarantee the defence of the safe areas', that the shortfall in troops would necessitate a 'credible air-strike capability', and that deterrence through presence would succeed only with the 'consent and cooperation of the parties'.[96] Boutros-Ghali also implied that the light option might be more in tune with the spirit of 836 if it were seen as 'an initial approach [with] limited objectives'.[97] Regardless of these concerns, states decided to choose the 'light option'.

Despite the steps taken to operationalize the safe areas concept in May and June 1993, the process remained incomplete. Though resolution 836 paved the way in theory to NATO air power being used as a credible dissuasive force against attacks on the safety zones, no concrete procedures were established at this time as to how UNPROFOR and NATO would work together if air power were needed. This omission was an important and telling oversight for a number of reasons. Along with the preference for the 'light option', it suggested that states underestimated, perhaps wilfully, the extent to which it would be difficult to implement an effective safe area policy: they seemed to be under the illusion that if the Security Council proclaimed the existence of safe areas, local belligerents would take them seriously. Further, it indicated the ambivalence with which states viewed air power. On the one hand, as there never was any intention to defend the safe areas through deployment of ground troops, air power was the only enforcement mechanism available to deter attacks against them. On the other hand, except perhaps for the US which did not have troops on the ground, the implementing states were not entirely committed to using it. For, if air power were used in a manner consistent with a strong interpretation of 824 and 836, it would undoubtedly target Bosnian Serb forces that might, in retribution, retaliate against UNPROFOR troops, including those of key implementing states. This potentially untenable situation, of which the UN Secretariat seemed the most aware and wary, was disregarded in the short-term by states as they responded to the immediate and urgent demands the war in Bosnia placed upon their competing state and community interests.[98] The crises from August 1993 onwards would test their willingness to follow through on what they had agreed upon in theory.

The safe area crises, August 1993–July 1995

In the months following the expansion of the safe area concept and the demise of the VOPP, the international community moved toward a different approach to the peace negotiations on Bosnia. The complexity of the VOPP was put aside first in favour of a simpler formula that recognized that ethnic partition – a three-way split – might be the only solution to the conflict.[99] However, with a three-party negotiation hindering progress, the Clinton administration pursued a diplomatic strategy designed to bring the Bosnian Muslims and Bosnian Croats together, which bore fruit at the end of February 1994: the two parties signed a comprehensive ceasefire, which solidified into an agreement in principle on the creation of the Federation of Bosnia and Herzegovina on 1 March 1994.[100] Another development was the creation in April 1994, at the initiative of the United States, of the Contact Group on Bosnia, consisting of the US, the UK, France, Russia and Germany. Its intent was to achieve greater coordination among the major external actors, to help smooth relations between them, and to encourage more important negotiating roles for the Americans and the Russians. From here emerged the main outlines of the second alternative settlement to the VOPP, key provisions of which would be carried through to the Dayton Accords of November 1995. The Contact Group plan of 4 July 1994 proposed a 51–49 per cent territorial split between the Croat-Muslim Federation and the Bosnian Serb Republic. Although the Contact Group did threaten various coercive gestures, such as strengthening the sanctions regime, the Bosnian Serbs remained recalcitrant, mainly because it would require them to relinquish territory they held on the battlefield. There was some talk of a bombing campaign to impose the settlement on the Serbs, but insufficient political will existed to follow through.[101] Hence, despite sustained diplomatic effort to end the war, fighting continued, leading to a number of crises relating to the safe areas. For the sake of clarity, these will be presented factually prior to an analysis of the role state and community interests played in their resolution.

The safe area crises

Overall, the safe areas encountered many problems on a daily basis. They remained under siege, enduring regular mortar attacks by the Bosnian Serbs and not receiving sufficient amounts of humanitarian assistance. Zepa and Srebrenica, the most eastern of the enclaves and deep within Bosnian Serb-held territory, were the worst affected in this regard as convoys were regularly prevented from reaching them. Boutros-Ghali admitted that 'living conditions in the safe areas remain[ed] appalling'.[102] Another key problem was the continued military activity within the safe areas. Bosnian government forces used them, in a number of instances, to regroup and rest in preparation for either smaller night-time raids or more important attacks against the Bosnian Serbs, as was the case in Gorazde in April 1994.

Sarajevo[103]

Although Sarajevo was under constant siege by the Bosnian Serbs, the city's main crisis began on 5 February 1994 when a single mortar blast killed 68 people in the crowded Markele marketplace.[104] Two parallel, and eventually coordinated, responses occurred. First, UNPROFOR commander in Bosnia Michael Rose worked to convince the two factions to reach an agreement on 9 February that provided for a ceasefire, a withdrawal of heavy weapons to a distance of 20 km from Sarajevo, and an interposition of UNPROFOR troops along the confrontation line in the city. Meanwhile, the North Atlantic Council (NAC) built consensus around the need for air strikes against the Serbs. Once Rose communicated the details of his deal to NATO, its ambassadors coalesced around the idea of a ten-day ultimatum to the belligerents demanding that they withdraw their heavy weapons from a Sarajevo exclusion zone or place them under the control of UNPROFOR. Any such armament not handed over or removed would be subject to air strikes. Little progress was made during the first week of the ultimatum period: the Serbs were reluctant to remove their heavy weapons as they argued this would give an unfair advantage to the Bosnian infantry within Sarajevo. PM Major flew to Moscow to encourage the Russians to become involved. After intensive negotiations with Russian envoy Vitaly Churkin, Mladic complied with the ultimatum by removing some weapons from the exclusion zone and placing others in concentration points that could be inspected by UNPROFOR. As part of the deal, Russia immediately deployed 400 troops to help UNPROFOR monitor confrontation lines within Sarajevo: the Serbs could trust both that a 'neutral' force now monitored Bosnian army activity and that NATO would not be able to act in the future without Russian consent. Although this conclusion to the crisis did improve conditions in Sarajevo and enabled the West to claim a victory of sorts, it did nothing to lift the siege and left air power untested.

Gorazde[105]

Gorazde was larger, more important strategically and better defended than the other two eastern enclaves of Zepa and Srebrenica. Its defenders had been well enough supplied to launch important raids from the pocket that inflicted heavy losses upon the Serbs. At the time of the April 1994 Gorazde crisis, UNPROFOR had as yet not positioned peacekeepers within the enclave due to a persistent shortage of troops,[106] and was reliant on the presence of a small number of military observers within the town. For the first ten days of the Serb assault on the enclave, Rose approached the problem from a peacekeeping perspective, endeavouring to establish a voluntary cease-fire between the parties and a negotiated solution as to the parameters of the enclave. When this strategy failed and the severity of the assault on the civilian population became known to the outside world, a more concerted effort was made by the UN to end the Serb advance. Close air support was called in by Rose on 10 April, but the strike did not succeed in stopping Serb bombardment of the town for more than a few hours. A second NATO pinprick mission occurred the next day, resulting in relative quiet for the next few days. However, on 14 April the Serbs took 150 UN personnel as hostages, most of them peacekeepers stationed at heavy weapons collection sites around Sarajevo, in retaliation for the bombings. The Serbs then advanced unabated, capturing the strategic heights around Gorazde on the 16th and moving in toward the town. Again Rose called aircraft in, but the Serbs managed to down a British jet, further deepening the international crisis.

Fearful an escalating crisis might trigger serious NATO involvement,[107] the Russians moved to intervene. Churkin obtained promises from Mladic that the Serbs would stop shelling the enclave, withdraw from it and release the UN hostages if NATO would call off its strikes, to which Yasushi Akashi, the UN Secretary-General's envoy to the Former Yugoslavia, agreed.[108] However the Serbs failed to honour their side of the bargain and persisted with the attack, moving into the town itself on 20 April. NAC issued two ultimatums. The first demanded that the attack against Gorazde cease immediately, that the Serbs pull back to a distance of 3 km from the centre of the city by 24 April, and that humanitarian convoys be permitted to enter the town. The second demanded that a heavy-weapons exclusion zone be established within a 20 km radius around Gorazde, and specified that similar zones could be erected around any of the other safe areas if deemed necessary by UNPROFOR. Defiance of these measures, including heavy weapons attacks against the other safe areas, would be met with air strikes. The NAC also requested that the Bosnian government not undertake any further action from within the safe areas. These arrangements were endorsed by Security Council resolution 913 of 22 April 1994. When the Serbs failed to comply with the immediate cease-fire demand, NATO moved to turn its key, only to be foiled by Akashi who believed the Serbs were making progress in other areas. Angered by their earlier betrayal by the Serbs, the Russians now moved to end the impasse by making clear that they no longer unequivocally opposed air strikes. The Serbs thus reached an agreement with UNPROFOR on 24 April, bringing into effect the conditions of the NATO ultimatum and permitting the deployment of British and Ukrainian peacekeepers into the enclave. The threat of a more determined NATO response seemed to help dissuade the Serbs, indicating that air power could play an important role in deterring attacks against the safe areas. However, the taking of UN hostages and the downing of the NATO aircraft highlighted the extent to which the NATO air power deterrent might not be credible.

Bihac[109]

The Bosnian army launched a major assault out of the pocket at the end of October 1994. The Serbs counterattacked at the start of November, shelling the enclave repeatedly and preventing humanitarian convoys from reaching the city. With its population swollen to 170,000, conditions within the enclave quickly deteriorated. In contravention of the no-fly zone over all of Bosnia, Serb aircraft from the Ubdina airfield situated in the Serb-controlled Krajina region of Croatia bombed the enclave on 18 and 19 November, terrorizing the population and putting at risk the lives of the 1,300 lightly armed Bangladeshi peacekeepers. In the largest operation so far in support of the safe areas, NATO aircraft targeted the Ubdina airstrip on 21, 23 and 25 November. Although NATO wished to neutralize the airport entirely, UNPROFOR scaled the bombing down so as to avoid retribution on the part of the Serbs.[110] Still 50 Canadian peacekeepers were taken hostage, again undermining the value of the NATO deterrent. The confrontation over Bihac eventually resolved itself when the Serb attack faltered in early December.

Sarajevo again[111]

Intensified fighting erupted around Sarajevo in March 1995, resulting in both sides removing some of their heavy weapons from UNPROFOR-supervised collection sites in May. Rupert Smith, the new UNPROFOR commander for Bosnia, warned them they would be attacked from the air unless they either removed their heavy weapons from the exclusion zone or returned them to the depots. When Serb forces failed to comply, air strikes hit two ammunition bunkers in the vicinity of Pale on 25 May. In retaliation, the Serbs not only continued bombarding Sarajevo, but also intensified their mortar attacks on the other enclaves, particularly in Tuzla where 71 civilians were killed and 200 others injured with a single round. NATO again responded, hitting six more ammunition bunkers near Pale. The Serbs took revenge by taking 400 UN personnel hostage, chaining some of them to possible NATO targets as human shields. Evincing the degrading quality of the NATO deterrent, further air strikes were called off as the top priority became the security of UN personnel. Eventually, after a series of meetings between Mladic and Bernard Janvier, Force commander in former Yugoslavia as a whole, the hostages were released between 2–18 June.

Srebrenica and Zepa[112]

On 6 July 1995, the Bosnian Serbs began shelling the Srebrenica safe area, also targeting one of UNPROFOR's observation posts, Foxtrot, which they finally overran on the 8th. Although the Dutch tried to set up a blocking position to keep the Serbs from advancing any further, this tactic proved unsuccessful, and gradually over the next two days 5 more observation posts were captured and Dutch peacekeepers taken hostage. During this time, several requests for close air support were turned down by Janvier, acting UNPROFOR commander in Bosnia in Rupert Smith's absence: Janvier did not believe the Serbs intended to take the enclave, and he was wary of using air power in the wake of what had occurred in Sarajevo in May. The Serbs eventually threatened, if further strikes were forthcoming, to kill the Dutch hostages and to shell the Dutch base at Potacari just North of Srebrenica where fleeing refugees began to gather for protection. The Dutch ministry of defence requested at this time that close air support be discontinued because its troops were in jeopardy, to which Akashi conceded. With Srebrenica in their hands, the Serbs moved to ethnically cleanse the area, evacuating women and children to Tuzla, but drawing aside at least 7,336 individuals, mostly men, whom they mass executed at several sites within eastern Bosnia.[113] The Serbs began their assault on Zepa three days after they captured Srebrenica, again targeting UNPROFOR observation posts. Due to the remote location of the enclave and poor communications with it, no effort was made to use air power to deter the attack. Zepa fell on the 19th, and its civilian inhabitants were deported. To avoid capture, its Bosnian defenders fled through the forest.

As these crises reveal, implementation of the safe area policy was plagued by difficulties and serious setbacks from the start. UNPROFOR could not keep the areas safe from the Bosnian Serbs,[114] was vulnerable to retaliation if it tried, and was at serious risk of losing its impartiality, by failing to prevent Bosnian government military activity within the safe areas and by opposing Serbian military activity outside of them. Yet, the safe areas forged unparalleled cooperation between the UN and NATO, opened the door toward unprecedented NATO military action in support of humanitarian objectives, and endured for more than two years despite their grave drawbacks. An examination of community and state interests can explain why all of this was so, again revealing the extent to which the safe areas satisfied states' compulsion to take sides when faced with ethnic cleansing and other violations of humanitarian law, while avoiding that very same taking of sides that would have catapulted them into full-fledged warfare in defence of the Bosnian government.

Community interests and the safe area crises

The safe area crises served as both focal points and catalysts for NATO and the Security Council. Each episode would cause a spurt in international activity to do something in order to respond to the violation of the safe area in question. Horrified by the indiscriminate bombing of civilians as well as at the blatant disregard the Serbs demonstrated toward common values and Security Council resolutions, key international actors showed a willingness to overcome some of their differences, notably on air strikes. Despite their persistent fears about bombing due to the vulnerability of their troops on the ground, the Europeans bowed, albeit reluctantly on the part of the British, to American pressure for stronger action in this regard, in part because they all opposed barbaric and inhuman behaviour. Likewise, community norms as focal points kept the Russians on board, despite their initial opposition to too great a NATO involvement in the conflict. Although they did not approve of the NATO ultimatum to the Serbs in February 1994, the Russians agreed that something had to be done about the indiscriminate attacks against civilians so stepped in to make the exclusion zone possible. And, in Gorazde, they eventually dropped their opposition to air strikes as they grew disgruntled with Serb disregard for both human life and the authority of the UN.

In relation to community interests as enablers, violations of humanitarian law and blatant disrespect of Security Council resolutions allowed permissive conditions to arise that eventually culminated in the operationalization of the NATO air power deterrent for the safe areas. That NATO action outside of its members' territory in contexts other than those associated with strict self-defence would be tolerated by other important international actors was not self-evident, particularly in the case of Russia and China. Yet, NATO's involvement in ultimatums against and eventually bombings of the Serbs eventually received widespread acceptance by states. This assent resulted from two beliefs on the part of states.

First, they accepted that NATO action occurred within a broader UN authorized involvement in former Yugoslavia, so was legitimate despite its novelty. Second, they considered NATO engagement to be carefully constrained through resolution 836 and the guidelines eventually agreed upon by UNPROFOR and NAC. NATO bombing would occur in close consultation with UN forces on the ground, for specific reasons as elaborated through the UN safe areas policy, and according to appropriate guidelines in terms of the extent and severity of bombing.[115] The UN and NAC facilitated the development of these arrangements.

Manfred Wörner, NATO Secretary-General, and Boutros-Ghali, worked out the dual-key arrangement, which ensured that NATO could not take unilateral action.[116] UNPROFOR further refined the criteria under which air power could be used, i.e. 'in self-defence; in reply to bombardment against the safe areas; in response to armed incursions into the safe areas; and to neutralize attempts to obstruct the freedom of movement of UNPROFOR forces or humanitarian convoys'.[117] NAC further elaborated a three step escalatory approach to air strikes on 9 August 1993, from 1) strikes against targets that were militarily significant and visibly impeding or preventing the implementation of Security Council resolutions;[118] to 2) strikes against a wider set of targets associated with the siege; to 3) strikes against targets outside the immediate areas of the siege.[119] The 5 February 1994 Sarajevo marketplace mortar attack further encouraged the UN Secretary-General to press NATO to reach a decision that would 'authorize the Commander-in-Chief of NATO's Southern Command to launch air strikes, at the request of the United Nations, against artillery or mortar positions in or around Sarajevo' deemed suitable by UNPROFOR.[120] Eventually overcoming mainly Britain's resistance to air strikes,[121] NAC followed through on this request, delegating authority to military commanders so that it could be carried out if and when necessary. The Gorazde crisis further broadened the authority of NATO's Southern Commander, Leighton Smith, to turn the NATO key in response to violations of any of the safe area exclusion zones. NATO's relative restraint and basic adherence to this agreed blueprint for action in each successive crisis meant that it slowly worked its way toward a more coercive, yet internationally acceptable, response in August 1995.

The constraining effect of community interests explains why the NATO deterrent was operationalized despite state ambivalence toward it discussed in the last section, and why in fact the safe area policy was maintained despite its evident shortcomings in practice. First, public pressure to do something would increase every time the Serbs perpetrated another major atrocity, making it difficult for states to leave the Muslims in the enclaves to their fate.[122] The need to do something to help a besieged Sarajevo in the summer of 1993 even propelled Clinton to contemplate seriously using ground troops to lift the strangulation of the city, although no such action was ever taken.[123] Other states were also so burdened by an imperative to act that they demanded that 836 be regarded as the peace-enforcement resolution they considered it to be, suggesting that it authorized the use of force not just in UNPROFOR's self-defence, but also to 'protect' the safe areas.[124] In a meeting of troop-contributing countries assembled by Boutros-Ghali in the midst of the Bihac crisis in the hope of ascertaining whether or not these countries wished to move from a peacekeeping to a peace-enforcement mandate, eight of 17 responded in the affirmative, expressing 'their inability to understand why more robust action was not being taken'.[125] Several members of the Council also considered that other areas in Bosnia deserved equal international attention, successfully incorporating a clause into Security Council resolution 900 of 4 March 1994 that requested the UN Secretary-General to 'report on the feasibility and modalities for the application of the protection, defined in resolutions 824 and 836, to Maglaz, Mostar and Vitez'. Once engendered by the UN, safe areas could not so easily be abandoned.

This constraining effect of community norms was further reinforced by the continued problem of the arms embargo, which many states considered a violation of Article 51 of the Charter. If the Security Council hindered the right of the Bosnian government to defend itself, then it was even more so the responsibility of the Council to ensure a successful deterrence against attacks on the safe areas. With each passing crisis, the non-aligned became more and more adamant that the UN either had to do something more or lift the arms embargo,[126] their Islamic members going so far as to issue a joint declaration in the Security Council that, in their view, the embargo no longer applied to the Republic of Bosnia and Herzegovina.[127] An unacceptable response in this context would have been to forsake the safe areas, but to neither intervene nor lift the embargo. Since both intervention and lifting the embargo were ruled out for state interest reasons, a continued safe area policy, backed by a NATO air power deterrent, seemed the only way forward.

As in the earlier phases of the safe area process, the UN and UNPROFOR were primarily responsible for the implementation of the safe area policy on a daily basis. Although the UN acts on behalf of member states, it often exercises discretion in carrying out its mandate. Confronted with ambitious, but ambiguous Security Council resolutions in relation to the safe areas and with the problematic consequences of the use of the NATO air power deterrent, the Secretary-General grew doubtful that the safe area policy could be implemented as envisioned by the Council; and key personnel in Bosnia came to the conclusion that the only way to proceed in theatre was with peacekeeping rules of engagement.[128] These institutional preferences affected how crises were resolved.

Although the UN established the safe areas in Bosnia, it remained sceptical about the potential for successful implementation. As crisis after crisis erupted in and around the safe areas, as UN personnel were taken hostage, and as both the Bosnians and the Serbs lost respect for the peacekeeping force, the Secretary-General's reservations began to coalesce around three key interrelated issues. First, the Secretary-General emphasized that the insufficient resources allocated to UNPROFOR implied that it would be unable to carry out the more robust policies envisioned by the Council. Instead, reliance upon the consent and cooperation of the parties would have to be recognized as the underlying foundation of activity in Bosnia.[129] Second, Boutros-Ghali highlighted the vulnerability of UN peacekeepers on the ground and the resulting ineffectiveness of air power to provide adequate deterrence against attacks on the safe areas.[130] Third, the Secretary-General worried about the potential loss of UNPROFOR's impartiality as a result of the peace-enforcement activities it was expected to undertake against the Bosnian Serbs. A UN force perceived as taking sides would be hindered in its other humanitarian objectives.[131] As a result of these problems, Boutros-Ghali gradually moved toward a position by mid-1994 that only peacekeeping rules of engagement were viable in Bosnia. For instance, he stated unequivocally:

> ...I do not believe that UNPROFOR should be given the mandate to enforce compliance with the safe area regime. The use of force that would be necessary to implement such a mandate would, as I have already stated, prevent UNPROFOR from carrying out its overall mandate in the former Yugoslavia, turn it into a combatant and further destabilize the situation in Bosnia and Herzegovina. In short, such a mandate would be incompatible with the role of UNPROFOR as a peace-keeping force.[132]

The Secretary-General also believed that the French, British and Dutch RRF that was to go into Bosnia in July 1995 should remain within the peacekeeping mandate of UNPROFOR. If it was envisioned for peace enforcement, then it and UNPROFOR should be replaced entirely by a multinational force.[133] Finally, by May 1995, Boutros-Ghali considered UNPROFOR's position in Bosnia to be untenable and recommended a scaling-down of the force and a return to 'those tasks that a peacekeeping operation can reasonably be expected to perform in the circumstances prevailing in Bosnia...'[134]

This institutional belief[135] in the culture and effectiveness of peacekeeping was also echoed by UN personnel in the field. Akashi was a seasoned veteran of UN peacekeeping operations, and did not fully appreciate how air power worked until General de LaPresle explained it to him in several lengthy sessions.[136] He believed that negotiation should take precedence over the use of force, as demonstrated by his refusal to authorize air strikes in several of the crises above, most notably in Srebrenica, for fear of taking sides and endangering UNPROFOR's mission. He also supported the Secretary-General's preference that the RRF remain within the confines of a peacekeeping mission, even suggesting that it be called a theatre reserve force so as to avoid provoking the Serbs.[137] And in a meeting in Paris on 12 May, Akashi concurred with Boutros-Ghali that 'the costs of a more robust use of force [were] high', also recommending that it might be best if a 'drastic reduction' were sought in the size and mandate of UNPROFOR.[138]

In tandem, General Rose feared crossing the 'Mogadishu line' in Bosnia and reiterated 'the dangers of abandoning peacekeeping for war fighting'.[139] In Sarajevo, he worked to establish an agreement between the parties on a heavy weapons exclusion zone prior to the issuance of the NATO ultimatum so as to avoid the necessity for air strikes: 'we had to act quickly if we were to prevent this "we must do something" reaction of the international community from dragging the UN mission into war in Bosnia'.[140] Likewise, in Gorazde he exhausted all possible means of negotiation before resorting to air power, at least partially because he believed that the Bosnian Muslims, as well as the relief workers in the town, were exaggerating the level of casualties and the strength of the Serb attack in the hope of provoking a foreign intervention.[141] He was concerned that taking sides would jeopardize UNPROFOR's humanitarian mission.[142] Rose also worked to keep the British position toward air strikes lukewarm at best.[143]

Janvier also opposed any move toward peace-enforcement, both before and after the May 1995 Sarajevo crisis. He spoke to the Security Council on 24 May 1995, telling members they needed to become more honest toward the mission in Bosnia and change the mandate in such a way as to minimize vulnerability.[144] In a meeting with Akashi and Rupert Smith on 9 June 1995, which aimed to harmonize their perspectives on the use of force, Janvier remarked, 'the main point of the situation is that we are a peacekeeping force, whether we want it or not'.[145] Finally, Janvier proved so reluctant to authorize air strikes during the Srebrenica crisis of July 1995 that it was suggested that he had struck a secret deal with Mladic and Milosevic involving the release of hostages for the non-use of air strikes.[146] However, neither the UN inquiry into the fall of Srebrenica nor the Netherlands Institute for War Documentation study on the Srebrenica safe area ever found any evidence to suggest a special arrangement had occurred;[147] and Janvier vehemently denies any involvement in a deal.[148] It is more likely that the hostage crisis of May–June 1995 encouraged both greater caution in the use of force and greater faith in pursuing the principles of peacekeeping.

State interests and the safe area crises

A number of state interests affected how the safe area policy was implemented, both at times enabling measures designed to strengthen the policy and at times constraining states in their ability to fulfill the requirements of 819, 824 and 836. Preserving Atlantic unity remained an important focal point, as well as enabler, allowing the Americans and the Europeans to work out the operationalization of the NATO air power deterrent. The British in particular went along with several of the NATO ultimatums against their better judgement to avoid a major confrontation with the Americans.[149]

American state interests played an important role in encouraging the use of air strikes during the crises, though Clinton's persistent lack of resoluteness regarding Bosnia meant that the US took no initiative during the Srebrenica catastrophe. The Principals Committee remained divided throughout the implementation phase of the safe areas policy in which the US had acquiesced, with Clinton vacillating between pushing for stronger involvement when the Serbs engaged in particularly nefarious behaviour and containment when it seemed that Bosnia was a morass to be avoided at any cost.[150] However, one constant remained a key part of American thinking on the war: air power. The US believed that air strikes, if used, could be the most important factor in coercing the Serbs into relinquishing territory and accepting a settlement. In this respect, the safe area policy provided the US with a spring board to more assertive NATO action from the air, which would also demonstrate the Alliance's continued worth in the post-Cold War world. As a result, the US was a key instigator behind NAC's decisions to fully operationalize the air power deterrent and to issue ultimatums to the Bosnian Serbs. Yet, the US did not want to jeopardize its NATO expansion policy, which was launched by Christopher at a NATO ministerial meeting in December 1993, so it gradually adopted the more cautious European stance when it became obvious UNPROFOR troops were in danger on the ground in May–July 1995.

Russian state interests proved salutary for the safe area policy on at least two occasions. Throughout the crisis in Bosnia, Russia was being pulled in opposite directions. On the one hand, due to domestic political constraints in the form of the rise of nationalists demanding a stronger stance in favour of the Serbs, Russia could not support the West unequivocally on Bosnia, most notably in relation to the use of NATO air strikes. On the other hand, Russia wished to avoid a confrontation with its post-Cold War international partners and a potential slide into renewed East-West hostility.[151] One way for Russia to overcome this dilemma was to limit NATO air action as much as possible, while capitalizing on its special access to and relationship with the Serbs so as to bring about an end to crises in a manner that nonetheless adhered to the general framework of Western policy in the Balkans. Also important to the Russians was Western acceptance of their status as a key international player that needed to be consulted on decisions, as opposed to being ignored as it was when NATO issued its ultimatum against the Serbs in February 1994.[152] Rendering their participation indispensable to the process, through either their presence on the ground or their influence over the Bosnian Serbs, was one way to ensure this inclusion in the decision-making process. In the 1994 Sarajevo and Gorazde crises, Russia intervened in instrumental ways: it sent peacekeepers into Sarajevo to quiet Bosnian Serb mistrust of a UNPROFOR interposition force along confrontation lines; and it played a vital diplomatic role in the resolution of the Gorazde hostage-taking incident. Both intercessions contributed to the safe area policy overcoming its first two hurdles.

In the end, however, the safe area policy was inadequately enforced due to the strong constraining effect of state interests. Implementing states were simply unwilling to incur the costs that stronger implementation would have entailed. They were willing to go only so far in support of their community interests – establishing a presence in the safe areas and using NATO air power as a means to deter attacks against them – but no further. The main problem was that these methods of enforcement did not work in the long run, thus requiring, if the safe areas were to be effectively implemented, the one more step states refused to take. The eastern enclaves were isolated and deep within Serb territory, so any peacekeepers stationed there were automatically vulnerable, which explains why Boutros-Ghali found it difficult to obtain troops for these remote locations.[153] A Nordic Battalion refused to follow General Jean Cot's order to take up a temporary assignment in Srebrenica in October 1993.[154] Overall Force Commander Bertrand de Lapresle almost resigned in April 1994 when his order to a French battalion to take up position in Gorazde was countermanded by Mitterrand a few hours before the British and French troops were to depart for the enclave.[155] The Dutch did not intend to remain in Srebrenica beyond August 1995 when their last troop rotation ended. In addition, the fact that the underlying intent of the safe areas policy, as demonstrated by the wording of resolutions 824 and 836, was a stance against the Bosnian Serbs meant that enforcement action through air power would be perceived by them as hostile, increasing the likelihood of their retaliation against both UNPROFOR and the safe areas. The means foreseen to deter that retribution, NATO dissuasive force from the air, faltered because its use threatened the lives of lightly-armed and dispersed UNPROFOR soldiers, something states simply would not risk. This fundamental unwillingness to take sufficient measures to deter attacks against the safe areas culminated in the tragic events of the summer of 1995.

Safe area endgame

The immediate international response to the fall of Srebrenica was to convene a meeting in London on 21 July 1995, at which a number of important decisions were made. First, it was agreed that 'any attack against Gorazde [would] be met with a substantial and decisive response, including the use of air power'. Second, participants 'underlined [their] determination to ensure access to Sarajevo for the delivery of provisions to the civilian population and resupply of the United Nations forces, and support for early use of the rapid reaction force to protect UNPROFOR in maintaining access for these deliveries'.[156] This line in the sand was then rendered operational through the NAC. On 25 July, NATO authorized the use of air strikes for as long as was considered necessary by NATO and UNPROFOR commanders 'to support the defence of Gorazde within a wider geographic area, including against any concentration of troops, if NATO and the United Nations commanders both judged that they posed a serious threat to the safe area'.[157] These arrangements were extended to the three other remaining safe areas of Tuzla, Bihac and Sarajevo on 1 August 1995. In order to streamline the UN chain of command in relation to requests for air strikes, the Secretary-General also delegated his key directly to the theatre-wide Force Commander Bernard Janvier.

The opportunity to put these new arrangements to the test occurred on 28 August when a Bosnian Serb-fired mortar landed in the Markele marketplace, killing 37 civilians and wounding 90 others. Prior to turning his key, Rupert Smith, UNPROFOR commander in Bosnia and acting Force Commander in Janvier's absence, made certain that all UNPROFOR personnel were removed from vulnerable locations. Air strikes then began on 30 August in what NATO called *Operation Deliberate Force*. For three days NATO attacked targets throughout Bosnia, with the RRF playing a key role in silencing Serbian heavy weapons around Sarajevo. A pause in the bombing occurred on 1 September, giving Mladic an opportunity to consider an end to attacks on the safe areas and a theatre-wide ceasefire in return for a cessation of air strikes. When his reply proved unsatisfactory, bombing resumed on 5 September for a further nine days. When another pause to *Operation Deliberate Force* on 14 September bore fruit, with the Serbs signing a Framework for a Cessation of Hostilities and removing their heavy weapons from the Sarajevo exclusion zone, NATO's operation was permanently halted. With the siege of Sarajevo lifted and humanitarian convoys now moving about freely, conditions within the safe areas improved dramatically. The Dayton Accords officially ended hostilities in November 1995, and allocated the territory of the four remaining safe areas to the Bosnian-Croat Federation.

Operation Deliberate Force did not bring the Serbs to the bargaining table on its own: parallel developments in the ground war brought about an approximate 51–49 split in territory held on the battlefield. A Croatian government offensive in the Krajina in early August rapidly defeated the Croatian Serbs. Now freed to fight in Bosnia, the Croatian army, alongside Bosnian armed forces, mounted an offensive in September that liberated first the Bihac pocket and then areas further east. By 22 September 1995, their objective of getting the Serbs to relinquish enough territory to be willing to end the war had been achieved. A new configuration of state and community interests enabled this remarkable change in the turn of events.

Community interests and the safe area endgame

The fall of Srebrenica and the horrific massacres that followed galvanized the international community to draw a line in the sand at Gorazde and to implement a tougher policy involving air strikes to shield the remaining safe areas from further infringement. Richard Holbrooke, America's chief negotiator at Dayton, explained that the air campaign had not been part of a master plan in American strategy, but rather that it had taken 'outrageous Bosnian Serb action to trigger *Operation Deliberate Force*'.[158] There was also a significant outcry in international public opinion,[159] and a growing sense of obligation among key players that something had to be done at last. The French especially were aghast. When the newly-elected President Chirac spoke to Clinton on the phone on 13 July, he was 'full of moral indignation'.[160] He even proposed to Clinton that French troops using American helicopters retake the enclave.[161] Although Clinton speculated that this was just a bluff, Chirac had in fact requested that the Ministry of Defence present serious options regarding Srebrenica in the wake of its capture.[162] If the US had agreed, it is probable that France would have followed through; its key problem was that it did not have enough military strength to act alone, so always favoured multilateralism.[163] Clinton was also shocked by what had happened in Srebrenica: 'This can't continue. We have to seize control'.[164] Finally, although the US had contemplated proposing a settlement that traded Gorazde, along with Srebrenica and Zepa, to the Serbs in return for territory in central Bosnia, it changed its stance: 'After Srebrenica, we cannot propose such a thing'.[165]

The other critical change provoked by the fall of Srebrenica was an institutional move away from peacekeeping rules of engagement in Bosnia. The first element of this shift was the altered chain of command respecting air strikes. With the UN key now solely in the hands of the UN Force Commander, it would now be much easier to initiate NATO bombing. Such a decision could be made without having to coordinate with the civilian branch of the UN. The second element of this shift was on the one hand, Boutros-Ghali's acquiescence in it, and on the other, Rupert Smith's clear adoption of peace-enforcement rules of engagement. When Akashi expressed concern to the UN Secretary-General that the new arrangements for air strikes were too robust, providing an automatic triggering mechanism that might be problematic, Boutros-Ghali replied that he nonetheless supported the NAC agreements and would be delegating authority to Janvier.[166]

Rupert Smith's behaviour was more surprising. As far back as the 9 June meeting with Akashi and Janvier, Smith had been advocating a more robust UNPROFOR stance in Bosnia. He stated then, for example: 'Mladic won't treat us as an enemy as long as we do everything on his terms. If we try to do our job our way, then we are his enemy and he will treat us that way. If we bring in force behind us, he will make concessions, but if we do things on his terms he will succeed in neutralizing us'.[167] Similarly, Smith believed the RRF needed to be used as a fighting force: 'We are in danger of reverting to the *status quo* minus, of operating in the mode of supplicant. This is why I keep returning to the question of the Rapid Reaction Force: are we going to use them to fight? If not, I am not sure I want them – they will just be more mouths to feed, and create expectations that I cannot meet'.[168] Starting in August 1995, Smith put his words into action: he began behaving as a commander in charge of war-fighting as opposed to peacekeeping.[169] With Janvier away when the Serbs hit the marketplace on 28 August, he was the one who turned the UN key for air strikes. He then removed UN personnel from vulnerable locations, all the while keeping it purposefully secret from Mladic and the international media that he had turned the key.[170] Then when NATO bombing began, he ordered the RRF to shell Serb positions around Sarajevo. Smith also decided unilaterally to open a land route into Sarajevo, thus ending its siege, simply informing the Serbs on 2 September that the road from the airport to the capital would be open to civilian traffic as of the next day. When the Serbs threatened retaliation, Smith replied that any Serb interference would be met with 'disproportionate force'.[171] In a series of unauthorized statements to the press through a UNPROFOR spokesperson, Smith made it clear the peacekeepers were now a fighting force: 'We're into peace enforcement here. Peace enforcement is not negotiating...We've seen that; it has failed over the years here. We are saying, "If you do not do this, no conditions, you continue to get bombed"'.[172] When the secretariat complained, requesting a formal explanation of these remarks, Smith replied that 'as a result of our enforcement action, UNPROFOR abandoned its peacekeeping mission – at least in the Sarajevo area. We remain, for the time being, in the position of combatants: coercing and enforcing our demands on the BSA'.[173] These instrumental changes to the way UNPROFOR operated on the ground enabled a more robust approach in Bosnia.

State interests and the safe area endgame

Although the fall of Srebrenica provided states with the impetus they needed to switch to a more coercive policy in Bosnia, changes in state interests had been afoot for a while. There were two elements to this change. First, the US acknowledged that the 'do something but do not actually do it' middle-ground approach was a failure. It recognized that state interests as constraints needed to be accepted fully and incompatible community interests abandoned. Or, as an alternative, new ways needed to be found to overcome the constraining effect of state interests so that community interests could be upheld. Both tactics were eventually adopted by the American government. Second, new interests as enablers appeared in both the US and Europe, opening the door to an alteration of the *status quo*.

If state interests would not permit military intercession on behalf of the Bosnian government or in defence of the remaining safe areas, then another method was needed to stop the war in Bosnia and satisfactorily conclude the safe area policy: NATO air strikes, but this time coupled with a ground offensive by a proxy army. To put this plan in motion, the US government began covertly arming and unofficially training first the Croat army, and later a coordinated Bosnian-Croat army, with the help of retired generals.[174] At the same time, the Clinton administration admitted that if their state interests would not allow the pursuit of certain moral imperatives, then these would have to be set aside to end the conflict. Key here was the acceptance that if the Bosnians would not live together intermixed, and if the West had insufficient will to preserve the multiethnic fabric of Bosnia, then partition and territory swaps were in order. There is no suggestion here that the US or any other western state found the atrocities that occurred in Srebrenica tolerable,[175] but only that the fall of the enclave eliminated a sticky problem that had hindered peace negotiations for over two years. It was likely for this reason that no effort whatsoever was made by key players to halt the Serb takeover of Zepa.

In terms of state interests as enablers, a new picture emerged in America in the spring of 1995 that made the US far more likely to engage in a Balkan and thus safe area endgame. Foremost was the fact that Clinton's indecision and vacillation on the Balkans was beginning to take its toll on American foreign policy as a whole. Echoing Albright's sentiments,[176] Lake wrote a memo to Clinton stating that the administration's 'weak, muddle-through strategy in Bosnia was becoming a cancer on Clinton's entire foreign policy – spreading and eating away at its credibility'.[177] The administration's work on NATO rejuvenation and expansion risked being undermined: NATO credibility was being called into question daily due to the continued fighting, UN hostage-taking and menacing of the safe areas. If this dismal state of affairs were to be addressed before Clinton entered an election year, then the US would have to adopt a leadership role and bring about a cessation of hostilities in the Balkans. To accomplish this goal, Lake set about building unity within the Principals Committee and developing an endgame strategy.[178] Whereas the administration had until this point focused on crisis management, it was now time to think about an end result and present it to the Europeans as American policy.

Moreover, as conditions under which UNPROFOR operated in Bosnia deteriorated in May–July 1995, it became a distinct possibility that the Europeans would consider withdrawing their troops. It was even conceivable they would have to extract them in hostile circumstances. Such a prospect implied the deployment of American ground troops for a rescue operation, as Clinton had made a promise as far back as the Joint Action Programme of 22 May 1993 to assist UNPROFOR troops in Bosnia if the need arose.[179] The US was doubly-obligated to go to the assistance of its NATO partners' troops if so requested. For some time, NATO military staff had been working on the tactical arrangements for just such an extraction plan, labelled OPLAN 40104 and already approved by the NAC.[180] In the event of such an extrication request, the President's choices would be seriously curtailed. Clinton was worried enough that this eventuality might take place that he decided to prepare the public in advance through a nationally broadcast radio address:

> I want to make it clear again, what I have said about the use of our ground forces. We will use them only if, first, there is a genuine peace with no shooting and fighting...Second, if our allies decide they can no longer continue the UN mission and decide to withdraw, but they cannot withdraw in safety, we should help them to get out...[Third,] I have decided that if a UN unit needs an emergency extraction, we would assist, after consulting with Congress. This would be a limited, temporary operation, and we have not been asked to do this.[181]

The US administration wanted to avoid this possibility at all costs: it would be political suicide to undertake such a risky mission and then have nothing to show for it – finally deploy ground troops in Bosnia in the context of a UN and NATO failure to end the conflict.[182] State interests dictated that another solution be pursued actively.

Not only did the President have to find a way to avert the above scenario, he was also under a severe time restraint to do so. This pressure resulted from the fact that the US Senate was moving forward with binding legislation unilaterally lifting the arms embargo, which it indeed adopted on 26 July 1995 by 69 votes to 29.[183] The problem with 'lift' was that it would almost assuredly guarantee Allied withdrawal from Bosnia, leading directly to the very circumstances Clinton was trying to avoid. Clinton vetoed the bill, but was not certain his veto would not be overturned. As the Senate went on summer recess, Clinton now had until September to come up with an alternative peace package for Bosnia. At a 7 August Principals Committee meeting, Clinton made clear that settlement needed to be reached 'within the next few months'.[184]

Finally, French state interests, with the British and Dutch in tow, had also shifted by the spring of 1995 toward a greater willingness to use force against the Bosnian Serbs. On 27 May 1995, Bosnian Serb forces, dressed in French uniforms, overran a UNPROFOR checkpoint near the Vrbanja bridge in downtown Sarajevo and captured the French peacekeepers stationed there. The UNPROFOR Commander of Sector Sarajevo, also a Frenchman, counterattacked, successfully retaking the bridge but losing two soldiers.[185] One important element that sparked this change in French behaviour was the election of Jacques Chirac to the Presidency of the Republic. He felt enough was enough: French forces had suffered so much humiliation in Bosnia that it was time to redress their image.[186] Military credibility and reputation were on the line.[187] It was for this reason that the French spearheaded the RRF, which operated under national command and was well-armed as well as mobile. After the fall of Srebrenica and renewed international commitment to Sarajevo and Gorazde, these states consented to the 15,000 strong RRF that played such a key role in the defence of Sarajevo during *Operation Deliberate Force*. Despite this new willingness to use force, the Europeans remained wary of taking sides in favour of the Bosnian government or of using their troops for anything other than the opening of Sarajevo: they pushed for the second, ultimately permanent NATO bombing pause and insisted that the Bosnian-Croat army not be encouraged to reclaim territory far beyond the 51–49 split envisaged by the Contact Group Plan.[188]

Conclusion

The Bosnian safe areas were an ideal solution for states not willing to intervene militarily in a meaningful way but nonetheless under pressure to do something to halt ethnic cleansing and the seizure of territory by force. If the safe areas had indeed been but an interim measure, in place perhaps for a few months until the transatlantic rift was overcome and a final settlement reached, then they might have worked. However, the ambiguities inherent in their creation and the ambivalence with which states implemented them meant that the safe areas could not withstand two and half years of continued warfare. Not willing to defend them on the ground, states envisaged NATO air power as the primary mechanism through which attacks against them could be deterred. Yet, as resolutions 824 and 836 made clear, the intent behind the safe area concept was to favour Bosnian government forces already disadvantaged by the UN arms embargo. Whereas the

Bosnian Serbs were required to treat the safe areas as inviolable, the Muslims not only retained their weapons inside the zones but also launched attacks out of them.[189] Not surprisingly UNPROFOR's impartiality *vis-à-vis* the Serbs was thus irrevocably tarnished; and NATO enforcement action on behalf of the safe areas, though UN authorized, was perceived by them as hostile. When the Bosnian Serbs retaliated against lightly-armed and vulnerable UNPROFOR troops, state willingness to enforce the safe area concept gradually eroded, eventually culminating in the tragic events of July 1995.

4
A Decision Not to Act: Proposed UN Secure Humanitarian Areas in Rwanda

Figure 4.1 Rwanda[1]

From April to July 1994, between 500,000 and 800,000 persons, mostly Tutsis and politically moderate Hutus, were massacred in Rwanda in a genocide organized and perpetrated by Hutu extremists who took control of the Rwandan government and armed forces. Unlike in Iraq and Bosnia where international humanitarian involvement was mobilized with relative speed, the state response to genocide in Rwanda was so inadequate and sluggish as to be almost insignificant in terms of the course of the genocide and the protection of targeted civilians. Nonetheless, the Security Council did eventually authorize a more vigorous military humanitarian operation in Rwanda. Safety zones again featured prominently as a possible means to guarantee the protection of a threatened civilian population. Security Council Resolution 918 of 17 May 1994, which authorized the expansion of UNAMIR from 270 to 5,500 troops, mandated the peacekeeping force to 'contribute to the security and protection of displaced persons, refugees and civilians at risk in Rwanda, including through the establishment and maintenance, where feasible, of secure humanitarian areas (SHAs)'. Since the UN Secretary-General was unable to find enough troops for this force until August after the genocide was over, such SHAs were never created, although it might be possible to consider the Amahoro Stadium location as a *de facto* prototype. It was there where a skeletal UNAMIR left in Rwanda after downsizing by the Security Council on 21 April[2] protected 15,000 Tutsis throughout the genocide.

This chapter aims to understand why states contemplated the possibility of creating safety zones in Rwanda, and to assess why they never implemented these proposed SHAs despite the Security Council's authorization for their use. An examination of how state and community interests interacted in complex, unforeseen ways will be helpful in trying to understand why states, when they shied away from involvement in Rwanda, nonetheless felt the need to play down the fact that a genocide was occurring; and why states were pressured into considering a safety zone policy when they were strongly opposed to further humanitarian engagement on African soil in the wake of the debacle in Somalia.

Rwanda prior to the genocide

Origins of the Rwandan genocide

Prior to the genocide of 1994, Rwanda's population numbered 7.7 million persons: 84 per cent Hutu, 15 per cent Tutsi and one per cent Twa. Though the origins of the Rwandan genocide date back to the period of Belgian colonial rule after WWI in which differences between Hutu and Tutsi were exaggerated and institutionalized to facilitate administration of the country,[3] we will focus on its more specific and recent causes. When the majority Hutu population finally revolted in the late 1950s and achieved independence from Belgium in 1962, they were not kindly predisposed to the Tutsis, whom they viewed as their oppressors and whom they envied for their socio-political status.[4] Thousands of Tutsis were massacred and tens of thousands more were sent into exile. A pattern then emerged in the *de facto* single-party Rwandan state in which Tutsi guerrilla incursions from neighbouring Uganda, Tanzania, Zaire and Burundi were followed by massacres of the Tutsi population in Rwanda, engendering ever greater numbers of refugees. For example, 10,000 Tutsis were slaughtered in the Bugasera area following the 1963 incursion.[5] By the 1990s, it was estimated that the total number of Tutsi refugees outside the country equaled 600,000 to 700,000 persons.[6]

Though conditions in Rwanda improved somewhat following Juvénal Habyarimana's coup d'état in 1973 – massacres of Tutsis ceased for a time and his first decade in power was characterized by unprecedented stability and economic development[7] – his initial success did not endure. Habyarimana's rule became more and more corrupt and authoritarian in an attempt to buttress the ethno-regional fortress that guaranteed his hold over Rwanda. Gradually power came to be concentrated in the hands of the President's *Akazu* ('little house') or clan, which later conceived of and implemented the genocide.[8] This favouritism of the northwestern region from which he came alienated the southern parts of Rwanda, leading to the emergence of discontented opposition groups, even among the Hutu population. Habyarimana also failed to make any substantial progress toward reconciliation with the Tutsis, preventing them from acquiring positions of power or influence and refusing unequivocally to let Tutsi refugees in exile return home.

This last repudiation prompted Tutsis in Uganda to take more concerted action toward enabling their re-entry into Rwanda. Although many of these refugees had lived in Uganda for decades – some even playing a prominent role in Yoweri Museveni's National Resistance Army (NRA) as it gained control of Uganda in 1986, they were never granted full settlement rights there.[9] Convinced by 1987 that returning to Rwanda thus represented their best option, the Tutsis formed the Rwandan Patriotic Front (RPF),[10] which developed an eight-point agenda reflecting the organization's belief that only a profound alteration of Rwandan government and society would permit those in exile to return to and settle in Rwanda.[11] Although this programme aimed for genuine change in Rwanda, it was unclear how such policies could be implemented considering the degree of ethnic disharmony in the country and the weakness of the link between the RPF and Rwanda.[12]

It was difficult for the RPF to build a firm political base in Rwanda where most political parties, including the new opposition ones, emphasized the overriding importance of ethnicity, to the detriment of the Tutsis.

Though the initial October 1990 RPF attack was not an overriding success,[13] the RPF was gradually able to 'liberate' a significant part of the North over the next three years, putting increasing pressure on the Habyarimana regime. These RPF advances coincided with the international community's growing insistence on 'good governance' as an important criterion for receiving aid following the end of the Cold War, a tactic which succeeded in pressuring Rwanda toward democratization and liberalization. In June 1991, a new Rwandan constitution was adopted which granted freedom of speech and the right to organize political parties. Within a year, the *Parti Libéral*, the *Parti Social Démocrate* and the *Mouvement Démocratique Républicain* joined Habyarimana's ruling *Mouvement Révolutionnaire National pour le Développement* (MRND) on the political scene. Habyarimana's regime was also under great international pressure to resolve its conflict with the invading RPF peacefully so as to avoid a large scale civil war as well as destabilization in the region.[14]

As a result, Habyarimana was forced to negotiate a settlement with the RPF. Signed in Tanzania on 4 August 1993, the Arusha Peace Agreement constituted a significant victory for the RPF and laid the groundwork for a 'virtual political revolution in Rwanda',[15] fundamentally altering its form of government and the composition of its army, as well as resolving the refugee problem in favour of the RPF.[16] The Agreement specified the precise composition of the Broad-Based Transitional Government (BBTG) that was to replace the existing regime until elections occurred. The RPF and the ruling MRND party were each to hold 5 portfolios of 21 in the cabinet, and 11 seats out of 70 in the transitional national assembly. The rest of the seats were distributed amongst other parties. The *Coalition pour la Défense de la République* (CDR), the movement Hutu extremists had founded in response to the growing number of threats to the *status quo* in Rwanda, was excluded from this power-sharing formula. Although Habyarimana insisted that the CDR be included in the BBTG, the RPF refused since the CDR was both fanatic and not a political party.[17]

The Arusha Agreement also delineated the method of integration of the armed forces of the government of Rwanda and the RPF. Here again, the RPF proved the significant victor in the negotiations. The Habyarimana regime was forced to accept a 50–50 split in the joint-command, and a 60–40 ratio in its favour in the distribution of regular troops. This last agreement meant that more than 20,000 government soldiers would be demobilized, twice as many as in RPF ranks.[18]

The Arusha Peace Accord was thus a 'perfect' agreement on paper, resolving the refugee problem, ending the civil war and setting up democratic institutions,[19] but this seeming flawlessness belied a fundamental discordance between the RPF and the Habyarimana regime. Growing RPF military superiority on the battlefield put pressure on the Rwandan government to accept Arusha provisions with which it did not agree.[20] In contrast to the unity, strength and preparedness of the RPF at Arusha,[21] the Rwandan government's negotiating team was factionalized and ineffective, and it did not represent the true nucleus of power in Rwanda. Habyarimana's executive power had been sufficiently eroded by the 1991 constitution that Rwanda's negotiating team was led by a member of the opposition and mainly represented the voice of the Liberal Hutus in the country.[22] Extremist factions in Rwanda limited their own involvement in negotiations, believing that they had nothing to gain and everything to lose by participating in the peace process.[23]

Thus locked into a settlement that would deprive them of power, Hutu extremists devised other means to prevent the implementation of the Arusha Peace Agreement. They made every attempt to delay the creation of the BBTG that was supposed to be in place before the end of December 1993. The extremists launched an unofficial and clandestine political and social movement called 'Hutu Power', which sought to split, sometimes through infiltration, all the parties that had been accorded portfolios in the cabinet and seats in the transitional National Assembly by playing upon fears that the RPF would seek to impose a dictatorship upon the majority Hutu. Consequently, parties now comprised both members sympathetic to 'Hutu Power' and members willing to proceed with the BBTG arrangements. Endless disputes ensued, preventing party leaders from reaching agreement on representation in the interim institutions. Complicating matters were the political assassinations carried out by extremists.[24]

Yet, the Arusha Peace Agreement would not disappear from the political landscape. The closer it came to being actualized, the more frantic grew the extremists,[25] and the more willing they became to implement a radical plan that would prevent, once and for all, the RPF from gaining political power.[26] The Hutu extremists thus set in motion a genocidal policy that was carefully planned by a core group consisting of Habyarimana's close entourage (the *Akazu*), the army leadership and key members of the CDR and the MRND.[27] They proceeded to create a climate of fear and paranoia in Rwanda by using *Radio Télévison Libre des Mille Collines* (RTLM) to broadcast explicit hate propaganda against the Tutsis and to incite Hutus to murder them in preemptive strikes. To carry out the actual killing, the extremists utilized Rwanda's administrative infrastructure as well as the *Interahamwe* work groups,[28] which were transformed into militias. Rwanda's centrally appointed local authorities, *préfets*, *sous-préfets* and *bourgmestres*, would order the commencement of massacres in towns and villages when the time came. The Presidential Guard, drawn almost exclusively from Habyarimana's home region, would murder key political figures in Kigali, including moderate Hutus, and then assist the *Interahamwe* militias in the killing of ordinary Tutsis.[29]

UN involvement in Rwanda prior to and during the genocide

Although the Security Council agreed to sponsor a peacekeeping mission to Rwanda to monitor the proper implementation of the Arusha Agreement, its general reluctance to oversee the peace process with any degree of commitment or determination foreshadowed its later response to the genocide. Both parties consented to the creation of a neutral international force to be under the responsibility and command of the United Nations.[30] The Accord suggested an appropriate mandate for the force, which was considerably broader in scope in certain key respects than the mandate eventually agreed upon by the Security Council in Resolution 872 of 5 October 1993, which established UNAMIR for an initial six-month period. Both mandates required that peacekeeping troops assist with humanitarian aid and mine clearance, as well as monitor observance of the ceasefire. However, the mandate contained in the Arusha Accord also specified that a neutral international force could contribute to the overall security of Rwanda, by 'catering for the security of civilians', by assisting 'in the tracking of arms caches and neutralization of armed gangs throughout the country', and by aiding 'in the recovery of all weapons distributed to or illegally acquired by the civilians'.

In contrast, UNAMIR's mandate relied explicitly on the consent and cooperation of the parties involved, and focused mainly on monitoring observance of the Arusha Accord. Except for the provisions relating to the coordination of humanitarian assistance and to the monitoring of the process of repatriation of refugees, all references to ensuring the security of civilians and to preventing the propagation of para-statal militarization were dropped. UNAMIR was authorized to 1) contribute to the security of the city of Kigali; 2) monitor the security situation during the final period of the transitional government, leading up to elections; 3) 'investigate at the request of the parties, or on its own initiative, instances of alleged non-compliance with the provisions of the Protocol of Agreement on the Integration of the Armed Forces of the Two Parties, and to pursue any such instances with the parties responsible and report thereon as appropriate to the Secretary-General'; and 4) investigate and report on incidents involving the gendarmerie and the police. This disjunction between the proposed mandate of the Arusha Agreement and the one approved by the Security Council suggests that Rwanda was expected to require far greater international assistance and commitment than the United Nations was willing to provide.[31]

From the beginning of its deployment, UNAMIR was unable to react decisively to flagrant violations of the disarmament provisions of the Arusha Agreement and to clear indications that premeditated violence on a large scale was being planned.[32] One particular incident stands out in this regard. UNAMIR was made aware of the likelihood of massive bloodshed by an informant who belonged to the upper echelons of Habyarimana's security entourage. Involved in the military training of the *Interahamwe*, the informant said that he suspected that these militias were being formed for the purpose of exterminating Tutsis. He offered to provide UNAMIR with the location of a major weapons cache in Kigali in return for political asylum.[33] However, when UNAMIR Force Commander Roméo Dallaire sought permission from New York to conduct a cordon and search operation, his request was refused, as this action risked exceeding both the force's mandate and rules of engagement.[34] Instead he was instructed to share the information with the President of Rwanda and the representatives of three western Embassies in Kigali, including the American and French Ambassadors. Dallaire was also advised to entreat Habyarimana to investigate the matter and ensure that further transgressions of the Arusha Accord be disallowed.[35] Despite the fact that these warnings of future calamity reached the Secretariat, their significance seemed to be downplayed by senior officials, and no form of contingency planning was set in motion to deal with a massive outbreak of violence.[36]

What became clear to Dallaire was that Rwanda and the UNAMIR operation were classified as low-priority by the Security Council.[37] The force never received all the equipment it needed, and it was 'constrained by numerous shortcomings in personnel, equipment and disbursable funds, and even basics such as ammunition'.[38] Whenever progress in the implementation of the Arusha Peace Agreement stalled, the Security Council threatened not to renew UNAMIR's mandate, thus indicating its unwillingness to enforce compliance if it became necessary to do so. UNAMIR's basic ineffectiveness had grave consequences. On the one hand, its presence reassured and gave a sense of security to opponents of Habyarimana's regime, who were more willing to speak out and were thus more identifiable when the killing started.[39] On the other hand, its lack of decisiveness and credibility indicated that the international community was unable, or simply unwilling, to respond to growing extremism in Rwanda, thus sending a message to the genocide plotters that they could proceed with their plan without fear of hindrance.

On 6 April 1994, the presidential plane carrying Habyarimana back to Rwanda from a summit meeting in Dar-es-Salaam was shot down at 8:30 PM.[40] By 9:15 PM *Interahamwe* roadblocks littered the capital, at which Tutsis were detained so they could be slaughtered; and RTLM was inciting Hutus to murder Tutsis in order to avenge the death of Habyarimana. The first people to be killed, mainly by the Presidential Guard, were leaders of the political parties that opposed extremism, and their families, regardless of whether or not they were Tutsi. At first, the killing was restricted to Kigali, only spreading throughout the country after 21 April when it became clear the UN would not intervene effectively to halt the genocide.[41] Within a few weeks, hundreds of thousands lay dead. When the genocide came to an end soon after the RPF gained victory over government forces on 18 July, between 500,000 and 800,000 had been slaughtered, in what has been widely recognized as genocide.[42]

The international community's initial reaction to the Rwandan genocide was two-fold: 1) evacuation of foreign nationals, and 2) the reduction in size of UNAMIR. In the first days following the outbreak of violence, a combined French, Belgian and Italian initiative called *Opération Amaryllis* was launched to evacuate foreign nationals from Rwanda. Within five days of the downing of the plane carrying Habyarimana, 2,000 Europeans had been flown out.[43] Ordered not to intervene in the local political or security situation, European soldiers on occasion became spectators to the slaughter of the Tutsis and the moderate Hutus.[44] On 7 April 1994, 10 Belgian UNAMIR peacekeepers were brutally murdered by the Presidential Guard as they endeavoured to prevent the moderate Hutu Prime Minister, Agathe Uwilingiyiamana, from being assassinated. It is almost certain that the killing of these Belgian peacekeepers was a deliberate act by extremists to weaken UNAMIR or to provoke its withdrawal. Their plan worked. Within ten days of the murder of these soldiers, the Belgium Ambassador informed the Security Council 'that the Belgian battalion in UNAMIR [would] be withdrawn...without delay'.[45] If UNAMIR's power and authority in Rwanda was feeble to begin with, the loss of the Belgian contingent undermined the peacekeeping force further. Its size dropped automatically from 2,486 to 1,705 personnel, but most importantly, UNAMIR lost its best equipped, best trained and most experienced contingent.

Faced with the *de facto* weakening of UNAMIR, the Secretary-General nonetheless presented three options for the future of UNAMIR to the Security Council: 1) an immediate and massive reinforcement of UNAMIR and an alteration of its mandate so that it could become a Chapter VII peace enforcement operation capable both of restoring law and order and of putting an end to the killings; 2) a scaling down of UNAMIR, leaving a small contingent stationed in Kigali that would act as an intermediary between the parties so as to encourage a ceasefire; and 3) a complete withdrawal of UNAMIR.[46] Through Resolution 912 of 21 April 1994, the Security Council selected the second option, reducing UNAMIR's size to 270 soldiers.

Only on 17 May 1994 did the Security Council reconsider this decision, authorizing the expansion of UNAMIR to 5,500 troops and altering its mandate so that it could 'contribute to the security and protection of displaced persons, refugees and civilians at risk in Rwanda, including through the establishment and maintenance, where feasible, of secure humanitarian areas'.[47] However, since it proved difficult to find properly equipped troops for the mission, UNAMIR II would only enter Rwanda in August when the genocide was over and when SHAS were no longer needed.

International inaction following the start of the genocide

Although it is tempting to interpret the state failure to respond adequately to the Rwandan genocide as a clear indication that community interests play second fiddle to state interests – that states only pay attention to their community interests when their state interests are also at stake – an analysis of the behaviour of states and the UN following the start of the genocide reveals a more complex course of events and interrelationship between these two different types of interests. The fact that the enabling effect of state interests was almost non-existent for most states in relation to Rwanda does indeed lie at the heart of any understanding of why nothing, or very little, was done in Rwanda in April 1994. However, the absence of state interests that would have encouraged intervention is only part of the picture. Two other factors, with reference to community interests, were also vitally important to the international community's inaction when faced with genocide. First, legitimate confusion existed, on the part of states, the UN Secretary-General and the media, as to whether or not what was occurring in Rwanda was genocide, thus dampening the enabling and constraining quality of this community norm for a critical two to three week period at the start of the calamity. Second, once it did become clear that the horrific killings taking place in Rwanda were in fact genocide, states in general, but the US in particular, deliberately embarked upon a campaign to minimize, avoid or deny outright the impact that the violation of this community norm would have on the Security Council. States did not simply do nothing in Rwanda. Rather, they were forced to counter the pressure their community interests placed upon them, instigating intentionally a policy of inaction toward Rwanda that resulted in a tardy, ineffective response to the genocide and a defunct safety zone approach.

State interests and international inaction

Very few states had strategic interests in Rwanda that might have bolstered their response to the genocide. With the end of the Cold War, American interest in sub-Saharan Africa declined drastically. Its monetary aid to the continent had fallen off, and its focus had shifted to eastern Europe and the Balkans. Although supportive of the Arusha peace process, the US was not keen on being drawn into an active role in helping to implement it, as evidenced by its unwillingness to contribute troops to UNAMIR and its reluctance to provide this peacekeeping force with a strong mandate. When James Woods, Deputy Assistant Secretary for African Affairs at the Department of Defense from 1986 to 1994, attempted to include Rwanda on a Pentagon list of potential trouble spots in the spring of 1993 as Clinton took over the presidency, he was told to leave it off since no American interests were at stake there.[48] The UK did not have an embassy in Rwanda, and it was neither actively involved in the Arusha peace process nor interested in taking any initiative in terms of Security Council decision-making regarding the humanitarian emergency. Both the British and the Americans considered Rwanda to be situated in France's sphere of influence, so deferred to the French on the matter.[49] French state interests in Rwanda, as further discussed in the next chapter, were predicated on allegiance to the Habyarimana regime, thus predisposing France to support the interim government. Only the Belgians, with their historical links to

Rwanda, had any impetus to become involved, and had shown their commitment to the country over long-months of propping up the UNAMIR force. However, the loss of their peacekeepers on 7 April 1994 generated, within public opinion, a strong knee-jerk opposition to continued participation within UNAMIR.[50] As a result, Belgium withdrew its troops almost immediately. When Boutros-Ghali begged its Foreign Minister Willy Claes to at least leave Belgium's military equipment behind so that UNAMIR would not be left stranded without adequate means, Belgium refused.[51] The only enabling state interest that seemed to exist at this time was the desire to protect foreign nationals, which explains the rapid response of the French, Belgian and American military operations in this regard.

Unlike in both Bosnia and Iraq, the refugee problem resulting from the genocide did not figure prominently in calculations of state interest. Over the course of the genocide and its aftermath, millions of Rwandans fled to neighbouring countries. However, this exodus did not affect either Europe or North America. Since refugees in Africa did not pose a threat to the state interests of key Security Council members, there was no self-interested inclination to prevent their movement out of Rwanda into Zaire, Tanzania and Burundi.

Not only did state interests not enable intercession in the Rwandan genocide, their constraining effect was so severe as to render involvement almost an impossibility. Although Western states' general unwillingness to suffer casualties amongst their troops, as evinced by the attitude of the Belgians to the killing of their troops, eclipsed any commitment they may have had toward fulfilling their community interests, American state interests most strongly dictated the international policy of inaction toward the genocide. Traumatized by its experience in Somalia, the US was determined to avoid any direct military involvement in Rwanda.[52] When the genocide erupted in Rwanda, the US had only just withdrawn their last soldiers from Somalia on 31 March 1994. Having kept them there that long despite domestic political opposition so as to ensure an orderly hand-over to the UN, the Clinton administration believed that neither public opinion nor Congress would tolerate even the suggestion of American military involvement in Rwanda.[53] On 10 April, well-respected and powerful Republican Senate Leader Bob Dole indicated he opposed any American role in Rwanda as no vital national interest was at stake there. Two influential democrats soon followed suit: Harry Johnston, Chairman of the Subcommittee on Africa of the House Committee on Foreign Affairs, and Robert Byrd, Chairman of the Senate Appropriations Committee, agreed that the US should not intervene in Rwanda.[54] America's response to the Somalia debacle also incorporated a strong element of UN scapegoating: it blamed the international organization for what had been predominantly a US disaster,[55] thus amplifying its conviction that American interests were not served by involvement in UN peacekeeping. Hence, when it appeared as if conditions in Rwanda were no longer ripe for a peacekeeping presence and risked degenerating into a Somalia-type situation, the US pushed for the complete and total withdrawal of UNAMIR in the informal sessions of the Security Council, and was the only state to do so.[56] Although it was persuaded to acquiesce in the consensus on adopting resolution 912, which left a skeletal UNAMIR of 270 soldiers in Rwanda, its position certainly created an insurmountable obstacle to a more forceful response.

Community interests and international inaction

Despite the presence of strong state interests that discouraged humanitarian intervention and led to the downsizing of UNAMIR on 21 April, it is nonetheless surprising that states failed so absolutely to respond to the Rwandan emergency in a timely fashion, given the enormity of the crime of genocide as well as the magnitude of suffering of the victims. The enabling and constraining quality of the 1948 Genocide Convention is perhaps the most potent of any of the community norms states recognize. States party to the convention 'confirm that genocide is a crime under international law which they undertake to prevent and to punish'.[57] Although it is not clear to what extent the crime of genocide imposes a legal obligation on states to intervene to halt it,[58] it certainly generates a very strong moral imperative, or 'obligation of a kind', to do so. Since the crime of genocide is particularly heinous, the moral duty to prevent it carries exceptional weight.[59] Eventually by mid-May the pressure that this violation of an important community norm imposed upon states did become too strong to ignore, dominating Security Council deliberations[60] and triggering the UN and French safety zone plans. However, in the interim, little was done at first because states, the UN Secretariat and the media failed to recognize that genocide was taking place. Once this misperception was finally corrected three weeks into the genocide and pressure mounted for intervention, states feared being pushed in a direction at odds with their self-interests, so they deliberately sought to pursue a policy of inaction in Rwanda and to deny the importance of their community interests.

Two important sets of factors explain why the community interest in preventing the crime of genocide did not reach the full force of its potential to affect state behaviour. Wary of the growing permissiveness of the UN toward humanitarian intervention in general, the US came to take a restrictive view of the circumstances in which the use of peacekeeping or other forces was justified; and confusion on the part of the UN Secretary-General and members of the Security Council regarding whether or not the killings in Rwanda constituted genocide hindered an effective institutional response to the calamity.

The United States must take considerable responsibility for the Security Council's inability to address the calamity in a meaningful way. Its reluctance to see the Council authorize intervention in Rwanda stemmed in part from its growing belief in the wake of Somalia that the permissive conditions prevalent in the Council toward becoming involved in humanitarian emergencies were dangerous to American strategic interests. The US considered that this tendency within the Security Council created a heightened risk that America would be pressured into involvement in foreign lands, either as the main saviour, the provider of key logistical support or the only power capable of rescuing other states' troops when disaster struck.[61] To combat this enabling effect of community norms, which meant that there were approximately 70,000 peacekeepers in the field by 1994[62] at an annual cost of $3 billion,[63] the Clinton administration unveiled a new policy labeled PDD 25 on 5 May 1994. Its main intention was to create rigorous standards of review for US support for or participation in UN peacekeeping operations, and its basic premise stated that the United Nations could not intervene in every conflict so involvement had to be selective and effective.[64] To facilitate the establishment of criteria for US participation, PDD 25 divided peacekeeping into two categories: peace operations that required US support in terms of Security Council backing and financial assistance; and peace operations that would also involve American soldiers. Stricter standards for participation would be applied in the second scenario.

In order for a peacekeeping mission to be sponsored by America, it would have to advance US interests, and there would have to exist a community of interest for dealing with the problem on a multilateral basis. Clear objectives would be necessary, as well as an understanding of where the mission fitted on the spectrum between traditional peacekeeping and peace enforcement. For a Chapter VI operation, a cease-fire would have to be in place, and the consent of the parties guaranteed; for a Chapter VII enforcement operation, the threat to international peace and security would have to be significant. The means to accomplish the mission would have to be available before authorization could occur, and its duration as well as an exit strategy would have to be anticipated. Finally, for a mission to proceed, the consequences of inaction would have to be deemed unacceptable.[65] In addition to these guidelines, if American troops were to participate directly in the peacekeeping operation, the mission would have to have clear objectives and advance American interests at acceptable risk. Congressional and domestic support would be prerequisites, as well as a definite endpoint for US involvement.[66]

In accordance with PDD 25, the US proved unwilling to support the idea of a strengthened UNAMIR.[67] Not just a restraint on American behaviour, PDD 25 also trapped the UN in a vicious circle with regard to intervention into Rwanda. The United States would refuse any new peacekeeping deployment unless all the necessary conditions were fulfilled, but these requirements could not be met without the support of the superpower. This problem was amplified by the fact that the US was assessed 31 per cent of the cost of all peacekeeping operations, thereby rendering American consent to and patronage of a mission necessary in order for it to proceed and be a success.[68] Further, if the United States chooses not to act or to delay the proceedings of the Security Council so as to hinder the decision-making process, then its weight is felt in the direction of non-involvement, and very little can be achieved.[69] In sum, America's inordinate influence upon peacekeeping activities and its unwillingness to engage in the Rwandan crisis greatly delayed and weakened the UN response to genocide.

Community interests can only enable and facilitate action if there is clear and relatively undisputed recognition that a grave violation of the norm in question is in fact occurring. In the case of Rwanda, clarity and consensus about events on the ground constituting genocide were not present. During the first two or three weeks of the genocide, bewilderment and confusion plagued the UN. Boutros-Ghali only declared it was genocide publicly on 4 May in a TV interview for *Nightline* in the US. Only at the end of April did the Security Council, mainly at the behest of the Czech Republic and New Zealand,[70] adopt a presidential statement that condemned the acts of genocide in Rwanda.[71] Even states with intelligence on the ground, such as the US, France and Belgium, were not fully aware genocide was underway in Rwanda for ten to fourteen days.[72] Though these states had a state interest in denying it was genocide so as to avoid intervention, it nevertheless seems safe to assume that there was some legitimate initial confusion for one to two weeks, and that this misperception delayed and hindered an effective UN response.

One clear factor militated against immediate recognition of genocide: the perpetrators' strategy of concealing their crime and sowing sufficient confusion in the international community to prevent intervention.[73] They presented the chaos in Rwanda as the consequence of both the resumption of the civil war and an outbreak of random tribal violence resulting from a spontaneous desire on the part of the Presidential Guard and Hutus in general to enact vengeance for the death of Habyarimana.[74] The new interim government also portrayed itself as a victim of circumstance and claimed that the RPF's violation of the ceasefire prevented it from restoring law and order. It also contended that the massacres were being perpetrated by the RPF.[75] Rwanda's presence on the Security Council throughout this period, which gave it a more powerful position than usual in the non-aligned and African caucuses, meant that the extremist interim government enjoyed exceptional influence.[76] Its constant and explicit support for UNAMIR, justified as a means to end the violence, rendered its stance regarding the killings and the RPF more credible. The international community's confidence in the Arusha Agreement's ability to resolve Rwanda's democratization and human rights problems encouraged a certain wishful thinking regarding events in Rwanda that facilitated the extremists' subterfuge.[77]

The Secretary-General of the United Nations (SG), whose job it is to inform and make recommendations to the Security Council based upon an accurate assessment of the facts, seemed unable to perform his job adequately in the particular context of Rwanda. Whether this lapse in performance stemmed from Boutros-Ghali's distance from the Council at this time,[78] from his disbelief that a genocide could actually take place in a poor African state[79] or from his previous cordial association with the Habyarimana regime as a Minister of State for Foreign Affairs in Egypt,[80] its result was a failure to characterize the events in Rwanda as genocide and to apply sustained pressure upon the Security Council for a forceful response. On 8 April, Boutros-Ghali sent a letter from Geneva to Colin Keating, President of the Council during the month of April, suggesting that the evacuation of UNAMIR from Rwanda might be unavoidable.[81] UNAMIR Force Commander Dallaire was in fact told to consider and plan a withdrawal on at least three separate occasions by senior UN officials, including the Secretary-General himself on 12 April.[82] Since he was loath to put the UN into another Somalia-type situation,[83] Boutros-Ghali never conveyed to the Council that Dallaire, instead of wanting to pull out, had recommended to the UN Secretariat that UNAMIR be expanded: the UNAMIR commander considered that 5,000 soldiers deployed under a Chapter VII mandate, with air support and proper logistics and communications, might have been able to achieve a great deal.[84]

After considerable delay,[85] in his first report to the Security Council on the crisis, in which the three possible future strategies for UNAMIR were presented, the Secretary-General stated that the death of Habyarimana 'set off a torrent of widespread killings' and that 'the violence appear[ed] to have political and ethnic dimensions'.[86] Yet the explanation for these massacres provided by the Rwandan interim government was also accepted: 'reliable reports strongly indicate that the killings were started by unruly members of the Presidential Guard'.[87] Regarding the three options presented in this report, total withdrawal, partial withdrawal and the strengthening of UNAMIR, Boutros-Ghali made clear his preference for the last option via a statement of his spokesman to the press, but did not adequately convey the advantages of this position to the Council.[88] It should also be noted that UNAMIR Force Commander Dallaire did not start using the term genocide in his own reports to UN headquarters until sometime after 24 April,[89] contributing in some measure to the Secretary-General's non-use of the word in the early phases of the genocide.

Even at the end of April, when the Secretary-General decided that more forceful action was needed by the Security Council in order to curtail the slaughter of civilians, there was still no mention of genocide. The massacres were associated with timeless ethnic hatred and a break down of law and order. The 29 April letter the Secretary-General addressed to the Security Council, claimed, for instance, that the assassination of Habyarimana 'reawakened deep-rooted ethnic hatreds, which have plagued Rwanda in the past and which have again led to massacres of innocent civilians on a massive scale'.[90] Responsibility for these killings was assigned as follows:

> Some of these [massacres] have been the work of uncontrolled military personnel but most of them have been perpetrated by armed groups of civilians taking advantage of the complete breakdown of law and order in Kigali and many other parts of Rwanda.[91]

In the Secretary-General's 13 May report to the Security Council, which proposed that the UNAMIR mission be expanded and strengthened, there was still no mention of genocide.[92] Only in his 31 May report did Boutros-Ghali at last label the events occurring in Rwanda as genocide and lay blame for their perpetration upon the extremists.[93]

The Secretary-General's initial distorted account of circumstances in Rwanda helped form the views of the members of the Security Council, thereby rendering an appropriate response to the crisis unlikely since it was not fully realized that a genocide was being perpetrated. Resolution 912, reducing the size and strength of UNAMIR, mentioned that the Security Council was 'appalled at the ensuing large-scale violence in Rwanda' and 'deeply concerned by the continuing fighting, looting, banditry and the breakdown of law and order, particularly in Kigali'.[94] The term 'genocide' was only employed in Security Council Resolution 925 of 8 June 1994, which altered UNAMIR II's deployment schedule. In addition, the first outright denunciations of Rwanda's continued presence on the Security Council and of its lies concerning events in Rwanda came only on 16 May 1994.[95] Smaller states, which did not have access to their own intelligence reports and which later complained that the Secretariat failed to provide them with correct and adequate information, were especially vulnerable to poor performance on the part of the UN Secretary-General.[96] The Czech Ambassador voiced his criticism of the Secretary-General's performance in the Security Council:

> My delegation is troubled that it has taken so long for the Secretary-General to use this description (i.e. 'genocide') in his reports, on which the Security Council bases its work so heavily...My delegation feels that if these facts [regarding the genocide] had been forcefully communicated to the Security Council as soon as the Secretariat became aware of them we might have been a step further today.[97]

The assumption that the massacres in Rwanda were a result of a breakdown of law and order, rendered more complicated by the resumption of the civil war, was vastly different from the reality of a planned genocide. It made the Rwandan crisis resemble the Somali one more than the Bosnian one, thus reinforcing institutional attitudes of restraint that emerged following the debacle in Somalia.[98] Yet Rwanda was not a failed state, but one capable of conducting genocide. Once this misperception had contributed to the decision-making process and UNAMIR was downsized, it was difficult to reverse momentum and retake the initiative, as the subsequent problems with the deployment of UNAMIR II revealed.[99] Moreover, this error in judgement was amplified by the swiftness of the genocide. Most of its victims succumbed in the first 6 to 8 weeks; therefore, even a two week delay in determining that genocide was taking place meant that the enabling effect of this community norm did not emerge in time to make a difference to the unfolding of the genocide.[100]

Not only did the early non-recognition of genocide mean that international action would not be enabled, it also implied that a clear imperative to do something on behalf of a population targeted for elimination would also not arise in a timely manner. Two factors are noteworthy here. First, Rwanda's coverage in the media did not bode well for the possibility of an effective international response, as it failed to generate a sufficiently strong public outcry that might have propelled governments to action. The media's initial coverage of the genocide presented the events taking place in Rwanda either as uncontrollable tribal violence or as simply a civil war. The first *New York Times* stories about Rwanda following the assassination of Habyarimana depicted the carnage in Rwanda exactly as the extremists would have wished. The headline on 8 April 1994 read 'Troops Rampage in Rwanda', and the article to which it was attached stated that Kigali 'had been given over to anarchy, with gangs of young men wielding clubs and machetes rampaging through the streets, looking to settle old scores'.[101] The next day, the same reporter claimed that 'rival tribal factions waged vicious street battles'.[102] This focus upon drunken soldiers, tribal warfare and anarchy continued in the *New York Times* until approximately 21 April when stories began to allude to the fact that Hutus were targeting Tutsis, and that the violence was politically motivated and not spontaneous.[103] An editorial on 22 April pronounced that 'what looks very much like genocide has been taking place in Rwanda'.[104] Yet even the acknowledgement of such a horrific crime did not alter the general stance of the American media, which regularly resurrected the ghost of Somalia and continued to oppose American involvement of any kind in a peacekeeping mission.[105] TV coverage mirrored that of the print media, in terms of content and timeframe analysis.[106]

British press content on Rwanda resembled that found in America, typifying the killings as spontaneous and as centuries' old tribal warfare.[107] *The Economist* came to the conclusion that 'whoever began it, soldiers were soon joined in their brutalities by drunken mobs and street gangs'.[108] On 23 April, it raised the question of whether or not the killings might have been premeditated and suggested that Tutsis were the main targets of the violence,[109] but only, on 21 May, labeled the events taking place in Rwanda as genocide.[110]

Moreover, compared with press coverage of the Bosnian crisis which depicted ethnic cleansing as a sophisticated military strategy dedicated to achieving clear political aims, the news about Rwanda conveyed the impression of a primitive, timeless conflagration of chaos and savagery, although both conflicts were similar in many respects.[111] The media thus reinforced stereotypical perceptions about Africa being both uncivilized and plagued by ancient tribal conflicts that were without remedy, and thus did not bring into play the constraining effect of the genocide norm in domestic politics. The amount of US press coverage on each conflict is also illustrative. Bosnia, on a monthly basis, was likely to feature in 25 times more articles than Rwanda did, indicating that the media, and most probably the political elite, deemed Bosnia to be a vastly more important story than Rwanda.[112] In sum, the failure of the media to report accurately and adequately on the genocide in Rwanda enhanced the confusion within the Security Council, augmented public indifference toward the calamity, and meant that there was little pressure on states to act in line with community interests.

Second, when it was becoming apparent that the killing in Rwanda was genocide, several western states sought to deny or call into question this reality in order to avoid the strong moral imperative or 'obligation of a kind' that recognition would have entailed.[113] The worst culprit in this regard was the US. Due in part to its large Jewish population and the widespread American public empathy toward the tragedy of the Holocaust,[114] the US government sought to reduce the usage of the term genocide in relation to Rwanda. The State Department sent out directives to diplomats in the field to the effect that, whereas the slaughter they were witnessing might seem to resemble genocide, it was not to be described as genocide but as 'acts of genocide'.[115] The following interchange on 10 June 1994 between reporters and Christine Shelley, the US State Department spokeswoman at the time of the genocide, is also telling:

Shelley: We have every reason to believe that acts of genocide have occurred.

Reporter: How many acts of genocide does it take to make a genocide?

Shelley: That's not a question I'm in a position to answer.

Reporter: Is it true that you have specific guidance not to use the word 'genocide' in isolation, but always preface it with the words 'acts of'?

Shelley: I have guidance...which I try to use as best as I can...There are formulations that we are using that we are trying to be consistent in our use of. I don't have an absolute categorical prescription against something, but I have the definitions. I have a phraseology which has been carefully examined and arrived at...

Reporter: So you say genocide happens when certain acts happen, and you say that those acts have happened in Rwanda. So why can't you say that genocide has happened?

Shelley: Because, Alan, there is a reason for the selection of words that we have made, and I have – perhaps I have – I am not a lawyer. I don't approach this from the international legal and scholarly point of view. We try, best as we can, to accurately reflect a description in particularly addressing that issue...There are obligations which arise in connection with the use of the term.[116]

It was also felt that to call it genocide and then be seen to be doing nothing about it would impact negatively on Congressional elections scheduled for November 1994.[117]

Clearly aware that a legal requirement to prevent genocide existed and that such an obligation might be powerful enough to impose a drastic change of course in UN policy, the United States delegation, with Britain alongside, also lobbied to have Security Council Resolutions contain somewhat ambiguous language. This fact explains the curious formulation present in the presidential statement of 30 April, which informed that 'the killing of members of an ethnic group with the intention of destroying such a group in whole or in part constitutes a crime punishable under international law'.[118] Colin Keating, whose tenure as President of the Council ended the following day, pressed for inclusion of the term 'genocide' in this statement, even threatening to table a resolution if wording could not be agreed, which would have forced the US, the UK and other members on the Council to explain publicly their dislike of the word 'genocide'. In the end the British delegation came up with the ambiguous solution above.[119] Security Council Resolution 925, extending the duration of the UNAMIR II mission, noted with concern 'the reports indicating that acts of genocide have occurred in Rwanda...',[120] as opposed to acknowledging that a full-fledged genocide was underway. In this manner, states endeavoured to circumvent the constraining effect of norms by denying, to a considerable extent, that genocide was taking place in Rwanda.

Though the US, the UK and others were not keen to intervene militarily in Rwanda, they were nonetheless concerned that non-involvement might tarnish their identities as upstanding, 'civilized' members of the international community. As a result, these states sought to frame their behaviour regarding Rwanda in such a way as to not jeopardize too severely the image they wished to portray or themselves. First, the portrayal of what was going on in Rwanda as ancient tribal violence and chaotic killing generated an identity chasm between a 'barbaric' Rwanda on the one hand and a 'civilized' international community on the other. This depiction of Rwanda amplified the sense that it lay outside the boundaries of international society and therefore not within the scope of states' community interests. The marginalization of Africa in international affairs following the end of the Cold War further contributed to Western states' sense of disassociation from Rwanda.

So as not to tarnish its reputation as a respectable UN partner that does not leave a peacekeeping mission in the lurch at the first sign of trouble, Belgium also sought to depict its withdrawal from Rwanda in a manner consistent with community interests. Not only did Belgium unilaterally withdraw it peacekeepers from Rwanda, it also launched a diplomatic campaign in the Security Council for the complete suspension of the UNAMIR mission.[121] Foreign Minister Claes contacted his counterparts in Washington, Paris and London,[122] and rang the foreign ministries of several of the other smaller members of the Council in order to convince them to do the same.[123] If Belgium could gloss the circumstances of its departure by suggesting that a UN mission to Rwanda was now pointless in the face of renewed fighting and a deteriorating security situation, then it could preserve its reputation as a solid UN partner. Its self-interested behaviour would then be diluted by the broader international abandonment of Rwanda. Having learned from bitter experience in Somalia about the importance of a 'face-saving way of extracting its soldiers' from far away places, the 'US was able to accommodate [the Belgians] in the Security Council by calling for the draw-down of the peacekeeping force in Rwanda'.[124]

Likewise, though the US and the UK did not want to become directly involved in ending the genocide in Rwanda, they nevertheless did not want to appear as if they were entirely indifferent to the calamity. As a result, they cultivated a kind of subterfuge that enabled them to maintain their honour whilst abandoning Rwanda to genocide. Avoiding labeling the killings genocide was one such tactic, since western states would 'have made fools of themselves' if they had called events in

Rwanda genocide and then done nothing about it.[125] When such a denial eventually became indefensible in the face of mounting evidence to the contrary, the only other option for preserving an upright international identity was to justify inaction in ethical, reputable terms. For example, the US advanced the idea that the future of UN credibility and peacekeeping depended on not becoming embroiled in a disastrous situation in Rwanda.[126] US diplomats could adopt a 'discourse of acting in the best interests of the international community', saying they were putting peacekeepers' lives and the needs of the UN first, while ignoring hundreds of thousands of deaths in Rwanda.[127] Madeleine Albright, US Permanent Representative to the UN at the time of the genocide, warned that the UN's reach should not exceed its grasp, or 'we will only further undermine UN credibility and support'.[128] PDD 25 was often rationalized in this way as well: if the ideal of effective multilateral peacekeeping were to have a future, then systematic, practical criteria were needed to ascertain whether the US would or would not become involved in peacekeeping missions, even if this meant ignoring cases where intervention was absolutely necessary but not in conformity with pre-established principles.[129]

The adoption and non-implementation of secure humanitarian areas (SHAs)

As the crisis deepened in Rwanda and it was acknowledged that the massacres there constituted genocide, pressure mounted on the UN and its member states to do something, which eventually culminated in Security Council authorization of the expansion of UNAMIR and of the creation of SHAs as a means to provide security for civilians and displaced persons in Rwanda. Safety zones were again envisaged, as they had been in Iraq and Bosnia, because they would allow states to counter the genocide in some way, without having to take sides militarily in the Rwandan civil war or be drawn into a Somalia-type peace-enforcement operation. The proposed SHAs were thus in keeping with the Security Council's preferences for obtaining a ceasefire and returning to the settlement framework of Arusha as the most effective ways of ending the slaughter and solving Rwanda's long-term problems without the need for direct international military intervention.[130] Despite the fact that the proposed SHAs were a compromise solution that offered states the opportunity to reconcile competing, and in this case opposing, interests, these safety zones were never implemented: state interest concerns advocating caution and restraint remained too strong, especially on the part of the US as it sought to avoid another Somalia.

Community interests and the adoption and non-implementation of SHAs

As the death-toll reached a staggering level, it became increasingly difficult for states to ignore that a universal and fundamental international crime was being perpetrated: the constraining effect of community interests became too strong to disregard. African states in particular played a prominent role in advocating the need for a decisive UN response to the genocide. Nigeria, then on the Security Council, spurred both the non-aligned and African caucuses to play an active role in the crisis and tabled on their behalf a draft resolution as early as 13 April calling for the strengthening of UNAMIR.[131] Dismayed at the level of indifference of the Security Council, several African countries also reminded the international community that the fabric of civilization was at stake in Rwanda.[132] At the 8 June 1994 meeting of the Security Council, almost every state voiced its commitment to halting the genocide since it was intolerable that such a travesty of justice and humanity should continue. The representative from Spain spoke the most forcefully:

> The international community cannot stand idly by when faced with these facts, particularly in view of the binding terms of the 1948 Convention on the Prevention and Punishment of the Crime of Genocide, which can be considered to form part of general international law.[133]

Similarly, the SG came under increasing pressure from African countries and NGOs to demonstrate that the United Nations did not have two standards for measuring human lives, one for Europeans and one for Africans. This outcry was supported by the mounting evidence that genocide was taking place throughout Rwanda, while the international community's attention and resources were directed toward the crisis in Bosnia. The Secretariat, headed by the Secretary-General, was the first UN institutional body to reverse course and call for a more forceful response to the violations of international law occurring in Rwanda. An April 29 letter to the Security Council urged it to reconsider its decision to reduce the size and strength of UNAMIR.[134]

On 13 May 1994, the Secretary-General recommended that the Security Council increase the size of UNAMIR to 5,500 troops and expand its mandate so that it would be possible to provide both 'safe conditions for displaced persons and other groups in Rwanda' and 'security assistance to humanitarian organizations in their programmes for distribution of relief supplies'.[135] The Secretary-General also warned that the deployment would have to occur rapidly or the chances of the project's success would be in jeopardy.[136] The Secretary-General proposed that the troops take control of Kigali airport and then fan out across the country, establishing a definite presence in Rwanda within 31 days.[137]

Four days later, in fulfillment of a moral duty to do something in the face of genocide, the Security Council mounted a more forceful response to the crisis. Implementing the Secretary-General's recommendations, it authorized the expansion of UNAMIR to 5,500 troops and altered its mandate to include the following task:

> (a) To contribute to the security and protection of displaced persons, refugees and civilians at risk in Rwanda, including through the establishment and maintenance, where feasible, of secure humanitarian areas...[138]

Security Council Resolution 918 sanctioned UNAMIR II to use force, but only in self-defense 'against persons or groups who threaten protected sites and populations, United Nations and other humanitarian personnel or the means of delivery and distribution of humanitarian relief...'[139]

State interests and the adoption and non-implementation of the SHAs

Despite the fact that the need to uphold an important community norm had finally compelled states to contemplate the possibility of creating safety zones to protect targeted civilians, the envisaged SHAs were never established, mainly because state interests that discouraged such a course of action were too strong. The United States again played an important role in impeding the effectiveness of the SHAs approach, though Boutros-Ghali found it exceedingly difficult to find and properly equip potential troops for the UNAMIR II mission.

The United States continued to counteract deliberately the pressure community norms were exerting upon the Security Council. Although isolated in its position favouring non-intervention in Rwanda, the US still forced at least a week's delay of the vote on Resolution 918. Concluding that it could not postpone a vote forever, the United States attempted to bring the plan more in line with PDD 25 before it reached the resolution phase.[140] Since the US was concerned about peacekeepers landing in Kigali in the middle of a conflict situation and did not want to be called upon to help facilitate such a risky proposition, it promulgated an 'outside-in' plan,[141] in which a 'protective zone' would be created 'along the Rwandan borders [with Tanzania and Zaire] with an international force to provide security to populations'.[142] Washington sent military experts to see Kofi Annan, then UN Under-Secretary-General for Peacekeeping, to persuade him of the merits of the plan: they informed him they would be willing to contribute to the costs of such a mission since it would require fewer soldiers and thus be cheaper to implement.[143] This proposal drew criticism from Dallaire in a cable dated 12 May: he wondered how Tutsis could get to such a zone without being killed.[144] It is not clear to what extent the Clinton administration was serious about this proposal: it never found its way into a draft resolution sponsored by the US, for instance, or into a modified version of 918. Senior officials in the US administration were unable to provide any details of the plan, of who instigated it, or of why it was never pursued more forcefully: they only suggested it was a worthwhile endeavour that should have been looked into further.[145] In an interview for the *Frontline* documentary *The Triumph of Evil*, Tony Marley, Political and Military Advisor for the State Department from 1992 to 1995, stated that the discussions about the safe havens were only feasibility discussions, and 'never went anywhere'.[146] According to Dallaire, a formal UN request to the US government to have it put an end to RTLM broadcasts, through jamming, airstrikes or covert operations, was turned down, due to cost and legal inconveniences.[147]

The US also played its part along with other states in hampering the deployment of UNAMIR for three months and in rendering the SHA policy defunct. Although resolution 918 called for the full UNAMIR II contingent to be stationed within Rwanda in 31 days, delays mounted, to the extent that the peacekeeping force only reached full capacity at the end of August. Even states such as New Zealand and the Czech Republic, so active in trying to get the Council to respond more effectively to the genocide, did not volunteer any of their own soldiers for UNAMIR II.[148] Boutros-Ghali solicited assistance from the Secretary-General and the Chairman of the Organization of African Unity (OAU) in his efforts to find troops, and asked 30 African states to contribute troops.[149] Eventually Ethiopia, Ghana, Senegal, Malawi, Mali, Zambia and Tunisia volunteered troops to UNAMIR II, but they lacked the equipment and logistical support necessary to carry out a successful mission. Although the permanent members of the Security Council and other developed states offered some financial and technical assistance, their contributions were minimal, and stalling tactics were common.[150] Donors of equipment alleged, for example, that African countries used peacekeeping operations to outfit their forces and worried that their equipment would be misused.[151]

Bureaucratic procedures also produced long delays, especially as there existed no political will to overcome them. The United States agreed to lease fifty APCs to the UNAMIR II mission on 19 May, but disagreement over remuneration, painting and transport meant they only arrived in Rwanda on 30 July 1994.[152] Instead of driving this type of issue through the Pentagon bureaucracy, the Clinton administration allowed it to 'proceed in its slowest, most torturous manner'.[153] The Secretary-General pursued every avenue to quicken the process, and even proposed an alternative deployment schedule for UNAMIR II, which the Security Council ratified on 8 June,[154] all to no avail. Frustrated by the continued inaction of the Security Council, Boutros-Ghali excoriated the international community for its continued failure to respond to the 'agony of Rwanda'.[155]

Conclusion

Though SHAs were never established in Rwanda mainly because strong state interests constrained state behaviour in relation to the calamity, the violation of such an important community norm as the prohibition against genocide could not be ignored outright. Powerful states such as the US and Britain found it difficult to do nothing with complete indifference. Hence, they opted for a policy of inaction that sought deliberately to diminish the pressure community interests placed on them. These states were often required to justify their non-action in terms that were of some acceptability to the rest of the international community. Despite their reluctance, these states also acquiesced eventually in the adoption of resolution 918, which enlarged UNAMIR and proposed the establishment of SHAs, though they would not become involved directly in its implementation. This unwillingness to institute the SHAs was unfortunate because there is some indication that such zones might have proved effective at saving lives, considering that much of the killing was carried out by militias armed with nothing but machetes. A sufficient deterrent threat and the silencing of RTLM, the propaganda radio station that incited Hutus to murder Tutsis, might have been enough to ensure adequate enforcement of the safety zones. If what occurred at the Amahoro football stadium and a few other UN-protected sites, such as the King Faisal Hospital, was indicative of what could be achieved with a small number of poorly equipped peacekeeping troops under Chapter VI rules of engagement, then the SHAs might indeed have been a realistic and effective response to the genocide. At Amahoro stadium, UNAMIR, through their presence alone, sheltered some 15,000 people, a large proportion of whom were orphaned children.[156] Dallaire reckons that over 30,000 Rwandans from both sides and behind both belligerent lines were protected in these reasonably defensible sites.[157]

5

Too Little, Too Late: France's *Zone Humanitaire Sûre* in Rwanda

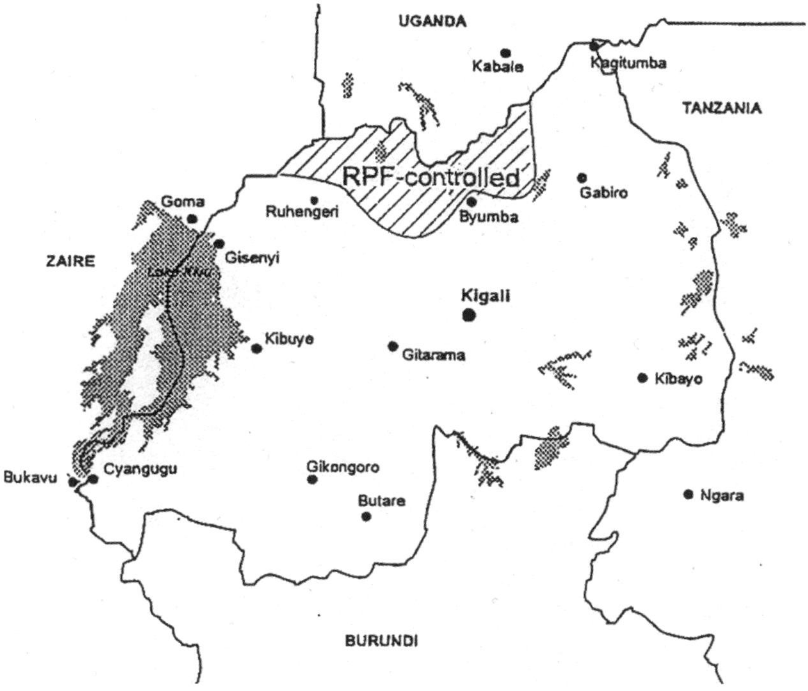

Figure 5.1 Approximate location of RPF/FAR front line, September 1993[1]

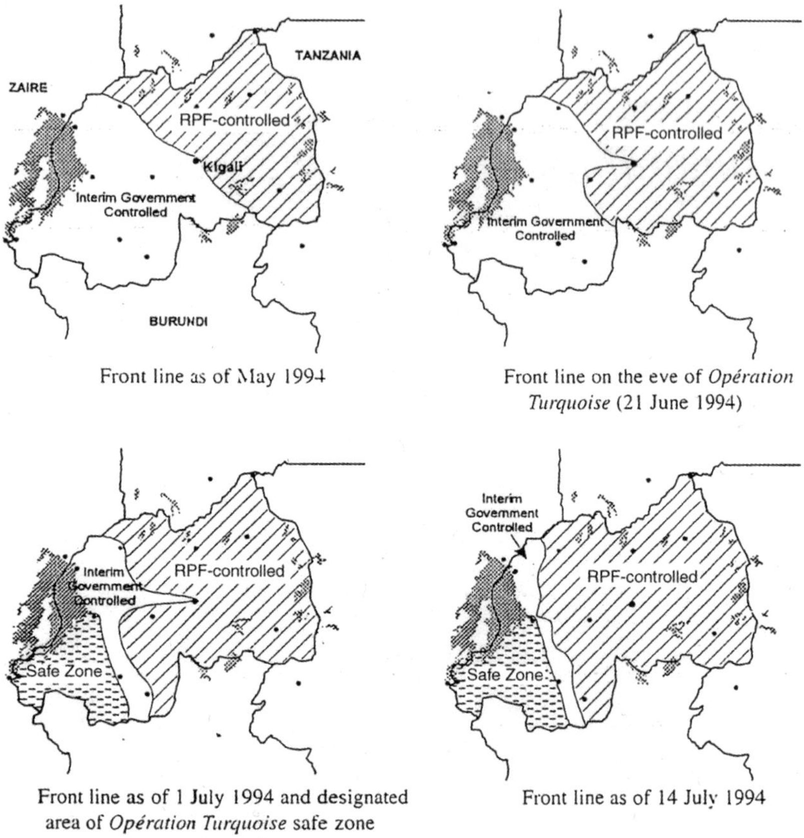

Figure 5.2 Approximate location of RPF/FAR front line, May–July 1994[2]

In the midst of the UN Secretary-General's desperate efforts to speed the deployment of UNAMIR II, the French government announced unexpectedly its decision to launch a military mission to Rwanda so as to maintain a humanitarian presence in the country pending the arrival of UNAMIR II. Assuring the international community that the operation's humanitarian purpose excluded 'any interference in the development of the balance of military forces between the

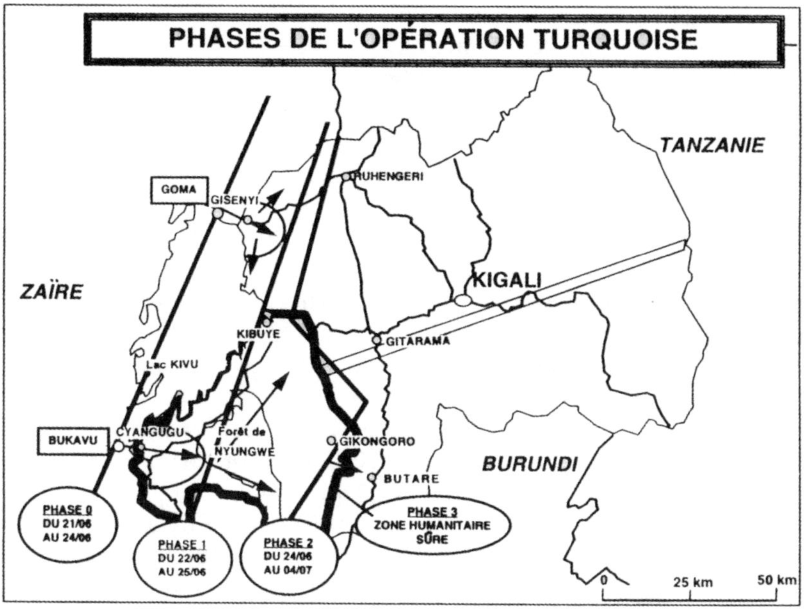

Figure 5.3 Opération Turquoise's Deployment into the ZHS[3]

parties involved in the conflict',[4] the French tabled a resolution in the Security Council on 22 June 1994, which was adopted by ten votes, with five abstentions. Acting under Chapter VII of the Charter of the United Nations, the Security Council stressed the 'strictly humanitarian character' of the mission and welcomed 'the establishment of a temporary operation under national command and control aimed at contributing, in an impartial way, to the security and protection of displaced persons, refugees and civilians at risk in Rwanda...'[5] *Opération Turquoise* was accorded a similar mandate to UNAMIR II in terms of the tasks it was allowed to perform, including the 'establishment and maintenance, where feasible, of secure humanitarian areas',[6] but was authorized to use all necessary means to achieve these humanitarian objectives. The resolution limited the French intervention to two months, unless the Secretary-General determined that an even earlier pull-out date was preferable in order to accommodate the deployment of UNAMIR II. The French military operation was launched the following day, on 23 June 1994. Over the next few days the first detachments out of a total estimated contingent of 3,000 troops, including 500 African soldiers from seven different countries, fanned out into Rwanda from bases in neighbouring Zaire.[7]

As *Opération Turquoise* moved further east into Rwanda, it became likely that the French might encounter the front between the RPF and the *Forces Armées Rwandaises* (FAR), as well as RPF forces desirous of continuing their march westwards. In order to avoid these eventualities that might endanger French troops and make the humanitarian objectives of the mission more difficult to achieve, France decided to use its prerogative under resolution 929 and create a *zone humanitaire sûre* (ZHS) informing the UN Secretary-General of this fact on 2 July 1994 via a letter from the Permanent Representative of France, which was then transmitted to the Council.[8] The ZHS spanned one fifth of south-western Rwanda, covering the *préfectures* of Butare, Cyangugu, Gikongoro and Kibuye, and was kept in place until the French left Rwanda as scheduled on 22 August 1994. Although UNAMIR II moved into the zone after the French departure, its status was not preserved as a safety zone *per se* and the new RPF government, victorious in the civil war by 18 July, reclaimed control of the territory.

This chapter aims to understand why the French opted for a safety zone policy in Rwanda, and to assess how effectively they implemented the ZHS. The theoretical framework developed in Chapter 1 will again be employed to help illustrate that the safety zone was established because France sought to reconcile its competing interests, both of a state-centric and community-oriented nature. An analysis that incorporates a broadened conception of interest will shed light on why *Opération Turquoise* was so controversial, earning praise and suffering condemnation at one and the same time, and on why its results in Rwanda were so mixed and ambiguous.

France and Rwanda, 1990–1994

France had been extensively involved in providing aid to Rwanda since its independence in 1962. It considered Rwanda to be a worthwhile and exemplary candidate for financial and cooperative assistance, especially throughout the 1980s when its economy performed well.[9] In tandem with the general trend within international society after the Cold War to render aid conditional on democratic reforms, President François Mitterrand, at the Franco-African summit of June 1990 in La Baule, linked development assistance and cooperation with progress toward multi-party democracy and pluralism.[10] Habyarimana seemed to respond positively to this pressure.[11] In July 1990 he made clear that his MRND party no longer had an undisputed right to the presidency of the country. In November 1990, the system of ethnic quotas for scholarships and professional placements was abolished; and in June 1991 he ended one-party rule and brought into existence a new constitution that permitted some freedom of expression and association. France further urged Habyarimana to work toward some accommodation of the Tutsi minority within Rwanda and toward a resolution of the Tutsi refugee problem outside Rwanda.

In terms of the civil war with the RPF, France consistently maintained the position that the only long-term solution to the impasse was a negotiated settlement,[12] and it fully supported the Arusha peace process, being the one permanent Security Council member to advocate actively a UN peacekeeping force for Rwanda that would monitor the implementation of the Accord. One outcome the French government disfavoured, however, was a military take-over of Rwanda by the RPF, which constituted a real possibility considering the weakness of the FAR and the prowess of the RPF on the battlefield. In order to prevent such an eventuality and to promote a negotiated settlement with the RPF, France provided fairly extensive military assistance to Rwanda from 4 October 1990 until 15 December 1993. As the RPF threat to the integrity of the Rwandan state mounted, France became increasingly focused on military aid, to the detriment of concern about good governance and Habyarimana's treatment of minorities. France grew indifferent to growing extremism and grave violations of human rights in Rwanda during this period, which generated a sense within the *Akazu* faction that they could do as they pleased and France would remain supportive.[13]

When the RPF first invaded Rwanda on 30 September 1990, the speed and success of their advance took the FAR by surprise: the RPF was easily able to capture Gabiro, 90 kilometres from Kigali. In order to guarantee French support for his country, Habyarimana staged an attack against Kigali on the night of 4 October 1993,[14] which provoked sufficient concern in Paris for it to launch *Opération Noroît* the next day. Its publicly stated purpose was to protect the French Embassy and French foreign nationals residing in the country. In actuality, however, *Noroît* became but one component of a cluster of military assistance mechanisms the French instituted in Rwanda to stabilize it and buttress it against RPF invasion. As it had with almost all its other Francophone-African allies, France signed a military assistance treaty with Rwanda in 1975 under the presidency of Valéry Giscard d'Estaing. As part of this agreement, France already had a *Mission Militaire de Coopération* in Rwanda when the RPF invaded, consisting of 20 technical advisers to the *Gendarmerie* and FAR. This advisory service was boosted by a *Détachement d'Assistance Militaire et d'Instruction* (DAMI) in March 1991 following a quick, successful RPF strike in January against Ruhengeri, which was near the heart of Habyarimana's political base in the North. Under the DAMI, 30 more military experts were sent to Rwanda: their goals were to recruit and train new soldiers, to improve the fighting strategy and capacity of the FAR and to instruct the elite Presidential Guard on how to create and discipline an effective fighting force.[15] The Presidential Guard was subsequently responsible for training the *Interahamwe* militias.[16]

France thus contributed to the increase in size and amelioration in strength of the Rwandan army: between October 1990 and the August 1993 Arusha Peace Agreement, FAR forces increased from 5,000 to 50,000.[17] The DAMI, in conjunction with the *Direction Générale de la Sécurité Extérieure* (DGSE),[18] also advised the Commander-in-Chief of the Rwandan army, as well as other top-ranking military officials, about tactical strategy and battle plans.[19] From 1990 to 1993 France also supplied Rwanda with 20 million francs worth of military equipment per year,[20] and underwrote Rwandan purchases of arms from Egypt and South Africa.[21] Considerable debate exists as to whether or not the French continued to supply the FAR with ammunition and weapons during the genocide, via *Opération Amaryllis* in April 1994[22] and through Goma, Zaire thereafter.[23]

At first *Opération Noroît* contributed in small ways to the activities of the DAMI: its soldiers conducted one-day training sessions for the FAR and completed 24hr sorties to the front to provide technical instruction to the Rwandan army, most probably in the use of mortars and Gazelle helicopters.[24] They were under strict orders not to interfere directly in the civil war or engage in battle alongside the FAR.[25] Despite this assistance, the FAR were unable to halt the RPF's conquest of territory in the north-east. In February 1993, the RPF doubled the area it controlled and advanced to within 30 kilometres of Kigali. It is likely that without the deterrent effect of *Opération Noroît*, the RPF might have taken or tried to take the capital. This near RPF victory led to the increase in size of *Noroît* to its zenith of nearly 700 men,[26] and to the broadening of its mandate in Rwanda. French troops now helped put into place a sophisticated surveillance system of Kigali, manning roadblocks to the North and checking identity cards so as to prevent RPF infiltration of the capital.[27] The French also started providing tactical support from just behind the lines, and Prunier believes they may have participated in combat in select instances.[28] It is also possible *Noroît* soldiers began interrogating RPF prisoners.[29] Despite this increase in military assistance to Rwanda, the poor performance of the FAR during February 1993 convinced the French that the Rwandan government could not win a war against the RPF, and thus that it was more essential than ever for Habyarimana to reach a settlement at Arusha. Once the Accord was in place, France left Rwanda on 15 December 1993 as it was required to do, leaving in place 24 technical experts under its initial *Mission Militaire de Coopération*.

When Habyarimana's plane was shot down on 6 April 1994, France sent *Opération Amaryllis* to Rwanda within 48 hours. Its purpose was to evacuate French foreign nationals as well as other European citizens, which it did rapidly, departing Rwanda on 14 April. The French embassy sheltered many political figures from Habyarimana's entourage, including the wife and children of the deceased President and other members of the *Akazu*. Many of this select group of Rwandans were flown to France along with French nationals, whereas Tutsis who had worked in the Embassy for many years were left behind.[30] Until *Opération Turquoise* two and a half months later, the French did nothing to halt the horrific activities of the Rwandan interim government.

France's initial indifference to genocide

Although France propped up an increasingly extremist regime, did not use its leverage over the Rwandan government to greater positive effect, and failed to take any action to halt the genocide when it started, much of the French literature on France's involvement in Rwanda during the 1990s tends nonetheless to be condemnatory in an excessively melodramatic way.[31] An analysis of more objective literature as well as of primary documents unveils a more complex picture of France's role in Rwanda,[32] one that was as much motivated by community interests as it was by state interests. France's long-standing friendship and military alliance with Rwanda, as well as its distrust of the RPF, certainly predisposed France toward remaining loyal to the interim government despite mounting evidence of extremism.

However, particularly after its military withdrawal in December 1993, France also considered itself to be following a principled course in Rwanda, working toward a democratization process that sought to accommodate the Tutsis but precluded the idea of an RPF military victory, and emphasizing the importance of a ceasefire between the belligerents as the most effective means to end the slaughter and achieve a lasting peace.

State interests and France's initial indifference

Overpopulated, landlocked and devoid of mineral riches, Rwanda in and of itself was of very little strategic importance to France. Nevertheless a number of important state interests ensured that, at the start of the genocide, France would neither move against the Rwandan government nor attempt to put an end to the massacres. These state interests constrained French behaviour toward the genocide, and also reconfirmed important elements of France's identity.

First, French loyalty to Habyarimana's Rwanda stemmed in part from a long habit of defending its '*pré carré*' in Africa as a means to enhance its international status and military credibility. Rwanda's inclusion within the club of Francophone African states amplified its significance and ensured that France's response to events taking place there would be dictated by the broader logic of French foreign policy toward its reserved zone in Africa. Since the end of WWII, preserving a domain of influence and culture within Sub-Saharan Africa has been a key component of France's quest for grandeur. French Africa offered France its most important opportunity to prove its worth on the international stage, because it allowed France to preserve the appearance of being a great power.[33] In order to protect its *domaine réservé* from encroachment by the two superpowers during the Cold War, France ensured that the French African states under its tutelage remained stable and reasonably prosperous, often supporting authoritarian and military regimes in the process just as long as regional order was secured.[34]

Certainly, France's credibility in this regard rested on quelling internal or external threats to these allies, a task France took very seriously as illustrated by the military cooperation agreements it signed with each one of them and the numerous military interventions it undertook on their behalf in such states as Gabon, Chad and Zaire.[35] The domino theory also applied to the French-Africa context: France worried that if it failed to come to the aid of one of its allies, the remainder would look elsewhere for support.[36] In return for such dedicated commitment, France could count on the support of its African partners at the UN and in other international institutions,[37] and thus gain prestige and recognition at the international level.

The end of the Cold War led to the loss of France's perceived 'uniqueness' in international affairs, which was a key element of its identity as an independent voice in a bipolar system.[38] The new international context was an uncertain one for France, and it felt restricted and constrained in its foreign policy. It was becoming an ordinary power.[39] As a result, the perpetuation of the French sphere of influence in Africa became all the more important as a means to preserve France's status and prestige in the international system. The Rwandan crisis was the perfect chance to demonstrate both the solidity of the French-African connection and France's continued dominance in its *pré carré*. Intervention in favour of Habyarimana's regime showed that France could still act as a great power and be counted upon to uphold its alliance commitments.

Second, France was disinclined to abandon a state that was part of '*la francophonie*' when it faced what the French considered to be an Anglo-Saxon menace. Another important component of the special relationship between France and Africa was language and culture: ensuring that parts of Africa continued to speak French helped promote France's world-wide cultural and linguistic prestige. In this respect, Rwanda was quite important because it lay along a 'political and linguistic faultline between francophone and anglophone east Africa'.[40] Protecting *la francophonie* from Anglo-Saxon influence turned into one of Mitterrand's favourite themes, and it became embedded in the political discourse in France.[41] Labeled the Fashoda-syndrome, this thinking mythologized the French capitulation of the rest of East Africa to the British at Fashoda, Sudan in 1898. Following World War II and even more strongly after the Cold War when America became the sole remaining superpower, France transferred its angst about its loss of prestige and power in world politics into a perceived rivalry with the US. The global dominance of anglophone language and culture, as well as American economic might, came to be seen as threats to France's state interests. In order to buttress French political influence, commercial dynamism and ability to promote its language, France would have to counteract this Anglo-Saxon menace.

This worldview was then transposed onto the Rwandan context, despite significant information belying it, often leading to biased interpretations of events. Since the RPF was a predominantly anglophone organization due to its founders having grown up in English-speaking Uganda, France considered its attacks on Rwanda as a foreign invasion assisted by a neighbouring country, rather than as a liberation struggle by a repressed minority within Rwanda. Similarly, though only ten per cent of Rwandans speak French,[42] the RPF offensive was deemed a menace to the 'Frenchness' of Rwanda. Although American interest in Africa was in sharp decline after the Cold War,[43] as demonstrated by slashed aid spending and reduced trade, and although the USA never directly supported the RPF,[44] France equated the RPF onslaught with a Washington scheme to augment its prestige and authority in Africa.[45] Prunier claims that the Fashoda syndrome is 'the main reason – and practically the only one – why Paris intervened so quickly and so deeply in the growing Rwandese crisis', and remained so loyal to the dictatorship from 1990 throughout the start of the genocide.[46]

Third, the way France conducted its day-to-day diplomatic relationship with Rwanda also contributed to both its allegiance to the Habyarimana regime and its indifference to genocide. As '*la Françafrique*' established a non-accountable family affair as the main format of association between France and its African allies, it meant that policy formation and implementation was controlled by very few persons, thus diminishing accountability and transparency *vis-à-vis* the French public and the *Assemblée Nationale*. France's constitution under the Fifth Republic concentrated unprecedented powers in the hands of the President. One of the President's *domaines réservés* was decolonization policy, which developed into a personalized bilateralism between the French President and his cronies, whether military or civilian, on the one hand, and each French African state on the other.[47] Unlike other aspects of foreign policy, *la Françafrique* was kept out of the purview of the Ministry of Foreign Affairs[48] and escaped all the normal channels of accountability and checks and balances. The President could, for example, dispatch regular troops overseas, as with *Opération Noroît,* without requiring Parliamentary approval. The only ministry involved in African affairs was the Ministry of Cooperation, which tended to toe the presidential line: it worked hand in hand with the *Élysée's* Africa office to carry out Presidential policy in Africa.[49] The *Assemblée Nationale* also gave the President free rein,[50] and there was surprising little public discussion of Africa and French involvement there.

The French President was thus largely responsible for setting the tone of policy toward French Africa. Mitterrand, deeply attached to the *pré carré* and to the preservation of *la francophonie*, was rarely contested in his policy toward Rwanda.[51] His influence was rendered even more pervasive by the very personalized, literally familial, nature of the relationships between the French President and his African counterparts. Mitterrand was often referred to as *l'Africain* in the African press, and it was not unusual for African heads of state to call him *'Papa'*. In the case of Rwanda, there were substantial personal links between the ruling *Akazu* of Habyarimana's regime and French President Mitterrand. Mitterrand's son and head of the Africa cell in the *Élysée* from 1986 until September 1992, Jean-Christophe, developed a strong personal friendship with Habyarimana and his son. He not only encouraged support for the Rwandan regime, but also bestowed lavish gifts on Habyarimana, such as the Falcon 50 jet and crew which served as Rwanda's presidential aircraft. When the RPF attacked in 1990, Habyarimana phoned Jean-Christophe directly and requested French assistance, which proved forthcoming.[52] In August 1993, Habyarimana received red carpet treatment when he was entertained by President Mitterrand in Paris. During *Opération Amaryllis* France evacuated several high ranking members of the *Akazu* to Paris, including Habyarimana's wife and family.

This familial attitude and inclination toward camaraderie with African leaders meant that criticism of domestic affairs was 'not the norm'.[53] The intimacy of such relationships made it difficult to know when to draw the line, and almost impossible to maintain an objective stance toward decision-making.[54] It is alleged that Mitterrand brought up the subject of human rights abuses in Rwanda, but accepted, at face value, Habyarimana's explanation that he had no power over the extremists that were responsible for them. Loyal to an old friend, Mitterrand preferred to ignore evidence of growing extremism in Rwanda.[55] France would often qualify as rumour any reports or testimonies alluding to gross violations of human rights taking place in Rwanda. Colette Braeckman also suggests that the 'clientelism' that pervaded most of the bilateral relationships with French-Africa meant that France was relatively ignorant of Rwanda's past and let itself be taken in by appearances: Rwanda was a well-administered country by African standards and gave an overall impression of progressive development and peace.[56] Habyarimana, a highly cultured man, was perceived as being wise and trustworthy. He gave every formal indication of supporting movement toward democratization, praising Mitterrand's La Baule speech, altering Rwanda's constitution to permit the formation of political parties and participating in the Arusha peace process.

The French military was often given a relatively free hand in French-African affairs.[57] Most of the French military stationed in Rwanda, either through belief or obedience to the command structure, adhered to the two central intertwined thrusts of French policy in Rwanda: 1) that there was an Anglo-Saxon menace to be countered; and 2) that the Hutus were the rightful rulers of Rwanda because they constituted a majority of the population and 'spoke' French. The RPF thus constituted a prime target due to its preference for the English language and to the fact that it was predominantly a Tutsi organization. Several top-ranking French army officers also reinforced the ethnic aspect of the conflict. It was common for General Jean-Pierre Huchon, head of the *Mission Militaire de Coopération* in Rwanda from 1993 to 1995, to refer to Tutsis as the enemy.[58] General Christian Quesnot, the President's chief of staff, also voiced this view within the Presidential circle, reinforcing the French executive's perceptions of what its state interests were in Rwanda.[59] Bureaucrats and field operatives of the DGSE regularly labeled the RPF as the '*Khmers Noirs*'. Such negative jargon trickled down to low-ranking soldiers who came to support the efforts of the FAR in their opposition to the RPF. Naturally, once French forces became so enmeshed in Rwandan military affairs, they acquired an interest in sustaining their ally, mainly out of an instinctive sense of solidarity for their fellow Rwandan combatants, a fact that would contribute to the later ambiguities of *Opération Turquoise*.[60]

Community interests and France's initial indifference

Not only did state interests contribute to France's support of the Habyarimana regime and later reluctance to address the genocide in a meaningful way, but community interests played their part as well. Believing in norms about democracy and about the peaceful resolution of disputes, but also wanting to be recognized as part of an international community that values highly these same principles, France worked to achieve democratization in Rwanda, as well as a negotiated settlement of the civil war.[61] France's pursuit of these community-oriented interests prevented an early response to the genocide for two important interrelated reasons.

First, France considered that a legitimate democracy in Rwanda would be one in which the Hutu majority retained a substantial hold over the institutions of government. This preference for Hutu majority rule did serve as a means to legitimate France's state interests. Despite the fact that Habyarimana came to power in a *coup d'état* in 1973, for instance, President Mitterrand and other French officials often made clear that they believed his rule to be lawful because he represented the Hutus, which constituted 80 per cent of the Rwandan population.[62]

However, by the 1990s and as demonstrated by its support for the Arusha Peace Agreement, France had accepted that the Tutsi minority deserved better representation within the Rwandan political system. The problem was that this accommodation was not a straightforward process. Ethnic divisiveness and extremism in Rwanda, as well as the Tutsi minority status, made it difficult for the RPF to achieve its aims of changing Rwandan society and bringing about democratization, except through a military take-over of Rwanda, at least in the short term.[63] France deemed this possibility to be unacceptable as it would violate the democratic rights of a large portion of the population. As a result, France was disinclined to abandon the Rwandan government when the RPF offensive resumed following the start of the genocide.

Second, France regarded a ceasefire followed by a negotiated settlement as the most appropriate and acceptable means to enable a greater Tutsi role in government but avoid an RPF military conquest of Rwanda. France was thus fully prepared to accept the RPF as a legitimate party at the negotiating table. The problem for France here, however, was that the Habyarimana regime needed to be propped up militarily if it were to be a viable party at Arusha, partially explaining France's military involvement in Rwanda from 1990 to 1993 and its indifference to growing extremism. Likewise, once the fighting resumed in April 1994, France would not risk weakening the Rwandan interim government further in the wake of the assassination of Habyarimana, for fear that the RPF would obtain their military victory over Rwanda. The result of this French principled commitment to a return to negotiations between the belligerents was that France took no action in relation to the genocide.

France's change of course and the launch of *Opération Turquoise*

Although France's decision to send a humanitarian military operation to Rwanda is a striking departure from its earlier indifference toward the genocide, the configuration of its state and community interests remained surprisingly constant in this new phase of France's involvement in Rwanda. Certainly new factors emerged, such as the public outcry against France's inaction in the face of genocide, but the shift in French policy was more a change of tactics, given the severity of the calamity engulfing Rwanda, than a fundamental rethinking of France's relationship with the country and its government. Despite France's sudden willingness to 'do something' in Rwanda to help protect civilians and displace persons at risk, it remained committed to preventing an RPF military victory through the establishment of a ceasefire and the negotiation of a permanent settlement. From its inception, France's *Opération Turquoise* was thus imbued with a certain ambivalence that did not inspire automatic international acceptance. France's change of course in Rwanda is but one part of the story of the launch of *Opération Turquoise*; the other is how this humanitarian mission came to be approved of by the Security Council despite France's previous questionable record of involvement in Rwanda.

State interests and the launch of *Opération Turquoise*

Opération Turquoise did not represent a drastic reconfiguration of France's state interests. To the contrary, to a large extent France acted now for the same reasons it had not acted earlier in the crisis. It still sought to preserve its *pré carré* and *la francophonie*, but the tactic to achieve these ends changed. As the crisis in Rwanda grew steadily worse and the death toll reached staggering proportions, France's actual power in French-Africa and its capacity to restore order in Rwanda were put into question. Its military credibility within the *pré carré* was at risk: France needed to prove its ability to respond to grave calamities in its sphere of influence in Africa. A decisive humanitarian intervention in Rwanda that stabilized the country would convince both African states and the international community as a whole that France could effectively project its power onto the African continent, thus justifying its pre-eminent place on the Security Council. By proceeding with an operation under national command and control, the French were able to guarantee that they would take much of the credit for the intervention.[64] Bringing about conditions that might generate a renewed ceasefire between the RPF and the FAR was also viewed as a way to prevent Rwanda from losing its 'French' identity.

One clear factor contributing to the change of French strategy toward the genocide was the appearance of a chink in the armour of *la Françafrique*. The magnitude of the tragedy underway in Rwanda and the seeming bewilderment as to what to do about it on the part of the executive branch offered the French Foreign Ministry, under the skilful direction of Alain Juppé, the opportunity to recapture the foreign policy initiative from President Mitterrand.[65] It viewed *Opération Turquoise* as a means to distance itself from the policies of the *Élysée* and to take control of an area of foreign policy previously monopolized by the executive branch. Whereas previous French policy in Rwanda was determined almost exclusively by Mitterrand and his executive entourage, *Opération Turquoise* was devised in a crisis council that included not only the President and his aides, but also Juppé, Prime Minister Édouard Balladur, Defence Minister François Léotard, and the Commander-in-Chief of the French military, Amiral Jacques Lanxade.[66] Two important developments emerged from this new collaboration. First, the inclusion of new voices in the decision-making process incorporated novel ideas into the spectrum of possible responses to Rwanda. As a result, new options, of which the ZHS was one, appeared on the horizon. Second, Rwanda policy was now subjected to the same checks and balances, as well as accountability requirements, as other aspects of French foreign policy. Mitterrand's views about French Africa were balanced against those of the more forward-looking Léotard and Juppé. Bringing members of the legislature into the decision-making process also meant that they were ultimately accountable to the public for the successes and failures of the project. Caution was therefore in order, and the military would no longer have free rein.

Community interests and the launch of *Opération Turquoise*

France's community interest in a negotiated settlement as the best long-term solution to Rwanda's problems remained constant during this new phase of France's involvement in the country and helped enable *Opération Turquoise*. Since it became unacceptable for France to maintain ties with the interim government once it was clear the killing in Rwanda was genocide,[67] France's new tactic to bring about conditions for renewed negotiations between the RPF and a more moderate Hutu leadership[68] was to stabilize the country through an intervention. Preventing an RPF victory remained high on the French agenda, but not in the very critical sense some have suggested. It was not that *Opération Turquoise* was intended to oppose or defeat the RPF militarily.[69] Rather, France wished to create conditions for a negotiated settlement without intervening directly in the conflict, thus making more difficult an RPF military take-over of Rwanda, which France viewed as anathema to true democratization in the country.[70]

It is not surprising that Mitterrand's initial preference for *Opération Turquoise* was as an interposition force between the RPF and FAR, which would land in Kigali and then spread out through the entire country.[71] However, fearing that such an action would be both too dangerous and misinterpreted by the international community, Foreign Affairs Minister Juppé and Prime Minister Balladur teamed up against Mitterrand to insist on a purely humanitarian mission that would penetrate Rwanda in the east from Zaire.[72] In essence, the underlying political aim of *Opération Turquoise* was to stabilize Rwanda in the hope of re-establishing a ceasefire and returning to the Arusha framework,[73] in the process bringing the genocide to an end and channeling humanitarian aid into the country. The fact that the aim of the mission was more to generate enough stability in Rwanda for a ceasefire than to stop the genocide against the Tutsis *per se* can be seen by the initial choice of an entry point from Zaire. At first, Gisenyi was selected, where most Tutsis had been driven out or massacred decades before.[74] Prunier, who was asked to join the crisis unit planning the intervention, managed to convince the government that an arrival at Gisenyi would both risk the wrath of the RPF and draw the criticism of the international community.[75]

The moral, and possibly legal, duty to prevent genocide, played an important role in pressuring France into a change of course in Rwanda: France could no longer ignore the violation of such an important community norm as the prohibition against genocide. Juppé, in particular, took the lead in reasserting France's adherence to the idea of *un devoir d'ingérence*.[76] Speaking to the Security Council in New York, PM Balladur confirmed that his country felt an obligation to do something: 'France...believed it had a moral duty to act without delay to stop the genocide and provide immediate assistance to the threatened populations'.[77] In a letter to Mitterrand on 21 June 1994, Balladur wrote that one of the reasons he and the President were able to reach agreement on the need for an intervention was because there existed '*un devoir moral*'.[78] Not unlike the coalition's sense of responsibility toward the Kurds following the 1991 Gulf War, France also felt an added pressure to act in Rwanda to make amends for any part it had played in either failing to avert genocide or in involuntarily facilitating it.

Moreover, once the Rwandan genocide captured the public imagination, both the media and powerful NGOs began to exert influence upon the French government, pressuring it to halt the slaughter. Like its Anglophone counterparts in America and the UK, the French media initially portrayed the Rwandan genocide with confusion, misinformation and indifference. For example, a *Le Monde's* 9 April story said that the Presidential Guard was engaging in aimless violence in its desire to avenge the President's death.[79] Only at the end of May did *Le Monde* begin referring, on a regular basis, to what was occurring in Rwanda as genocide. In June, as the RPF controlled part of the country was opened to journalists, eye witness accounts of the slaughter, as well as detailed descriptions of the remains found at massacre sites, became the focus of many articles. *Le Monde*, for instance, ran stories, on five separate occasions, dealing with tales of the killings and bodies left behind by the killers. Numerous editorials and signed articles emerged, analyzing France's role in the affair and accusing the French government of complicity in the genocide.[80] TV reporting also followed this same general pattern of reporting.[81]

Made aware that its government had played an important role in facilitating the genocide in Rwanda, the French public demanded that its leaders now take the initiative to make amends for the past. Public pressure became so great Foreign Ministry officials could not tolerate it.[82] President Mitterrand was acutely affected by this intense public criticism. Wishing to avoid the possibility of having his second and final seven-year term end in scandal, he sought to distract attention from the *Élysée's* previous policy and thus ensure his continued popularity amongst a disenchanted public.[83] A successful French intervention in Rwanda would enable him to achieve this goal. Influential NGOs, such as *Médecins Sans Frontières* and the *Fédération Internationale des Ligues des Droits de L'Homme*, exploited this policy window generated by increased public awareness: they blitzed the TV, radio and print media for two solid months in May and June 1994. They sent an open letter to President Mitterrand stating that what was going on in Rwanda was a clear case of genocide and that an international armed intervention was necessary to stop it.[84]

Finally, the identity variable contributed to the constraining effect of the genocide convention upon France. The President, the Prime Minister and the Minister of Foreign Affairs each saw humanitarian intercession in Rwanda as an opportunity to restore France's international image which had become badly tarnished as knowledge regarding France's involvement in arming and training the *génocidaires* grew. France needed a new policy that would enable it to atone for past errors and show its respect for fundamental norms of international law.[85] For the *Quai D'Orsay* as well as *l'Élysée*, restoring France's identity as a trusted, upstanding member of international society, capable of defending community interests, was of paramount importance.

Community interests also enabled *Opération Turquoise* from another perspective. Without the legitimacy the UN Secretary-General and the Security Council conferred on the intervention, it would have been far more difficult for France to launch a humanitarian mission. In the first instance, the genocide had such a strong impact on states that they were willing to cast aside concern about France's past onerous role in Rwanda and support its proposed humanitarian intervention. The impetus to prevent genocide acted as a focal point for the Security

Council, uniting a somewhat divided and dubious Council behind the French plan. Even so, the French offer was greeted with some degree of suspicion and skepticism: resolution 929 was passed by only ten votes, with five abstentions.[86] Despite these misgivings, *Opération Turquoise* was approved, mainly due to the unconditional support professed by the Secretary-General and a sense of shame on the part of the international community that it had failed to uphold its community interests by taking any sort of decisive action in Rwanda.[87]

The fact that France chose to launch its initiative within the confines of the Security Council reveals the extent to which it recognized the legitimating effect of community norms and institutions. Only through the mechanism of Security Council authorization could France hope to render permissible and appropriate actions that otherwise might have been opposed or frowned upon, considering France's previous support of the genocidal dictatorship. For the French plan to receive widespread acceptance, France would have to lay bare its mission to the scrutiny and assessment of the international community, which Prime Minister Balladur understood clearly. He imposed on the French government's executive branch five conditions needing to be satisfied before the mission could proceed: 1) necessity of a UN mandate; 2) a clear time-limit; 3) no in-depth penetration of Rwanda;[88] 4) purely humanitarian objectives; and 5) allied troops should be involved.[89] The Security Council obliged where it was able: acting under Chapter VII of the Charter of the United Nations, it stressed the 'strictly humanitarian character' of the mission and welcomed 'the establishment of a temporary operation under national command and control aimed at contributing, in an impartial way, to the security and protection of displaced persons, refugees and civilians at risk in Rwanda...'[90] *Opération Turquoise* was accorded a similar mandate as UNAMIR II in terms of the tasks it was allowed to perform, but was authorized to use all necessary means to achieve these humanitarian objectives. 929 also limited the French intervention to two months.

The transparency and legitimacy the Security Council conferred on *Opération Turquoise* reassured states: not one member of the Security Council publicly questioned France's humanitarian motives or suggested that France's history in Rwanda might be problematic. Instead, concern rested with two pragmatic issues. It was suggested that *Opération Turquoise* would detract from the efforts to deploy UNAMIR II since resources, both military and financial, would be expended on the French mission at a crucial time. There were also reservations about having two simultaneous missions in the field, with differing capabilities to fulfill their mandates – one being under a Chapter VII resolution and the other being under a Chapter VI resolution. It was felt that this duality might jeopardize the success of both missions.[91] It is also likely that the RPF would have opposed more vehemently French humanitarian intercession in Rwanda if it had not been for the legitimacy bestowed upon the operation by Security Council authorization limiting the nature and scope of France's intervention.

The implementation and effectiveness of the ZHS

A week after launching *Opération Turquoise*, France informed Boutros-Ghali that it would be establishing a ZHS in Rwanda. As with the Iraqi safe haven and the Bosnian safe areas, a safety zone approach was chosen because it enabled France to reconcile a number of competing, and contradictory, interests. The ZHS allowed France to do something in the face of genocide, while simultaneously avoid entering the Rwandan civil war and risking casualties amongst its troops. The ZHS also granted France the opportunity to rekindle ceasefire negotiations, without either propping up militarily the discredited Rwandan government or awakening too seriously the hostility of the RPF. The ZHS in fact provided France with the most legitimate excuse possible for keeping the RPF out of a significant section of Rwandan territory – namely to preserve the humanitarian character of the zone – in the hope of delaying an RPF military conquest of Rwanda until all opportunities for a peaceful settlement had been exhausted.

As a result of these inherent ambiguities, the ZHS produced mixed results and much controversy. The ZHS enabled France to reconcile such diverse interests that at least one commentator actually labeled the intervention 'schizophrenic',[92] being at one and the same time humanitarian and intensely political.[93] On the one hand, France's humanitarian mission to Rwanda was relatively effective. It is estimated that the French saved approximately 13,000 to 15,000 Tutsi lives, mainly by protecting Tutsis in camps surrounded by hostile *gendarmerie*.[94] The general improvement of security in the south-west enabled a dramatic increase in humanitarian assistance activities of NGOs and other international organizations, and thus greatly ameliorated living conditions for the Hutu and Tutsi civilians who found themselves in the ZHS.[95] By 12 August 1994, there were over 1.5 million internally displaced persons in camps in the ZHS.[96] France claims to have provided the equivalent of 11,000 days of hospitalization, 14,000 medical consultations, and 78,000 acts of care, as well as distributing 700 tons of aid to Rwandans.[97] The fact that many NGOs that had initially refused to work with France later changed their stance, even requesting that France remain beyond the mandated 22 August date set for its withdrawal, reflected their acknowledgement of the generally successful performance of the French troops. The French were also able to ensure the inviolability of the zone. As the French had assembled an impressive deterrent fighting force, with 12 Jaguar planes stationed in Goma, Zaire and an abundance of armoured cars, they deterred the RPF from moving into the zone. Likewise, their earlier friendship with the Hutus meant that French soldiers were able to conduct the humanitarian operation without too much difficulty.

On the other hand, the implementation of the ZHS was not without serious problems. Not in sufficient numbers to control the entire territory of the ZHS and having brought the wrong kind of military equipment with them, the French were unable to save most of the Tutsis still hiding in the bush and atop hillsides: they lacked the small transport vehicles needed to collect dispersed Tutsis and bring them to central locations.[98] Upon finding small pockets of hunted Tutsis, the French would promise to return the next day to deliver them to safety, but either failed to do so altogether or did so only after it was too late,[99] once again putting into question their humanitarian motives. The ZHS also provided refuge to members of the Rwandan interim government, to the FAR and to the *Interahamwe* militias, which France also failed to disband and disarm. Following the RPF's victory in the civil war on 18 July 1994, many of these groups came to fear RPF reprisals and fled unhindered to Zaire in a massive refugee exodus that France proved unable to prevent.

Despite requests from humanitarian organizations and UN agencies to remain in Rwanda longer than the mandated two-month period so as to perpetuate a stabilizing presence in the south-west, the French honoured the time-frame set out in Security Council resolution 929 and departed on 22 August 1994, while a less well equipped and organized UNAMIR II contingent entered the area. Their departure created considerable fears amongst the Hutu population regarding its safety, leading to another major exodus of refugees to Zaire. As France proved unwilling to resolve the problem of the continued military activity of the Hutu extremists, this militant extremism was transferred to Zaire when the French pulled out. Upon the termination of *Opération Turquoise*, the south-west of Rwanda remained the most insecure part of the country for several years, mainly because of the concentration of former Hutu soldiers and militia along the Rwandan-Zairian border. This enduring insecurity resulted in a number of tragic incidents, including an RPF massacre of thousands of internally displaced persons at the Kibeho camp in 1995.[100] The fact that the FAR and *Interahamwe* launched attacks into Rwanda from Zaire led to two wars between Rwanda and Zaire (now the Democratic Republic of Congo [DRC]), in 1996 and 1998. According to the International Rescue Committee, the 1998 conflict has caused, either directly or indirectly, the death of 3.8 million persons in the DRC to date.[101]

The ambiguities inherent in the implementation of the ZHS provoked accusations that France sought to buttress the nearly defunct, extremist interim government and prevent outright an RPF victory in the civil war; or that France purposefully let the FAR and *Interahamwe* escape into Zaire so they could return to fight another war against the RPF.[102] However, such accusations are too simplistic. Although the French could and should have done more than they did – three examples being silencing the broadcasts of RTLM that encouraged the exodus to Zaire,[103] distancing themselves from the interim government sooner than they did,[104] and seeking to disarm the *Interahamwe* in a systematic way – there were no great conspiracies at work in Rwanda. The French found themselves in challenging circumstances; and their behaviour during *Opération Turquoise* was constrained by both their state and community interests.

State interests and the implementation of the ZHS

Although France sought to prevent an RPF military victory in the hope of rekindling ceasefire negotiations, it certainly did not go to Rwanda either to fight a direct war against the RPF or to act as a force of interposition between the two belligerents. Such courses of action were far too risky, militarily and politically, to even contemplate. The French government hoped, rather, that its presence in the country might generate conditions that would be ripe for renewed talks between the parties. France was so concerned about the possibility of confronting the RPF in battle that it created the ZHS as a means to prevent contact between its forces and those of the RPF: it informed the Security Council on 2 July that if the creation of the zone was opposed by the international community, it would withdraw its soldiers from Rwanda immediately.[105]

Though France was accorded a Chapter VII authorization by the Security Council, mainly so as to ensure it could defend its troops with all necessary means, in most respects *Opération Turquoise* actually operated on the ground under Chapter VI rules of engagement.[106] In practice, the proper functioning of the ZHS relied on the consent of the parties on the ground. The French only had 1,200 active soldiers in the ZHS.[107] In order mainly to deter a confrontation with the RPF, these French soldiers on the ground were buttressed by an important deterrent force: 12 Jaguar and Mirage F-1 aircraft, as well as artillery fire, light armour vehicles and helicopters, were at the ready at bases in Zaire.[108] Although the RPF chose not to challenge the zone in the short-run, France remained concerned that its deterrent force might be undermined in the long-run. It therefore attempted to establish a *modus vivendi* with the RPF: General Lafourcade, the French commander in the ZHS, worked diligently with General Paul Kagame, commander of the RPF, to manage their conflictual relationship by communicating via the offices of UNAMIR.[109] Nonetheless France believed that if it overstayed its welcome in Rwanda, the RPF might consider retaking the ZHS forcefully.[110] This menacing possibility was the key reason why France left Rwanda so promptly on 22 August 1994 when its Security Council mandate ended. Although the UN might have been willing to extend its authorization for a longer period, the RPF opposed a continued French presence and pushed hard to have UNAMIR II replace *Opération Turquoise*, something the French took very seriously.[111] At any rate, it was clear to the French by this point that the RPF would not return to the negotiating table and would not be dissuaded from their military take-over of Rwanda.

In terms of the Hutus within the zone, France was initially guaranteed the support of the government side, considering its past links with the Habyarimana regime and the warm welcome shown French troops.[112] France was thus able to control the military activity of the FAR within the zone by issuing clear directives in this regard to the interim government leaders.[113] Although the French did not wish initially to weaken the Hutu side further through a disarmament campaign of the FAR,[114] it is also true that such an undertaking would have required *Opération Turquoise* to move to Chapter VII enforcement action. General Lafourcade voiced his concern at the fact that Hutu militias and civilians had turned decidedly cool toward the French military because of its persistent neutrality in the face of pleas for support from the FAR and the interim government.[115] In these circumstances in which its troops could be at risk, it was unlikely that France would undertake enforcement operations, such as the active disarming of FAR and *Interahamwe* soldiers through systematic cordon and search operations.[116] Challenging the FAR and *Interahamwe* would have required strong political and military resolve, which state interests did not permit. The French were not alone in this lost opportunity to seize perpetrators of the genocide. No attempt was made by UNHCR or any other state involved in the refugee camps in Zaire either to screen refugees entering neighbouring countries or to arrest those suspected of genocidal acts. General state unwillingness to tackle through force the FAR and *Interahamwe* problem persists to this day, with the UN's peacekeeping mission in the DRC, Mission de l'Organisation des Nations Unies au Congo (MONUC), never having being mandated to forcibly disarm or repatriate to Rwanda these combatants, who continue in 2005 to prey upon the civilian populations of eastern DRC.

Likewise, the French did not purposefully let the FAR escape into Zaire weapons intact, or hide members of the interim government in the ZHS.[117] French troops found it difficult to deal with FAR and militias numbering in the tens of thousands and blending into the civilian population.[118] The extremist interim government was in the ZHS for all of one day, before it discovered it was not welcome in the zone and fled to Zaire.[119] Many of the allegations directed toward France about not capturing the *génocidaires* and letting the FAR slip through its fingers into Zaire only arose because of France's previous ties to the Habyarimana regime. France's failings are suspicious only in this context of past wrongs. The fact that some of *Opération Turquoise's* soldiers considered themselves to be operating in a pre-genocide context was also highly problematic: Dallaire and Prunier tell the story of a certain Colonel Thibault, a former DGSE-man who had advised the FAR prior to the genocide, who made incendiary comments about the RPF while serving in the ZHS. He was immediately reprimanded by General Lafourcade and recalled to France.[120]

Finally, both the *Élysée* and the *Quai D'Orsay* designed *Opération Turquoise* as a public relations vehicle to bolster their popularity, to allay criticism for the government's previous association with the Habyarimana regime and to show their commitment to preventing the international crime of genocide. A swift, successful mission that was shown to save Tutsi lives would be sufficient for this purpose.[121] In fact, the French ensured TV cameras access to the ZHS, and French soldiers were instructed to evince great understanding toward journalists.[122] Getting bogged down in Rwanda and risking casualties would defeat this objective and negatively affect public popularity ratings. In sum, important state interests encouraged a quick and tidy humanitarian foray into Rwanda, and discouraged a risky enforcement operation that might have resolved the problem of the militarized extremists once and for all.

Community interests and the implementation of the ZHS

France's behaviour in relation to the ZHS was also guided by its community interests. If France was to halt the killing, rectify its international image as a trusted member of the international community and allay suspicions regarding its true intentions in Rwanda, then it needed both to succeed in Rwanda and to remain strictly within the confines of the mandate it was given by the UN. France worried a lot about how other states would perceive its intentions in Rwanda, and thought consciously about what was the best way to proceed without feeding suspicion. The most obvious way for France to convince the international community of the propriety of its behaviour in Rwanda during *Opération Turquoise* was to not overstep the parameters of the authorization it had received from the Security Council. French officials repeated often that France was bound by resolution 929, which restricted it from interfering in the military conflict between the RPF and the FAR; and the highest echelons of the French government denied any intention of preventing the RPF from gaining a military victory in Rwanda.[123] France also made clear that it was entitled to establish a ZHS in the country, under resolution 929, in order to ensure the 'security and protection of displaced persons, refugees and civilians at risk in Rwanda'.[124] France's departure from Rwanda exactly two months after its arrival was also in strict conformity with resolution 929.

Documents contained within *Tome II, Annexes* of the French *Assemblée Nationale* inquiry also indicate the extent to which the French were concerned about keeping their activities under *Opération Turquoise* legitimate. The government stressed the absolute importance of neutrality and impartiality in Rwanda, again confirming its Chapter VI approach to the peacekeeping mission, as well as its determination to pursue humanitarian objectives designed to assist all civilians, whether they be Hutu or Tutsi.[125] Also apparent in French policy is an unwillingness to step beyond the confines of a classical peacekeeping operation, since such a movement might easily be misinterpreted, blemishing the humanitarian nature of its intervention or leading to allegations of interference in the internal affairs of Rwanda. As a result a number of possible activities were officially discounted. Although French troops were given clear authority to disarm Rwandan soldiers or militiamen threatening human lives,[126] systematic disarmament was discounted as it would be an abrogation of impartiality, seeing as the French could not take weapons away from the RPF.

In relation to the extremist interim government and the possible arrest of its members by French soldiers, General Lafourcade requested instructions from Paris as to what to do, believing that house arrest might be the best option until such time as the international community made a decision regarding their fate.[127] Although France would not tolerate any political activity on the part of this interim government,[128] it decided that it did not have a mandate to take further action against its members and requested guidance from the United Nations.[129] Likewise, France did not consider that arresting or detaining genocide perpetrators within the FAR or the militias as legitimate actions under Security Council resolution 929. The enabling effect of the Genocide Convention meant that France could probably have proceeded unilaterally in this matter without violating international law.[130] However, widespread arrests of this kind by a third party state had never been undertaken before. Although a Commission of Experts was set up by the Security Council on 1 July 1994 to examine information on grave violations of international humanitarian law and possible acts of genocide in Rwanda,[131] the international criminal tribunal for Rwanda (ICTR) was only established by the Security Council on 8 November 1994.[132] Arrests and trials were not at the forefront of the international community's agenda in the summer of 1994: by August international attention had shifted to the plight of the refugees in Goma, Zaire. In sum, community interests also encouraged restraint on the part of the French in their humanitarian involvement in Rwanda.

Conclusion

Given its military capacity in the region and its potential to influence the interim government, France was perhaps the best placed of any of the major powers to intervene in Rwanda once the genocide started. However, due to its special friendship with the Habyarimana regime and its distrust of the RPF, France remained initially indifferent to the genocide, just as it had turned a blind eye to the government's growing extremism in the early 1990s. Only when conditions in the country became desperate and the death toll reached over 500,000 was France eventually compelled, through public pressure and a sense of a *devoir d'ingérence*, to launch *Opération Turquoise*. A ZHS fitted very well into this humanitarian operation because it allowed the French to save lives and do something in the face of genocide, while also creating the opportunity for a ceasefire and a return to negotiations between the belligerents without having to take sides in the civil war or oppose the RPF militarily. The underlying ambivalence with which the safety zone was thus implemented meant that the French left themselves open to criticism and controversy, even though they endeavoured to stay within the parameters of Security Council Resolution 929 and succeeded in implementing the zone with relative effectiveness while they remained in Rwanda. Once the genocide was over and once it became clear the RPF was in control of the country, France's rationales for being in Rwanda disappeared. The French thus departed Rwanda abruptly, but as scheduled on 22 August 1994, leaving behind the problems that continue to plague parts of the African Great Lakes region to this day, namely militarism and extremism.

6
Conclusion

This book sought to provide a comprehensive account of the safety zones of the 1990s, explaining the reasons for their creation, detailing their implementation, and shedding some light on why each succeeded or failed. Moreover, in an attempt to help bridge the gap between schools of thought in International Relations that are sometimes needlessly polarized, this book developed the idea of a broadened conception of interest, suggesting that states neither operate in a purely strategic, material environment, nor in a uniquely normative and legal one. Rather, states exist in both milieus simultaneously. They seek to satisfy, in some measure, both interests associated with more narrow, short-term and self-regarding concerns, which we have labeled state interests, and interests associated with broader, long-term and international society-oriented considerations, which we have called community interests. Each of the case studies then sought to show the extent to which community and state interests interacted in complex, dynamic ways, producing particular outcomes, namely safety zones, which neither type of variable could explain on its own.

In tandem with these main focuses of the book, two general themes concerning the various experiences of safety zones in the 1990s in Iraq, Bosnia and Rwanda are explored here. First, an attempt is made to assess the roles of state and community interests across the cases, with particular attention paid to the relative importance of, and complex interrelationships between, these two types of interests in the creation and implementation of the safety zones. Second, some observations are offered about the extent to which the safety zones of the 1990s met the objectives set for them. These observations have implications for any future attempts to establish safety zones in areas of conflict, but the main focus of this study in general, and of this chapter, is the attempt to understand the zones that were created in the 1990s, rather than to adopt a more prescriptive approach.

State and community interests and the creation and implementation of safety zones in the 1990s

As a result of the consistency of investigation across the cases, it is now possible to make a certain number of overall observations regarding the interplay between state and community interests and the creation and implementation of safety zones in the 1990s. The most important observation to be made, and one that appears to be true in each case, is that the presence of both community and state interests was necessary for a safety zone policy: neither type of interest was sufficient on its own to explain why states pursued this option. In other words, both state and community interests that enabled or encouraged action had to rise above a certain threshold for safety zones to be created. If either type of interest was absent or discouraged involvement, then a safety zone was not created. In Iraq, Bosnia and Rwanda, community interests were without doubt at stake; and in the case of the Iraqi safe haven, the Bosnian safe areas and the French ZHS, important state interests, such as preserving military credibility or keeping an alliance strong, factored into why states intervened in the way that they did. The one instance where a safety zone was not fully pursued seems to confirm this hypothesis. In Rwanda, the UN SHAs were contemplated as a possibility only when the mistaken perception that what was occurring in Rwanda was random violence and not genocide was rectified, thus setting the triggering and enabling effect of community interests into motion. However, these SHAs never moved beyond the declaratory phase, mainly as a result of the strong constraining effects of several key players' state interests. In addition to this bottom level threshold, we also seem to see an upper level one at work: if state and community interests both become very strong enablers – if they both demanded action – then the response of the international community moved beyond a safety zone policy to a more unambiguous intervention, such as was the case with *Operation Deliberate Force* in Bosnia. In sum, for safety zones to occur, both community and state interests needed to be present, but their configuration needed to involve conflicting tendencies. In fact, safety zones were selected precisely because they were ambiguous: they allowed states to do something, either when they might have preferred to do nothing at all or when they could not agree amongst themselves on what needed to be done.

There were at least four variations on this theme of ambiguity, which occurred simultaneously or at different times within and across the case studies. First, and as seen in all the cases, safety zones allowed states to reconcile opposing state and community interests. Whereas strong pressures existed for states to halt Iraqi repression, Bosnian Serb ethnic cleansing and the Rwandan genocide in fulfillment of their community interests, important state interests meant that states were unwilling to take all necessary measures to do so. The absence of important strategic interests, the fear of casualties amongst their troops, and the unwillingness to get drawn into civil wars were some of the most important factors that constrained state responses to these calamities. Safety zones were an effective response to such contradictory impulses: they let states do something to protect civilians targeted for slaughter, but let them simultaneously avoid becoming intentionally embroiled in the military conflict underway.

Second, safety zones allowed states to reconcile competing state interests within their own governments. For instance, with the ZHS in Rwanda, France wanted to demonstrate that it had the capacity to project its military power into French Africa to put an end to a horrific situation, and yet at the same time find a way to avoid a clash with the RPF while still preventing its military take-over of Rwanda. A safety zone was an ideal means to achieve all these objectives. In tandem, the US vacillated on Bosnia between more assertive action against the Serbs from the air, mainly to affirm NATO's capacity to respond to new types of threats in the post-Cold War context, and non-involvement in the conflict until a peaceful settlement could be reached. Safe areas allowed the Americans to reconcile, in some measure, these competing inclinations.

Third, safety zones facilitated harmonization of differences between countries in the pursuit of their state interests. The most glaring example of safety zones as a means to overcome discord occurred in Bosnia: safe areas were endorsed by the Europeans and acquiesced in by the Americans because they could not see eye to eye on either a diplomatic solution to the conflict or on what constituted the most appropriate means to achieve it. Since both the VOPP and 'lift and strike' were unacceptable to one or the other faction, a safe area policy was advanced to bridge the transatlantic rift. The Europeans averted an outcome they considered would greatly threaten their peacekeeping troops on the ground, and the Americans were granted an opening on air power.

Fourth, safety zones helped reconcile competing community interests, as seen in Iraq. The safe haven was an ideal compromise solution to both the problems of Iraqi repression of the Kurdish population and of the preservation of state sovereignty. Since Security Council Resolution 688 and the subsequent safe haven broke new ground in terms of humanitarian intervention, there were legitimate concerns about the possible violation of sovereignty, of which Iraq's invasion of Kuwait had been a prime example. A safe haven could allay these fears because the western military presence on the ground in Iraq would be temporary and for a clear humanitarian purpose.

In terms of the significance of the relative weight of state and community interests with respect to the creation and implementation of safety zones, two comments seem relevant. First, community interests seemed somewhat more influential than state interests in terms of the timing of the adoption of both the initial concept and subsequent new implementation measures. It was not that state interests were not present at these moments: they did in fact play vitally important roles, a point already clarified above. State interest concerns over massive refugee flows, for example, certainly impelled states to find ways to prevent them, most evidently in Iraq and Bosnia. Likewise, concern that NATO's credibility was being tarnished encouraged the eventual adoption of tougher measures against the Serbs in Bosnia. However, it was often the transgression of important community interests that created the sense of urgency within the US, the UK or France, enhanced by media and public pressure, that finally tipped the balance in favour of doing something on behalf of beleaguered civilians or of improving what was already being done. TV images of thousands of Kurds dying along the Turkish border, which sparked a public outcry and persistently negative editorial comments about American government inaction, finally altered Bush's stance against a safe haven in Iraq. Declaring Srebrenica a safe area in April 1993 occurred at that moment because public awareness about the plight of the town rose dramatically as a result of Morillon's impromptu visit a few weeks earlier, thus increasing pressure upon states to do something. In addition, each new international initiative to better implement the safe areas in Bosnia, mainly involving a more effective use of NATO air power, was preceded by Bosnian Serb atrocities against the civilians within the zones. France's sudden *volte-face* over Rwanda occurred once the French public made it clear they would no longer tolerate the government's abdication of its international responsibilities. In sum, without denying the important role played by state interests throughout the process, it is nonetheless possible to suggest that community interests were somewhat more instrumental in terms of timing: they tipped the scale toward initial adoption or toward a new implementation approach. The one significant exception to this argument seems to be the expansion of the safe area concept in Bosnia in May 1993, the main impetus for which was the alleviation of the transatlantic rift.

As for state interests, their relative weight was stronger in relation to effectiveness of implementation. This observation makes sense in light of the fact that state resources, both civil and military, were essential to adequate constitution of the zones. When hard choices had to be made regarding what actually ought to be done to carry through with the safety zone policies, it was normal that more self-regarding concerns would affect decisions in vital ways. The greater the enabling effect of state interests and the lesser the constraining effect, the more likely the safety zones would be effectively implemented. In Rwanda, the constraining effect was so severe as to render the SHAs approach a non-starter. For both the Iraqi safe haven and the French *Opération Turquoise*, state interests as enablers were relatively strong; and state interests as constraints relatively weak. As Chapters 2 and 5 indicated, reasonably strong state interests encouraged action in both cases. Moreover, both France and the coalition against Iraq were able to muster a strong military force, as well as an adequate deterrent, at relatively little cost to themselves: France had troops and equipment at the ready for action in French Africa, and the coalition headed by America was easily prepared for further involvement in Iraq, having just won the 1991 Gulf War. In addition, as compared to Bosnia, the conditions on the ground were far more favourable in Rwanda and in Iraq: France was unlikely to be severely menaced by its former Hutu allies; Saddam Hussein was militarily spent, so not prone to renewed attack against the coalition; and the Kurdish uprising would not resume as a result of how severely it had been crushed. Nonetheless, in both cases, the configuration of state interests was such that, once the worst of the calamity receded, the intervening states withdrew, simply because the constraining effect of their state interests reasserted itself, which explains the hasty departure from Rwanda in the case of France and the state of uncertainty left behind in northern Iraq in the case of the coalition. Until September 1995 in Bosnia, the constraining effect of state interests made effective implementation of the safe areas sporadic at best, although the enabling effect of state interests did ensure some attempts at adequacy, namely through the passage of Security Council Resolution 836 and the subsequent use of NATO air strikes on several key occasions. The main problem in Bosnia was that conditions on the ground were highly problematic: UNPROFOR was operating in a war zone; the Bosnian Serbs were particularly intransigent in their aims; and suitable deterrence was not achieved through air power. Tougher action against the Bosnian Serbs was made possible only when the threat to UNPROFOR troops was reduced after retrenchment and when the emergence of a better trained Croat-Muslim army made it unnecessary for the US to have to consider entering the war on the ground.

Overall assessment of the safety zones of the 1990s

Despite the fact that none of the safety zones of the 1990s was a resounding success, it would be premature to dismiss the concept as entirely unhelpful. The safety zones of the 1990s represented a movement away from the more traditional zones outlined in humanitarian law: they were innovative attempts by states to 'protect', within their countries of origin, civilians who were directly targeted for slaughter or ethnic cleansing. As such, they offer us the opportunity to reflect upon whether and how well third parties can guarantee, during a conflict or complex emergency, the security of civilians within their own country, although one must be careful not to over-generalize or to make naïve assumptions considering the extent to which contexts differ. Nonetheless, as it is both likely that wars involving significant violations of humanitarian law will continue to occur, and that the international community will be called upon to 'do something' in the face of the massacre of civilians, safety zones may yet again prove a temptation that states cannot resist. Although tentative and incomplete, what is learned here may be of some use in the future.

The Iraqi safe haven was likely the most effective safety zone of the 1990s: it enabled the Kurds to return to their homes in Iraq and remain there in relative safety; its military protection by well-armed coalition troops, further enhanced through the no-fly zone, successfully deterred any reprisals from Saddam Hussein; and its implementation went smoothly, allowing unhindered access by humanitarian organizations and causing little international controversy. Its main drawback was that its establishment and termination were not integrated into a broader, long-term solution to the overall Iraqi problem, thus leading to a situation where this region's status within the country remained unclear and its border with Turkey, pregnable and problematic. In some respects France's ZHS in Rwanda proved equally effectual: it saved a number of the remaining Tutsis in the zone; it allowed humanitarian agencies to reach a part of the country that had been hitherto inaccessible; and its inviolability was ensured by sufficient deterrent force. However, France's effort was marred, partially as a result of its previous involvement in Rwanda and partially due to implementation problems on the ground. Its most blatant drawback was its tardy creation: one can only wonder what France might have been able to achieve if its configuration of state and community interests had permitted its immediate intercession at the outset of the genocide. As it turned out, France set up the zone only near the end of the calamity; it did not always protect the targeted group effectively; it failed either to adequately disarm the *génocidaires* and FAR or to prevent their exodus to Zaire; and it withdrew rather abruptly despite signs that its continued presence in the country would be a stabilizing factor, so as to avoid further international reproach regarding the validity of its humanitarian motives. Although France's ZHS was not an unequivocal success, it nonetheless accomplished more than the proposed UN SHAs, which were never implemented. If the *ad hoc, de facto* sites that the small UNAMIR contingent managed to oversee in Kigali had been strengthened or expanded throughout the country, they might also have served as examples of how internationally implemented safety zones can protect beleaguered civilians.

Finally, despite the collapse of the Srebrenica and Zepa safe areas in the summer of 1995 and the tragic massacres that ensued, the safety zones in Bosnia were not without merit. Sarajevo was supplied with humanitarian assistance throughout the siege and its exclusion zone kept both Muslim and Serb heavy weapons silent for long stretches of time. Although it was not applied consistently and faltered badly during the fall of Srebrenica and Zepa, NATO's air power deterrent did play a role in keeping the safety zones out of Bosnian Serb hands on a number of occasions, and thus in preventing ethnic cleansing. In addition, as an accompaniment to the Muslim-Croat ground offensive, *Operation Deliberate Force* helped relieve the siege of Sarajevo and helped keep Gorazde within government-held territory. At times sporadic, humanitarian assistance nonetheless filtered into the eastern enclaves, thanks in part to the continued peacekeeping presence there. However, the failings of the Bosnian safe areas surpassed those of the Iraqi safe haven or the Rwandan ZHS, perhaps because the safe areas were not as clearly demarcated, endured for a much longer period of time and were utilized in the context of continued warfare. On the one hand, failure to fully disarm Muslim fighters within the safe areas, as well as the use of NATO's deterrent force in a manner that seemed to favour the Bosnian government side, damaged UNPROFOR's impartiality in the eyes of the Bosnian Serbs and thus contributed to the latter's growing hostility toward both the safe areas and the peacekeepers. On the other hand, not-insignificant disarmament of Srebrenica, as well as inadequate international military protection on the ground and from the sky, left it and Zepa open to invasion when UNPROFOR proved unable to deter the Bosnian Serb attacks.

A discussion of two broad interrelated points permits us to grapple with some of these successes and failures across the case studies: the improvised nature of these safety zones and the level of state willingness to enforce them. First, all of the instances of safety zone usage in the 1990s were *ad hoc* responses to immediate crises involving grave violations of humanitarian law. Pressured by their community and state interests, states were looking for ways to get through the day, and very often decisions taken regarding the zones were reactive in nature, not to mention hasty and sometimes rash. With little to draw upon from past experience due to the distinctiveness of the 1990s zones, states were also exploring uncharted territory. As a result, the safety zone policies were neither adequately thought-through nor clearly associated with a broader solution to the problems at hand,[1] mainly because there was no solution or too little international consensus to achieve one. The coalition implementing the Iraqi safe haven, most particularly the United States, did not stop to consider what would happen to the Kurds once their repression ceased and *Operation Provide Comfort* closed down. Once the killing stopped in Rwanda and it became clear that it would be impossible to resurrect a legitimate Hutu side for negotiations with the RPF, France had little idea as to how to deal with the FAR and *génocidaires*: it was caught off-guard by their massive flight to Zaire. The safe areas in Bosnia were viewed initially as an interim measure, needed for a few months at most so as to avoid a transatlantic rift, further ethnic cleansing and the loss of more territory to the Bosnian Serbs until a final settlement could be reached. Instead, the UN and implementing states were saddled with a policy that was never designed to cope with two and half years of continued conflict. As a result, the safe area policy eventually buckled under the strain of unforeseen consequences, such as peacekeepers being taken as hostages or the low effectiveness of the arrangements for close air support.

In certain ways, the Iraqi safe haven and the ZHS were more coherent territorially and militarily than the safe areas, perhaps partially explaining their greater level of effectiveness. In both Iraq and Rwanda, once the political decision was made to create a safety zone, a clear military chain of command, disposing of sufficient resources and deterrent capability, made the decisions regarding precisely where the zone should be, how large it should be and what use of force should be allowed in relation either to self-defense or the protection of the zone. In addition, in each of these cases there was a clearer idea than in Bosnia about the time frame needed to reach the proposed objectives, as well as a more definite intention to avert an open-ended commitment. The safe haven and the ZHS also covered a contiguous piece of land that bordered a friendly state, thus making implementation by land and air more feasible than in Bosnia where the safe areas were dispersed and isolated.[2]

In contrast, the Bosnian safe areas were far less coherent in a number of respects, perhaps due to the greater complexity of the situation. At the most basic level, the safe areas were never properly demarcated, leading to confusion as to their size and precise location, although 20 km heavy weapons exclusion zones were eventually erected around Sarajevo and Gorazde in February and April of 1994 respectively. The safe areas were also brought into being by the Security Council prior to confirmation that sufficient resources would be made available for their effective implementation, resulting ultimately in a scenario where UNPROFOR's duties exceeded its means to carry them out.[3] Although intended as an interim measure, the safe areas became part of the UN's relatively open-ended humanitarian commitment to Bosnia: as just another element of UNPROFOR's mandate, they were no longer invested with the kind of sharpness of purpose that was seen in Iraq or in Rwanda. Finally, Security Council resolutions regarding the safe areas, in terms of the forcefulness of language and stance against the Serbs, were at odds with the more cautious approach taken by UNPROFOR on the ground. As a result, the safe area policy seemed to be a mixture of both a conciliatory approach, in which the zones' existence was reliant upon the consent of the belligerents, and an assertive approach, in which enforcement would be necessary, particularly as the Bosnian Serbs were disadvantaged in the equation.[4] This combination of incompatible elements helped engender a whole plethora of problems and miscalculations that eventually proved too much of a burden for the safe areas to bear. On the one hand, perhaps due to an underestimation of the intransigence of the Bosnian Serbs,[5] the safety of the zones was to be guaranteed through cooperation: it was assumed that Security Council resolutions alone would be sufficient to ensure compliance; and that a relatively small number of lightly-armed peacekeepers, backed by a limited NATO deterrent capability,[6] would be adequate for the job. On the other hand, the safety zones were being imposed on the Bosnian Serbs: Security Council resolutions openly favoured the Muslims, allowing them to keep their weapons in all but Srebrenica; and when NATO air strikes were eventually used on several occasions, they were mostly directed at Bosnian Serb targets. Such strong, seemingly biased, enforcement measures angered the Serbs, who then retaliated against the safe areas and against an ill-prepared and spread-too-thin UNPROFOR.

Second, the level of state willingness to enforce the safety zones was also influential in terms of why they succeeded or failed. As already explained, states were attracted to safety zones precisely because they were ambiguous, halfway measures that enabled them to reconcile competing, and often contradictory, interests. Almost by definition, therefore, the willingness to enforce them was to varying degrees ambivalent and erratic, which in turn had major repercussions for the level of deterrence provided and the amount of protection accorded. Certainly, there existed some will to enforce the zones in Iraq, Bosnia and Rwanda, seeing as they were implemented without the clear and unequivocal consent of the belligerents and at some risk, political and military, to the states that intervened. Yet, this willingness was difficult to sustain over long periods of time, mainly as a result of how events on the ground affected state interests, which we have already identified as being the weightier factor in terms of explaining effectiveness of implementation. Often, even though a direct threat to state interests was rectified through the creation of a safety zone, its implementation would trigger other state interest concerns, such as the fear of suffering casualties amongst their troops, of being drawn into a quagmire or of it costing too much politically. For example, preventing refugee flows was one state interest that enabled the pursuit of the safe haven in Iraq and the safe areas in Bosnia. However, once the tide had been stemmed and once the displaced had either been returned home or effectively prevented from leaving, then there was less willingness to follow through with adequate protection in their country of origin, a reality that must be kept in mind as a potentially significant drawback of 1990s safety zones in general.[7] The declining will to enforce is what predominantly lies behind the hasty departures from Iraq and Rwanda, as well as the growing resistance to the use of NATO air strikes in Bosnia.

Another aspect which affected state willingness to enforce the safety zones was the level of opposition encountered, and the extent to which that opposition could be thwarted without significant cost. As several have pointed out and as the case studies in this book indicate, enforcement action almost always has underlying political motivations:[8] it is state interests that more often than not play a key role in deciding when and to what extent force is to be used. For instance, coalition monitoring of the no-fly zone over Iraq was just as much about keeping Iraq weak as it was about facilitating the enforcement of Security Council Resolutions 687 and 688. As a consequence, rarely can enforcement action be undertaken to manipulate the behaviour of one or all of the warring parties, even if in strict conformity with a UN mandate, without incurring their wrath.[9] It is very likely that the targeted party will consider itself designated as an enemy and respond accordingly. In the cases of Iraq, Bosnia and Rwanda, these factors played themselves out rather differently, which helps explain their differing degrees of success. In the case of the safe haven, although Saddam Hussein considered coalition action to be a violation of Iraqi sovereignty and thus an attack against Iraq, the 1991 Gulf War had weakened the country to the extent that it dared not oppose anew American, British and French troops. In addition, the coalition had at its disposal, as a residue of the 1991 Gulf War, an impressive deterrent capability, so knew that the following syllogism would prove correct: 'we rattle the saber, and they withdraw'.[10] The presence of such a strong deterrent also meant that the Kurds had little to fear from Saddam Hussein. Such circumstances created excellent conditions for a safety zone; and, since the Kurds were not in a position to resume their insurrection due to how severely it had been crushed, the issue of their disarmament within the zone was somewhat inconsequential.

Opération Turquoise operated initially in a somewhat similar environment, but France nonetheless abdicated responsibility for enforcement when it appeared that all might not end well. As seen in Chapter 5, the underlying state interest behind France's enforcement action was to avoid an RPF military victory through a ceasefire and renewed negotiations between Hutus and Tutsis over Rwanda's future. Even though the creation of the ZHS also had humanitarian motivations and was endorsed by the Security Council, the RPF perceived France's presence as hostile and thus threatened retaliation. However, they were successfully deterred by French military capability. The more worrisome problems in some respects for the French were the FAR and the *génocidaires* who sought refuge in the zone: they initially welcomed *Opération Turquoise*, but when they realized that France would not be using force on their behalf they became decidedly cooler, though they dared not attack their former ally. So again, in terms of the inviolability of the safety zone *per se*, disarmament was not a strong issue of concern during the existence of the ZHS. The main problem, however, was that

France was not willing to risk encountering serious opposition on the part of the FAR or the *génocidaires*, or for that matter the RPF, through a prolonged, stabilizing presence in the country, so it neither disbanded the FAR nor prevented their escape into Zaire, and left Rwanda as fast as its humanitarian conscience would allow. As the French proved unwilling to solve permanently the problem of Hutu extremists' continued military activity, it was transferred to Zaire, leading to two wars between Rwanda and Zaire (now the DRC).

The Bosnian Serbs also perceived UN mandated enforcement action in relation to the safety zones as hostile to them, which it was in its political undertones of support for the Bosnian government. Unlike the Iraqi army or the RPF though, the Bosnian Serbs were not deterred and retaliated, putting UN and western state willingness to enforce the safe areas to the test. This fact alone – that the Bosnian Serbs took military action against intervening peacekeeping troops – meant that Bosnia became the toughest theatre of the three in which to enforce safety zones, so it is not surprising that implementation here was problematic and not altogether successful. Two factors permitted Bosnian Serb retribution: they were a determined, well-trained and well-armed Eastern European fighting force capable of inflicting damage; and the will to enforce the safe areas was insufficiently strong to mount a successful, prolonged deterrent, especially as the West was not acting from a position of unity and feared being drawn into a full-fledged war with the Bosnian Serbs. Although some will existed, especially through April 1994, to deter attacks against the safe areas through a NATO air power deterrent, its dissuasive force degraded over time as implementing states, as well as the UN itself, grew wary of putting their soldiers' lives at risk. The US, Britain and France were simply not willing to incur the costs more aggressive enforcement action would have entailed. What remaining will to enforce disappeared almost entirely following the hostage-taking episode of May 1995, resulting in nefarious Serb behaviour in Srebrenica being unchallenged two months later. The issue of disarmament became a problem in this context of insufficient willingness to enforce: as UNPROFOR could neither adequately monitor the safe areas nor ensure their safety, it encouraged both Bosnian military activity within them, further enraging the Serbs, and Serb attacks against them because they were not properly defended, particularly in the case of Srebrenica which had been demilitarized. In the wake of such a potentially undermining failure for NATO and the UN, and in response to the gravity of the crimes committed by the Bosnian Serbs following their capture of Srebrenica, the will to enforce reasserted itself; and Serb ability to take revenge receded as a result of their accumulating losses to the Muslim-Croat army on the battlefield.

It appears from the above assessments that many complex and interrelated factors played a role in determining when and why the safety zones of the 1990s worked and when and why they did not, so any future attempts to employ such similarly distinctive methods to 'protect' civilians in their country of origin are best considered cautiously on a case by case basis. Very broadly however, safety zones seem more likely to succeed, especially when established without the consent of the belligerents, if they are coherent in design;[11] if they are implemented as a clear interim measure by a single state or a multinational force with clear rules of engagement;[12] if they are integrated into a broader conflict resolution approach; and if they are buttressed by a sufficiently strong state willingness to enforce them so as to ensure credible deterrent threat against possible retributive measures.

Beyond the 1990s

Given the difficulties of implementation and enforcement, it is perhaps not surprising that no further attempts to create safety zones of a 1990s variety have occurred since *Opération Turquoise*. However, less ambitious efforts to provide a semblance of protection to civilians within combat zones have arisen. For example, the Interim Emergency Multinational Force, the European-sponsored and French-led temporary force that went into the town of Bunia in eastern DRC in June 2003, was authorized by Security Council Resolution 1484 of 30 May 2003, to 'ensure the protection of the internally displaced persons in the camps in Bunia'. However, there was no formal recognition that these camps constituted safety zones. The UN peacekeeping mission in the DRC (MONUC) also created, for a few weeks, a temporary buffer zone between belligerents in eastern DRC in December 2004 to enable humanitarian assistance to reach displaced persons, but this zone, which was never delineated, was not designed to protect civilians *per se*, who at any rate were not specifically targeted for slaughter although some killings occurred.

Finally, on 5 August 2004, the UN and the Sudanese government agreed to a Plan of Action to resolve the humanitarian crisis in Darfur. This plan involved the establishment by the Sudanese government of 'secure safe areas' for the displaced civilian population.[13] In these areas – camps and areas around certain towns and villages with a high concentration of local population – the Sudanese government was to prevent further military activity on the part of its own soldiers, of rebel factions, and of armed militias. Such safe areas cannot be compared to the zones of the 1990s because they do not involve outside intervention and external military force as a means to guarantee the security of civilians at risk, although an African Union force was sent to Darfur separately as mainly an observer mission due to its small size and lack of capacity. These secure safe areas were rather aimed at forcing the Sudanese government to assume responsibility for the safety of the country's citizens, and were conceived as a stepping stone toward achieving security in Darfur as a whole. They resemble in certain respects the neutralized zones of Article 15 of the 1949 Geneva Convention IV or the demilitarized zones of Article 60 of the 1977 Geneva Protocol I, particularly as Sudanese government forces were required to reach agreement with rebel forces not to engage in clashes in some of the zones concerned.[14] Only limited progress has so far been made in rendering these areas, and Darfur as a whole, truly safe. Nevertheless, the use of the safe area concept in the new millennium – albeit the more traditional kind as laid out in international humanitarian law rather than the 1990s variety – indicates continued interest in safety zones as a means to protect beleaguered civilians in warfare.

Notes

Chapter 1 Introduction

1. SC Res. 824, 6 May 1993. Srebrenica was designated a safe area in SC Res. 819 of 16 April 1993.
2. SC Res. 918, 17 May 1994.
3. SC Res. 929, 22 June 1994.
4. On safety zones, see Comité International de la Croix Rouge (CICR), 'Zones Sanitaires et Zones de Sécurité', *Revue Internationale de la Croix Rouge* 82 (1951), 442–483 and 628–662; Karin Landgren, 'Safety Zones and International Protection: A Dark Grey Area', *International Journal of Refugee Law* 7 (1995), 436–458; Jean-Philippe Lavoyer, 'International Humanitarian Law, Protected Zones and the Use of Force', in *UN Peacekeeping in Trouble: Lessons Learned from the Former Yugoslavia: Peacekeepers' Views on the Limits and Possibilities of the United Nations in a Civil War-Like Conflict*, eds, Wolfgang Biermans and Martin Vadset (Aldershot: Ashgate, 1998), pp. 262–279; Yves Sandoz, 'The Establishment of Safety Zones for Persons Displaced within Their Country of Origin', in *International Legal Issues Arising under the United Nations Decade of International Law*, eds, Najeeb Al-Nauimi and Richard Meese (London: Martinus Nijhoff Publishers, 1995), pp. 899–927; and Maurice Torrelli, 'Les Zones de Sécurité', *Revue Générale de Droit International Public* (1995) 99, 787–849.
5. See CICR, 'Zones Sanitaires et Zones de Sécurité', 442.
6. Ibid., 443.
7. 1907 Hague Convention IV Respecting the Laws and Customs of War on Land, in Adam Roberts and Richard Guelff, eds, *Documents on the Laws of War*, 3rd ed. (Oxford: Oxford University Press, 2000), p. 78.
8. See CICR, 'Zones Sanitaires et Zones de Sécurité', 446–448.
9. See Ibid., 448–455, and Jean S. Pictet, ed., *Commentary I Geneva Convention for the Amelioration of the Condition of the Wounded and Sick in Armed Forces in the Field* (Geneva: ICRC, 1952), pp. 208–209.
10. On 9 September 1939, the ICRC sent a memorandum to belligerents, in which it encouraged them to consider the possibility of setting up hospital zones and neutralized zones, and declared itself willing to administer them as a neutral party. Only Germany responded affirmatively to the idea, but on condition of reciprocity by the other parties to the conflict. A second attempt by the ICRC in March 1944 to convince belligerents to create such zones also failed. See CICR, 'Zones Sanitaires and Zones de Sécurité', 460–468.
11. See Ibid., 455–457.
12. See Ibid., 458–459 and Bernard Wasserstein, *Secret War in Shanghai* (London: Profile Books, 1998), pp. 18, 61.
13. See CICR, 'Zones Sanitaires and Zones de Sécurité', 469–483.

14. All articles pertaining to safety zones in humanitarian law are merely permissive and not mandatory, as states proved reluctant to accept stronger commitments. See Geoffrey Best, *War & Law Since 1945* (Oxford: Clarendon Press, 1994), pp. 116–117.
15. 12 August 1949 Geneva Convention I for the Amelioration of the Condition of the Wounded and Sick in Armed Forces in the Field, in *Documents on the Laws of War*, eds, Roberts and Guelff, p. 206. See also Pictet, *Commentary I*, pp. 206–216.
16. Annex I to 1949 Geneva Convention I, Draft Agreement Relating to Hospital Zones and Localities, in Pictet, *Commentary I*, pp. 415–429.
17. 12 August 1949 Geneva Convention IV Relative to the Protection of Civilian Persons in Time of War, in *Documents on the Laws of War*, eds, Guelff and Roberts, pp. 306–307. See also Jean S. Pictet, ed., *Commentary IV Geneva Convention Relative to the Protection of Civilian Persons in Time of War* (Geneva: ICRC, 1958), pp. 118–133.
18. Annex I to 1949 Geneva Convention IV, Draft Agreement Relating to Hospital and Safety Zones and Localities, in Pictet, *Commentary IV*, pp. 627–639.
19. 8 June 1977 Geneva Protocol I Additional to the Geneva Conventions of 12 August 1949, and Relating to the Protection of Victims of International Armed Conflicts, in *Documents on the Laws of War*, eds, Roberts and Guelff, pp. 454–456. See also Yves Sandoz, Christophe Swinarski and Bruno Zimmermann, eds, *Commentary on the Additional Protocols of 8 June 1977 to the Geneva Conventions of 12 August 1949* (Geneva: Martinus Nijhoff Publishers, 1987), pp. 697–712.
20. For similar lists, see Landgren, 'Safety Zones and International Protection', 440; and Sandoz et al., eds, *Commentary on the Additional Protocols*, p. 697.
21. The text of this ICRC mediated agreement can be found in Marco Sassoli and Antoine A. Bouvier, eds, *How Does Law Protect in War? Cases, Documents and Teaching Materials on Contemporary Practice in International Humanitarian Law* (Geneva: International Committee of the Red Cross, 1999), pp. 1055–1056.
22. In the case of Sri Lanka, the zones were administered by the United Nations High Commissioner for Refugees (UNHCR).
23. It must also be noted that the zones provided for in international humanitarian law relate only to international armed conflict. There is no mention of a safety zone concept in either the 1977 Geneva Protocol II Relating to the Protection of Victims of Non-International Armed Conflict or in Common Article 3 of the 1949 Geneva Conventions that outlines minimum provisions for protection of civilians and other non-combatants in an armed conflict not of an international character. However, there is no reason why they should not serve as a model applied by analogy to internal wars, since they are only permissible courses of action for states, not obligatory ones. See Lavoyer, 'Protected Zones and the Use of Force', p. 266.
24. Safety zones differ from what one might refer to as protectorates, as established in Kosovo and East Timor in 1999, in which the international community takes over the administration of an entire territory following the cessation of a conflict.
25. See for instance Hans J. Morgenthau, *Politics Among Nations: The Struggle for Power and Peace*, 5th ed. (New York: Alfred A. Knopf, 1978), p. 236.

26. Many theorists have called for an exploration of the connection between rationality and norms, suggesting that a simple dichotomy is far too naive. See Martha Finnemore and Kathryn Sikkink, 'International Norm Dynamics and Political Change', in *Exploration and Contestation in the Study of World Politics*, ed. Peter J. Katzenstein et al., (Cambridge, Massachusetts: The MIT Press, 1999), pp. 271–272; S. Neil MacFarlane and Thomas Weiss, 'Political Interest and Humanitarian Action', *Security Studies* 10 (Autumn 2000), 113–115; and Andrew Hurrell, 'Conclusion: International Law and the Changing Constitution of International Society', in *The Role of Law in International Politics: Essays in International Relations and International Law*, ed. Michael Byers (Oxford: Oxford University Press, 2000), p. 328.
27. William E. Connolly, *The Terms of Political Discourse* (Oxford: Martin Robertson & Company, 1983), pp. 55–56.
28. Jane J. Mansbridge, 'The Rise and Fall of Self-Interest in the Explanation of Political Life', in *Beyond Self-Interest*, ed. Jane J. Mansbridge (Chicago: The University of Chicago Press, 1990), pp. 3–22.
29. Hedley Bull, *The Anarchical Society: A Study of Order in World Politics* (London: the Macmillan Press, 1977), p. 13.
30. Ibid., pp. 67–73.
31. Andrew Hurrell, 'International Society and the Study of Regimes: A Reflective Approach', in *Regime Theory and International Relations*, ed. Volker Rittberger (Oxford: Clarendon Press, 1999), p. 59.
32. Michael Byers, *Custom, Power and the Power of Rules: International Relations and Customary International Law* (Cambridge: Cambridge University Press, 1999), p. 6.
33. Robert O. Keohane, 'Empathy and International Regimes', in *Beyond Self-Interest*, ed. Mansbridge, p. 230.
34. Ibid., p. 229.
35. For a similar point, see Louis Henkin, *How Nations Behave: Law and Foreign Policy*, 2nd ed. (New York: Columbia University Press, 1979) p. 331.
36. Finnemore and Sikkink also suggest that such a synthesis might be useful so as to understand which logic (rationalist or normative) applies to what kind of actors under what circumstances. Finnemore and Sikkink, 'International Norm Dynamics', p. 273.
37. See Morgenthau, *Politics Among Nations*; E. H. Carr, *The Twenty Years' Crisis, 1919–1939: An Introduction to the Study of International Relations*, 2nd ed. (London: Macmillan, 1946), and Kenneth Waltz, *Theory of International Politics* (Reading, Massachusetts: Addison-Wesley, 1979).
38. Stephen D. Krasner, 'Structural Causes and Regime Consequences: Regimes as Intervening Variables', in *International Regimes*, ed. Stephen D. Krasner (Ithaca: Cornell University Press, 1983), p. 1.
39. Arthur A. Stein, 'Coordination and Collaboration: Regimes in an Anarchic World', in *International Regimes*, ed. Krasner, p. 123.
40. Hurrell, 'International Society and the Study of Regimes', p. 55.
41. See Byers, ed., *Role of Law in International Politics*; Abram Chayes and Antonia Chayes, *The New Sovereignty: Compliance with International Regulatory Agreements* (Cambridge, Massachusetts: Harvard University Press, 1995); Henkin, *How Nations Behave*; and Anne-Marie Slaughter et al., 'International Law and International Relations Theory: A New Generation of Interdisciplinary Scholarship', *American Journal of International Law* 92, 3 (July 1998), 367–397.

42. For a description of how this process works, see Chayes and Chayes, *The New Sovereignty*, pp. 118–123. These authors explain that, as a 'matter of international legal practice, questionable action must be explained and justified' in terms that are understood by other states. Therefore, 'the reasons adduced in explanation of justifications cannot be merely self-regarding, but must have an objective appeal to the interlocutor' so that good legal arguments can be distinguished from bad.
43. Martha Finnemore, 'Constructing Norms of Humanitarian Intervention', in *The Culture of National Security: Norms and Identity in World Politics*, ed. Peter J. Katzenstein (New York: Columbia University Press, 1996), p. 157.
44. Martha Finnemore, *National Interests in International Society* (Ithaca: Cornell University Press, 1996), pp. 1–2.
45. Robert O. Keohane, *International Institutions and State Power: Essays in International Relations Theory* (London: Westview Press, 1989), pp. 128, 161.
46. Ronald Jepperson, Alexander Wendt and Peter J. Katzenstein, 'Norms, Identity, and Culture in National Security', in *Culture of National Security*, ed. Katzenstein, p. 41.
47. Paul Kowert and Jeffrey Legro, 'Norms, Identity, and Their Limits: A Theoretical Reprise', in *Culture of National Security*, ed. Katzenstein, p. 461.
48. I adapt similar definitions from Finnemore, *National Interests in International Society*, p. 22; and Charles Kegley, Jr. and Gregory A. Raymond, *When Trust Breaks Down: Alliance Norms and World Politics* (Columbia: University of South Carolina Press, 1990), p. 14.
49. Judith Goldstein and Robert O. Keohane, 'Ideas and Foreign Policy: an Analytical Framework', in *Ideas and Foreign Policy: Beliefs, Institutions and Political Change*, ed. Judith Goldstein and Robert O. Keohane (London: Cornell University Press, 1993), p. 9.
50. Krasner, 'Structural Causes and Regime Consequences', p. 2.
51. Byers, *Custom, Power and the Power of Rules*, p. 32.
52. Krasner, 'Structural Causes and Regime Consequences', p. 2.
53. A similar idea is presented in Goldstein and Keohane, 'Ideas and Foreign Policy', pp. 17–20.
54. Friedrich Kratochwil, *Rules, Norms, and Decisions: On the Conditions of Practical and Legal Reasoning in International Relations and Domestic Affairs* (Cambridge: Cambridge University Press, 1991), p. 11.
55. Adam Roberts, 'Humanitarian War: Military Intervention and Human Rights', *International Affairs* 69 (1993), 243.
56. Kegley and Raymond, *When Trust Breaks Down*, p. 18.
57. Finnemore, 'Constructing Norms of Humanitarian Intervention', p. 158. Chayes and Chayes put it this way: 'Norms…are not predictive. Since they are prescriptions for action in situations of choice, the actor may or may not choose to obey them', in *The New Sovereignty*, p. 113.
58. See Richard Price and Nina Tannenwald, 'Norms and Deterrence: The Nuclear and Chemical Weapons Taboos', in *Culture of National Security*, ed. Katzenstein p. 149; Goldstein and Keohane, 'Ideas and Foreign Policy', pp. 16–17; and Kratochwil, *Rules, Norms, and Decisions*, p. 69.
59. These rules are called *erga omnes* in scope. The International Court of Justice, in the Barcelona Traction case, compared regular rules with *erga omnes* rules thus: 'an essential distinction should be drawn between obliga-

tions of a State towards the international community as a whole, and those arising *vis-à-vis* another State in the field of diplomatic protection. By their very nature the former are the concern of all States. In view of the importance of the rights involved, all States can be held to have a legal interest in their protection; they are obligations *erga omnes*', in *Case Concerning the Barcelona Traction, Light and Power Company, Limited,* 5 February 1970, *ICJ Reports*, 1970, p. 32.

60. See Michael Byers, 'Conceptualising the Relationship between Jus Cogens and Erga Omnes Rules', *Nordic Journal of International Law* (1997), 211–212, 239. Theodor Meron explains this point clearly: 'When a state breaches an obligation *erga omnes*, it injures every state, including those not specifically affected. As a victim of a violation of the international legal order, every state is therefore competent to bring actions against the breaching state. When an obligation erga omnes, in whose fulfilment all states have a legal interest, is breached, the breaching state's responsibility is engaged *vis-à-vis* all of the other members of the international community', in *Human Rights and Humanitarian Norms as Customary Law* (Oxford: Clarendon Press, 1989), p. 191.
61. 1948 Convention on the Prevention and Punishment of the Crime of Genocide, in *Documents on the Laws of War*, ed. Roberts and Guelff, pp. 180–184.
62. 'The High Contracting Parties undertake to respect and ensure respect for this Protocol in all circumstances'.
63. See Frits Kalshoven, 'The Undertaking to Respect and Ensure Respect in All Circumstances: From Tiny Seed to Ripening Fruit', *Yearbook of International Humanitarian Law* 2 (1999), 3–61.
64. See *Prosecutor vs. Zoran Kupreskic* et al., ICTY Trial Chamber II, Judgement, The Hague, 14 January 2000, Case No. IT-95-16, pars. 518–519. See also Roberts and Guelff, 'Introduction', *Documents on the Laws of War*, pp. 33–34.
65. Author interview with Sir Franklin Berman, Legal Advisor to the UK Foreign and Commonwealth Office from 1991 to 1999, Oxford, 6 June 2001.
66. See 'The Establishment of Protected Zones for Endangered Civilians in Bosnia-Herzegovina', *ICRC Position Paper*, distributed on 30 October 1992 to the governments concerned, the Co-chairmen of the London Conference on the former Yugoslavia and the Office of the High Commissioner for Refugees, in *How does Law Protect in War?*, ed. Sassoli and Bouvier, pp. 1127–1129.
67. Oran Young, 'Regime Dynamics: the Rise and Fall of International Regimes', in *International Regimes*, ed. Krasner, p. 94.
68. In legal terminology, these rules are *jus cogens*. See Article 53 of the 1969 Vienna Convention on the Law of Treaties, in *International Legal Materials* 8 (1969), p. 679. Although there is not complete agreement about which norms are considered to *be jus cogens*, the *International Law Commission* lists, in the commentary illuminating Article 53 of the 1969 Vienna Convention, three examples of rules, which if implemented, would be violations of existing *jus cogens* rules: 'a) a treaty contemplating an unlawful use of force contrary to the principles of the Charter; b) a treaty contemplating the performance of any other act criminal under international law; and c) a treaty

contemplating or conniving at the commission of acts, such as the trade in slaves, piracy or genocide', in the suppression of which every state is called upon to co-operate', in *United Nations Report of the International Law Commission*, covering the work of its fifteenth session, May 6–July 12 1963, in *The American Journal of International Law* 58 (1964), 265.

69. See generally Hermann Mosler, *The International Society as a Legal Community* (Alphen aan den Rijn, The Netherlands: Sijthoff & Noordhoff, 1980); and George Abi-Saab, 'Introduction', in *Conference on International Law: The Concept of Jus Cogens in International Law, Papers and Proceedings. Lagonissi, April 3–8, 1966* (Geneva: Carnegie Endowment for International Peace, 1967), pp. 7–15.
70. Author interview with Sir Franklin Berman, Oxford, 6 June 2001. Some scholars, mainly in reference to the recent practice of the Security Council in the humanitarian emergencies of the 1990s, contend that Common Article 1 of the Geneva Conventions does entail a legal obligation on the part of states to act to ensure respect for humanitarian law. See Laurence Boisson de Chazournes and Louigi Condorelli, 'Common Article 1 of the Geneva Conventions revisited: Protecting collective interests', *International Review of the Red Cross* 82 (2000), 67–87; Hans-Peter Gasser, 'Ensuring Respect for the Geneva Conventions and Protocols: the Role of Third Party States and the United Nations', in *Armed Conflict and the New Law Volume II: Effecting Compliance*, ed. Hazel Fox and Michael A. Meyer (London: the British Institute of International and Comparative Law, 1993), pp. 15–49.
71. Kalshoven explains this point well: 'I believe that when it comes to reading into Article I [of the Geneva Conventions] any effect beyond the sphere of the state's internal affairs, it lies in adding to the state's *right* as a Party to the Conventions to make other states respect their terms, *a moral incentive* or "obligation" to do so...It implies that in weighing the admittedly many factors involved in the process of decision making, the moral duty to "ensure respect" for international humanitarian law carries particular weight: the graver the "situation of apparent disregard", the heavier the weight of this factor', in 'The Undertaking to Respect and Ensure Respect', 60.
72. Stephen D. Krasner, 'Regimes and the Limits of Realism: Regimes as Autonomous Variables', *International Regimes*, ed. Krasner, p. 358.
73. Finnemore, *National Interests in International Society*, p. 30.
74. The capacity of particular moral arguments to influence government policies stems directly from the political influence of domestic and transnational moral entrepreneurs. See Ethan Nadelmann, 'Global prohibition regimes: the evolution of norms in international society', *International Organization* (1990) 44, 483.
75. MacFarlane and Weiss, 'Political Interest and Humanitarian Action', 128.
76. See Andrew P. Cortell and James W. Davis, Jr., 'How Do International Institutions Matter? The Domestic Impact of International Rules and Norms', *International Studies Quarterly* (1996) 40, 452–454; and McElroy, *Morality and American Foreign Policy*, p. 44.
77. See generally Katzenstein, 'Introduction: Alternative Perspectives on National Security'.
78. Jepperson, Wendt and Katzenstein, 'Norms, Identity, and Culture in National Security', p. 62.

79. See Audie Klotz, *Norms and International Relations: The Struggle Against Apartheid* (London: Cornell University Press, 1995), p. 28.
80. For a discussion of why reputation is so important to states and why its loss affects their relations with others, see McElroy, *Morality and American Foreign Policy*, pp. 46–53.
81. As will be apparent in later chapters, the process of state interest formation within states is complex and at times disharmonious. However, from an international perspective, it is more straightforward because the pursuit of state interests does not involve the same dynamics of coordination and collaboration that the pursuit of community interests entails.
82. One of the purposes of MacFarlane's and Weiss' article on 'Political Interest and Humanitarian Action', is to reveal the extent to which state interests expand humanitarian space. See generally, and 112–116.
83. These peacekeeping operations were intended to enable the superpowers to disentangle themselves from a number of conflict areas, including Afghanistan (UNGOMAP, 1988), Iran/Iraq (UNIIMOG, 1988), Angola (UNAVEM I, 1988), Namibia (UNTAG, 1989) and Central America (ONUCA, 1989).
84. SG report, *An Agenda for Peace: Preventive Diplomacy, Peacemaking and Peace-keeping*, UN Doc. A/47/277–S/24111, 17 June 1992, pars. 3, 15.
85. Ibid., par. 17.

Chapter 2 Providing Comfort at Home: Safe Haven in Iraq

1. Map No. 3835, Rev. 4, January 2004, United Nations Cartographic Section.
2. James L. Jones, 'Operation Provide Comfort: Humanitarian and Security Assistance in Northern Iraq', *Marine Corps Gazette* 75, 11 (November 1991), 107. The area covered by the no-fly zone was yet larger.
3. The no-fly zone was erected on 10 April 1991.
4. On why Iraq invaded Kuwait, see Fred Halliday, 'The Gulf War and its Aftermath: First Reflections', *International Affairs* 67, 2 (1991), 223–234; and Efraim Karsh, 'Survival at All Costs: Saddam Hussein as Crisis Manager', in *The Gulf Crisis and its Global Aftermath*, ed. Gad Barzilai et al., (London: Routledge, 1993), pp. 51–66.
5. By taking over Kuwait's oil fields, Iraq would have controlled more than a fifth of the world's oil production. See John Pimlott, 'The Gulf Crisis and World Politics', in *The Gulf War Assessed*, ed. John Pimlott and Stephen Badsey (London: Arms and Armour Press, 1992), p. 40.
6. The bombing of Iraq began on the night of 16–17 January 1991.
7. Ahmed Hashim, 'Iraq, the Pariah State', *Current History* 91, 561 (1992), 12. This fundamentalist takeover made the Americans more wary of the insurrection as a whole. See Michael Sterner, 'Closing the Gate: the Persian Gulf War Revisited', *Current History* 96, 606 (1997), 18.
8. On Kurdish history, see David McDowall, *A Modern History of the Kurds* (London: I. B. Tauris, 1997); Gerard Chaliand, ed., *A People Without a Country: the Kurds of Kurdistan*, trans. Michael Pallis (London: Zed Books, 1993); and Michael M. Gunter, *The Kurds of Iraq: Tragedy and Hope* (New York: St. Martin's Press, 1992).

9. Towards the end of the Iran–Iraq war, repression of the Kurds worsened as a form of punishment for their assistance to the Iranian side. In what is referred to as the *Anfal* campaign, the Governor of Northern Iraq conducted a campaign of systematic elimination of the Kurds, culminating in March 1988 when Iraq launched chemical attacks, killing approximately 150,000 to 200,000 Kurds and causing thousands more to flee to Turkey. See McDowal, *A Modern History of the Kurds*, pp. 357–361.
10. Elizabeth N. Offen, 'Migrants and Refugees: the Human Toll', in *The Gulf Crisis and Its Global Aftermath*, ed. Barzilai, pp. 117–118. UNHCR's numbers at the end of April 1991 were 400,000 amassed on the Iraqi side of Turkish border and another 400,000 just inside Turkey. See David Keen, *The Kurds in Iraq: How Safe is Their Haven Now?* (London: Save the Children Fund, 1993), p. 7.
11. Helena Cook, *The Safe Haven in Northern Iraq: International Responsibility for Iraqi Kurdistan* (University of Essex: Human Rights Centre & London: Kurdistan Human Rights Project, 1995), p. 36.
12. UN Doc. S/22435, 3 April 1991.
13. See Lawrence Freedman and David Boren, '"Save havens" for Kurds in Post-War Iraq', in *To Loose the Bands of Wickedness: International Intervention in Defence of Human Rights*, ed. Nigel S. Rodley (London: Brassey's, 1992), p. 52.
14. See 'Excerpts from Bush's News Conference: Relief Camps for Kurds in Iraq', *New York Times*, 17 April 1991, p. 12.
15. Freedman and Boren, '"Safe Havens"' for Kurds', p. 59. Colin Powell explained, 'One Sunday afternoon,...we sketched out a "security zone", a sector around Kurdish cities in Iraq that Saddam's troops would not be allowed to enter', in *My American Journey*, with Joseph E. Persico (New York: Random House, 1995), p. 808.
16. Jones, 'Operation Provide Comfort', 107.
17. Over 20,000 military personnel from 13 different nations, though more than half were American, helped implement the safe haven; and thirty states contributed material and financial assistance to *Operation Provide Comfort*. See Michael E. Harrington, 'Operation Provide Comfort: A Perspective in International Law', *Connecticut Journal of International Law* 8, 2 (1993), 650–653.
18. See Iraq's letter to the SG, UN Doc. S/22513, 21 April 1991.
19. UN Doc. S/22513, 22 April 1991.
20. Iraq consented to the presence of no more than 500 UN Guards. See *Annex to the Memorandum of Understanding*, UN Doc. S/22663, 31 May 1991.
21. Turkey provided asylum to 150,000 Kurdish refugees in 1988. See Lawrence Freedman and David Boren, '"Safe havens" for Kurds in post-war Iraq', 49.
22. See Mahmut Bali Aykan, 'Turkey's Policy in Northern Iraq, 1991–95', *Middle Eastern Studies* 32, 4 (1996), 346–347.
23. James Baker said that 'we did, however, hope and believe that Saddam Hussein would not survive in power after such a crushing defeat', in *The Politics of Diplomacy, Revolution, War, and Peace, 1989–92,* with Thomas M. DeFrank (New York: G. P. Putnam's Sons, 1995), p. 435. See also p. 437. Powell confirmed: 'What we hoped for, frankly, in a postwar Gulf region

was an Iraq still standing, with Saddam overthrown', in *My American Journey*, p. 745.

24. Baker, *Politics of Diplomacy*, p. 435
25. Ibid., p. 437. Powell also said, 'But our practical intention was to leave Baghdad enough power to survive as a threat to an Iran that remained bitterly hostile toward the United States', in *My American Journey*, p. 808.
26. These negotiations led to SC Res. 687, 2 April 1991. Bush also feared more generally that pursuing goals to which the coalition had not agreed would fracture it and prevent further cooperation on ceasefire terms. See Baker, *Politics of Diplomacy*, p. 437.
27. Bush said, 'I don't think that support would last if it were a long drawn-out conflagration. I think that support would erode, as it did in the Vietnam conflict', quoted in John Mueller, *Policy and Opinion in the Gulf War* (Chicago: The University of Chicago Press, 1994), p. 121. See also Baker, *Politics of Diplomacy*, p. 436.
28. Ibid., p. 439.
29. George Bush, quoted in Maureen Dowd, 'After the War: Bush Stands Firm on Military Policy in Iraq', *The New York Times*, 13 April 1991, p. 1.
30. Baker explained: 'From the start of the crisis, we had said repeatedly that the United States had no motives beyond forcing compliance with UN resolutions and ejecting Iraq from Kuwait. We had argued that we had no grand design for a substantial permanent military presence in the region. The simplest way to establish our credibility on this score with all parties was to remain true to our word and withdraw promptly from Iraq', in *Politics of Diplomacy*, p. 436.
31. According to testimony by General Schwarzkopf before the Senate Armed Services Committee, coalition armed forces suffered 466 dead and 350 wounded. Iraqi military casualties have been estimated 'between 25,000 and 50,000 dead and 100,000 wounded', although higher figures exist. In terms of the Iraqi civilian population, the dead may have reached as high as 250,000. See discussion in Adam Roberts, 'The Laws of War in the 1990–91 Gulf Conflict', *International Security* 18, 3 (Winter 1993/94), 158, 170.
32. See Baker, *Politics of Diplomacy*, p. 436.
33. China abstained on Security Council resolution 678, for example, but did not veto it.
34. Zimbabwe, Cuba and Yemen voted against the resolution; India and China abstained; and the UK, the US, the USSR, France, Côte D'Ivoire, Romania, Ecuador, Zaire, Austria and Belgium voted in favour of it.
35. See generally S/PV.2982, 5 April 1991, and interventions by India (p. 62) and Zimbabwe (p. 31) in particular.
36. Reference to 2(7) was inserted into 688 by France once it became clear that an earlier draft resolution not containing such a clause was not supported by nine Security Council members. See Simon Chesterman, *Just War or Just Peace? Humanitarian Intervention and International Law* (Oxford: Oxford University Press, 2001), p. 132.
37. Ibid. Statements in the deliberations of the Council also made clear that it was the repercussions of Iraq's repression – refugee flows – that justified the need for a resolution, and not the repression *per se*. See S/PV.2982, 5 April 1991.

38. See Nicholas Wood and Martin Fletcher, 'Coming to the Rescue of the Kurds', *The Times*, 18 April 1991, p. 2.
39. Even states that did not vote in favour of resolution 688 still alluded to Iraq's horrific conduct toward its civilian population. See comments by Yemen, in S/PV. 2982, 5 April 1991, p. 26.
40. See S/PV.2982, 5 April 1991.
41. Ibid., pp. 65–66.
42. Author interview with Sir David Hannay, UK Permanent Representative to the UN from 1990 to 1995, London, 13 June 2001.
43. George Bush, 'After the War; Excerpts from Bush's News Conference: Relief Camps for the Kurds in Iraq', *New York Times*, 17 April 1991, p. 12 (section A).
44. John Major, *House of Commons Parliamentary Debates (Hansard), 1990–1991,* Vol. 189, Col. 159, 16 April 1991. Douglas Hurd also said that the presence of foreign troops in Iraq was 'wholly consistent with Security Council resolution 688...' in Ibid., Vol. 189, Col. 423, 17 April 1991. The Foreign and Commonwealth Office (FCO) believed that the more appropriate legal justification for the safe haven was extreme humanitarian need, but nonetheless agreed that the operation was consistent with 688. Author interview with Sir Franklin Berman, Oxford, 6 June 2001; and with Sir David Hannay, London, 13 June 2001. See also testimony by a Legal Counsellor for the FCO before the House of Commons Foreign Affairs Committee on 2 December 1994, in *House of Commons Parliamentary Papers 1992–93*, H.C.235-iii, pp. 59, 85; and discussion in Chesterman, *Just War or Just Peace?*, p. 204.
45. UN Doc. S/22513, 22 April 1991.
46. Freedman and Boren, '"Safe Havens" for Kurds in post-war Iraq', p. 48.
47. Quoted in *International Herald Tribune*, 9 April 1991, p. 52. Baker said in his memoirs that 'the desperation and deprivation of the scene was literally unbearable', and that it made him 'determined to see us do all we could do to prevent this from becoming even more of a humanitarian catastrophe than it already was', in *Politics of Diplomacy*, pp. 432, 433.
48. Ibid., pp. 433, 434. According to Baker, he secured Bush's and Dick Cheney's support with this proactive approach. The US Senate also recognized a 'moral obligation to provide sustained humanitarian relief for Iraqi refugees', US Senate Resolution 99 Concerning the Protection of Refugees in Iraq, 11 April 1991, quoted in Chesterman, *Just war or Just Peace?*, p. 200
49. George Bush, 'Address to Joint Session of Congress, 6 March 1991', in *Encyclopedia of the Persian Gulf War,* ed. Mark Grossman (Santa Barbara: ABC-CLIO, 1995), p. 443. Margaret Thatcher, PM of Britain when Saddam Hussein invaded Kuwait, also echoed similar sentiments. See Dan Keohane, 'British Policy in the Conflict', in *International Perspectives on the Gulf Conflict, 1990–1991*, eds. Alex Danchev and Dan Keohane (London: The Macmillan Press, 1994), pp. 152.
50. Sir David Hannay, S/PV.2982, 5 April 1991, p. 63.
51. Douglas Hurd, *House of Commons Parliamentary Debates (Hansard) 1990–1991*, Vol. 189, Col. 26, 15 April 1991.
52. Bush, 'Remarks to the American Association for the Advancement of Science', quoted in *Financial Times*, 16 February 1991.

53. Powell said that 'President Bush's rhetoric urging the Iraqis to overthrow Saddam, however, may have given encouragement to the rebels', in *My American Journey*, p. 808.
54. Schwarzkopf agreed to this concession because these flights posed no threat to American troops. See Baker, *Politics of Diplomacy*, p. 439.
55. Ibid., p. 440.
56. See Ibid.
57. McElroy cites a number of studies which have shown a significant causal relationship between sustained public and elite preferences on foreign policy matters and the making of American foreign policy. See McElroy, *Morality and American Foreign Policy*, pp. 43–44.
58. 'Injustice pour les Kurdes', editorial in *Le Monde*, 2 April 1991, p. 1.
59. See 'Devoir d'ingérence', editorial in *Le Monde*, 7–8 April 1991, p. 1, which affirmed the right of humanitarian intervention in the internal affairs of states.
60. 'Desert Temptation', editorial, *Times*, 2 April 1991.
61. Beginning on 5 April, the *Times* began running editorials, OP-ED pieces and letters to the editor, all in favour of intervention. See editions on the 6, 9, 10 and 18 April.
62. See 'Want another war?', *The Economist*, 13 April 1991, p. 16.
63. See 'Sanctuary for the Kurds', *The Economist*, 20 April 1991, p. 14.
64. For example, see 'The Allies' New Duty: Refugees', editorial, *New York Times*, 5 April 1991, p. 24; A. M. Rosenthal, 'America at the Vistula', *New York Times*, 9 April 1991, p. 25; and Anthony Lewis, 'Abroad at Home: Politics and Decency', *New York Times*, 15 April 1991, p. 17.
65. See 'At Last, the Kurds Find Friends', editorial, *New York Times*, 18 April 1991, p. 24.
66. See *Le Monde*, 4 April 1991, p. 4.
67. Douglas Kellner, *The Persian Gulf TV War* (Boulder: Westview Press, 1992), p. 417
68. Ibid., p. 418.
69. US Senate Resolution 99 Concerning the Protection of Refugees in Iraq, 11 April 1991, quoted in Chesterman, *Just War or Just Peace?*, p. 200.
70. Bush for instance publicly called for a new world order 'freer from the threat of terror, stronger in the pursuit of justice, and more secure in the quest for peace', quoted in George Bush, 'Address Before a Joint Session of Congress on the Persian Gulf Crisis and the Federal Budget Deficit, 11 September 1990', in *Public Papers of the Presidents of the United States: George Bush 1990, Book II – July 1 to December 31, 1990* (Washington: United States Government Printing Office, 1991), p. 1219. See also deliberations in the Security Council, in S/PV.2982, 5 April 1991.
71. These words were voiced by the American Ambassador to the UN. S/PV.2982, 5 April 1991, p. 58.
72. See generally Rikka Kuusisto, 'Framing the Wars in the Gulf and in Bosnia: the Rhetorical Definitions of the Western Power Leaders in Action', *Journal of Peace Research* 35, 5 (1998), 603–620.
73. Barney Dickson, 'From Emperor to Policeman – Britain and the Gulf War', in *The Gulf War and the New World Order*, eds. Haim Bresheeth and Nira Yuval-Davis (London: Zed Books, 1991), p. 43.

74. Mario Bettati and Bernard Kouchner, *Le Devoir D'ingérence: Peut-On Les Laisser Mourir?* (Paris: Denoël, 1987).
75. Ian Johnstone, *Aftermath of the Gulf War: an Assessment of United Nations Action* (London: Lynne Rienner Publishers, 1994), p. 18. Author interview with Sir David Hannay, London, 13 June 2001.
76. The *non-refoulement* principle enshrined in this convention, which prohibits the return of refugees to a territory where their lives are threatened, is rendered moot when there are significant grounds for believing that the refugees in question comprise a security threat to the country of asylum. See Article 33 of 28 July 1951 Convention relating to the Status of Refugees. See also Guy S. Goodwin-Gill, *The Refugee in International Law*, 2nd ed. (Oxford: Clarendon Press, 1996), pp. 139–141; and Kemal Kirisci, '"Provide Comfort" and Turkey: Decision-Making for Refugee Assistance', *Low Intensity Conflict & Law Enforcement* 2, 2 (1993), 233.
77. Quoted in Kirisci, '"Provide Comfort" and Turkey', 231.
78. Philip Webster, 'Major Gives Aid to Kurds after MPs Urge Action', *Times*, 4 April 1991, p. 22.
79. Ilan Greilsammer, 'European Reactions to the Gulf Challenge', in *The Gulf Crisis and Its Global Aftermath*, ed. Barzilai et al., p. 215. On France's role in the Gulf crisis, see Jolyon Howorth, 'French Policy in the Conflict', in *International Perspectives on the Gulf Conflict*, eds. Danchev and Keohane, pp. 175–200.
80. See Friedemann Buettner and Martin Landgraf, 'The New World Order of Europe and the Gulf Crisis', in *The Gulf War and the New World Order: International Relations of the Middle East*, eds. Tareq Y. Ismael and Jacqueline S. Ismael (Gainsville: University Press of Florida, 1994), pp. 94–95.
81. Ibid., p. 94.
82. George Bush, 'Address to Joint Session of Congress, 6 March 1991', quoted in *Encyclopedia of the Persian Gulf War*, ed. Grossman, p. 442.
83. See Sterner, 'Closing the Gate', 15–16. See also Transcript of David Frost interview with General Norman Schwarzkopf, *Talking with David Frost*, 27 March 1991, in *Encyclopedia of the Persian Gulf War*, ed. Grossman, p. 447.
84. Sterner, 'Closing the Gate', 16.
85. Several commentaries in the *New York Times* questioned why a state as predominant as the US did not take military action against the gunships. See William Safire, 'Bush's Moral Crisis', 1 April 1991, p. 17; and Leslie H. Gelb, 'Foreign Affairs: White House Guilt?', 14 April 1991, p. 19.
86. Between January 1992 and July 1994, there were over one hundred incidents of violence perpetrated against the Kurds. See Cook, *The Safe Haven in Northern Iraq*, p. 48.
87. Between August 1991 and May 1992, for example, there were 9 Turkish incursions into Iraq. See Kemal Kirisci and Gareth M. Winrow, *The Kurdish Question and Turkey: An Example of a Trans-State Ethnic Conflict* (London: Frank Cass, 1997), p. 162.
88. 'The Kurds Still Need Protectors', editorial, *New York Times*, p. 16.
89. See Freedman and Boren, '"Safe Havens" for Kurds', pp. 70–71.
90. UN Doc. S/22663, 31 May 1991.
91. Keen, *How Safe is Their Haven Now?*, p. 18.

Chapter 3 Only So Far but No Further: Safe Areas in Bosnia

1. Reprinted from Ivo H. Daalder, *Getting to Dayton: The Making of America's Bosnia Policy* (Washington D.C.: Brookings Institution Press, 2000), p. 65.
2. Reprinted from Jan Willem Honig and Norbert Both, *Srebrenica: Record of a War Crime* (London: Penguin Books, 1996), p. xiv. Copyright © Jan Willem Honig and Norbert Both, 1996. Reproduced by permission of Penguin Books Ltd.
3. See Annex to UN Doc. S/25829, 24 May 1993.
4. See Noel Malcolm, *Bosnia: A Short History* (London: Papermac, 1996); Misha Glenny, *The Fall of Yugoslavia: the Third Balkan War*, 3rd ed. (New York: Penguin Books, 1996); Dusko Doder, 'Yugoslavia: New War, Old Hatreds', *Foreign Affairs* 91 (Summer 1993), 3–23; and Steven L. Burg and Paul S. Shoup, *The War in Bosnia-Herzegovina: Ethnic Conflict and International Intervention* (New York: M. E. Sharpe, 1999), pp. 15–61.
5. These figures are from the 1991 census and are quoted in Burg and Shoup, *War in Bosnia-Herzegovina*, p. 26.
6. The International Conference on the Former Yugoslavia described the problem thus: 'the population of Bosnia and Herzegovina is inextricably intermingled. Thus there appears to be no viable way to create three territorially distinct States based on ethnic or confessional principles', quoted in *Report of the Secretary-General pursuant to General Assembly resolution 53/35: The fall of Srebrenica*, UN Doc. A/54/549, 15 November 1999, par. 30.
7. SC Res. 743, 21 February 1992.
8. *Keesing's Record of World Events*, 38 (March 1992), p. 38832.
9. Burg and Shoup, *War in Bosnia-Herzegovina*, p. 117.
10. *Keesing's* 38 (March 1992), p. 38833.
11. See Burg and Shoup, *War in Bosnia-Herzegovina*, pp. 134–136.
12. 'Bosnia & Herzegovina: Yellow card', *The Economist*, 25 April 1992, p. 54.
13. See Honig and Both, *Srebrenica*, p. 73.
14. 'Europe: When will they call it peace?', *The Economist*, 1 August 1992, p. 29.
15. For more about the marked imbalance in weaponry between the Bosnian Muslims and Serbs, see Tadeusz Mazowiecki, UN Special Rapporteur on the Situation of Human Rights in the Territory of the Former Yugoslavia, *Report on the situation of human rights in the territory of the former Yugoslavia*, UN Doc. A/47/666, 17 November 1992, p. 7. On the links between the Bosnian Serb militia and Serbia, see Honig and Both, *Srebrenica*, pp. 74–75.
16. See generally David C. Gompert, 'The United States and Yugoslavia's Wars', in *The World and Yugoslavia's Wars*, ed. Richard H. Ullman (New York: Council on Foreign Relations, 1996), pp. 122–144; and Wayne Bert, *The Reluctant Superpower: United States' Policy in Bosnia, 1991–1995* (New York: St. Martin's Press, 1997), pp. 127–163.
17. SC Res. 757, 30 May 1992.
18. See SG report, S/23900, 12 May 1992, pars. 25–26.
19. SC Res. 758, 8 June 1992.
20. SC Res. 761, 29 June 1992.

21. SG Report, *Fall of Srebrenica*, par. 42.
22. The text of the Vance-Owen Peace Plan is annexed to the SG report, S/25479, 26 March 1993. See also Burg and Shoup, *War in Bosnia-Herzegovina*, pp. 215–262; and James Gow, *Triumph of the Lack of Will: International Diplomacy and the Yugoslav War* (London: Hurst & Company, 1997), pp. 223–253.
23. Its pre-war population was 9,000. David Rohde, *Endgame: The Betrayal and Fall of Srebrenica: Europe's Worst Massacre since World War II* (Boulder, CO: Westview Press, 1998), p. 45. See also See Burg and Shoup, *War in Bosnia-Herzegovina*, p. 140.
24. For an account of Morillon's involvement in the March–April 1993 Srebrenica crisis, see his interview for *Death of Yugoslavia,* BBC documentary, transcript, Liddell Hart Centre for Military Archives, ref: 3/56, pp. 28–44.
25. *Rapport D'information déposé par la Mission D'information Commune sur les Événements de Srebrenica. Tome I (Rapport et Annexes)* (Paris: Assemblée Nationale, 11 décembre 2001), p. 17.
26. Quoted in Rohde, *Endgame*, p. XV.
27. On April 12, an intense artillery attack killed 56 people in less than an hour. See Laura Silber and Allan Little, *The Death of Yugoslavia*, rev. ed. (London: Penguin Books, 1996), p. 269.
28. Tadeusz Mazowiecki wrote in a November 1992 report, 'the greater prevalence of ethnic cleansing in Serbian occupied territories is undoubtedly related to political objectives formulated and pursued by the Serbian nationalists, namely ensuring Serbian control over all territories inhabited by significant numbers of Serbs, as well as adjacent territories assimilated to them owing to logistic and military considerations'. See UN Doc. A/47/666, 17 November 1992, pp. 6–7.
29. Ethnic cleansing, as defined in a UN report prepared by Tadeusz Mazowiecki, is 'the elimination by the ethnic group exercising control over a given territory of members of other ethnic groups', – it is a strategy of consolidating and legitimising territorial control on the basis of ethnic homogeneity. See Ibid, p. 6.
30. Since Bosnia was technically a civil war, it was not clear whether Article 2(4) of the Charter was applicable. Nonetheless the repeated international condemnations of the Bosnian Serb seizure of territory though the use of force suggest that states considered this behavioural to be reprehensible and something about which the Security Council needed to be concerned.
31. Other Security Council resolutions condemning violations of international humanitarian law were 771 (13 August 1992), 779 (6 October 1992) and 780 (6 October 1992).
32. See S/PV. 3106, 13 August 1992; S/PV 3119, 6 October 1992; S/PV. 3134, 13 November 1992; S/PV. 3175, 22 February 1993; S/PV. 3200, 18 April 1993; S/PV 3201, 19 April 1993; and S/PV. 3203, 20 April 1993.
33. S/PV. 3134, 13 November 1992. Other states which decried the violations of international law taking place in Bosnia at this same meeting were China, France, Japan, Malaysia, India, Russia, the UK and the US.
34. See in particular comments by Madeleine Albright in S/PV. 3175, 22 February 1993. Several other states, including Ecuador (S/PV. 3106),

Venezuela (S/PV. 3119), Jordan (S.PV/3203) and Hungary (S/PV. 3175) drew parallels between the Bosnian crisis and Nazi brutality during World War II.

35. See comments by Austria in the Security Council on 13 August 1992 (S/PV. 3106).
36. See *Rapport D'information (Srebrenica), Tome I*, pp. 113, 123.
37. ICRC, Position Paper, 'The Establishment of Protected Zones for Endangered Civilians in Bosnia-Herzegovina', in *How does Law Protect in War?*, ed. Sassoli and Bouvier, pp. 1127–1128.
38. See for example UN Doc. E/CN.4/1993/50, 10 February 1993, p. 57.
39. S/PV. 3143, 13 November 1992.
40. Morillon deliberately used the media to convince the international public that something had to be done to help the people of Srebrenica. He granted several interviews while in Srebrenica via a network of short-wave radios. See his interview for *Death of Yugoslavia*, transcript, Liddell Hart Centre for Military Archives, ref: 3/56, pp. 37–38.
41. Several prominent intellectuals in France, including Bernard-Henri Lévy (*Le Monde*, 5 January 1993), André Glucksmann (*Figaro*, 29 June 1992), and Bernard Kouchner had been pressing the French government to intervene in Bosnia for humanitarian reasons. See generally Jolyon Howorth, 'The Debate in France over Military Intervention in Europe', in *Military Intervention in European Conflicts*, ed. Lawrence Freedman (Oxford: Blackwell Publishers, 1994), pp. 113–121.
42. See David C. Hendrickson, 'The Recovery of Internationalism', *Foreign Affairs* 73, 5 (1994), 26–28.
43. *New York Times* columnist Anthony Lewis likened acceptance of the VOPP to Munich. See *New York Times*, 7 January 1993, p. A23 and 8 January 1993, p. A25. Other influential political figures, such as President Jimmy Carter's National Security Adviser, Zbigniew Brzezinski, condemned the plan 'as appeasement'. See Burg and Shoup. *War in Bosnia-Herzegovina*, pp. 230, 247.
44. UNHCR head Ogata was concerned enough about this dilemma that she began to refer to the principle of the 'right to stay' in her public speeches, which she interpreted as the 'right to be allowed to remain in one's home in safety and dignity, and not to be forced out by ethnic cleansing'. See Bill Frelick, '"Preventive Protection" and the Right to Seek Asylum: A Preliminary Look at Bosnia and Croatia', *International Journal of Refugee Law* 4, 4 (1992), 447.
45. UN Doc. S/25519, 3 April 1993.
46. Author interview with Diego Arria, Permanent Representative of Venezuela to the UN from 1990–1993, New York, 28 September 1999. The other non-aligned members sitting on the Council were Cape Verde, Djibouti and Morocco: they also supported the Srebrenica safe area.
47. This deviation led Diego Arria to say that there were in fact two United Nations. See Ibid.
48. SC Res. 819, 16 April 1993.
49. Silber and Little, *Death of Yugoslavia*, p. 272. Many within the UN came to see the Srebrenica talks as a surrender. Diego Arria likened Srebrenica to a concentration camp with UN peacekeepers subservient to Serbian orders, in interview with author, New York, 28 September 1999.

José-Maria Mendiluce, Coordinator for UNHCR in former Yugoslavia, said the talks at Sarajevo airport 'succeeded in humiliating the UN [and] transforming a resolution of the Security Council into pure rubbish', in interview for *Death of Yugoslavia*, transcript, Liddell Hart Centre for Military Archives, ref: 3/52, pp. 10–11, 14.

50. The text of the agreement for the demilitarization of Srebrenica is annexed to SG report, S/25700, 30 April 1993.
51. SG report, S/25700, 30 April 1993, par. 16.
52. See James Gow, 'British Perspectives', in *International Perspectives on the Yugoslav Conflict*, ed. Alex Danchev and Thomas Halverson (London: Macmillan Press, 1996), pp. 89–90.
53. The 1993 *White Paper on Defence* justified Britain's extensive involvement in UN peacekeeping on the grounds that such participation was expected of a permanent member of the Security Council. See Philip Towle, 'The British Debate About Intervention in European Conflicts', in *Military Intervention in European Conflicts*, ed. Freedman, p. 95; and Alex Macleod, 'La Grande-Bretagne: faire face à ses responsabilités internationales', *Relations Internationales et Stratégiques* 19 (Automne 1995), 94.
54. See Hubert Védrine, *Les Mondes de François Mitterrand: À l'Élysée 1981–1995* (Paris: Fayard, 1996), p. 669. France had also issued a *Livre blanc sur la défense* in 1994, which reconfirmed its responsibilities at the international level with respect to peace and security, and emphasized the significance of being treated as an equal by the major powers. See Alex Macleod, 'La France: à la recherche du leadership international', *Relations Internationales et Stratégiques* 19 (Automne 1995), 76. France also believed initially that it would better preserve its ability to influence the Serbs if it did not support a too antagonistic use of force against them. See Pia Christina Wood, 'France and the Post Cold War Order: the Case of Yugoslavia', *European Security* 3, 1 (Spring 1994), 141.
55. Quoted in Védrine, *Les Mondes de François Mitterrand*, p. 642.
56. See Susan Woodward, *Balkan Tragedy: Chaos and Dissolution After the Cold War* (Washington D.C.: The Brookings Institution, 1995), pp. 6–20.
57. Author interview with Sir David Hannay, London, 13 June 2001. According to Hannay, the prime reason for the failure of the policy was that it was 'sustained for too long'.
58. Elizabeth Drew, *On the Edge: The Clinton Presidency* (New York: Simon & Schuster, 1994), p. 138.
59. See Powell, *My American Journey*, p. 877.
60. See generally Drew, *On the Edge*, 141–151; and Daalder, *Getting to Dayton*, pp. 12–14.
61. EU foreign ministers had unequivocally backed the VOPP on 1 February 1993.
62. See Warren Christopher, *In the Stream of History: Shaping Foreign Policy for a New Era* (Stanford: Stanford University Press, 1998), p. 344, note 1. See also Daalder, *Getting to Dayton*, p. 10; and Burg and Shoup, *War in Bosnia-Herzegovina*, p. 410.
63. Quoted in Drew, *On the Edge*, p. 151.
64. On the demise of the VOPP, see Silber and Little, *Death of Yugoslavia*, pp. 276–290; and James Gow, *Triumph of the Lack of Will*, pp. 246–248.

65. Delegates from France, Hungary, New Zealand, Pakistan, Russia and Venezuela travelled to Srebrenica on 23 April 1993. See UN Doc. S/25700, 30 April 1993.
66. See Annex to UN Doc. S/25829, 24 May 1993.
67. See Christopher, *In the Stream of History*, pp. 345–347.
68. For this same reason, the UN Secretary-General also opposed 'lift and strike'. See Boutros Boutros-Ghali, *Unvanquished: A U.S.-U.N. Saga* (New York: I. B. Tauris, 1999), p. 90.
69. See Douglas Hurd interview for *Death of Yugoslavia*, transcript, Liddell Hart Centre for Military Archives, ref: 3/31, p. 4.
70. UNPROFOR peacekeepers referred to 'lift and strike' as 'stay and pray'. See General Sir Michael Rose, *Fighting for Peace: Bosnia 1994* (London: The Harvill Press, 1998), p. 9.
71. Polling data from the US suggests that there was consistently relatively strong public support for multilateral intervention in Bosnia, but little public willingness for the US to take unilateral action. See Richard Sobel, 'Portraying American Public Opinion toward the Bosnia Crisis', *Harvard International Journal of Press Politics* 3, 2 (1998), 19–24.
72. Christopher, *In the Stream of History*, p. 346.
73. See Ibid., pp. 129–130.
74. Drew, *On the Edge*, p. 157; and Daalder, *Getting to Dayton*, p. 17.
75. Christopher, *In the Stream of History*, p. 347.
76. British Foreign Secretary Douglas Hurd believed that, compared to the British, the French 'were less concerned with military advice and practicalities', so tended to take action whenever there was a dramatic worsening of the situation. See his interview for *Death of Yugoslavia*, transcript, Liddell Hart Centre for Military Archives, ref: 3/31, p. 13.
77. *Rapport D'information (Srebrenica), Tome I*, p. 124.
78. The British also concurred that 'they couldn't go on having this very serious division over how to handle the war in the Balkans between the Americans and the Europeans', so 'cobbled a safe area policy together'. Author interview with Lord David Owen, Co-Chairman of the Steering Committee of the International Conference on the former Yugoslavia from 1992 to 1995, London, 17 July 2001.
79. Author interview with Stuart Seldowitcz, Political Advisor in the US Mission to the UN from 1991–1996, Washington D.C., 9 October 1999. According to Seldowitcz, the US remained 'very sceptical about safe areas after the Joint Action Plan – that was our concession to what we thought were European concessions on other points'. And author interview with Robert Hunter, US Ambassador to NATO, 1993–1998, Washington D.C., 12 October 1999: 'we formally accepted that deal (i.e. NATO's involvement in deterring attacks against the safe areas), but informally of course that's not what we wanted'.
80. David Owen, *Balkan Odyssey* (London: Indigo, 1996), p. 175.
81. Annex to UN Doc. S/25800, 19 May 1993.
82. 'Bosnia: Filibuster here, terror there', *The Economist*, 27 March 1993, p. 52.
83. The Slovenian government had initially proposed the creation of four safe haven zones in July 1992 to 'avert new flows of refugees and

displaced persons from Bosnia and Herzegovina'. See *Proposals Concerning the Measures for Voluntary Return Home of the Displaced Persons and Refugees from Bosnia and Herzegovina*, Ministry of Foreign Affairs, Republic of Slovenia, 29 July 1992, quoted and discussed in Bill Frelick, '"Preventive Protection"', 443.

84. UNHCR estimated that as of the start of 1993 there were 581,425 refugees from the former Yugoslavia scattered throughout Europe. Most EU countries made visas mandatory in November 1992, putting a halt to the flow. See Michael Dewar, 'Intervention in Bosnia – the case against', *The World Today* 49, 2 (February 1993), 32–33. For a complete table of year-end statistics 1991–95 on populations of concern to UNHCR in relation to the former Yugoslavia, see Amir Pasic and Thomas G. Weiss, 'The Politics of Rescue: Yugoslavia's Wars and the Humanitarian Impulse', *Ethics and International Affairs* 11 (1997), 120.
85. *Keesing's*, 38 (July 1992), p. 39012.
86. According to one senior UN official, the pressure of the non-aligned became 'juggernaut-like. There was no stopping them and they would not listen to reason'. Author interview with Alvaro de Soto, Senior Advisor to the UN Secretary-General from 1992–1994 and Assistant Secretary-General of the UN for Political Affairs from 1995–1999, New York, 4 October 1999.
87. See Judge Ad Hoc Lauterpacht's dissenting opinion in the case before the International Court of Justice, in which Bosnia charged Serbia with failing to prevent the crime of genocide: '[I]t is not to be contemplated that the Security Council would ever deliberately adopt a resolution clearly and deliberately flouting a rule of *jus cogens* or requiring a violation of human rights. But the possibility that a Security Council resolution might inadvertently or in an unforeseen manner lead to such a situation cannot be excluded...On this basis, the inability of Bosnia-Herzegovina sufficiently strongly to fight back against the Serbs and effectively to prevent the implementation of the Serbian policy of ethnic cleansing is at least in part directly attributable to the fact that Bosnia-Herzegovina's access to weapons and equipment has been severely limited by the embargo. Viewed in this light, the Security Council resolution can be seen as having in effect called on Members of the United Nations, albeit unknowingly and assuredly unwillingly, to become in some degree supporters of the genocidal activity of the Serbs and in this manner and to that extent to act contrary to a rule of *jus cogens*.' In *Bosnia v. Serbia II*, ICJ Reports 1993, at 436–41, quoted in Craig Scott et al., 'A Memorial For Bosnia: Framework of Legal Arguments Concerning the Lawfulness of the Maintenance of the UN Security Council's Arms Embargo on Bosnia and Herzegovina', *Michigan Journal of International Law* 16, 1 (Fall 1994), 14–15.
88. Clinton often raged about the arms embargo and its effect on the fighting capability of the Bosnian government. See Drew, *On the Edge*, p. 150.
89. The Europeans were dismayed to discover that, in the middle of the drafting process of 836, the Americans no longer supported a wording that would have called for more general disarmament of the safe areas.

Author interview with Lord David Owen, London, 17 July 2001. The US sided with the non-aligned mainly at the request of the Bosnian government which insisted its soldiers in the safe areas be able to keep their weapons: 'we were trying to define what was meant by safe areas in a pro-Bosnian way, so that we could show the Bosnians we hadn't sold them out'. Author interview with Stuart Seldowitcz, Washington D.C., 9 October 1999.

90. Author interview with Sir David Hannay, London, 13 June 2001.
91. Author interview with Diego Arria, New York, 28 September 1999. On the non-aligned view of the safe area concept, see Honig and Both, *Srebrenica*, p. 113.
92. Prior to the adoption of resolution 836, the Secretariat's fears were communicated to the Council in a non-paper, but states proceeded with the safe area concept anyway. Author interview with Boutros Boutros-Ghali, UN Secretary-General from 1991–1996, Paris, 18 June 1999; Diego Arria, New York, 28 September 1999; Shashi Tharoor, Special Assistant to the UN Under-Secretary-General for Peacekeeping Operations, 1989–1996, New York, 30 September 1999; and Alvaro de Soto, New York, 4 October 1999. Boutros-Ghali writes, in *Unvanquished*, that he believed 70,000 additional troops would have been necessary for effective implementation of the safe areas (p. 84).
93. SG report, S/25939, 14 June 1993.
94. Ibid.
95. Boutros-Ghali says he was reluctant to present a light option to the Council as it would 'be able to do nothing without the consent and cooperation of the warring parties', but was pressed to do so by the UK permanent representative (see *Unvanquished*, p. 86) and because he 'was at the service of the Security Council' (interview with author, Paris, 18 June 2001).
96. SG report, S/25939, 14 June 1993
97. Ibid.
98. The Netherlands Institute for War Documentation identified the 'trial and error character' of the international intervention in Bosnia – the 'tendency to "muddle through"' – as one of the main reasons why the safe areas policy was so problematic. In 'Epilogue', *Srebrenica: a 'safe' area – Reconstruction, background, consequences and analyses of the fall of a safe area*, 10 April 2002 (www.srebrenica.nl), part. 2.
99. This was the *HMS Invincible* package of September 1993 which proposed a union of three republics, one Croatian, one Bosnian and one Serbian. The country would be roughly split in three. This plan came close to fruition on a number of occasions, but eventually proved barren as the Bosnians would not exchange Srebrenica and Zepa for Serb-controlled territory around Sarajevo. See Burg and Shoup, *War in Bosnia-Herzegovina*, pp. 280–286.
100. This was finalized through the Washington agreement of 10 May 1994.
101. See SG report, *Fall of Srebrenica*, par. 155.
102. SG report, S/1994/291, 11 March 1994, par. 5.
103. See Burg and Shoup, *War in Bosnia-Herzegovina*, pp. 145–146.
104. It was never conclusively determined by UNPROFOR which side was responsible for the attack. The Americans and international public

opinion blamed the Serbs, whereas UNPROFOR was more sceptical, believing that the Bosnian army might have attacked its own citizens in an attempt to trigger foreign intervention. See David Binder, 'Anatomy of A Massacre', *Foreign Policy* 97 (Winter 1994), 70–78; Rose, *Fighting for Peace*, pp. 43–44; and SG report, *Fall of Srebrenica*, par. 119.

105. See generally Burg and Shoup, *War in Bosnia-Herzegovina*, pp. 146–151; and SG report, *Fall of Srebrenica*, pars. 131–145.
106. SG report, S/1994/555, 9 May 1994. By March 1994, still only 5,200 of 7,600 troops needed for the safe areas had been deployed.
107. NATO would indeed have to move to the next level of bombing – strategic air strikes designed to hit targets further afield from the fighting – since it could no longer risk close air support and the possibility that more of its planes would be shot down.
108. The dual key arrangement for the authorization of NATO air strikes meant that both the UN and NATO had to agree to air strikes before they could happen. Initially, the UN key was in the Secretary-General's hands, but he delegated his authority to his special envoy, Akashi.
109. See generally Burg and Shoup, *War in Bosnia-Herzegovina*, pp. 154–158; and SG report, *Fall of Srebrenica*, pars. 157–163.
110. SG report, *Fall of Srebrenica*, par. 159.
111. See generally Burg and Shoup, *War in Bosnia-Herzegovina*, pp. 167–168.
112. See generally Rohde, *Endgame*; Honig and Both, *Srebrenica*; SG report, *Fall of Srebrenica*, pars. 239–393; and Netherlands Institute for War Documentation, *Srebrenica: a 'safe' area*.
113. SG report, *Fall of Srebrenica*, par. 409.
114. Up to 20,000 people, mostly Muslim, perished in and around the safe areas. See Ibid., par. 3.
115. Robert Hunter confirms that a major concern of NATO states during the deliberations on air strikes related to the importance of ensuring that 'the use of force was tightly controlled' and that 'any use of force had to be proportionate and related in part to the kind of non-interference quality and we're just there for humanitarian purposes quality'. Interview with author, Washington D.C., 12 October 1999.
116. See SG report, *Fall of Srebrenica*, pars. 107–108, 110.
117. Ibid., pars. 111–113. Boutros-Ghali communicated to the council that air power capability was fully operative on 18 August 1993, UN Doc. S/26335, 20 August 1993.
118. This type of air power is what is commonly referred to as close air support.
119. SG report, *Fall of Srebrenica*, par. 109.
120. Annex to UN Doc. S/1994/131, 6 February 1994.
121. Britain considered that NATO's use of force in support of the UN mission would destroy UNPROFOR's impartiality. See Silber and Little, *Death of Yugoslavia*, p. 315; Rose, *Fighting for Peace*, p. 46; and SG report, *Fall of Srebrenica*, par. 118.
122. Several officials I interviewed, including Stuart Seldowitcz; Jim O'Brien, Legal Advisor to US State Department from 1993–1994 and Advisor to Madeleine Albright from 1994 onwards; Bill Woodward, Speechwriter for Madeleine Albright from 1993–1999; Sir David Hannay; and Lord David

Owen all concurred that the public outcry following each major Bosnian Serb violation of a safe area engendered a more concerted effort on the part of states to do something to deter further attacks against the safe areas.

123. Clinton asked the Pentagon to provide him with an assessment of the number of ground troops needed to lift the siege. Although Powell's figure of 70,000 was subsequently lowered to 25,000, Clinton still considered this number far beyond anything Congress would accept. See Drew, *On the Edge*, pp. 273–274.
124. See comments by New Zealand and Austria in the Security Council, S/PV. 3336, 14 February 1994. At the same meeting, France and Britain vehemently opposed any such interpretation of 836. Such persistent confusion in the Council led the Czech ambassador to 'realize that certain delegations found it impossible to make the distinction between the military view of what is safe and the political and psychological view of what it means to declare an area safe', S/PV. 3356, 31 March 1994.
125. See SG report, *Fall of Srebrenica*, pars. 162–163. Of the nine who opposed such a transformation of the mandate, three were permanent members of the Security Council, i.e. Britain, France and Russia.
126. See interventions by the non-aligned states in relevant Security Council debates, as well as remarks by Pakistan, Nigeria, Oman and Malaysia in S/PV. 3336, 14 February 1994.
127. See statement by Pakistan, S/PV. 3370, 27 April 1994.
128. The Assemblée Nationale inquiry into Srebrenica criticizes harshly the UN in this regard, condemning key figures for not being able to see beyond traditional peacekeeping practices. See *Rapport D'information (Srebrenica), Tome I*, p. 64.
129. See the following SG reports: S/1994/300, 16 March 1994, par. 32; S/1994/555, 9 May 1994, pars. 12–13; S/1994/1389, 1 December 1994, pars. 41, 54.
130. When the Bosnian Serbs acquired anti-aircraft capability in April 1994, pinprick bombing (close air support) became less feasible, as it was considered too dangerous for NATO pilots. Heavier air strikes encouraged Serb retaliation. Author interview with Bertrand de Lapresle, UN Force Commander in former Yugoslavia from March 1994–February 1995, Paris, 21 June 2001. See also Netherlands Institute for War Documentation, 'Epilogue', in *Srebrenica: a 'safe' area*, part. 7; and the following SG reports: S/1994/555, 9 May 1994, par. 10; S/1994/1389, 1 December 1994, pars. 30–31, 56; and S/1995/444, 30 May 1995, par. 58.
131. See SG reports: S/1994/300, 16 March 1994, par. 34; S/1994/555, 9 May 1994, pars. 14–15, 24; S/1994/1067, 17 September 1994, par. 43; S/1995/444, 30 May 1995, par. 62.
132. S/1994/1389, 1 December 1994, par. 57.
133. SG report, S/1995/444, 30 May 1995, par. 76.
134. Ibid., par. 78. See also Boutros-Ghali, *Unvanquished*, pp. 233, 236.
135. Shashi Tharoor said: 'we, having been schooled in peacekeeping and having as an institution already done peacekeeping, did not see ourselves as a body that could have actually gone and taken sides in a war and won it'. Interview with author, New York, 30 September 1999.

136. Author interview with Bertrand de Lapresle, Paris, 21 June 2001.
137. Transcript of 9 June meeting between Akashi, Janvier and Smith, in *Rapport D'information (Srebrenica), Tome I*, p. 299.
138. SG report, *Fall of Srebrenica*, par. 186.
139. Rose, *Fighting for Peace*, p. 46.
140. Ibid., p. 43.
141. Burg and Shoup, *War in Bosnia-Herzegovina*, pp. 150–151.
142. Rose, *Fighting for Peace*, p. 241.
143. See Ibid., p. 179.
144. Intervention of Bernard Janvier before the Security Council, in *Rapport D'information (Srebrenica), Tome I*, pp. 268–269.
145. Ibid, p. 299.
146. See Rhode, *Endgame*, pp. 359–362. In a directive of 2 July 1995, Janvier writes that UNPROFOR must avoid any action that might trigger the use of air power. See *Rapport D'information (Srebrenica), Tome I*, p. 67.
147. SG report, *Fall of Srebrenica*, par. 197; and Netherlands Institute for War Documentation, 'Epilogue', in *Srebrenica: a 'safe' area*, part. 6.
148. Testimony of Bernard Janvier, in *Rapport D'information (Srebrenica), Tome II (Auditions)*. Boutros-Ghali also denies any such deal was reached, in *Unvanquished*, p. 237.
149. Robert Hunter said that the British were almost always the last to concede in the NAC deliberations on air strikes. Interview with author, Washington D.C., 12 October 1999.
150. The war in Bosnia was 'framed' in different ways, depending on whether or not more forceful action was sought at the time. Comparing the Serbs to the Nazis, for example, encouraged tougher action against them; whereas equating Bosnia to a Vietnam- or WWI-like morass discouraged involvement. See generally Kuusisto, 'Framing the Wars in the Gulf and in Bosnia', 603–620; and K. M. Fierke, 'Multiple Identities, Interfacing Games: The Social Construction of Western Action in Bosnia', *European Journal of International Relations* 2, 4 (1996), 467–477.
151. See quotation by Vitaly Churkin, Russian envoy to the Balkans, in Silber and Little, *Death of Yugoslavia*, p. 328. On the Russian dilemma more generally, see Paul A. Goble, 'Dangerous Liaisons: Moscow, the Former Yugoslavia, and the West', in *The World and Yugoslavia's Wars*, ed. Ullman, pp. 182–197.
152. Andrei Kozyrev, 'The Lagging Partnership', *Foreign Affairs* 73, 3 (1994), 66.
153. It took until June 1994 for the full 7,600 troops authorized under the 'light option' to reach Bosnia. SG report, S/1994/1389, 1 December 1994, par. 2.
154. See angry correspondence on this issue by Cot, the UN Secretariat and the Swedish government responsible for the battalion, in *Rapport D'information (Srebrenica), Tome I*, pp. 243–248.
155. Interview with author, Paris, 21 June 2001.
156. Statement made after the meeting, quoted in SG report, *Fall of Srebrenica*, par. 406.
157. Quoted in Ibid., par. 412.
158. Richard Holbrooke, *To End a War* (New York: Random House, 1998), p. 104.
159. The *Washington Post* and the *New York Times* published 70 reports on Bosnia between them from 11–18 July 1995; and the CBS *Evening News* ran the fall of Srebrenica as its headline story from 11–14 July. See Piers

Robinson, 'The Policy-Media Interaction Model: Measuring Media Power During Humanitarian Crisis', *Journal of Peace Research* 37, 5 (2000), 619.

160. Bob Woodward, *The Choice* (New York: Simon & Schuster, 1996), p. 259. The moralistic tone and principled content of Chirac's message to Clinton was confirmed by his advisor, Jean-David Levitte. See *Rapport D'information (Srebrenica), Tome I*, p. 107.
161. Woodward, *The Choice*, p. 260.
162. 12 July 1995 French Ministry of Defence Note on possible options for Bosnia, in *Rapport D'information (Srebrenica), Tome I*, pp. 372–375.
163. See French PM Alain Juppé's comments, in Ibid., p. 111. See also p. 112; and author interview with Laurent Klein, Chef du Secrétariat de la Commission des Affaires Étrangères, Assemblée Nationale and person in charge of drafting the *Rapport D'information (Srebrenica)*, Paris, 15 June 2001.
164. Woodward, *The Choice*, p. 260.
165. Holbrooke, *To End a War*, p. 75.
166. SG report, *Fall of Srebrenica*, par. 413.
167. Transcript of 9 June meeting between Akashi, Janvier and Smith, in *Rapport D'information (Srebrenica), Tome I*, p. 299.
168. Ibid., p. 301.
169. A lot of speculation has arisen regarding Rupert Smith's behaviour since he seemed to act without the clear consent of his UN military or civilian superiors. He is not listed as having been interviewed for the Secretary-General's report into the fall of Srebrenica, a strange omission, and he declined to talk to the French National Assembly inquiry, saying he could not get permission to do so from the UK Ministry of Defence. Author interview with Laurent Klein, Paris, 15 June 2001. It also seems fortuitous that Smith happened to be holding the UN key in Janvier's absence just when the Serbs decided to shell Sarajevo, but no evidence exists to suggest that this was anything other than an authentic Serb provocation. See SG report, *Fall of Srebrenica*, pars. 438–439.
170. Ibid., par. 440.
171. Ibid., par. 450.
172. Ibid., pars. 453–454.
173. Ibid., par. 454.
174. It seems likely the US turned a blind eye from summer 1994 onwards to third party shipments of arms to the Bosnian government, and began covertly arming the Bosnians itself in 1995. Through a private military contractor, *Military Professional Resources, Inc.*, the US apparently sent a former US Army chief of staff and a former commander of US forces in Europe, among other retired officers, to assist Croatia and later Bosnia with military planning and combat preparedness. See Burg and Shoup, *War in Bosnia-Herzegovina*, pp. 308–309, 313, 339–340; and Netherlands Institute for War Documentation, 'Epilogue', in *Srebrenica: a 'safe' area*, part 8.
175. No evidence suggests that a conspiracy was afoot – that governments purposefully let Srebrenica fall so that they could resolve the Bosnian conflict. See SG report, *Fall of Srebrenica*, par. 485; and *Rapport D'information (Srebrenica), Tome I*, p. 114.
176. Albright had been lobbying for a more aggressive policy in Bosnia for several months prior to Lake's decision to forge an endgame strategy.

Author interview with Jim O'Brien, Washington D.C., 10 October 1999; and Bill Woodward, Washington D.C., 9 October 1999.

177. Woodward, *The Choice*, p. 253.
178. See Daalder, *Getting to Dayton*, pp. 85–116. Unity was possible to achieve after Srebrenica because 'basically people had got to the point where advocates of a stronger reaction won...Other fronts had just run out of arguments, like we can wait'. Author interview with Bill Woodward, Washington D.C., 9 October 1999.
179. The US informed Boutros-Ghali of this decision on 8 December 1994 in a *démarche* stating that 'the US would participate with ground troops in a NATO-led withdrawal of allied forces in Bosnia', quoted in Boutros-Ghali, *Unvanquished*, p. 215. See also Burg and Shoup, *War in Bosnia-Herzegovina*, p. 323.
180. Daalder, *Getting to Dayton*, pp. 56–57.
181. 'Radio Address by the President to the Nation' (White House, Office of the Press Secretary, June 3, 1995), in Ibid., pp. 54–55.
182. Withdrawal was not the best option for Britain and France either: it would be an admission of defeat and might generate casualties in the context of failure.
183. The timing of the vote in the Senate was linked to the Srebrenica tragedy because members who had previously been sitting on the fence on the issue of the arms embargo suddenly felt that this was the 'last straw', revealing 'the total inefficacy of Clinton's policy'. Author interview with Mira Baratta, Legislative Assistant to Senator Bob Dole for Arms Control and Foreign Policy from 1989–1996, New York, 4 October 1999.
184. See Woodward, *The Choice*, p. 265.
185. See SG report, *Fall of Srebrenica*, par. 192.
186. Thierry Tardy, 'Le président Chirac et la Bosnie-Herzégovine: les limites d'une politique', *Relations Internationales et Stratégiques* 25 (Printemps 1997), 145–146.
187. The British also came to adopt this view of the 'need to show determination...Here was a clear attempt at intimidation, and either you reacted by talking all the time about withdrawal, or you show determination'. See Douglas Hurd's interview for *Death of Yugoslavia*, transcript, Liddell Hart Centre for Military Archives, ref: 3/31, p. 24.
188. Burg and Shoup, *War in Bosnia-Herzegovina*, p. 357. According to Boutros-Ghali, the Russians were so alarmed by the prospect that the bombing campaign might lead to a Bosnian Serb military defeat that they requested the Secretary-General to retrieve the key he had delegated to Janvier. See *Unvanquished*, p. 245.
189. Srebrenica was the one exception, although its disarmament was only partially completed.

Chapter 4 A Decision Not to Act: Proposed UN Secure Humanitarian Areas in Rwanda

1. Map No. 3717, Rev. 9, January 2004, United Nations Cartographic Section.
2. SC Res. 912, 21 April 1994.

3. On the origins of the Rwandan genocide, see Gérard Prunier, *The Rwanda Crisis 1959–1994: History of a Genocide* (London: Hurst & Company, 1995); Linda Melvern, *A People Betrayed: The Role of the West in Rwanda's Genocide* (London: Zed Books, 2000); Alison Des Forges, *Leave None to Tell the Story: Genocide in Rwanda* (New York: Human Rights Watch and Paris: Fédération Internationale des Ligues des Droits de L'Homme, 1999); and Howard Adelman and Astri Suhrke, eds, *The Path of a Genocide: The Rwanda Crisis from Uganda to Zaire* (New Brunswick, New Jersey: Transaction Publishers, 1999).
4. Gérard Prunier, *La Crise Rwandaise: Structures et Déroulement* (WRITENET, July 1994), p. 2.
5. Howard Adelman and Astri Suhrke, 'Early Warning and Response: Why the International Community Failed to Prevent the Genocide', *Disasters: The Journal of Disaster Studies and Management* 20 (1996), 296.
6. Ibid.
7. Prunier, *La Crise Rwandaise*, p. 3.
8. *Who Is Killing; Who Is Dying; What Is to Be Done* (London: African Rights, May 1994), p. 8.
9. Howard Adelman and Astri Suhrke, *The International Response to Conflict and Genocide: Lessons from the Rwandan Experience. Study II: Early Warning and Conflict Management* (Copenhagen: DANIDA Joint Evaluation of Emergency Assistance to Rwanda, 1996), p. 18.
10. The RPF included some dissident Hutus amongst its ranks. For more on the RPF and its Ugandan origins, see Gérard Prunier, 'The Rwandan Patriotic Front', in *African Guerrillas,* ed. Christopher Clapham (Oxford: James Currey, 1998), pp. 119–133; and Cyrus Reed, 'Exile, Reform, and the Rise of the Rwandan Patriotic Front', *The Journal of Modern African Studies* 34, 3 (1996), 479–501
11. The agenda called for the creation of a genuine democracy in Rwanda in which all Rwandans could feel secure and be equally integrated. See Patrick Mazimhaka, 'Background and the Current Situation', in *Genocide in Rwanda: Background and Current Situation,* ed. Napoleon Abdulai (London: Africa Research & Information Centre, 1994), p. 28.
12. Prunier, 'Rwandan Patriotic Front', pp. 128–129. Many of the Tutsis within the RPF had never set foot in Rwanda, or had left the country as children during the massacres of the late 1950s and 1960s.
13. See Prunier, 'Rwandan Patriotic Front', pp. 130–132; and Reed, 'Exile, Reform', 487–497.
14. See generally Christopher Clapham, 'Rwanda: The Perils of Peacemaking', *Journal of Peace Research* 35, 2 (1998), 193–210.
15. Adelman and Suhrke, *Study II*, p. 28.
16. For a copy of the Arusha Agreement, see UN. Doc. A/48/824–S/26915, 23 December 1993.
17. Adelman and Suhrke, *Study II*, p. 25.
18. Ibid.
19. Adelman and Suhrke, *Study II,* p. 26. For further commentary on the problematic nature of the Arusha Peace Agreement and process, see Clapham, 'Perils of Peacemaking'; Alan J. Kuperman, 'The Other Lesson of Rwanda: Mediators Sometimes Do More Damage than Good', *SAIS Review* 16, 1 (1996), 221–240.

20. Adelman and Suhrke, *Study II*, p. 26.
21. Ibid, p. 25; and Clapham, 'Perils of Peacemaking', 203.
22. Prunier, *La Crise Rwandaise*, p. 6.
23. The Arusha Peace Agreement would have deprived the *Akazu*, the extremist clan around Habyarimana which controlled the Presidential Guard and the FAR, of the Presidential favouritism it enjoyed, and reduced their influence over the army. Clapham, 'Perils of Peacemaking', 201–03.
24. Prunier, *History of a Genocide*, p. 10.
25. Adelman and Suhrke, *Study II*, p. 35.
26. For an analysis of how the extremists became convinced that genocide represented the most logical solution to their problem, see Prunier, *History of a Genocide*, p. 226.
27. Linda Melvern has done extensive research on who masterminded and directed the genocide. See generally Linda Melvern, *Conspiracy to Murder: The Rwandan Genocide* (London: Verso, 2004).
28. These had been set up by Habyarimana during the 1980s as a means to further the development of Rwanda.
29. Adelman and Suhrke, 'Early Warning', p. 298.
30. See Article 52 of the Arusha Peace Agreement (annexed to UN Doc. A/48/824–S/26915, 23 December 1993).
31. Adelman and Suhrke, *Study II*, p. 39. See also Conclusion 2 of the *Report of the Independent Inquiry into the Actions of the UN into the Actions of the United Nations During the 1994 Genocide in Rwanda*, 15 December 1999 (www.un.org/News/ossg/rwanda_report.htm).
32. For an assessment of the overall weakness of UNAMIR and its effect on the genocide, see Conclusions 2 and 3 of the *Report of the Independent Inquiry into the Actions of the UN*.
33. This information is contained in a cable UNAMIR Force Commander Roméo Dallaire sent to the Department of Peacekeeping Operations (DPKO) on 11 January 1994. See Roméo Dallaire, *Shake Hands with the Devil: the Failure of Humanity in Rwanda*, with Brent Beardsley (Toronto: Random House, 2003), pp. 141–144.
34. For a detailed account of how the UN responded to this informant, see the section untitled 'The 11 January Cable' in *Report of the Independent Inquiry into the Actions of the UN* and Conclusion 3.
35. Boutros Boutros-Ghali, 'Introduction', in *The United Nations and Rwanda, 1993–1996*, The United Nations Blue Book Series, Vol. X (New York: United Nations Department of Public Information, 1996), pp. 31–32.
36. Adelman and Suhrke, 'Early Warning', 299. Justifying UN caution, Boutros-Ghali explained that alarming reports from the field are not uncommon during peacekeeping operations. See 'Introduction' in *The United Nations and Rwanda, 1993–1996*, p. 31. The *Report of the Independent Inquiry into the Actions of the UN* also criticizes the UN's lack of contingency planning despite countless warning signals that the Arusha process was breaking down. See Conclusion 9.
37. Roméo A. Dallaire, 'The Changing Role of United Nations Peacekeeping Forces: The Relationship between UN Peacekeepers and NGOs in Rwanda', in *After Rwanda: The Coordination of United Nations Humanitarian Assistance,*

eds. Jim Whitman and David Pocock (London: Macmillan Press, 1996), p. 208.

38. Adelman and Suhrke, *Study II*, pp. 36–37. See also Roméo A. Dallaire, 'The End of Innocence: Rwanda 1994', in *Hard Choices*, ed. Moore, p. 76.
39. Adelman and Suhrke, *Study II*, p. 40.
40. It is still not known who shot down Habyarimana's plane. Some believe that Habyarimana was assassinated by desperate members of his Akazu circle who considered him a traitor for signing the Arusha Peace Agreement and needed a plausible explanation for the violence that was to engulf Rwanda following the downing of the plane. See Prunier, *History of a Genocide*. In contrast, others believe the RPF shot down the plane, since it deemed a military victory the only way to transform Rwandan politics. See Filip Reyntjens, *Rwanda. Trois Jours qui ont fait basculer l'histoire* (Paris: Éditions l'Harmattan, 1995).
41. Melvern, *A People Betrayed*, p. 171; and Nicholas J. Wheeler, *Saving Strangers: Humanitarian Intervention in International Society* (Oxford: Oxford University Press, 2000), p. 223.
42. John Borton, Emery Brusset and Alistair Hallam, *The International Response to Conflict and Genocide: Lessons from the Rwandan Experience. Study III: Humanitarian Aid and Effects* (Copenhagen: DANIDA Joint Evaluation of Emergency Assistance to Rwanda, 1996), p. 5. The Rwandan government has revised this estimate to over 1 million killed in the genocide.
43. Larry Minear and Philippe Guillot, *Soldiers to the Rescue: Humanitarian Lessons from Rwanda* (Paris: Development Centre of the Organization for Economic Cooperation and Development, 1995), p. 58.
44. Prunier, *History of a Genocide*, pp. 234–235.
45. See UN Doc. S/1994/446, 15 April 1994.
46. SG report, S/1994/470, 20 April 1994.
47. SC Res. 918, 17 May 1994,
48. James Woods, interview for *The Triumph of Evil*, *Frontline* PBS Documentary, transcript, 26 January 1999 (www.pbs.org/wgbh/pages/frontline/shows/evil/).
49. Jean-Claude Willame, *L'ONU au Rwanda (1993–1995): La 'Communauté Internationale' à L'Épreuve d'un Génocide* (Paris: Éditions Labour, 1996), p. 35.
50. Author interviews with Alain Destexhe and Philip Mahoux, Belgian Senators and members of the Belgian Parliamentary Commission of Inquiry into Rwanda, Brussels, 20 June 2001. See *Rapport fait au nom de la Commission D'enquête Parlementaire Concernant les Événements du Rwanda* (Brussels: Sénat de Belgique, 6 Décember 1997), pp. 563–566.
51. Author interview with Boutros-Ghali, Paris, 18 June 2001; and in *Unvanquished*, p. 132.
52. A US led UN attempt to capture Somali warlord Mohammed Farah Aidid in August 1993 led to 18 American rangers being killed and 84 being wounded. The bodies of US servicemen were dragged through the streets of Mogadishu. See Matthew Bryden, 'Somalia: The Wages of Failure', *Current History* 94 (1995), 145–151; and Walter Clarke and Jeffrey Herbst, eds, *Learning from Somalia: The Lessons of Armed Humanitarian Intervention* (New York: Westview Press, 1997).

53. Author interview with Anthony Lake, President Clinton's National Security Advisor, 1993–1997, Washington D.C., 13 October 1999. See also Bill Clinton, *My Life* (New York: A. Knopf, 2004), p. 593.
54. Arthur Jay Klinghoffer, *The International Dimension of Genocide in Rwanda* (Chippenham, Wiltshire: Antony Rowe, 1998), pp. 95–96.
55. *Rwanda: The Preventable Genocide, Report of the International Panel of Eminent Personalities to Investigate the 1994 Genocide in Rwanda and the Surrounding Events* (Organization of African Unity, 29 May 2000), chap. 12, par. 37.
56. See interventions by the US on 12 and 13 April, in a leaked hand-written document containing a verbatim account of the informal sessions of the Security Council from 4 April to 17 May, in possession of Linda Melvern and consulted in London on 2 July 2001.
57. Article I, 1948 Genocide Convention, in *Documents on the Laws of War*, eds. Roberts and Guelff, p. 181.
58. See discussion in Chapter 1; and author interview with Sir Franklin Berman, Oxford, 6 June 2001.
59. *The Independent Inquiry into the Actions of the United Nations* describes the 'obligation' thus: 'The parties to the 1948 Convention took upon themselves a responsibility to prevent and punish the crime of genocide. This is not a responsibility to be taken lightly. Although the main action required of the parties to the Convention is to enact national legislation to provide for jurisdiction against genocide, the Convention also explicitly opens the opportunity of bringing a situation to the Security Council. Arguably, in this context, the members of the Security Council have a particular responsibility, morally if not explicitly under the Convention, to react against a situation of genocide. See Conclusion 5b.
60. Leaked hand-written document containing a verbatim account of the informal sessions of the Security Council from 4 April to 17 May, in possession of Linda Melvern, consulted in London on 2 July 2001; and author interview with Andrew Gilmour, Political Affairs Officer and Note-taker for the UN Secretary-General in the Security Council, New York, 28 September 1999.
61. Author interview with Boutros Boutros-Ghali, Paris, 18 June 2001. Tony Marley, Political Military Advisor for the US State Department from 1992–1995, interview for *Triumph of Evil*, transcript, 26 January 1999.
62. See Conclusion 6 of the *Report of the Independent Inquiry into the Actions of the United Nations*.
63. Melvern, *A People Betrayed*, p. 77.
64. *The Clinton Administration's Policy on Reforming Multilateral Peace Operations*, Department of State Publication 10161, May 1994, p. 1.
65. Ibid., p. 4.
66. Ibid., p. 5. See also Holly J. Burkhalter, 'The Question of Genocide: The Clinton Administration and Rwanda', *World Policy Journal* 11 (1994–1995), 48–49.
67. Nicholas Wheeler argues that it would have been possible to make the case for an intervention in Rwanda using the strict criteria of PDD 25, but leadership and public persuasion would have been necessary, something Clinton would not risk in the wake of Somalia. See *Saving Strangers*, p. 224.

68. Adelman and Suhrke, *Study II*, p. 51.
69. Ibid.
70. Karel Kovanda, Permanent Representative of the Czech Republic to the UN from 1993–1997, said he knew it was genocide within a week or two of 6 April. Interview with author, Brussels, 19 June 2001. Keating received a letter from the President of Human Rights Watch, Kenneth Roth, on 19 April, stating that this organization believed that what was going on in Rwanda was genocide and urging the Council to act. See Melvern, *A People Betrayed*, p. 169.
71. UN Doc. S/PRST/1994/21, 30 April 1994.
72. James Wood, Deputy Assistant Secretary for African Affairs at the Department of Defence from 1986–1994, believes the Pentagon figured out it was genocide 'within 10 to 14 days' after the first week of conflicting reports. Interview for *Triumph of Evil*, transcript, 26 January 1999. A senior state department official says that although everyone knew it was 'something bad' from the moment Habyarimana's plane went down, it 'was a matter of at least a couple of weeks' before it became clear that it was a 'massive, organized, carefully orchestrated genocide'. Interview with author, Washington D.C., 12 October 1999. Albright states it was 'weeks before most of us understood the nature and the scale of the violence', in Madeleine Albright, *Madame Secretary. A Memoir* (New York: Miramax Books, 2003), p. 148. According to the *Rapport fait au nom de la Commission D'enquête Parlementaire*, the Belgian government was in possession by January 1994 of an important amount of information suggesting that a genocide, or at least massacres on a grand scale, was being plotted in Rwanda, p. 506. Both Dallaire and Melvern believe that states such as Belgium, France, the US and Germany had sufficient resources on the ground prior to April 1994 to know exactly what was being planned. See respectively *Shake Hands*, p. 90 and *Conspiracy*, p. 119.
73. *Who is Killing*, p. 36.
74. See letter that Rwanda sent to the Security Council, UN Doc. S/1994/428, 13 April 1994.
75. See Rwanda's statement in the deliberations of the Security Council, S/PV.3368, 21 April 1994, p. 5.
76. Author interview with Karel Kovanda, Brussels, 19 June 2001. The *Report of the Independent Inquiry into the Actions of the UN* also states that 'the Rwandan presence hampered the quality of the information that the Secretariat felt it possible to provide to the Council and the nature of the discussion in that body'. See Conclusion 18.
77. Adelman and Suhrke, *Study II*, p. 70.
78. The Secretary-General was out of New York during most of the time of the genocide, and he often relied on indirect contact with the Council via his senior political advisors. The *Report of the Independent Inquiry into the Actions of the UN* faults this distancing, saying that the role of the Secretary-General can only to a limited extent be performed by proxy and that Boutros-Ghali's absence reduced his influence upon the Council. See conclusions 14 and 15. In *Unvanquished*, Boutros-Ghali claims he did not find out about the 11 January cable from Dallaire until three years after the genocide, p. 130.

79. Boutros-Ghali explained his initial misperception: 'Genocide is what happened in Germany. We never imagined that you could have genocide with machetes...It needs a certain kind of imagination to see that this could happen with light arms...' Interview with author, Paris, 18 June 2001.
80. Boutros-Ghali was instrumental in helping Rwanda secure a desperately needed arms deal with Egypt in October 1990. See Melvern, *A People Betrayed*, pp. 30–33.
81. Considerable controversy surrounds this letter. The copy Colin Keating gave Linda Melvern states: 'it is quite possible that the evacuation of UNAMIR and other UN personnel might become unavoidable...'; whereas the copy the Secretary-General had printed in *The United Nations and Rwanda, 1993–1994* (p. 255) says: 'it is quite possible that the evacuation of civilian staff from the United Nations system, as well as other foreign nationals, might become unavoidable...' See Melvern, *A People Betrayed*, p. 139 and p. 149, footnote 9.
82. Dallaire refused on each occasion. See Melvern, *A People Betrayed*, p. 146.
83. Iqbal Riza, Assistant Secretary-General in the UN Department for Peacekeeping Operations from 1993–1996, interview for *Triumph of Evil*, transcript, 26 January 1999.
84. This was the initial military assessment made by Dallaire, and since confirmed by a Carnegie Commission conference of military experts set up to study his proposal. See S. R. Feil, *Preventing Genocide: A Report to the Carnegie Commission on Preventing Deadly Conflict* (New York: Carnegie Corporation, 1998). Kuperman disagrees with this analysis, claiming only a much larger peace enforcement operation would have succeeded in mitigating the genocide. See Alan J. Kuperman, 'Rwanda in Retrospect', *Foreign Affairs* 79, 1 (2000), 94–118. See also Wheeler, *Saving Strangers*, pp. 222–223.
85. It took until 20 April for the Secretary-General to produce recommendations for the Council. Members claimed that his sluggishness, indecision and bewilderment hindered the decision-making process. Boutros-Ghali on the other hand contends it took him so long to present his initial report because he was trying very hard to convince members of the Council to authorize a new force. See Melvern, *A People Betrayed*, pp. 153–162.
86. SG report, S/1994/470, 20 April 1994, par. 2.
87. Ibid., par. 3.
88. See Conclusion 5a, *Report of the Independent Inquiry into the Actions of the UN*.
89. Dallaire, *Shake Hands*, p. 333.
90. UN Doc. S/1994/518, 29 April 1994.
91. Ibid.
92. See SG report, S/1994/565, 13 May 1994.
93. See SG report, S/1994/640, 31 May 1994, par. 6 and 36.
94. SC Res. 912, 21 April 1994.
95. See interventions by New Zealand and the United Kingdom, S/PV.3377, 16 May 1994.
96. Adelman and Suhrke, *Study II*, p. 89 (footnote 70).
97. S/PV.3388, 8 June 1994.

98. Adelman and Suhrke, *Study II*, p. 43.
99. Ibid.
100. On why only an early intervention might have halted the genocide, see Feil, *Preventing Genocide*, pp. 26–27; and Wheeler, *Saving Strangers*, p. 223.
101. William E. Schmidt, 'Troops Rampage in Rwanda; Dead Said to Include Premier', *New York Times*, 8 April 1994, p. A1.
102. William E. Schmidt, 'Terror Convulses Rwandan Capital as Tribes Battle', *New York Times*, 9 April 1994, p. A1.
103. See, for example, Donatella Lorch, 'The Massacres in Rwanda: Hope is also a Victim', *New York Times*, 21 April 1994, p. A3.
104. Editorial, 'Cold Choices in Rwanda', *New York Times*, 22 April 1994.
105. See, for example, editorial, 'Look Before Plunging Into the Rwanda', *New York Times*, 18 May 1994, p. A22.
106. See Steven Livingston and Todd Eachus, 'US Policy and Television Coverage', in *Path to a Genocide*, ed. Adelman and Suhrke, pp. 209–228.
107. For more on the British media in particular, see Judith Murison, *Fleeing the Jungle Bloodbath: the Method in the Madness. British Press Reporting of Rwanda* (Edinburgh: Centre of African Studies, 1996).
108. 'The Bleeding of Rwanda', *The Economist*, 16 April 1994, p. 77.
109. See 'Rwanda: No end in sight', *The Economist*, 23 April 1994, p. 66.
110. See 'Saving Rwandan Lives: It is too late for many, but the UN can protect others if it acts quickly', *The Economist*, 21 May 1994, p. 15.
111. For a detailed comparison of the American newspaper coverage of both conflicts, see Garth Myers, Thomas Klak and Timothy Koehl, 'The Inscription of Difference: News Coverage of the Conflicts in Rwanda and Bosnia', *Political Geography*, 15, 1996, 32–36. This study assessed numerically the words used to describe both conflicts.
112. Ibid., 30–31.
113. Every person interviewed about Rwanda by the author, including all American officials, agreed that the Genocide Convention places a very strong moral, if not legal, obligation to act to prevent it.
114. The United States Holocaust Memorial Museum was inaugurated in Washington D.C. in May 1993. At the ceremony, Clinton added his voice to the ritual chorus of 'never again': 'The evil represented in this museum is incontestable, but as we are its witness, so must we remain its adversary in the world in which we live', quoted in *The Triumph of Evil*, transcript, 26 January 1999. See also Philip Gourevitch, *We wish to inform you that tomorrow we will be killed with our families: Stories from Rwanda* (New York: Picador USA, 1998); African Rights, *Rwanda: Death, Despair and Defiance*, rev. ed. (London: African Rights, 1995), pp. 151–152.
115. It is important to note that the term 'acts of genocide' does appear in Article VIII of the 1948 Genocide Convention, but the intent behind US insistence on the usage of this particular terminology does seem to have been to downplay the seriousness of events in Rwanda.
116. See *The Triumph of Evil*, transcript, 26 January 1999; and Gourevitch, *We wish to inform you*, pp. 152–153.
117. Tony Marley, Political Military Advisor for the US State Department from 1992–1995, interview for *Triumph of Evil*, transcript, 26 January 1999.
118. UN Doc. S/PRST/1994/21, 30 April 1994.

119. See Melvern, *A People Betrayed*, pp. 179–180.
120. SC Res. 925, 8 June 1994.
121. For detailed information about the Belgian campaign to get UNAMIR out of Rwanda, see *Rapport fait au nom de la Commission D'enquête Parlementaire*, pp. 546–553.
122. Melvern, *A People Betrayed*, p. 148.
123. Author interview with Karel Kovanda, Brussels, 19 June 2001.
124. Tony Marley, interview for *The Triumph of Evil*, transcript, 26 January 1999.
125. Author interview with Sir David Hannay, London, 13 June 2001. Hannay also suggested the 1948 Genocide Convention would have been discredited in such a circumstance.
126. For a first hand account and theoretical elaboration of how indifference was bureaucratized into a justifiable US position on Rwanda at the UN, see Michael N. Barnett, 'The UN Security Council, Indifference, and Genocide in Rwanda', *Cultural Anthropology* 12, 4 (1997), pp. 551–578. It is interesting to note that Clinton considers the 'failure to try to stop Rwanda's tragedies...one of the greatest regrets of [his] presidency', in *My Life*, p. 593.
127. Barnett, 'The UN Security Council', p. 562.
128. Quoted in editorial, 'Look Before Plunging Into Rwanda, *New York Times*, 18 May 1994, p. A22.
129. Both a senior state department official and Anthony Lake believed that PDD 25 was a good document designed to enable the US to continue to participate in multilateral peacekeeping missions in the wake of Somalia. Interviewed by author in Washington D.C. on 12 and 13 October 1999 respectively.
130. Throughout the genocide, the Security Council stressed the importance of establishing a ceasefire so that the Arusha Peace Agreement could be implemented. The *Report of the Independent Inquiry into the Actions of the UN* stated that they found it 'disturbing' that records of meetings between the UN and the Rwandan interim government showed 'a continued emphasis on a cease-fire, more than the moral outrage against the massacres'; and that they considered the Council's accommodation of this government 'a costly error in judgment' considering that it was carrying out a genocide. See Conclusion 8.
131. Leaked hand-written document containing a verbatim account of the informal sessions of the Security Council from 4 April to 17 May, in possession of Linda Melvern, consulted on 2 July 2001. See Nigerian interventions on 9, 12, 13 and 15 April. Nigerian Ambassador Ibrahim A. Gambari even pleaded personally with Boutros-Ghali to do all he could to arrest the Security Council's movement toward a decision to withdraw UNAMIR. See section entitled 'The Continued Role of UNAMIR', in *Report of the Independent Inquiry into the Actions of the UN*.
132. See the letter Tanzania addressed to the Security Council, UN Doc. S/1994/527, 2 May 1994.
133. See generally S/PV.3388, 8 June 1994.
134. UN Doc. S/1994/518, 29 April 1994.
135. SG report, S/1994/565, 13 May 1994.
136. Ibid., par. 23.
137. Ibid., pars. 18–21.

138. SC Res. 918, 17 May 1994.
139. Ibid.
140. Adelman and Suhrke, *Study II*, p. 51.
141. Burkhalter, 'Clinton Administration and Rwanda', 50.
142. See section entitled 'New Proposals on the mandate of UNAMIR', in *Report of the Independent Inquiry into the Actions of the UN*; and Dallaire, *Shake Hands*, p. 364.
143. Melvern, *A People Betrayed*, p. 199.
144. See section entitled 'New Proposals on the mandate of UNAMIR', in *Report of the Independent Inquiry into the Actions of the UN*.
145. Even Anthony Lake could not tell me who had instigated the proposal and whether or not it was actively pursued. Author interviews with Anthony Lake, another senior state department official and Michael Sheehan, Member of Madeleine Albright's staff at the US Mission to the UN from 1993–1995, Washington D.C., 12–13 October 1999. Officials tended to discuss this nebulous plan with a sense of pride, as if to prove that the US had at least done something.
146. Tony Marley, interview for *Triumph of Evil*, transcript, 26 January 1999.
147. Dallaire, *Shake Hands*, p. 375.
148. When asked if his country had ever considered contributing troops to UNAMIR II, Karel Kovanda said, 'No that was pretty much out of the question'. Interview with author, Brussels, 19 June 2001.
149. Klinghoffer, *International Dimension of Genocide in Rwanda*, p. 51.
150. The Secretary-General complained to this effect in a letter dated 1 August 1994 from him to the President of the Security Council, UN Doc. S/1994/923, 3 August 1994.
151. Adelman and Suhrke, *Study II*, p. 51.
152. Ibid.
153. James Woods, interview for *Triumph of Evil*, transcript, 26 January 1999.
154. SC Res. 925, 8 June 1994.
155. SG report, S/1994/640, 31 May 1994.
156. Des Forges, *Leave None to Tell the Story*, p. 630. The UN's job at the Amohoro stadium was facilitated after 9 April by the RPF, which had secured that part of Kigali, though mortar attacks by the extremists continued. See Dallaire, *Shake Hands*, p. 276.
157. Dallaire, 'End of Innocence', p. 79. On the protection of civilians at these sites, see also Dallaire, *Shake Hands*, pp. 262, 264, 270, 272, 276, 284, 291, 302, 405 and 441.

Chapter 5 Too Little, Too Late: France's *Zone Humanitaire Sûre* in Rwanda

1. Reprinted from Borton, Brusset and Hallam, *Study III*, p. 27.
2. Reprinted from Ibid.
3. Reprinted from Jacques Lanxade, 'L'opération Turquoise', *Défense Nationale*, February 1995 (Paris: Éditions CEDN), 10.
4. Letter dated 20 June 1994 from the Permanent Representative of France to the United Nations addressed to the Secretary-General, UN Doc. S/1994/734, 21 June 1994.

5. SC Res. 929, 22 June 1994. France was not specifically named in the resolution.
6. These words were not directly incorporated into Resolution 929, but the operation was authorized to achieve the same humanitarian objectives laid out in paragraphs 4 (a) and (b) of Resolution 925, which did use the precise wording above.
7. These bases were in Goma and Bukavu. Only 60 per cent of these troops actually entered Rwanda. The rest were stationed in Zaire as part of the rear-guard team ensuring logistics, transport, communications and supplies. See Lanxade, 'L'opération Turquoise', 9–10.
8. UN Doc. S/1994/798, 6 July 1994. The English version of this letter contains the wording 'safe humanitarian zone'.
9. Stephen Smith, 'France-Rwanda: Lévirat Colonial et Abandon dans la Région des Grands Lacs', in *Les Crises Politiques au Burundi et au Rwanda, 1993–1994: Analyses, Faits et Documents*, ed. André Gichaoua (Paris: Karthala, 1995), p. 448.
10. Adelman and Suhrke, 'Early Warning', 297.
11. Pascal Krop, *Le Génocide Franco-Africain: Faut-il Juger les Mitterrand?* (Paris: Éditions Jean-Claude Lattes, 1994), p. 72; and Jean-Pierre Chrétien, *Le Défi de l'Ethnisme: Rwanda et Burundi: 1990–1996* (Paris: Éditions Karthala, 1997), p. 127. Chrétien argues that Habyarimana believed that he would be able to carry out a superficial democratization without truly altering the extent of his power in Rwanda.
12. See letter President Mitterrand sent to Habyarimana on 30 January 1991, in *Rapport D'information déposé par la Mission D'Information de la Commission de la Défense Nationale et des Forces Armées et de la Commission des Affaires Étrangères, sur les Opérations Militaires Menées par la France, d'Autres Pays et l'ONU au Rwanda entre 1990 et 1994, Tome 1* (Paris: Assemblée Nationale, 1998), p. 160.
13. The *Rapport D'information (Rwanda), Tome I* acknowledges that France's indifference to the growing extremism right under its nose was its most direct contribution to the later genocide (p. 370).
14. *Rapport D'information (Rwanda), Tome 1*, p. 127.
15. Ibid., pp. 145–156.
16. Rony Brauman, *Devant le Mal: Un Génocide en Direct* (Paris: Arléa, 1994), p. 62. Here again the issue of French culpability and indifference arises. It is difficult to imagine that the French military did not have any idea of the purpose for which these militias were being trained. See Agnès Callamard, 'French Policy in Rwanda' in *The Path of A Genocide*, ed. Adelman and Suhrke, pp. 168–169.
17. Mel McNulty, 'France's Rwanda Débâcle', *War Studies Journal* 2, 2 (Spring 1997), 10.
18. The DGSE specializes in undercover operations, counter-espionage and discreet advisement. See Philippe Marchesin, 'Mitterrand l'Africain', *Politique Africaine* 58 (June 1995), 19.
19. See Patrick de St. Exupéry, *L'Inavouable: La France au Rwanda* (Paris: Éditions des Arènes, 2004), pp. 178–79.
20. François-Xavier Verschave, *Complicité de Génocide? La Politique de la France au Rwanda* (Paris: Éditions La Décourverte, 1994), p. 36.

21. Prunier, *History of a Genocide*, p. 113.
22. Colonel Luc Marchal, the commander of the Belgium contingent of UNAMIR, recounts that a peacekeeper informed him that he saw unidentified boxes being taken off a French jet at Kigali airport on 8 April 1994 and then driven away on FAR vehicles, interview for *Rwanda Bloody Tricolor, Panorama* BBC Documentary, 19 May 1995. In contrast, French Colonel Henri Poncet, who was responsible for what was on that flight, explains that no ammunition was aboard, and that he requisitioned the FAR vehicles to transport a detachment to the French embassy. See *Rapport D'information (Rwanda), Tome I*, p. 279.
23. Human Rights Watch says it has evidence to prove that the French delivered arms to the FAR via Goma five times in May and June. See Des Forges, *Leave None to Tell the Story*, pp. 661–662. The French *Assemblée Nationale* inquiry concludes in contrast that such deliveries never occurred. See *Rapport D'information (Rwanda), Tome I*, p. 186. Both Gouteux and St. Exupéry offer considerable, sometimes anecdotal, evidence that France supplied the FAR directly or indirectly with weapons during the genocide. See Jean-Paul Gouteux, *Un Génocide Secret D'État: La France et le Rwanda 1990–1997* (Paris: Éditions Sociales, 1998), pp. 74–75; and St. Exupéry, *L'Inavouable*, pp. 183–184, 202–203. Melvern claims that France destroyed the paper trail linking it to the Rwandan extremists. See Melvern, *Conspiracy to Murder*, p. 186.
24. *Rapport D'information (Rwanda), Tome I*, p. 161.
25. Ibid.
26. Ibid., p.165.
27. Ibid., pp. 172–176; and Colette Braeckman, *Rwanda: Histoire d'un Génocide* (Paris: Fayard, 1994), p. 258.
28. Prunier, *History of a Genocide*, p. 100. The French government has always categorically denied that its troops became involved in the civil war. See *Rapport D'information (Rwanda), Tome I*, p. 171.
29. According to the *Rapport D'information (Rwanda), Tome 1*, Colonel Bernard Cussac and Lieutenant-Colonel Gilles Chollet did interrogate some prisoners, but there were no systematic or coercive interrogations (pp. 176–177).
30. Krop, *Le Génocide Franco-Africain*, pp. 97–105.
31. See for instance St. Exupéry, *L'Inavouable*; Verschave, *Complicité de Génocide*; Brauman, *Devant le Mal*; Gouteux, *Génocide Secret D'État*; Mehdi Ba, *Rwanda, 1994: Un Génocide Français* (Paris: L'Esprit Frappeur, 1997); Krop, *Le Génocide Franco-Africain;* and Stephen Smith, 'France-Rwanda', pp. 447–453.
32. See for example Prunier, *History of a Genocide*; Sara Fyson, *French Intervention in Rwanda 1990–1994: Acting out Ideational Biases* (M.Phil. Thesis, University of Oxford, 2000); McNulty, 'France's Rwanda Débâcle', pp. 3–22; Chrétien, *Le Défi de l'Ethnisme*; and *Rapport D'information (Rwanda)*, especially *Tome II Annexes*, and *Sommaires des Comptes Rendus d'Auditions*.
33. Daniel Bourmaud, 'France in Africa: African Politics and French Foreign Policy', *Issue: A Journal of Opinion* 23, 2 (1995), 60.
34. Jean-François Bayart, 'Réflexions sur la Politique Africaine de la France', *Politique Africaine* 58 (June 1995), 48.

35. For details, see Fyson, *French Intervention in Rwanda*, p. 116.
36. See testimony by Hubert Védrine, in *Rapport D'information, Sommaire des Comptes Rendus D'Auditions du 24 mars au 5 mai 1998*.
37. Marchesin, 'Mitterrand l'Africain', 8.
38. See Asteris C. Huliaras, 'The "anglosaxon conspiracy": French perceptions of the Great Lakes Crisis', *The Journal of Modern African Studies* 36, 4 (1998), 598–599; and Rachel Utley, 'The New French Interventionism', *Civil Wars* 1, 2 (Summer 1998), 85.
39. Huliaras, 'The "anglosaxon conspiracy"', 598.
40. McNulty, 'France's Rwanda Débâcle', 5.
41. Fyson, *French Intervention in Rwanda*, pp. 31–32.
42. Smith, 'France-Rwanda', p. 448. Most Rwandans speak only Kinyarwanda.
43. See Huliaras, 'The "anglosaxon conspiracy"', 596–597, 603–604.
44. The United States was furnishing military aid to Uganda as a means to counteract the spread of Islamic fundamentalism from Sudan. Indirectly, therefore, some of its assistance likely benefited the RPF since Museveni supported its agenda. In addition, some RPF soldiers, including General Paul Kagame, received training in the US, but they were Ugandan soldiers at the time. See testimony of Herman Cohen, Advisor for African Affairs in the US Department of State from 1989–1993, in *Rapport D'information (Rwanda), Sommaire des Comptes Rendus D'Auditions du 30 juin au 9 juillet 1998*.
45. Brauman, *Devant le Mal*, p. 55.
46. Prunier, *History of a Genocide*, p. 105.
47. McNulty, 'France's Rwanda Débâcle', 6.
48. Callamard, 'French Policy in Rwanda', 172.
49. See Fyson, *French Intervention in Rwanda*, p. 55.
50. During the entire time that *Opération Noroît* was in Rwanda, not one single debate about it occurred in the *Assemblée Nationale*.
51. See Fyson, *French Intervention in Rwanda*, p. 58; and Marchesin, 'Mitterrand l'Africain', pp. 5–7.
52. Prunier, *History of a Genocide*, pp. 100–101.
53. Fyson, *French Intervention in Rwanda*, p. 59.
54. Jean-François Bayart, *La Politique Africaine de François Mitterrand* (Paris: Éditions Karthala, 1984), pp. 57–65.
55. Krop, *Le Génocide Franco-Africain*, p. 88.
56. Collette Braeckman, 'Le Rôle de la France au Rwanda: Un Génocide par Inadvertance?', in *Rwanda: Perspectives*, eds. Maryse Bray et al., p. 33.
57. See generally Fyson, *French Intervention in Rwanda*, pp. 63–68.
58. Gouteux, *Un Génocide Secret D'État*, p. 155.
59. On the extremism of Huchon and Quesnot, see St. Exépury, *L'Inavouable*, pp. 179–182.
60. Braeckman, *Histoire d'un Génocide*, p. 278.
61. French politicians believed strongly that they were pursuing appropriate, morally-acceptable policies in Rwanda, which explains why they considered much of the criticism against France in relation to Rwanda to be too harsh. Former PM Édouard Balladur, during his testimony to the *Assemblée Nationale* inquiry, said he hoped some light can be shed on why the French government has been so virulently attacked for its policies in Rwanda when it was the only state that tried to do anything for

Rwanda, both before and after Arusha. See *Rapport D'information (Rwanda), Sommaire des Comptes Rendus d'Auditions du 24 Mars au 5 Mai 1998*. Similarly, Hubert Védrine, private secretary to President Mitterrand for many years, expressed similar incredulity at the backlash against France. See *Les Mondes de François Mitterrand*, p. 703.

62. See Mitterrand, quoted in Smith, 'France-Rwanda', p. 449; and Védrine, *Les Mondes de François Mitterrand*, p. 698.
63. Unfortunately, due perhaps to this ethnic divisiveness and the fear of another genocide, the RPF has only been able to retain its hold over Rwanda by putting in place a *de facto* relatively benign dictatorship, lending greater credibility to France's views of the early 1990s about the prospects for democracy in Rwanda under the RPF. See, for example, Filip Reyntjens, 'Rwanda, Ten Years On: From Genocide to Dictatorship', *African Affairs* 103 (2004), pp. 177–210.
64. Adelman and Suhrke, *Study II*, p. 54.
65. Verschave, *Complicité de Génocide?*, p. 122.
66. For an account of what went on in these crisis councils, see Védrine, *Les Mondes de François Mitterrand*, p. 701.
67. By the end of the first week of July, France agreed that the interim government was too tarnished by the genocide to have any role whatsoever as a governing authority over Rwanda. See, 'Telegram from French ambassador to Kigali to the Ministry of Foreign Affairs', 6 July 1994 (p. 411); and 'Notes of the Minister of Foreign Affairs', 7 July 1994 (p. 447) in *Rapport D'information (Rwanda), Tome II, Annexes*.
68. Several notes between Paris and the French Embassy in Rwanda reveal a growing concern that France not be associated with those responsible for genocide, but work to find a more moderate leadership, perhaps at the local level. See notes on 6 July (p. 411), 7 July (pp. 412–413), 8 July (p. 414), and 10 July 1994 (p. 416), in *Rapport D'information (Rwanda), Tome II, Annexes*.
69. See testimony of Christian Quesnot, the President's military chief of staff, in *Rapport D'information (Rwanda), Sommaire Des Comptes Rendus D'Auditions du 6 mai au 3 juin 1998*.
70. A Ministry of Foreign Affairs 1 July 1994 note explains that an RPF military victory would spell the end of representative democracy in Rwanda and lead to future warfare. In *Rapport D'Information (Rwanda), Tome II, Annexes*, p. 433.
71. See the very telling letter from former PM Balladur to Bernard Debré, 9 June 1998, in *Rapport D'Information (Rwanda), Tome II, Annexes*, p. 378. Some authors believe that the Kigali landing scenario had been part of a conjectural plan developed in April by the General Huchon and the *Mission Militaire de Coopération* to 'save' the interim government from the RPF advance. See Des Forges, *Leave None to Tell the Story*, p. 664.
72. Védrine, *Les Mondes de François Mitterrand*, p. 701.
73. The second aim of Opération Turquoise, as listed in its rules of engagement, was to work towards renewed negotiations between belligerents. In *Rapport D'information (Rwanda), Annexes, Tome II*, p. 389. See also note from France's Rwandan Ambassador, Jean-Michel Marlaud, to French Ministry of Foreign Affairs, 27 June 1994, in Ibid., p. 427.

74. Prunier, 'Opération Turquoise: A Humanitarian Escape from a Political Dead End', in *Path of a Genocide*, ed. Adelman and Suhrke, p. 286.
75. Ibid., p. 287.
76. See Alain Juppé, 'La Responsabilité de tous', *Le Monde*, 2 July 1994, p. 4.
77. S/PV.3402, 11 July 1994. See also France's remarks in the Security Council, S/PV.3392, 22 June 1994.
78. In *Rapport D'Information (Rwanda), Tome II*, Annexes, p. 375.
79. Jean de la Guérivière, 'Le premier ministre a été assassiné lors des massacres qui ont suivi la mort du chef de l'État', *Le Monde*, 9 April 1994, p. 3.
80. Adelman and Suhrke, *Study II*, p. 47. Examples of pieces questioning French involvement in Rwanda are: 'La France perd la mémoire au Rwanda', *Le Canard Enchaîné*, 4 May 1994; 'Rwanda: les amitiés coupables de la France', *Libération*, 18 May 1994; 'Les responsabilités françaises dans le drame rwandais', *L'Humanité*, 20 May 1994.
81. See generally Danielle Birck, 'La Télévision et le Rwanda ou le Génocide Déprogrammé', *Les Temps Modernes* 50, 583 (1995), 181–197.
82. See Marie-Pierre Subtil, 'La France pourrait prendre l'initiative d'une intervention au Rwanda', *Le Monde*, 17 June 1994.
83. Verschave, *Complicité de Génocide?*, p. 122.
84. Jean-Hervé Bradol, 'Rwanda, Avril–Mai 1994, Limites et Ambiguïtés de L'Action Humanitaire', *Les Temps Modernes* 50, 583 (1995), 142.
85. Adelman and Suhrke, *Study II*, p. 55.
86. The five countries which abstained were: China, New Zealand, Pakistan, Nigeria and Brazil.
87. Verschave, *Complicité de Génocide?*, p. 133. Almost everyone interviewed about Rwanda by author, including Boutros-Ghali, Kovanda and Lake, confirmed that, although they had misgivings about France's motives, they were shamed into supporting the French plan because no one else had been willing to step forward with another option.
88. This condition was later abandoned when the ZHS was implemented.
89. Prunier, *History of a Genocide*, pp. 287–288. Other African countries eventually sent token numbers of troops as well, including Senegal, Chad, Congo, Guinea-Bissau, Mauritania and Niger.
90. SC Res. 929, 22 June 1994.
91. See generally S/PV.3392, 22 June 1994.
92. St. Exupéry, *L'Inavouable*, p. 102
93. Ibid, pp. 101–102 and 107; and Dallaire, *Shake Hands*, p. 451. Dallaire writes that, between the colonial tradition of military intervention and the modern tradition of humanitarianism, the 'French never did reconcile which attitude was supreme in Turquoise' (451).
94. Adelman and Suhrke, *Study II*, pp. 54–55.
95. Borton, Brusset and Hallam, *Study III*, p. 11.
96. Ibid., p. 43.
97. Lanxade, 'L'opération Turquoise', 12.
98. Prunier, 'Opération Turquoise', p. 292; and Dallaire, *Shake Hands*, p. 451.
99. See St. Exépury, *L'Inavouable*, p. 68.
100. Borton, Brusset and Halam, *Study III*, p. 55. See also Stephanie T. E. Kleine-Ahlbrandt, *The Protection Gap in the International Protection of*

Internally Displaced Persons: the Case of Rwanda (Geneva: Institut Universitaire de Hautes Etudes Internationales, 1996); and Adelman and Suhrke, *Study II*, pp. 62–65.

101. See *Mortality in the Democratic Republic of Congo: Results from a Nationwide Survey* (New York: International Rescue Committee, 2004).
102. Braeckman, 'Un Génocide par Inadvertance?', p. 35.
103. See *Rapport D'information (Rwanda), Tome I*, p. 348.
104. Ibid., p. 363.
105. UN Doc. S/1994/798, 6 July 1994.
106. See *Opération Turquoise's* rules of engagement, which stress an attitude of strict neutrality *vis-à-vis* both parties, as well as humanitarian character of the operation, in *Rapport D'information (Rwanda), Tome II, Annexes*, p. 389.
107. Lanxade, 'L'opération Turquoise', 9. The rest of the contingent was in Zaire, performing rear-guard duties.
108. Borton, Brusset and Hallam, *Study III*, p. 42.
109. Adelman and Suhrke, *Study II*, p. 56.
110. Dallaire mentions at least one incident where the RPF ambushed a French convoy returning from Butare. See *Shake Hands*, p. 458.
111. See Ministry of Foreign Affairs note on Rwanda, 8 August 1994, in *Rapport D'information (Rwanda), Tome II, Annexes*, p. 478. According to this note, the main reason why France did not stay longer, although it felt compelled to, was because of Kigali's opposition.
112. St. Exépury says *Opération Turquoise* was greeted by Hutus carrying French flags. See *L'Inavouable*, pp. 24–25.
113. See Letter dated 15 July from the Permanent Representative of France to the President of the Security Council, UN Doc. S/1994/832, 15 July 1994.
114. See Ministry of Foreign Affairs note, 7 July 1994, in *Rapport D'information (Rwanda), Tome II, Annexes*, p. 446.
115. See testimony of General Jean-Claude Lafourcade, in *Rapport D'information (Rwanda), Sommaire des Comptes Rendus D'Auditions du 9 juin au 25 juin 1998*. See also Ministry of Foreign Affairs note about the Rwandan situation, 28 June 1994, in *Rapport D'information (Rwanda), Tome II, Annexes*, p. 429; and *Rapport D'information (Rwanda), Tome I*, p. 345. The interim government was not happy about the small size of the ZHS, and informed France that they would not cooperate with the neutralization of the FAR and militias. See Ibid., pp. 340–342. Dallaire agrees that the extremists were growing desperate because France was 'seemingly doing nothing to help their cause', in *Shake Hands*, p. 451.
116. Adelman and Suhrke, *Study II*, p. 56. The strength and determination of the FAR and *Interahamwe* should not be underestimated. Even the RPF could not capture or disarm them in significant numbers during close to five years of military occupation of eastern DRC.
117. France was very worried about the interim government and members of the FAR seeking refuge in the ZHS. See notes between French embassy in Rwanda and Paris, 9 July (p. 415) and 15 July (p. 419), in *Rapport D'information (Rwanda), Tome II, Annexes*; and *Rapport D'information (Rwanda), Tome I*, p. 343.
118. A note prepared by Captain Marin Gillier about an intervention in Bisesero reveals the extent of this type of difficulty. He said that although

the rules of engagement were clear in theory, in practice it was difficult to implement them because it was impossible to tell Hutu and Tutsi apart, and because not speaking the same language as Rwandans posed a real problem. In *Rapport D'information (Rwanda), Tome II, Annexes*, p. 402. The *Rapport D'information (Rwanda), Tome I* concurs that it would have been impossible for a force the size of *Opération Turquoise* to deal with the FAR and *Interhamwe* effectively (p. 347). See also testimony of Alain Juppé, French Foreign Minister in 1994, in *Rapport D'information (Rwanda), Sommaire des Comptes Rendus D'Auditions du 24 mars au 5 mai 1998.*

119. *Rapport D'information (Rwanda), Tome I*, p. 344.
120. See Prunier, 'Opération Turquoise', p. 294; Dallaire, *Shake Hands*, p. 459; and *Rapport D'information (Rwanda), Tome I*, p. 323.
121. Verschave, *Complicité de Génocide?*, p. 142.
122. Adelman and Suhrke, *Study II*, p. 55.
123. See Mitterrand quoted in Jacques Isnard, 'La rébellion rwandaise n'entend pas affronter les forces françaises', *Le Monde*, 7 July 1994, p. 1.
124. SC Res. 929, 22 June 1994.
125. See Opération Turquoise's rules of engagement, in *Rapport D'information (Rwanda), Tome II, Annexes*, p. 389.
126. See *Rapport D'information (Rwanda), Tome I*, p. 345.
127. See notes from French Embassy in Rwanda to Paris, 15 July 1994, in *Rapport D'information (Rwanda), Tome II, Annexes*, pp. 418–419.
128. See 15 July 1994 public communiqué of French Ministry of Foreign Affairs, in *Rapport D'information (Rwanda), Tome I*, pp. 343–344.
129. In a communiqué issued on 16 July, the Ministry of Foreign Affairs said it would implement any decisions of the UN *vis-à-vis* the interim government, but made clear that *Opération Turquoise's* current mandate did not authorize France to arrest its members, in *Rapport D'information (Rwanda), Tome I*, p. 344.
130. The OAU inquiry into the Rwandan genocide condemns the French for not either changing the mandate or acting unilaterally under the Genocide Convention. See *Rwanda: the Preventable Genocide*, chap. 15, par. 73.
131. SC Res. 935, 1 July 1994.
132. SC Res. 955, 8 November 1994.

Chapter 6 Conclusion

1. As others have noted, humanitarian intervention itself was often used as substitute for a real policy aimed at ending the conflict in question. See Roberts, 'Humanitarian War', 442; and in relation to Bosnia, Woodward, *Balkan Tragedy*, generally and p. 325.
2. Honig and Both make a similar point. See *Srebrenica*, p. 100.
3. The *Report of the Panel on United Nations Peace Operations* makes the point that UN peacekeeping missions should not be authorized until such time as member states have committed the necessary troops and resources. See UN Doc. A/55/305–S/2000/809, 21 August 2000, pars. 59–60.

4. The safe areas seemed thus to be a mixture of both a more traditional safety zone concept as elaborated in humanitarian law, and a more ambitious safety zone concept as seen in the other cases of the 1990s.
5. The UN Secretary-General acknowledges that this failure to fully grasp Serb war aims was a critical mistake. See SG report, *Fall of Srebrenica*, pars. 488–493.
6. UNPROFOR commanders initially believed that close air support would be adequate to deter attacks against the safe areas. Author interview with Bertrand de LaPresle, Paris, 21 June 2001. The Secretary General's report on the *Fall of Srebrenica* also acknowledges that air power was viewed 'as a last resort'. See par. 483.
7. Many have suggested that safety zones, though said to provide safety within the country of origin, are in fact aimed almost exclusively at refugee containment, to the detriment of the refugees themselves. See Mikhael Barutciski, 'The Reinforcement of Non-Admission Policies and the Subversion of UNHCR: Displacement and International Assistance in Bosnia-Herzegovina (1992–1994), *International Journal of Refugee Law* 8, 1–2 (1996), 49–110; Landgren, 'Safety Zones and International Protection'; Frelick, '"Preventive Protection"'; Erin D. Mooney, 'Presence, ergo protection? UNPROFOR, UNHCR and the ICRC in Croatia and Bosnia and Herzegovina', *International Journal of Refugee Law* 7 (1995), 407–435; *Inter-Agency Expert Consultation on Protected Areas (Harvard University, 23–24 February 1999): Report*, 7 April 1999; and chapters by Leonardo Franco, Yves Sandoz and B. S. Chimni in *International Legal Issues*, eds. Al-Nauimi and Meese.
8. A similar point is made by Landgren, 'Safety Zones and International Protection', 455.
9. On the issue of peace enforcement, and whether it can in fact be distinguished from war, see Mats Berdal, 'Lessons Not Learned: the Use of Force in 'Peace Operations' in the 1990s', *International Peacekeeping* 7, 4 (Winter 2000), 55–62, 67–68; and Shashi Tharoor, 'Should UN Peacekeeping Go "Back to Basics"?', *Survival* 37, 4 (Winter 1995–96), 60–61.
10. Powell, *My American Journey*, p. 809.
11. See SG report, *Fall of Srebrenica*, par. 499, which suggests that if safety zones were to have a future role in protecting civilians in armed conflict, 'it is clear that either they must be demilitarized and established by the agreement of the belligerents, as in the case of the 'protected zones'...recognized by international humanitarian law, or they must be truly safe areas, fully defended by a credible military deterrent. The two concepts are absolutely distinct and must not be confused'.
12. The *Report of the Panel on United Nations Peace Operations* makes a similar point. See UN Doc. A/55/305–S/2000/809, 21 August 2000, par. 53.
13. See generally SG report, S/2004/703, 30 August 2004.
14. See SG report, S/2004/787, 4 October 2004, pars. 2–8.

Bibliography

Interviews

Conducted by author

Diego Arria, Permanent Representative of the Venezuela to the UN, 1990–1993. New York City, 28 September 1999.

Mira Baratta. Legislative Assistant to Senator Bob Dole for Arms Control and Foreign Policy, 1989–1996. New York, 4 October 1999.

Sir Franklin Berman. Legal Advisor to UK Foreign and Commonwealth Office, 1991–1999. Oxford, 6 June 2001.

Boutros Boutros-Ghali. UN Secretary-General, 1991–1996. Paris, 18 June 2001.

Christopher Coleman. Chief of Policy and Analysis Unit, UN Department of Peacekeeping Operations. New York, 4 October 1999.

Ivo H. Daalder. Director for European Affairs on US National Security Council, 1995–1996, and author of *Getting to Dayton: The Making of America's Bosnia Policy*. Washington D.C., 7 October 1999.

Alain Destexhe. Belgian Senator and Working Member of *Commission D'enquête Parlementaire Concernant les Événements du Rwanda*. Brussels, 20 June 2001.

John Fawcett. UN Special Coordinator for Sarajevo, 1992–1996. Washington D.C., 8 October 1999.

Andrew Gilmour. UN Political Affairs Officer and Secretary-General's Note-taker in the Security Council during the Bosnian conflict and Rwandan genocide. New York, 28 September 1999.

Sir David Hannay. Permanent Representative of the UK to the UN, 1990–1995. London, 13 June 2001.

Robert Hunter. American Ambassador to NATO, 1993–1998. Washington D.C., 12 October 1999.

Ian Johnstone. Assistant to Shashi Tharoor and First Officer, UN Secretary-General's Office. New York, 30 September 1999.

Laurent Klein. Chef du Secrétariat de la Commission des Affaires Étrangères, Assemblée Nationale and Chief Drafter of the Srebrenica Inquiry. Paris, 15 June 2001.

Karel Kovanda. Permanent Representative of the Czech Republic to the UN, 1993–1997. Brussels, 19 June 2001.

Anthony Lake. President Clinton's National Security Advisor from 1993–1997. Washington D.C., 13 October 1999.

Bertrand de LaPresle. UN Force Commander in former Yugoslavia, March 1994–February 1995. Paris, 21 June 2001.

Philip Mahoux. Belgian Senator, and Vice President and Rapporteur of *Commission D'enquête Parlementaire Concernant les Événements du Rwanda*. Brussels, 20 June 2001.

Jim O'Brien. Attorney in Legal Advisor's Office in US State Department, 1993–1994, and Advisor to Madeleine Albright, 1994–1995. Washington D.C., 10 October 1999.

Lord David Owen. Co-Chairman of the Steering Committee of the International Conference on the Former Yugoslavia, 1992–1995. London, 17 July 2001.

Tom Quiggen. Intelligence Officer for the Canadian government stationed in former Yugoslavia, 1991–1995. Telephone interview. Oxford, 28 May 2001.

David Rieff. Author of *Slaughterhouse: Bosnia and the Failure of the West*. New York, 1 October 1999.

Iqbal Riza. Assistant Secretary-General in the UN Department for Peacekeeping Operations, 1993–1996. New York, 28 September 1999.

Stuart Seldowitcz, Political Advisor for the US Mission to the UN, 1991–1996. Washington D.C., 9 October 1999.

Michael Sheehan. Member of the President Clinton's National Security Council, 1992–1993, and Member of Madeleine Albright's team at the US Mission to the UN, 1993–1995. Washington D.C., 12 October 1999.

Alvaro de Soto. Senior Advisor to the Secretary-General, 1992–1994, and Assistant Secretary-General for Political Affairs, 1995–1999. New York, 4 October 1999.

Shashi Tharoor. Special Assistant to the Under-Secretary-General for Peace-keeping Operations, 1989–1996. New York, 30 September 1999.

Bill Woodward. Speechwriter for Madeleine Albright from 1993–1999. Washington D.C., 9 October 1999.

Conducted by 'Death of Yugoslavia', BBC documentary

Diego Arria. Permanent Representative of Venezuela to the UN, 1990–1993. Liddell Hart Centre for Military Archives, Box No. 18, File 1.

Sir David Hannay. Permanent Representative of the UK to the UN, 1990–1995. 21 July 1995. Liddell Hart Centre for Military Archives, Box No. 18.

Douglas Hurd. British Foreign Secretary, 1991–1996. 20 July 1995. Liddell Hart Centre for Military Archives, Box No. 18, File 2.

José-Maria Mendiluce. Coordinator of UNHCR in former Yugoslavia during Bosnian conflict. 11 May 1995. Liddell Hart Centre for Military Archives, Box No. 3, File 3.

Phillipe Morillon. UNPROFOR Commander in Bosnia, September 1992–July 1993. 10 April 1995. Liddell Hart Centre for Military Archives, Box No. 18, File 3.

Lord David Owen. Co-Chairman of the Steering Committee of the International Conference on the Former Yugoslavia, 1992–1995. 11 April 19995. Liddell Hart Centre for Military Archives, Box No. 18, File 3.

Charles Redman. President Clinton's Special Envoy to the Peace Talks, 1993–1994. 10 May 1995. Liddell Hart Centre for Military Archives, Box No. 18, File 4.

Sir Michael Rose. UNPROFOR Commander in Bosnia, January 1994–January 1995. Liddell Hart Centre for Military Archives, Box No. 18, File 4.

Shashi Tharoor. Special Assistant to the Under-Secretary-General for Peacekeeping Operations, 1989–1996. 30 April 1995. Liddell Hart Centre for Military Archives, Box 18, File 4.

Conducted by 'The Triumph of Evil', Frontline PBS documentary www.pbs.org/wgbh/pages/frontline/shows/evil/.

Philip Gourevitch. Author of *We wish to inform you that tomorrow we will be killed with our families: Stories from Rwanda*. 26 January 1999.

Colonel Luc Marchal. Commander of the Belgian Contingent within UNAMIR, 1994. 26 January 1999.

Tony Marley. Political Military Advisor for the US State Department, 1992–1995. 26 January 1999.

Iqbal Riza. Assistant Secretary-General of the UN for Peacekeeping Operations, 1993–1996. 26 January 1999.

James Woods. US Deputy Assistant Secretary for African Affairs at the Department of Defense, 1986–1994. 26 January 1999.

Inquiries and special reports

Bosnia

The Betrayal of Srebrenica: Why Did the Massacre Happen? Will it Happen Again? Hearing before the Subcommittee on International Operations and Human Rights of the Committee on International Relations. House of Representatives, 105th Congress, Second Session, 31 March 1998: 49–268 CC 1998.

Report of the Secretary-General pursuant to General Assembly resolution 53/35: *The Fall of Srebrenica*. UN Doc. A/54/549, 15 November 1999.

Rapport D'information déposé par la Mission D'information Commune sur les Événements de Srebrenica. Tome I (Rapport et Annexes); Tome II (Auditions). Paris: Assemblée Nationale, 11 décembre 2001.

Netherlands Institute for War Documentation. *Srebrenica: a 'safe' area – Reconstruction, background, consequences and analyses of the fall of a safe area.* 10 April 2002. www.screbrenica.nl.

Rwanda

Rapport fait au nom de la Commission D'Enquête Parlementaire Concernant les Événements du Rwanda. Brussels: Sénat de Belgique, 6 Décembre 1997.

Rapport fait au nom de la Commission D'Enquête Parlementaire Concernant les Événements du Rwanda. Annexes 1–13. Brussels: Sénat de Belgique, 6 Décembre 1997.

Rwanda: Genocide and the Continuing Cycle of Violence. Hearing before the Subcommittee on International Operations and Human Rights of the Committee on International Relations. House of Representatives, 105th Congress, Second Session, 5 May 1998.

Rapport D'information par la Mission D'information de la Commission de la Défense Nationale et des Forces Armées et de la Commission des Affaires Étrangères, sur les opérations militaires menées par la France, d'autres pays et l'ONU au Rwanda entre 1990 et 1994. Tome I, Rapport; Tome II, Annexes; Sommaire des Comptes Rendus D'Auditions du 24 mars 1998 au 5 mai 1998; Sommaire des Comptes Rendus D'Auditions du 6 mai 1998 au 3 juin 1998; Sommaire des Comptes Rendus D'Auditions du 9 juin 1998 au 25 juin 1998; Sommaire des Comptes Rendus D'Auditions du 30 juin 1998 au 9 juillet 1998. Paris: Assemblée Nationale, 15 décembre 1998.

Report of the Independent Inquiry into the Actions of the United Nations During the 1994 Genocide in Rwanda. New York: United Nations, 15 December 1999.
Rwanda: The Preventable Genocide. Report of the International Panel of Eminent Personalities to Investigate the 1994 Genocide in Rwanda and the Surrounding Events. Addis Ababa: Organization of African Unity, 29 May 2000.

Secondary sources

Abi-Saab, George. 'Introduction', in *Conference on International Law: The Concept of Jus Cogens in International Law, Papers and Proceedings.* Lagonissi, April 3–8, 1966. Geneva: Carnegie Endowment for International Peace, 1967: 7–15
Adelman, Howard. 'Humanitarian Intervention: The Case of the Kurds.' *International Journal of Refugee Law* 4, 1 (1992): 4–38.
Adelman, Howard and Astri Suhrke. *The International Response to Conflict and Genocide: Lessons from the Rwandan Experience. Study II: Early Warning and Conflict Management.* Copenhagen: DANIDA Joint Evaluation of Emergency Assistance to Rwanda, 1996.
Adelman, Howard and Astri Suhrke. 'Early Warning and Response: Why the International Community Failed to Prevent the Genocide', *Disasters: The Journal of Disaster Studies and Management* 20, 4 (December 1996): 295–304.
Adelman, Howard and Astri Suhrke, eds, *The Path of a Genocide: The Rwanda Crisis from Uganda to Zaire.* New Brunswick, USA: Transaction Publishers, 1999.
Aykan, Mahmut Bali. 'Turkey's Policy in Northern Iraq, 1991–95', *Middle Eastern Studies* 32, 4 (October 1996): 343–366.
Albright, Madeleine. *Madam Secretary. A Memoir.* New York: Miramax, 2003.
Ba, Mehdi. *Rwanda, 1994: Un Génocide Français.* Paris: L'Esprit Frappeur, 1997.
Baker, James A. III. *The Politics of Diplomacy: Revolution, War, and Peace, 1989–92.* With Thomas M. DeFrank. New York: G.P. Putnam's Sons, 1995.
Barnett, Michael N. 'The UN Security Council, Indifference, and Genocide in Rwanda.' *Cultural Anthropology* 12, 4 (1997): 551–578.
Barutciski, Mikhael. 'The Reinforcement of Non-Admission Policies and the Subversion of UNHCR: Displacement and Internal Assistance in Bosnia-Herzegovina (1992–1994)', *International Journal of Refugee Law* 8, 1/2 (1996): 49–110.
Bayart, Jean-François. *La Politique Africaine de François Mitterrand.* Paris: Éditions Karthala, 1984.
Bayart, Jean-François. 'Réflexions sur la Politique Africaine de la France', *Politique Africaine* 58 (June 1995): 41–50.
Berdal, Mats. 'Lessons Not Learned: The Use of Force in "Peace Operations" in the 1990s', *International Peacekeeping* 7, 4 (Winter 2000): 55–74.
Bert, Wayne. *The Reluctant Superpower: United States' Policy in Bosnia, 1991–95.* New York: St. Martin's Press, 1997.
Best, Geoffrey. *War & Law Since 1945.* Oxford: Clarendon Press, 1994.
Bettati, Mario and Bernard Kouchner. *Le Devoir D'ingérence: Peut-On Les Laisser Mourir?* Paris: Denoël, 1987.
Binder, David. 'Anatomy of a Massacre.' *Foreign Policy* 97 (Winter 1994–95): 70–78.

Birck, Danielle. 'La Télévision et le Rwanda ou le Génocide Déprogrammé', *Les Temps Modernes* 50, 583: 181–197.

Boisbouvier, Christophe. 'Le Rôle de la France au Rwanda depuis 1990', in *Rwanda: Perspectives*, eds, Maryse Bray, Aline Cook and Samantha Neath. London: University of Westminster Francophone Africa Research Centre, 1994: 37–45.

Boisson de Chazournes, Laurence and Luigi Condorelli. 'Common Article 1 of the Geneva Conventions revisited: Protecting collective interests', *International Review of the Red Cross* 82, 837 (March 2000): 67–87.

Borton, John, Emery Brusset and Alistair Hallam. *The International Response to Conflict and Genocide: Lessons from the Rwandan Experience. Study III: Humanitarian Aid and Effects*. Copenhagen: DANIDA Joint Evaluation of Emergency Assistance to Rwanda, 1996.

Bourmaud, Daniel. 'France in Africa: African Politics and French Foreign Policy', *Issue: A Journal of Opinion* 23, 2 (1995): 58–62.

Boutros-Ghali, Boutros. 'Introduction', in *The United Nations and Rwanda, 1993–1996*. The United Nations Blue Book Series, Vol. X. New York: United Nations Department of Public Information, 1996: 3–111.

Boutros-Ghali, Boutros. *Unvaquished: A U.S.-U.N. Saga*. New York: I. B. Tauris, 1999.

Bradol, Jean-Hervé. 'Rwanda, Avril–Mai 1994, Limites et Ambiguïtés de L'Action Humanitaire', *Les Temps Modernes* 50, 583 (1995): 126–148.

Braeckman, Colette. *Rwanda: Histoire d'un Génocide*. Paris: Fayard, 1994.

Braeckman, Colette. 'Le Rôle de la France au Rwanda: Un Génocide par Inadvertance?', In *Rwanda: Perspectives*, eds, Maryse Bray, Aline Cook and Samantha Neath. London: University of Westminster Francophone Africa Research Centre, 1994: 32–36.

Brauman, Rony. *Devant le Mal. Rwanda: Un Génocide en Direct*. Paris: Arléa, 1994.

Bray, Maryse, Aline Cook and Samantha Neath, eds, *Rwanda: Perspectives*. London: University of Westminster, Francophone Africa Research Centre, 1994.

Bresheeth, Haim and Nira Yuval-Davis. *The Gulf War and the New World Order*. London: Zed Books, 1991.

Broadcasting Genocide: Censorship, Propaganda & State-Sponsored Violence in Rwanda 1990–1994. London: Article 19, 1996.

Brown, Michael E, ed. *Ethnic Conflict and International Security*. Princeton: Princeton University Press, 1993.

Brownlie, Ian. *Principles of Public International Law*. 4th ed. Oxford: Oxford University Press, 1990.

Bryden, Matthew. 'Somalia: The Wages of Failure', *Current History* 94 (1995): 145–151.

Buettner, Friedemann and Martin Landgraf. 'The New World Order of Europe and the Gulf Crisis', in *The Gulf War and the New World Order: International Relations of the Middle East,* eds, Tareq Y. Ismael and Jacqueline S. Ismael. Gainsville: University Press of Florida, 1994: 77–115.

Bull, Hedley. *The Anarchical Society: A Study of Order in World Politics*. London: the Macmillan Press, 1977.

Bulloch, John and Harvey Morris. *No Friends but the Mountains: the Tragic History of the Kurds*. London: Viking, 1992.

Burg, Steven L. and Paul S. Shoup. *The War in Bosnia-Herzegovina: Ethnic Conflict and International Intervention*. New York: M. E. Sharpe, 1999.

Burkhalter, Holly J. 'The Question of Genocide: The Clinton Administration and Rwanda.' *World Policy Journal* 11, 4 (Winter 1994/95): 44–54.

Byers, Michael. 'Conceptualising the Relationship between Jus Cogens and Erga Omnes Rules', *Nordic Journal of International Law* 66 (1997): 211–239.

Byers, Michael. *Custom, Power and the Power of Rules: International Relations and Customary International Law*. Cambridge: Cambridge University Press, 1999.

Byers, Michael, ed. *The Role of Law in International Politics: Essays in International Relations and International Law*. Oxford: Oxford University Press, 2000.

Callamard, Agnès. 'French Policy in Rwanda', in *The Path of A Genocide: The Rwanda Crisis from Uganda to Zaire*, eds, Howard Adelman and Astri Suhrke. New Brunswick, USA: Transaction Publishers, 2000: 157–183.

Carr, E. H. *The Twenty Years Crisis, 1919–1939: An Introduction to the Study of International Relations*, 2nd ed. London: Macmillan, 1946.

Chaliand, Gerard. *A People without a Country: The Kurds & Kurdistan*, trans. Michael Pallis. London: Zed Books, 1993.

Chayes, Abram and Antonia Chayes. *The New Sovereignty: Compliance with International Regulatory Agreements*. Cambridge, Massachusetts: Harvard University Press, 1995.

Chesterman, Simon. *Just War or Just Peace? Humanitarian Intervention and International Law*. Oxford: Oxford University Press, 2001.

Chimni, B.S. 'The Incarceration of Victims: Deconstructing Safety Zones', in *International Legal Issues Arising under the United Nations Decade of International Law*, eds, Najeeb Al-Nauimi and Richard Meese. The Hague: Martinus Nijhoff Publishers, 1995: 823–853.

Chrétien, Jean-Pierre. *Le Défi de l'Ethnisme: Rwanda et Burundi: 1990–1996*. Paris: Éditions Karthala, 1997.

Christopher, Warren. *In the Stream of History: Shaping Foreign Policy for a New Era*. Standford: Standford University Press, 1998.

Clapham, Christopher. 'Rwanda: The Perils of Peacemaking', *Journal of Peace Research* 35, 2 (March 1998): 193–210.

Clarke, Walter and Jeffrey Herbst. *Learning from Somalia: The Lesson of Armed Humanitarian Intervention*. New York: Westview Press, 1997.

Clinton, Bill. *My Life*. New York: A. Knopf, 2004.

The Clinton Administration's Policy on Reforming Multilateral Peace Operations (Presidential Decision Directive 25). Department of State Publication 10161. Washington, D.C.: Bureau of International Organization Affairs, May 1994.

Comité International de la Croix Rouge. 'Zones Sanitaires et Zones de Sécurité.' *Revue Internationale de la Croix Rouge* 82 (1951): 443–483 and 629–662.

Condorelli, Luigi and Laurence Boisson de Chazournes. 'Quelques remarques à propos de l'obligation des Etats de "respecter et faire respecter" le droit international humanitaire "en toutes circonstances"', in *Studies and Essays on International Humanitarian Law and Red Cross Principles in honour of Jean Pictet*, ed. Christophe Swinarski. The Hague: Martinus Nijhoff Publishers, 1984: 17–35.

Connolly, William E. *The Terms of Political Discourse*. 2nd ed. Oxford: Martin Robertson & Company, 1983.

Cook, Helena. *The Safe Haven in Northern Iraq: International Responsibility for Iraqi Kurdistan*. University of Essex: Human Rights Centre & London: Kurdistan Human Rights Project, 1995.

Cortell, Andrew P. and James W. Davis, Jr. 'How do International Institutions Matter? The Domestic Impact of International Rules and Norms', *International Studies Quarterly* (1996), 40: 451–478.

Daalder, Ivo H. *Getting to Dayton: The Making of America's Bosnia Policy*. Washington D.C.: Brookings Institution Press, 2000.

Dallaire, Roméo A. 'The End of Innocence: Rwanda 1994', in *Hard Choices: Moral Dilemmas in Humanitarian Intervention*, ed. Jonathan Moore. New York: Rowman & Littlefield Publishers, 1998: 71–86.

Dallaire, Roméo A. 'The Changing Role of United Nations Peacekeeping Forces: The Relationship between UN Peacekeepers and NGOs in Rwanda', in *After Rwanda: The Coordination of United Nations Humanitarian Assistance*, eds, Jim Whitman and David Pocock. London: Macmillan Press, 1996: 205–218.

Dallaire, Roméo A. *Shake Hands with the Devil: The Failure of Humanity in Rwanda*, with Brent Beardsley. Toronto: Random House Canada, 2003.

Danchev, Alex and Dan Keohane, eds, *International Perspectives on the Gulf Conflict, 1990–1991*. London: The Macmillan Press, 1994.

Danchev, Alex and Thomas Halverson, eds, *International Perspectives on the Yugoslav Conflict*. London: Macmillan Press, 1996.

Danreuther, Roland. *The Gulf Conflict: A Political and Strategic Analysis*. Adelphi Paper, no. 264. London: International Institute for Strategic Studies, 1992.

Des Forges, Alison. *Leave None to Tell the Story: Genocide in Rwanda*. New York: Human Rights Watch, 1999.

Des Forges, Alison L. 'Alas, We Knew.' *Foreign Affairs* 79, 3 (May/June 2000): 141–142.

Destexhe, Alain. 'The Third Genocide.' *Foreign Policy* 97 (Winter 1994–95): 3–17.

Destexhe, Alain. *Rwanda and Genocide in the Twentieth Century*. Trans. Alison Marschner. London: Pluto Press, 1995.

De Waal, Alex and Rakiya Omaar. 'The Genocide in Rwanda and the International Response', *Current History* 94, 591 (April 1995): 156–161.

Dewar, Michael. 'Intervention in Bosnia–the case against', *The World Today* 49, 2 (February 1993): 32–34.

Dickson, Barney. 'From Emperor to Policeman–Britain and the Gulf War', in *The Gulf War and the New World Order*, eds. Haim Bresheeth and Nira Yuval-Davis. London: Zed Books, 1991: 40–48.

Doder, Dusko. 'Yugoslavia: New War, Old Hatreds', *Foreign Policy* 91 (Summer 1993): 3–23.

Drew, Elizabeth. *On the Edge: The Clinton Presidency*. New York: Simon & Schuster, 1994.

Feil, S. R. *Preventing Genocide: A Report to the Carnegie Commission on Preventing Deadly Conflict*. New York: Carnegie Corporation, 1998.

Ferney, Jean-Christophe. 'La France au Rwanda: raison du prince, dé-raison d'État?' *Politique Africaine* 51 (October 1993): 170–175.

Fierke, K. M. 'Multiple Identities, Interfacing Games: The Social Construction of Western Action in Bosnia', *European Journal of International Relations* 2, 4 (1996): 467–497.

Finnemore, Martha. *National Interests in International Society*. Ithaca: Cornell University Press, 1996.

Finnemore, Martha. 'Norms, culture, and world politics: insights from sociology's institutionalism.' *International Organization* (1996), 50: 325–47.

Finnemore, Martha. 'Constructing Norms of Humanitarian Intervention', in *The Culture of National Security: Norms and Identity in World Politics*, ed. Peter J. Katzenstein. New York: Columbia University Press, 1996: 153–185.

Finnemore, Martha and Kathryn Sikkink. 'International Norm Dynamics and Political Change', in *Exploration and Contestation in the Study of World Politics*, eds, Peter J. Katzenstein, Robert O. Keohane and Stephen D. Krasner. Cambridge, Massachusetts: The MIT Press, 1999: 247–277.

Franck, Thomas M. *The Power of Legitimacy Among Nations*. Oxford: Oxford University Press, 1990.

Franco, Leonardo. 'An Examination of Safety Zones for Internally Displaced Persons as a Contribution toward Prevention and Solution of Refugee Problems', in *International Legal Issues Arising under the United Nations Decade of International Law*, eds, Najeeb Al-Nauimi and Richard Meese. London: Martinus Nijhoff Publishers, 1995: 871–897.

Freedman, Lawrence and David Boren. '"Safe Havens" for the Kurds in Post-War Iraq', in Nigel S. Rodley, ed. *To Loose the Bands of Wickedness: International Intervention in Defence of Human Rights*. London: Brassey's, 1992: 43–92.

Freedman, Lawrence and Efraim Karsh. *The Gulf Conflict: 1990–91*. London: Faber and Faber, 1994.

Freedman, Lawrence, ed. *Military Intervention in European Conflicts*. Oxford: Blackwell Publishers, 1994.

Freedman, Lawrence. 'Why the West Failed', *Foreign Policy* 97 (Winter 1994–95): 53–67.

Frelick, Bill. '"Preventive Protection" and the Right to Seek Asylum: A Preliminary Look at Bosnia and Croatia', *International Journal of Refugee Law* 4, 4 (1992): 439–453.

Fyson, Sara. *French Intervention in Rwanda 1990–1994: Acting Out Ideational Biases*. Oxford: M.Phil. Thesis, 2000.

Gasser, Hans-Peter. 'Ensuring Respect for the Geneva Conventions and Protocols: The Role of Third States and the United Nations', in *Armed Conflict and the New Law, Volume II: Effecting Compliance*, eds, Hazel Fox and Michael A. Meyer. London: The British Institute of International and Comparative Law, 1993.

Guichaoua, André, ed. *Les Crises Politiques au Burundi et au Rwanda, 1993–1994: Analyses, Faits et Documents*. 2nd Ed. Paris: Karthala, 1995.

Guillot, Philippe. 'France, Peacekeeping and Humanitarian Intervention', *International Peacekeeping* 3, 1 (Spring 1996): 30–43.

Goble, Paul A. 'Dangerous Liaisons: Moscow, the Former Yugoslavia, and the West', in *The World and Yugoslavia's Wars*, ed. Richard H. Ullman. New York: Council on Foreign Relations, 1996: 182–197.

Goldstein, Judith and Robert O. Keohane, eds, *Ideas and Foreign Policy: Beliefs, Institutions and Political Change*. London: Cornell University Press, 1993.

Goldstein, Judith and Robert O. Keohane. 'Ideas and Foreign Policy: an Analytical Framework', In *Ideas and Foreign Policy: Beliefs, Institutions and Political Change*, eds, Judith Goldstein and Robert O. Keohane. London: Cornell University Press, 1993: 3–30.

Gompert, David C. 'The United States and Yugoslavia's Wars', in *The World and Yugoslavia's Wars*, ed. Richard H. Ullman. New York: Council on Foreign Relations, 1996: 122–144.

Goodwin-Gill, Guy. *The Refugee in International Law*, 2nd ed. Oxford: Clarendon Press, 1996.

Goulding, Marrack. 'The Use of Force by the United Nations', *International Peacekeeping* 3, 1 (Spring 1996): 1–18.

Gourevitch, Philip. 'The Genocide Fax: The United Nations was Warned about Rwanda. Did Anyone Care?', *The New Yorker*, May 11 1998.

Gourevitch, Philip. *We wish to inform you that tomorrow we will be killed with our families: Stories from Rwanda*. New York: Picador USA, 1998.

Gouteux, Jean-Paul. *Un Génocide Secret D'État: La France et le Rwanda 1990–1997*. Paris: Éditions Sociales, 1998.

Gow, James. 'British Perspectives', in *International Perspectives on the Yugoslav Conflict*, eds, Alex Danchev and Thomas Halverson. London: Macmillan Press, 1996: 87–99.

Gow, James. *Triumph of the Lack of Will: International Diplomacy and the Yugoslav War*. London: Hurst & Company, 1997.

Gow, James and Lawrence Freedman. 'Intervention in a fragmenting state: the case of Yugoslavia', in *To Loose the Bounds of Wickedness: International Intervention in Defence of Human Rights*, ed. Nigel S. Rodley. London: Brassey's, 1992: 93–132.

Gowing, Nik. 'Real-time TV Coverage from War: Does it Make or Break Government Policy?', in *Bosnia by Television*, eds, James Gow, Richard Paterson and Alison Preston. London: British Film Institute, 1996: 81–91.

Greenwood, Christopher. 'Is there a right to humanitarian intervention?' *The World Today* 49, 2 (February 1993): 34–40.

Greilsammer, Ilan. 'European Reactions to the Gulf Challenge', in Gad Barzilai, Aharon Klieman and Gil Shidlo, eds, *The Gulf Crisis and Its Global Aftermath*. London: Routledge, 1993: 208–227.

Grossman, Mark, ed. *Encyclopedia of the Persian Gulf War*. Santa Barbara: ABC-CLIO, 1995.

Glenny, Misha. *The Fall of Yugoslavia: the Third Balkan War*. 3rd ed. New York: Penguin Books, 1996.

Gunter, Michael M. *The Kurds of Iraq: Tragedy and Hope*. New York: St. Martin's Press, 1992.

Gunter, Michael M. 'Foreign Influences on the Kurdish Insurgency in Iraq', *Conflict Quarterly* 12, 4 (Fall 1992): 7–24.

Gunter, Michael M. 'A De Facto Kurdish State in Northern Iraq', *Third World Quarterly* 14, 2 (1993): 295–319.

Haass, Richard N. 'The Squandered Presidency: Demanding More from the Commander-in-Chief', *Foreign Affairs* 79, 3 (May/June 2000): 136–140.

Hale, William. 'Turkey, the Middle East and the Gulf Crisis', *International Affairs* 68, 4 (1992): 679–692.

Halliday, Fred. 'The Gulf War and Its Aftermath: First Reflections', *International Affairs* 67, 2 (1991): 223–234.

Harrington, Michael E. 'Operation Provide Comfort: A Perspective in International Law', *Connecticut Journal of International Law* 8, 2 (1993): 635–655.

Hashim, Ahmed. 'Iraq, the Pariah State', *Current History* 91, 561 (January 1992): 11–17.

Henkin, Louis. *How Nations Behave: Law and Foreign Policy*, 2nd ed. New York: Columbia University Press, 1979.

Hendrikson, David C. 'The Recovery of International', *Foreign Affairs* 73, 5 (September/October 1994): 26–43.

Holbrooke, Richard. *To End a War*. New York: Random House, 1998.

Honig, Jan Willem and Norbert Both. *Srebrenica: Record of a War Crime*. London: Penguin Books, 1996.

House of Commons Parliamentary Debates (Hansard), 1990–1991, Vols. 189–190, April–May 1991.

Howorth, Jolyon. 'The Debate in France over Military Intervention in Europe', in *Military Intervention in European Conflicts*, ed. Lawrence Freedman. Oxford: Blackwell Pulblishers, 1994: 106–124.

Howorth, Jolyon. 'French Policy in the Conflict', in *International Perspectives on the Gulf Conflict, 1990–1991*, eds, Alex Danchev and Dan Keohane. London: The Macmillan Press, 1994: 175–200.

Huliaras, Asteris C. 'The "anglosaxon conspiracy": French Perceptions of the Great Lakes crisis', *The Journal of Modern African Studies* 36, 4 (1998): 593–609.

Hurrell, Andrew. 'International Society and the Study of Regimes: A Reflective Approach', in *Regime Theory and International Relations*, ed. Volker Rittberger. Oxford: Clarendon Press, 1999: 49–72.

Hurrell, Andrew. 'Conclusion: International Law and the Changing Constitution of International Society', in *The Role of Law in International Politics: Essays in International Relations and International Law*, ed. Michael Byers. Oxford: Oxford University Press, 2000: 327–347.

Ismael, Tareq Y. and Jacqueline S. Ismael, eds, *The Gulf War and the New World Order: International Relations of the Middle East*. Gainsville: University Press of Florida, 1994.

Jakobsen, Peter Viggo. 'National Interest, Humanitarianism or CNN: What Triggers UN Peace Enforcement After the Cold War?', *Journal of Peace Research* 33, 2 (1996): 205–215.

Jepperson, Ronald L., Alexander Wendt and Peter J. Katzenstein. 'Norms, Identity, and Culture in National Security', in *The Culture of National Security: Norms and Identity in World Politics*, ed. Peter J. Katzenstein. New York: Columbia University Press, 1996: 33–75.

Jones, Bruce D. 'Civil War, the Peace Process, and Genocide in Rwanda', in *Civil Wars in Africa: Roots and Resolution*, eds, Taisier M. Ali and Robert O. Matthews. Montreal & Kingston: McGill-Queen's University Press, 1999: 53–86.

Jones, James L. 'Operation Provide Comfort: Humanitarian and Security Assistance in Northern Iraq', *Marine Corps Gazette* 75, 11 (November 1991): 98–107.

Johnstone, Ian. *Aftermath of the Gulf War: An Assessment of United Nations Action*. London: Lynne Rienner Publishers, 1994.

Kalshoven, Frits. 'The Undertaking to Respect and Ensure Respect in All Circumstances: From Tiny Seed to Ripening Fruit', *Yearbook of International Humanitarian Law* 2 (1999): 3–61.

Karsh, Efraim. 'Survival at All Costs: Saddam Hussein as Crisis Manager', in *The Gulf Crisis and Its Global Aftermath*, eds, Gad Barzilai, Aharon Klieman and Gil Shidlo. London: Routledge, 1993: 51–66.

Katzenstein, Peter J. ed, *The Culture of National Security: Norms and Identity in World Politics*. New York: Columbia University Press, 1996.

Katzenstein, Peter J. 'Introduction: Alternative Perspectives on National Security', in *The Culture of National Security: Norms and Identity in World Politics*, ed. Peter J. Katzenstein. New York: Columbia University Press, 1996: 1–32.

Katzenstein, Peter J. 'Norms, Identity, and Culture in National Security', in *The Culture of National Security: Norms and Identity in World Politics*, ed. Peter J. Katzenstein. New York: Columbia University Press, 1996: 33–75.

Katzenstein, Peter J., Robert O. Keohane and Stephen D. Krasner, eds, *Exploration and Contestation in the Study of World Politics*. Cambridge, Massachusetts: The MIT Press, 1999.

Keen, David. *The Kurds in Iraq: How Safe is Their Haven Now?* London: Save the Children Fund, 1993.

Kegley, Charles Jr. and Gregory A. Raymond. *When Trust Breaks Down: Alliance Norms and World Politics*. Columbia: University of South Carolina Press, 1990.

Kellner, Douglas. *The Persian Gulf TV War*. Boulder: Westview Press, 1992.

Keohane, Dan. 'British Policy in the Conflict', in *International Perspectives on the Gulf Conflict, 1990–1991*, eds, Alex Danchev and Dan Keohane. London: The Macmillan Press, 1994: 145–174.

Keohane, Robert O. *International Institutions and State Power: Essays in International Relations Theory*. London: Westview Press, 1989.

Keohane, Robert O. 'Empathy and International Regimes', in *Beyond Self-Interest*. ed. Jane J. Mansbridge. Chicago: The University of Chicago Press, 1990: 227–236.

Kirisci, Kemal. '"Provide Comfort" and Turkey: Decision-Making for Refugee Assistance.' *Low Intensity Conflict & Law Enforcement* 2, 2 (Autumn 1993): 227–253.

Kirisci, Kemal and Gareth M. Winrow. *The Kurdish Question and Turkey: An Example of a Trans-State Ethnic Conflict*. London: Frank Cass, 1997.

Kleine-Ahlbrandt, Stephanie T. E. *The Protection Gap in the International Protection of Internally Displaced Person: the case of Rwanda*. Geneva: Institut Universitaire de Hautes Etudes Internationales, Université de Genève, 1996.

Klinghoffer, Arthur J. *The International Dimension of Genocide in Rwanda*. Chippenham, Wiltshire: Antony Rowe, 1998.

Klotz, Audie. *Norms in International Relations: The Struggle Against Apartheid*. London: Cornell University Press, 1995.

Kowert, Paul and Jeffrey Legro. 'Norms, Identity, and Their Limits: A Theoretical Reprise', in *The Culture of National Security: Norms and Identity in World Politics*, ed. Peter J. Katzenstein. New York: Columbia University Press, 1996: 451–497.

Kozyrev, Andrei. 'The Lagging Partnership.' *Foreign Affairs* 73, 3 (May/June 1994): 59–71.

Krasner, Stephen D., ed. *International Regimes*. Ithaca: Cornell University Press, 1983.

Krasner, Stephen D. 'Structural Causes and Regime Consequences: Regimes as Intervening Variables', in *International Regimes*, ed. Stephen D. Krasner. Ithaca: Cornell University Press, 1983: 1–21.

Krasner, Stephen D. 'Regimes and the Limits of Realism: Regimes as Autonomous Variables', in *International Regimes*, ed. Stephen D. Krasner. Ithaca: Cornell University Press, 1983: 355–368.

Kratochwil, Friedrich. *Rules, Norms, and Decisions: On the Conditions of Practical and Legal Reasoning in International Relations and Domestic Affairs*. Cambridge: Cambridge University Press, 1991.

Krop, Pascal. *Le Génocide Franco-Africain: Faut-Il Juger les Mitterrand?* Paris: Éditions Jean-Claude Lattes, 1994.

Kuperman, Alan J. 'The Other Lesson of Rwanda: Mediators Sometimes Do More Damage Than Good.' *SAIS Review: A Journal of International Affairs* 16, 1 (Winter–Spring 1996): 221–240.

Kuperman, Alan J. 'Rwanda in Retrospect.' *Foreign Affairs* 79, 1 (January/February 2000): 94–118.

Kuusisto, Rikka. 'Framing the Wars in the Gulf and Bosnia: the Rhetorical Definitions of the Western Power Leaders in Action', *Journal of Peace Research* 35, 5 (1998): 603–620.

Laizer, Sheri. *Martyrs, Traitors and Patriots: Kurdistan after the Gulf War*. London: Zed Books, 1996.

Landgren, Karin. 'Safety Zones and International Protection: A Dark Grey Area.' *International Journal of Refugee Law* 7, 3 (1995): 436–458.

Lanxade, Jacques. 'L'Opération Turquoise', *Défense Nationale*, February 1995. Paris: Éditions CEDN, 7–15.

Lavoyer, Jean-Philippe. 'International Humanitarian Law, Protected Zones and the Use of Force.' In *UN Peacekeeping in Trouble: Lessons Learned from the Former Yugoslavia: Peacekeeper's Views on the Limits and Possibilities of the United Nations in a Civil War-Like Conflict*, eds, Wolfgang Biermann and Martin Vadset. Aldershot: Ashgate, 1998: 262–279.

Legro, Jeffrey W. 'When norms matter: revisiting the 'failure' of internationalism', *International Organization* (1997), 51: 31–63.

Livingston, Steven and Todd Eachus. 'Rwanda: U.S. Policy and Television Coverage', In *The Path of A Genocide: The Rwanda Crisis from Uganda to Zaire*, eds, Howard Adelman and Astri Suhrke. New Brunswick, USA: Transaction Publishers, 2000: 209–228.

Lucarelli, Sonia. *Europe and the Breakup of Yugoslavia: A Political Failure in Search of a Scholarly Explanation*. The Hague: Kluwer Law International, 2000.

Marchesin, Philippe. 'Mitterrand L'Africain', *Politique Africaine* 58 (June 1995): 5–24.

Macleod, Alex. 'La France: à la recherche du leadership international', *Relations Internationales et Stratégiques* 19 (Automne 1995): 69–80.

Macleod, Alex. 'La Grande-Bretagne: faire face à ses responsabilités internationales', *Relations Internationales et Stratégiques* 19 (Automne 1995): 90–100.

MacFarlane, S. Neil and Thomas Weiss. 'Political Interest and Humanitarian Action', *Security Studies* (Autumn 2000) 10, 2: 112–142.

Malcolm, Noel. 'Bosnia and the West: A Study in Failure.' *The National Interest* 39 (Spring 1995): 3–14.

Malcolm, Noel. *Bosnia: A Short History*. London: Papermac, 1996.

Malone, David. *Decision-Making in the UN Security Council, 1990–1996: The Case of Haiti*. Oxford: Clarendon Press, 1998.

Mansbridge, Jane J. ed. *Beyond Self-Interest*. Chicago: The University of Chicago Press, 1990.

Mansbridge, Jane J. 'The Rise and Fall of Self-Interest in the Explanation of Political Life.' In *Beyond Self-Interest*, ed. Jane J. Mansbridge. Chicago: The University of Chicago Press, 1990: 3–22.

Mas, Monique. *Paris-Kigali 1990–1994: Lunettes Coloniales, Politique du Sabre et Onction Humanitaire pour un Génocide en Afrique*. Paris: L'Harmattan, 1999.

Mayall, James, ed. *The New Interventionism 1991–1994: United Nations Experience in Cambodia, Former Yugoslavia and Somalia*. Cambridge: Cambridge University Press, 1996.

Mazimhaka, Patrick. 'Background and the Current Situation', In *Genocide in Rwanda: Background and Current Situation*, ed. Napoleon Abdulai. London: Africa Research & Information Centre, 1994: 24–32.

McDowall, David. *A Modern History of the Kurds*. London: I. B. Tauris, 1997.

McElroy, Robert W. *Morality and American Foreign Policy: The Role of Ethics in International Affairs*. Princeton: Princeton University Press, 1992.

McNulty, Mel. 'France's Rwanda Débâcle', *War Studies Journal* 2, 2 (Spring 1997): 3–22.

Melvern, Linda. *The Ultimate Crime: Who Betrayed the UN and Why*. London: Allison & Busby, 1995.

Melvern, Linda. 'Genocide behind the Thin Blue Line.' *Security Dialogue* 28, 3 (September 1997): 333–346.

Melvern, Linda. *A People Betrayed: The Role of the West in Rwanda's Genocide*. London: Zed Books, 2000.

Melvern, Linda. *Conspiracy to Murder: The Rwandan Genocide*. London: Verso, 2004.

Meron, Theodor. *Human Rights and Humanitarian Norms as Customary Law*. Oxford: Clarendon Press, 1989.

Minear, Larry and Philippe Guillot. *Soldiers to the Rescue: Humanitarian Lessons from Rwanda*. Paris: Development Centre of the Organization for Economic Cooperation and Development, 1995.

Mooney, Erin D. 'Presence, ergo protection? UNPROFOR, UNHCR and the ICRC in Croatia and Bosnia and Herzegovia', *International Journal of Refugee Law* 7, 3 (1995): 407–435.

Moore, Jonathan, ed. *Hard Choices: Moral Dilemmas in Humanitarian Intervention*. New York: Rowman & Littlefield Publishers, 1998.

Morgenthau, Hans J. *Politics Among Nations: The Struggle for Power and Peace*. 5th ed. New York: Alfred A. Knopf, Inc., 1978.

Mosler, Hermann. *The International Society as a Legal Community*. Alphen aan den Rijn, The Netherlands: Sijthoff & Noordhoff, 1980.

Mueller, John. *Policy and Opinion in the Gulf War*. Chicago: The University of Chicago Press, 1994.

Murison, Judith. *Fleeing the Jungle Bloodbath: The Method in the Madness. British Press Reporting of Rwanda*. Edinburgh University: Centre of African Studies, 1996.

Murphy, Sean D. *Humanitarian Intervention: the United Nations in an Evolving World Order*. Philadelphia: University of Pennsylvania Press, 1996.

Myers, Garth, Thomas Klak and Timothy Koehl. 'The Inscription of Difference: News Coverage of the Conflicts in Rwanda and Bosnia', *Political Geography* 15 (1996): 21–46.

Nadelmann, Ethan A. 'Global prohibition regimes: the evolution of norms in international society', *International Organization* (1990), 44: 479–526.

Offen, Elizabeth N. 'Migrants and Refugees: the Human Toll', in *The Gulf Crisis and Its Global Aftermath*, eds, Gad Barzilai, Aharon Klieman and Gil Shidlo. London: Routledge, 1993: 103–126.

Olson, Robert. 'The Kurdish Question in the Aftermath of the Gulf War: Geopolitical and Geostrategic Changes in the Middle East', *Third World Quarterly* 13, 3 (1992): 475–499.

Owen, David. *Balkan Odyssey*. London: Indigo, 1996.

Pasic, Amir and Thomas G. Weiss. 'The Politics of Rescue: Yugoslavia's Wars and the Humanitarian Impulse', *Ethics and International Affairs* 11 (1997): 105–131.

Pictet, Jean S., ed. *Commentary I Geneva Convention for the Amelioration of the Condition of the Wounded and Sick in Armed Forces in the Field*. Geneva: International Committee of the Red Cross, 1952.

Pictet, Jean S., ed. *Commentary IV Geneva Convention Relative to the Protection of Civilian Persons in Time of War*. Trans. Major Ronald Griffin and Mr. C. W. Dumbleton. Geneva: International Committee of the Red Cross, 1958.

Pimlott, John. 'The Gulf Crisis and World Politics', In *The Gulf War Assessed*. John Pimlott and Stephen Badsey, eds. London: Arms and Armour Press, 1992: 35–56.

Powell, Colin. *My American Journey*. With Joseph E. Persico. New York: Random House, 1995.

Price, Richard and Nina Tannenwald. 'Norms and Deterrence: The Nuclear and Chemical Weapons Taboos', in *The Culture of National Security: Norms and Identity in World Politics*, ed. Peter J. Katzenstein. New York: Columbia University Press, 1996: 114–152.

Prunier, Gérard. 'La Crise Rwandaise: Structures et Déroulement.' UNHCR REF-WORLD – Country Information: WRITENET, July 1994.

Prunier, Gérard. *The Rwanda Crisis 1959–1994: History of a Genocide*. London: Hurst & Company, 1995.

Prunier, Gérard. 'The Rwandan Patriotic Front', in *African Guerillas*, ed. Christopher Clapham. Oxford: James Currey, 1998.

Prunier, Gérard. 'Opération Turquoise: A Humanitarian Escape from a Political Dead End,' in *The Path of A Genocide: The Rwanda Crisis from Uganda to Zaire*, eds, Howard Adelman and Astri Suhrke. New Brunswick, USA: Transaction Publishers, 2000: 281–305.

Public Papers of the Presidents of the United States: George Bush 1990. Book II – July 1 to December 31, 1990. Washington: United States Government Printing Office, 1991.

Reed, Cyrus. 'Exile, Reform, and the Rise of the Rwandan Patriotic Front', *The Journal of Modern African Studies* 34, 3 (September 1996): 479–501.

Reyntjens, Filip. *Rwanda. Trois Jours qui ont fait basculer l'histoire*. Paris: Éditions l'Harmattan, 1995.

Reyntjens, Filip. 'Rwanda: Ten Years On: From Genocide to Dictatorship', *African Affairs* 103 (2004): 177–210.

Rieff, David. *Slaughterhouse: Bosnia and the Failure of the West*. New York: Simon & Schuster, 1995.

Rittberger, Volker, ed. *Regime Theory and International Relations*. Oxford: Clarendon Press, 1989.

Roberts, Adam. 'Humanitarian War: Military Intervention and Human Rights', *International Affairs* 69, 3 (July 1993): 429–449.

Roberts, Adam. 'The Laws of War in the 1990–91 Gulf Conflict', *International Security* 18, 3 (Winter 1993/94): 134–181.

Roberts, Adam. *Humanitarian Action in War: Aid, Protection and Implementation in a Policy Vacuum*. Adelphi Paper, no. 305. London: International Institute for Strategic Studies, 1996.

Roberts, Adam and Richard Guelff, eds, *Documents on the Laws of War*. 3rd ed. Oxford: Oxford University Press, 2000.

Robinson, Piers. 'The Policy-Media Interaction Model: Measuring Media Power During Humanitarian Crisis', *Journal of Peace Research* 37, 5 (2000): 613–633.

Rodley, Nigel S., ed. *To Loose the Bands of Wickedness: International Intervention in Defence of Human Rights*. London: Brassey's, 1992.

Rohde, David. *Endgame: The Betrayal and Fall of Srebrenica: Europe's Worst Massacre since World War II*. Boulder, Colorado: Westview Press, 1998.

Rose, Michael. *Fighting for Peace: Bosnia 1994*. London: The Harvill Press, 1998.

Rwanda Bloody Tricolor, Panorama BBC Documentary, 19 May 1995.

Rwanda: Death, Despair and Defiance. rev. ed. London: African Rights, 1995.

Sandoz, Yves, Christophe Swinarski and Bruno Zimmermann, eds, *Commentary on the Additional Protocols of 8 June 1977 to the Geneva Conventions of 12 August 1949*. Geneva: International Committee of the Red Cross, 1987.

Sandoz, Yves. 'The Establishment of Safety Zones for Persons Displaced within Their Country of Origin', In *International Legal Issues Arising under the United Nations Decade of International Law*, eds, Najeeb Al-Nauimi and Richard Meese. London: Martinus Nijhoff Publishers, 1995, pp. 899–927.

Sassoli, Marco and Antoine A. Bouvier, eds, *How Does Law Protect in War? Cases, Documents and Teaching Materials on Contemporary Practice in International Humanitarian Law*. Geneva: International Committee on the Red Cross, 1999.

Scott, Craig, Abid Qureshi, Jasminka Kalajdzic, Peter Copeland, Francis Chang and Paul Michell. 'A Memorial for Bosnia: Framework of Legal Arguments Concerning the Lawfulness of the Maintenance of the UN Security Council's Arms Embargo on Bosnia and Herzegovina', *Michigan Journal of International Law* 16, 1 (Fall 1994): 1–140.

Sharp, Jane M.O. 'Intervention in Bosnia – the case for', *The World Today* 49, 2 (February 1993): 29–32.

Shaw, Martin. *Civil Society and Media in Global Crises: Representing Distant Violence*. New York: Pinter, 1996.

Silber, Laura and Allan Little. *The Death of Yugoslavia*, rev. ed. London: Penguin Books, 1996.

Sitbon, Michel. *Un Génocide sur la Consicence*. Paris: L'Esprit frappeur, 1998.

Slaughter, Anne-Marie, Andrew S. Tulumello and Stepan Wood. 'International Law and International Relations Theory: A New Generation of Interdisciplinary Scholarship', *American Journal of International Law* 92, 3 (July 1998): 367–397.

Smith, Stephen. 'France-Rwanda: L'Evirat Colonial et Abandon dans la Région des Grands Lacs', in *Les Crises Politiques au Burundi et au Rwanda, 1993–1994: Analyses, Faits et Documents*. 2nd ed., ed. André Guichaoua. Paris: Karthala, 1995: 447–453.

Sobel, Richard. 'Portraying American Public Opinion toward the Bosnia Crisis', *Harvard International Journal of Press Politics* 3, 2 (1998): 16–33.

St. Exépury, Patrick de. *L'Inavouable: La France au Rwanda*. Paris: Éditions des Arènes, 2004.

Stein, Arthur A. 'Coordination and Collaboration: Regimes in an Anarchic World', in *International Regimes*, Ed. Stephen D. Krasner. Ithaca: Cornell University Press, 1983: 115–140.

Sterner, Michael. 'Closing the Gate: The Persian Gulf War Revisited', *Current History* 96, 606 (January 1997): 13–19.

Tardy, Thierry. 'Le président Chirac et la Bosnie-Herzégovine: les limites d'une politique.' *Relations Internationales et Stratégiques* 25 (Printemps 1997): 141–156.

Telhami, Shibley. 'Between Theory and Fact: Explaining US Behaviour in the Gulf Crisis', in *The Gulf War and the New World Order: International Relations of the Middle East*, eds, Tareq Y. Ismael and Jacqueline S. Ismael. Gainsville: University Press of Florida, 1994: 153–183.

Thakur, Ramesh and Carlyle A. Thayer, eds, *A Crisis of Expectations: UN Peacekeeping in the 1990s*. Boulder, Colorado: Westview Press, 1995.

Tharoor, Shashi. 'Should UN Peacekeeping Go "Back to Basics"?', *Survival* 37, 4 (Winter 1995–96): 52–64.

The Triumph of Evil. Prods. Mike Robinson and Ben Loeterman. PBS *Frontline* Documentary, 26 January 1999. www.pbs.org/wgbh/pages/frontline/shows/evil/.

The United Nations and Rwanda, 1993–1996. The United Nations Blue Book Series, Vol. X. New York: United Nations Department of Public Information, 1996.

The United Nations and the Iraq-Kuwait Conflict, 1990–1996. The United Nations Blue Book Series, Vol. IX. New York: United Nations Department of Public Information, 1996.

Tiso, Christopher M. 'Safe Haven Refugee Programs: A Method of Combatting International Refugee Crises', *Georgetown Immigration Law Journal* 8 (Fall 1994): 575–601.

Torrelli, Maurice. 'Les Zones de Sécurité.' *Revue Générale de Droit International Public* 99 (1995): 787–849.

Towle, Philip. 'The British Debate about Intervention in European Conflicts', in *Military Intervention in European Conflicts*, ed. Lawrence Freedman. Oxford: Blackwell Pulblishers, 1994: 94–105.

Ullman, Richard H, ed. *The World and Yugoslavia's Wars*. New York: Council on Foreign Relations, 1996.

Utley, Rachel. 'The New French Interventionism', *Civil Wars* 1, 2 (Summer 1998): 83–103.

Védrine, Hubert. *Les Mondes de François Mitterrand: À L'Élysée 1981–1995*. Paris: Fayard, 1996.

Verschave, François-Xavier. *Complicité de Génocide? La Politique de la France au Rwanda*. Paris: Éditions La Décourverte, 1994.

Waltz, Kenneth. *Theory of International Politics*. Reading, Massachusetts: Addison-Wesley, 1979

Wasserstein, Bernard. *Secret War in Shanghai*. London: Profile Books, 1998.

Wheeler, Nicholas J. *Saving Strangers: Humanitarian Intervention and International Society*. Oxford: Oxford University Press, 2000.

When Good Men Do Nothing. Prod. Mike Robinson. BBC *Panorama* Documentary. 7 December 1998.

Whitman, Jim and David Pocock, eds, *After Rwanda: The Coordination of United Nations Humanitarian Assistance*. London: Macmillan Press, 1996.

'Who is Killing; Who is Dying; What is to be Done: Discussion Paper.' London: African Rights, May 1994.

Willame, Jean-Claude. *L'ONU au Rwanda (1993–1995): La 'Communité Internationale' à L'Épreuve d'un Génocide*. Paris: Éditions Labor, 1996.

Wood, Pia Christina. 'France and the Post Cold War Order: the Case of Yugoslavia', *European Security* 3, 1 (Spring 1994): 129–152.

Woodward, Bob. *The Choice*. New York: Simon & Schuster, 1996.

Woodward, Susan. *Balkan Tragedy: Chaos and Dissolution After the Cold War*. Washington D.C.: The Brookings Institution, 1995.

Young, Oran. 'Regime Dynamics: The Rise and Fall of International Regimes', in *International Regimes*, ed. Stephen D. Krasner. Ithaca: Cornell University Press, 1983: 93–113.

Young, Oran. 'International Regimes: Toward a New Theory of Institutions', *World Politics* (1986), 39: 104–122.

Index